Cerebral Signal Transduction

Contemporary Neuroscience

Cerebral Signal Transduction

From First to Fourth Messengers

Edited by

Maarten E. A. Reith

College of Medicine, University of Illinois, Peoria, IL

Humana Press
Totowa, New Jersey

This publication is printed on acid-free paper. ∞
ANSI Z39.48-1984 (American Standards Institute) Permanence of Paper for Printed Library Materials.

Cover design by Martin E. A. Reith, Karen Schultz, and Patricia F. Cleary.

For additional copies, pricing for bulk purchases, and/or information about other Humana titles, contact Humana at the above address or at any of the following numbers: Tel.: 973-256-1699; Fax: 973-256-8314; E-mail: humana@humanapr.com

Photocopy Authorization Policy:

Printed in the United States of America. 10 9 8 7 6 5 4 3 2 1

Library of Congress Cataloging in Publication Data

Cerebral signal transduction : from first to fourth messengers / edited by Maarten E. A. Reith.
        p. cm. -- (Contemporary neuroscience)
     Includes bibliographical references and index.
     ISBN 0-89603-608-1 (alk. paper)
     1. Molecular neurobiology. 2. Cellular signal transduction.
  3. Nervous system--Diseases--Molecular aspects. 4. Nervous system-
  Degeneration--Molecular aspects. 5. Second messengers
  (Biochemistry) I. Series.
     [DNLM: 1. Signal Transduction--physiology. 2. Apoptosis-
  -physiology. 3. Depressive Disorder--physiopathology. 4. Memory-
  -physiology. 5. Nerve Degeneration--physiopathology. 6. Signal
  Transduction--drug effects. QH 601 C4139 2000]
  QP356.2.C475 2000
  612.8--dc21
  DNLM/DLC   99-38476
  for Library of Congress   CIP

Preface

Since the pioneering discovery of cyclic AMP four decades ago, a multitude of signaling pathways have been uncovered in which an extracellular signal (first messenger) impacts the cell surface, thereby triggering a cascade that ultimately acts on the cell nucleus. In each cascade the first messenger gives rise to the appearance of a second messenger such as cyclic AMP, cyclic GMP, or diacylglycerol, which in turn triggers a third messenger, a fourth messenger, and so forth. Many advances in elucidating such pathways have been made, including efforts to link messenger molecules to brain processes operative in health or disease. However, the latter type of information, relating signaling pathways to brain function, is scattered across a variety of publication media, which makes it difficult to integrate the multiple roles of different signaling cascades into our understanding of brain function in health and disease.

The primary aim of *Cerebral Signal Transduction: From First to Fourth Messengers*, therefore, is to offer a comprehensive picture of the recent advances made in the signaling field as it relates to neuronal and cerebral function. The current state of progress provides an exciting opportunity for such a comprehensive focus because molecular tools have become available to selectively remove, reduce, or enhance specific components in the signaling pathways, e.g., by interfering with the genes encoding key proteins. In addition, the increased awareness of crosstalk between different signaling cascades has revealed many possibilities for changes in gene expression underlying long-term changes in brain function.

Normal cerebral functions, such as memory or apoptosis during development, may be compromised in disease, as seen in Alzheimer's, in such neurodegenerative diseases as Parkinson's, Huntington's, or amyotrophic lateral sclerosis, or in stroke and brain trauma. In addition, there has been recent progress in elucidating the role of signaling messengers in depression and in the action of drugs of abuse. Accordingly, *Cerebral Signal Transduction: From First to Fourth Messengers* is organized around four themes involving brain functions: memory, neurodegeneration/apoptosis, mood disorders, and drug dependence. This book advances understanding of the mechanistic underpinnings for complex behavioral processes and clinically relevant brain diseases, and will be of interest to scientists, graduate students, and advanced undergraduates seeking a comprehensive overview of the cerebral signaling field. Selected chapters will also be of interest to physicians carrying out postmortem measurements related to cerebral signaling and who wish to study in more detail the mechanisms underlying brain diseases and the actions of pharmacotherapeutics. Most therapeutic drugs target the effect of first messengers (neurotransmitters) by either interfering with or mimicking their receptor action or altering their levels by acting on enzymes involved in their synthesis, degradation, or storage. The future will undoubtedly see new drugs targeting events downstream in the cascade of second, third, and fourth messengers, and we believe that *Cerebral Signal Transduction: From First to Fourth Messengers* will contribute to progress towards such novel pharmacotherapeutics.

Each chapter in *Cerebral Signal Transduction: From First to Fourth Messengers* is not simply a review of the work carried out in the author's laboratory, but rather presents a critical survey and synthesis of achievements in that area. Chapter 1 offers an overview of the various signaling cascades and their crosstalk, with the intent to provide basic resource material for reading the more specialized subsequent chapters. Under the section *Memory*, Chapters 2–4 discuss cAMP/PKA, Ca^{2+}/calmodulin-dependent protein kinase, DAG/PKC, and NO/PKG signaling pathways operative in learning and memory. The coverage includes simpler model systems for learning and memory such as *Aplysia californica* (Chap. 2) and *Drosophila* (Chap. 3) as well as more complicated systems including the honeybee (Chap. 3) and the mammalian hippocampus (Chaps. 2 and 4). Under the section *Neurodegeneration and Apoptosis*, Chaps. 5–8 describe cAMP/PKA, DAG/PKC, NO/PKG, and neurotrophic factor signaling cascades involved in these processes. Chapter 5 focuses on receptor–G protein interactions in Alzheimer's disease and Chap. 6 on NO signaling involved in neural injury, neurological disorders, and aggression. Chapter 7 discusses pro- and antiapoptotic neurotrophic factor signaling pathways involving Ras, and Chap. 8 focuses on pathways in neurodegeneration that utilize Ca^{2+}. Both Chaps. 5 and 8 connect signaling messengers in neurodegeneration with clinical findings in or implications for humans. Under the section *Depression*, Chaps. 9–11 cover cAMP/PKA, DAG/PKC, and neurotrophic factor signaling pathways thought to be important in the development and treatment of mood disorders. Stress and the development of depression are linked through cAMP/PKA (Chap. 9) and neurotrophic factor pathways (Chaps. 9 and 10) potentially involved in the novel, nonconvulsive treatment of repeated transcranial magnetic stimulation (Chap. 10). A strong connection between signaling messengers in mood disorders and clinical findings is continued in Chap. 11 focusing on components of the cAMP/PKA and DAG/PKC cascades. In the section *Drug Dependence*, Chaps. 12–15 discuss DAG/PKC signaling pathways and other cascades regulating the production of transcription factors implicated in the development and expression of drug dependence. Various signaling pathways in opiate (Chap. 12) and psychostimulant (Chaps. 13–15) dependence are discussed involving cyclic AMP, protein kinases, and transcription factors. Chapters 12 and 13 review the wealth of information that has come from recent studies with knockout mice lacking genes for the production of various key signaling messengers or receptor proteins acted upon by messengers. Chapters 14 and 15 discuss the role of the dopamine transporter in regulating the first messenger dopamine involved in the action of psychostimulant drugs, in particular that of cocaine. Phosphorylation of the dopamine transporter by the DAG/PKC signaling pathway is described (Chap. 14) and the transcriptional regulation of the dopamine transporter is reviewed (Chap. 15). Additionally, the latter chapter links pharmacodynamic mechanisms operative in human cocaine dependence with those studied in animal models.

The choice of authors for each chapter reflects the editor's identification of investigators who have been instrumental in developing these new frontiers in neuroscience. I thank the authors for their patience, during the process of putting this book together. I deeply appreciate the opportunity offered by Paul Dolgert and Tom Lanigan at Humana Press to produce this book in recognition of the importance of cerebral signal transduction in both health and disease.

Maarten E. A. Reith

Contents

Contributors

TED ABEL • *Department of Biology, University of Pennsylvania, Philadelphia, PA*

AILEEN J. ANDERSON • *Institute for Brain Aging and Dementia, University of California, Irvine, CA*

STAVROULA ANDREOPOULOS • *Section of Biochemical Psychiatry, Clarke Division, Center for Addiction and Mental Health, Toronto, Canada*

CARL W. COTMAN • *Institute for Brain Aging and Dementia, University of California, Irvine, CA*

RICHARD F. COWBURN • *Department of Clinical Neuroscience, Occupational Therapy and Elderly Care Research (NEUROTEC), Alzheimer's Disease Resesarch Center, Karolinska Institute, Huddinge, Sweden*

TED M. DAWSON • *Departments of Neurology and Neuroscience, John Hopkins University School of Medicine, Baltimore, MD*

VALINA L. DAWSON • *Departments of Neurology, Neuroscience, and Physiology, John Hopkins University School of Medicine, Baltimore, MD*

JOHN F. DISTERHOFT • *Department of Cell and Molecular Biology, Northwestern University Medical School, Chicago, IL*

BAS R. K. DOUMA • *Department of Animal Physiology, University of Groningen, Haren, The Netherlands*

RONALD S. DUMAN • *Departments of Psychiatry and Pharmacology, Yale University School of Medicine, New Haven, CT*

AASE FRANDSEN • *Department Pharmacology, Royal Danish School of Pharmacy, Copenhagen, Denmark*

RENÉ HEN • *Center for Neurobiology and Behavior, Columbia University, New York, NY*

SARI IZENWASSER • *Department of Neurology, University of Miami School of Medicine, Miami, FL*

JETTE BISGAARD JENSEN • *Department of Pharmacology, Royal Danish School of Pharmacy, Copenhagen, Denmark*

LI CHEN KRAMER • *Department of Neurology, University of Miami School of Medicine, Miami, FL*

MATTHEW LATTAL • *Department of Biology, University of Pennsylvania, Philadelphia, PA*

PETER P. LI • *Section of Biochemical Psychiatry, Clarke Division, Center for Addiction and Mental Health, Toronto, Canada*

PAUL G. M. LUITEN • *Department of Animal Physiology, University of Groningen, Haren, The Netherlands*

RAFAEL MALDONADO • *Department of Neuropharmacology, Universidad Pompeu Fabra, Barcelona, Spain*

DEBORAH C. MASH • *Departments of Neuology and Molecular and Cellular Pharmacology, University of Miami School of Medicine, Miami, FL*

ULI MÜLLER • *Institute for Neurobiology, Freie Universitat Berlin, Berlin, Germany*

ROBERT M. POST • *National Institute of Mental Health, Bethesda, MD*

HAOYU QIAN • *Institute for Brain Aging and Dementia, University of California, Irvine, CA*

MAARTEN E. A. REITH • *Department of Biomedical and Therapeutic Sciences, University of Illinois College of Medicine, Peoria, IL*

MASAYUKI SASAKI • *Department of Neurology, John Hopkins University School of Medicine, Baltimore, MD*

KIMBERLY SCEARCE-LEVIE • *Center for Neurobiology and Behavior, Columbia University, New York, NY*

DAVID SEGAL • *Department of Neurology, University of Miami School of Medicine, Miami, FL*

ARNE SCHOUSBOE • *Department of Pharmacology, Royal Danish School of Pharmacy, Copenhagen, Denmark*

OLGA VALVERDE • *Department of Neuropharmacology, Universidad Pompeu Fabra, Barcelona, Spain*

EDDY A. VAN DER ZEE • *Department of Zoology, University of Groningen, Haren, The Netherlands*

ROXANNE A. VAUGHAN • *Department of Biochemistry and Molecular Biology, School of Medicine, University of North Dakota, Grand Forks, ND*

JERRY J. WARSH • *Section of Biochemical Psychiatry, Clarke Division, Center for Addiction and Mental Health, Toronto, Canada*

SUSAN R. B. WEISS • *National Institute of Mental Health, Bethesda, MD*

M. GUOQIANG XING • *National Institute of Mental Health, Bethesda, MD*

LI-XIN ZHANG • *National Institute of Mental Health, Bethesda, MD*

Part I

Introduction

From First to Fourth Messengers in the Brain
An Overview

Maarten E. A. Reith

INTRODUCTION

This volume attempts to review *cerebral* signal transduction in health and disease under four different topics: memory, neurodegeneration/apoptosis, depression, and drugs of abuse. Research on intracellular signaling was initiated four decades ago by the pioneering discovery of Sutherland, Rall, and Wosilait of cyclic adenosine monophosphate (cAMP) (*see* ref. *1*). Since then, a multitude of signaling pathways have been uncovered in a wide variety of cells and tissues from yeasts to humans. In general, these pathways are initiated by extracellular signals impacting the cell surface, triggering a cascade that ultimately acts on the cell nucleus. Each cell type expresses a subset of receptors and messenger proteins. *Neuronal* cells many times express unique cell-surface proteins that recognize specific extracellular signals. These surface proteins can be different from, for example, those in *yeast* cells, although they link to similar kinases and phophatases that make up the signaling cascade or effectors in a given subcellular compartment. Such differences have been amply documented. For example, photoreceptors along with their G-proteins and effectors reside in the same outer segment compartment of the rod photoreceptor cell *(2)*, whereas receptors and G-proteins are asymmetrically distributed in neuronal growth cones *(3)* and rat Sertoli cells *(4)*. There is also evidence that protein–tyrosine kinases produce distinct cellular responses as a function of subcellular location *(5)*. Not only is location important, but different cell types are known to express different complements of receptors, G-proteins, and effectors. With regards to G-proteins, there are more than 20 distinct α-subunits, 5 β-subunits, and 10 γ-subunits. These subunits are capable of producing a multitude of combinations to form heterotrimers linked to heptahelical receptors. Different cell types and different effector systems use different subunit combinations, and many receptor subtypes can be linked to more than one G-protein at a time *(2)*.

From: *Cerebral Signal Transduction: From First to Fourth Messengers*
Edited by: M. E. A. Reith © Humana Press Inc., Totowa, NJ

Other examples of cell-dependent differences in components of signaling systems can be found in the family of calmodulin (CaM)–phosphodiesterases (PDEs), which are involved in cyclic nucleotide breakdown. CaM–PDE1A is the only CaM–PDE isozyme expressed in kidney medulla, whereas brain cells can express CaM–PDE1A, B, and C with distinct regional patterns of distribution *(6)*. Within the Ras subfamily of small G-proteins, Rab3A is found only in cells with regulated secretion, including neuronal cells. Evidence suggests a role of Rab3A in Ca^{2+}-dependent exocytosis, in conjunction with rabphilin-3A, which is expressed only in the brain *(7)*. It is clear that cell specificity in signaling pathways can arise from different subtypes or subcellular locations of the individual signaling components. Crosstalk among various signaling cascades, and the cell-dependent differences in ligand, receptor, G-protein, and effector interactions adds to the complexity and yields many different cascading scenarios.

Despite the complex interactions, general signaling mechanisms used by many cell types can be described. For example, the growth-factor-triggered mitogen-activated protein kinase (MAPK) signaling cascade has a basic pattern in both the yeast *S. cerevisiae* and in vertebrates. However, different MAPK isoforms serve at a given level for each signaling pathway. MAPK isoforms serve in *S. cerevisiae* as FUS3 and KSS1 and in vertrebrates as extracellular regulated kinases ERK-1 and -2 *(8)*. The STE11 protein at the MAP3K (i.e., two kinase levels upstream from MAPK) level in *S. cerevisiae* is homologous to mitogen-activated, ERK-activating kinase (MEK) kinase (MEKK) in vertebrates *(8)*.

Growth-factor-initiated signaling pathways in the brain have been shown to be involved in the action of drugs of abuse *(9)*. In addition, signaling pathways in brain lead, via phosphorylation of the transcription factor cAMP response element-binding protein (CREB), to the expression of immediate early genes (IEGs) such as *c-fos (10–14)* (*see also* Chapters 2, 8, 9, 10, 12, and 13). Regulation of the expression of CRE-bearing genes through CREB binding is not limited to the brain *(15)*.

The purpose of this chapter is to briefly summarize the major signaling pathways. Signaling information will be compared from a variety of systems, mostly but not limited to the brain. It is hoped that this review will assist the reader to place the detailed signaling information in the following chapters in a larger context and help to explore potential interfaces between pathways. First, the messengers are treated in a "horizontal" manner, categorized by their placement as first, second, third, and fourth in the sequence of events across cascades. Second, the messenger cascades are described in a "vertical" way, for each cascade separately. Third, potential points of crosstalk between the cascades are presented as an important mechanism for multiple cellular responses to a given extracellular signal.

MESSENGERS USED IN SIGNAL TRANSDUCTION: HORIZONTAL APPROACH

First Messengers

First messengers are extracellular signals impacting the cell surface, thereby setting in motion a sequence of signaling events. Extracellular signals capable of triggering signaling cascades include neurotransmitters, neuromodulators, or hormones acting upon receptors. In addition, nerve impulses can serve as extracellular signals. Certain receptors can be ionotropic, so that activation by a first messenger triggers influx of a second messenger such as Ca^{2+} (Table 1). Receptors can also be linked to G-proteins coupled to an ion channel or to other second-messenger systems via adenylate cyclase, phospholipase C (PLC), nitric oxide (NO) synthase, or phospholipase A_2 (PLA_2) (Table 1). Examples of signaling pathways that differ from the classic synaptic point-to-point transmission are becoming more and more numerous. Neuromodulators, such as peptides and small molecules like adenosine, can act as first messengers in the brain. Hormones such as progesterone and adrenal steroids can also exert effects on plasma membrane receptors. Classically studied steroid hormone effects involve intracellular receptors (*see* Third Messengers). Finally, nerve impulses can act as first messengers by depolarization-induced Ca^{2+} channel opening allowing influx of the second messenger Ca^{2+} (Table 1).

Second Messengers

The mechanism by which the first messenger triggers the appearance of the second messenger depends on the event that links the two (Table 1). In the case of the second messengers cAMP, cyclic guanosine monophosphate (cGMP), diacylglycerol (DAG), inositol triphosphate (IP_3), and arachidonic acid (AA), the link between the activation of a receptor by the first messenger and the stimulation of a second messenger occurs via G-proteins. The G-proteins (G_s, G_i, G_o, $G_{i,o}$, G_q) make up a complex family with many different combinations of the various α-, β-, and γ-subunits (*see* Chapters 5 and 11 for G-protein changes in Alzheimer's disease and bipolar affective disorder, respectively). Stimulation of the G_s subfamily increases adenylate cyclase activity, inhibits Na^+ channels, and opens Ca^{2+} channels, whereas G_i subfamily activation has the opposite effect and can also promote cGMP phosphodiesterase. G_o protein activation closes Ca^{2+} channels, whereas $G_{i,o}$ proteins inhibit adenylate cyclase and stimulate the β isoform of PLC. G_q

Table 1
Messenger Pathways: A Horizontal View

Messengers with links	Ca²⁺/CaM pathways	cAMP/PKA pathways	DAG/PKC pathways[a]	NO/PKG pathways	AA pathways	Steroid receptor pathways
First messenger	Neurotransmitter or nerve impulse	Neurotransmitter, modulator, or hormone	Neurotransmitter, modulator, or hormone	Neurotransmitter	Neurotransmitter	Neurotransmitter
Link	Cation channel of NMDA receptor, receptor linked by G-protein to Ca²⁺ channel, or voltage-sensitive Ca²⁺ channel	G_s, G_i, or $G_{i,o}$ protein-coupled receptor → adenylate cyclase	G_q protein-coupled receptor → PLC	Cation channel of NMDA receptor → Ca²⁺/CaM → NO synthase	Cation channel of NMDA receptor → Ca²⁺ → PKC → PLA₂	G_s protein-coupled receptor → adenylate cyclase
Second messenger	Ca²⁺	cAMP	DAG, IP₃	NO	AA	cAMP
Link	CaM KI, II, IV	PKA	PKC (guanylate cyclase)[a]	Guanylate cyclase	K⁺ channels, cannabinoid receptors, glutamate transporters, dopamine transporters	PKA → phosphorylation steroid receptor
Third messenger	CREB, SRF, SIF	CREB, SRF, SIF, IRBP	IRBP, Raf (cGMP)[a]	cGMP		Phosphorylated steroid receptors as transcription factors
Link	Binding to CRE, SRE, SIE	Binding to CRE, SRE, SIE, IRE	Binding to IRE; MAPK	PKG		Binding to RE
Fourth messenger	Fos, prodynorphin	Fos, prodynorphin, Jun B, Zif/268, Fos B[b], ΔFosB[b], Fras[b], Jun[b]	Jun B, Fos[c], Jun[c]			

Note: In composing the table, the following sources were used: refs. *8,13,16–18,21–26,40–45,* and *71.* The information is not meant to be exhaustive, especially on the transcription factors, which are subject of active ongoing research.

[a] Activity/compound in parentheses are less well studied.

[b] Possible fourth messengers resulting from PKA stimulation but third messenger unknown.

[c] Expression known to be stimulated by PKC activation but third messenger unknown.

protein activation can enhance PLC activity, which results in an increase of IP_3 and DAG. DAG can stimulate protein kinase C (PKC) and increase cGMP, as does the unique messenger nitric oxide (NO) (Table 1). AA has been reported to modulate neurotransporter function for glutamate *(16–18)*, glycine *(19)*, γ-aminobutyric acid *(20)*, and dopamine *(21)*.

The second messenger Ca^{2+}, in addition to being generated by influx through G-protein-coupled Ca^{2+} channels, can be increased intracellularly by activation of the *N*-methyl-D-aspartate (NMDA) receptor. Voltage-dependent Ca^{2+} channels can also increase Ca^{2+} by opening upon depolarization. The NMDA receptor plays an additional role in the production of the second messenger NO by Ca^{2+}/CaM stimulation of NO synthase activity (Table 1).

Third Messengers

Third messengers are generally transcription factors that are phosphorylated by protein kinases, which are, in turn, under the influence of various second-messenger systems (*see* Table 1). For example, the second messenger Ca^{2+} (produced intracellularly by either Ca^{2+} influx through the NMDA receptor, Ca^{2+} channels coupled to G-proteins, or voltage-sensitive Ca^{2+} channels) stimulates Ca^{2+}/CaM-dependent kinase (CaM K) I, II, or IV, which phosphorylates the third messengers such as CREB, serum response factor (SRF), and *sis*-inducable factor (SIF). These third messengers bind to the cAMP response element (CRE), SRF response element (SRE), and SIF response element (SIE), respectively *(13,22)*. Another example is the second-messenger cAMP stimulation of protein kinase A (PKA), which phosphorylates CREB, SRF, SIF, or the inverted repeat element (IRE) binding protein (IRBP) *(23)*. DAG, which is formed by activation of receptors linked to the phospholipid signaling system, activates PKC. Endogenous PKC is available in the nucleus to affect the phosphorylation state and activity of several transcription factors. Activated PKC can also stimulate the Ras (small G-protein) pathway, which leads from Ras and Raf (a cytoplasmic serine/threonine protein kinase) to MAPK (Table 1). Finally, several members of the steroid receptor family (i.e., for progesterone or vitamin D, or the orphan receptor chicken ovalbumin upstream promotor [COUP-TF]) can be regarded as third messengers, as these receptors can be phosphorylated by PKA, which is stimulated by cAMP. These cytoplasmic receptors then function as ligand-regulated transcription factors upon translocation to the nucleus (Table 1).

Fourth Messengers

The third messenger CREB, itself a transcription factor, induces the expression of IEGs that encode transcription factors such as Fos (the protein of the *c-fos* gene). These transcription factors then act as a fourth messengers (*see*

Table 1). CREB also binds to the CRE of the promoter of prodynorphin, which plays an important role in the autofeedback system for inhibition of dopaminergic overactivity *(11,24)*. SRF and SIF regulate the expression of Fos by the *c-fos* gene, whereas the Jun B promoter bears an IRE for IRBP, the activity of which is inducable by both PKA and PKC *(23)*. The promoter of the *Zif/268* gene carries an SRE for SRF *(25)*. In contrast, more work is needed to identify the third messenger for the activation of Fos B, ΔFosB, the fos-related antigens (Fras), and Jun. These transcription factors are likely to result from PKA activation as suggested by the similarity between their response and that of Fos to stimulation of the dopaminergic system in the brain *(26)*. In addition, there is evidence implicating the cAMP/PKA pathway in the effect of indirect dopamine agonists *(27,28)*. Jun can be activated by forskolin, which increases cAMP *(25)*; however, the third messengers involved are not known. *c-Fos* and *c-Jun* can also be induced by PKC activation to express Fos and Jun *(25)*, but, again, the third messengers have not been identified in these pathways.

In the brain, much attention has been devoted to the role of IEGs in the dynamic changes that occur when the dopamine system is activated. Expression of Fos by *c-fos* is induced by the combined stimulation of dopamine D_1 and D_2 receptors. Activation of these receptors stimulates, via G_s, adenylate cyclase (enabling PKA to phosphorylate CREB). D_1 receptor antagonists can attenuate *c-fos* induction *(13)*. A complex D_1–D_2 interaction has also been reported for the induction of the IEG *zif/268* (giving the product Zif/268) by indirect dopamine agonists *(29,30)*. An additional effect of indirect dopamine agonism is the induction of *jun B*, occurring most likely through an inverted repeat element in the *jun B* promoter which responds to PKA and PKC *(23,31)*. Other transcription factors acting as third messengers are the Fras (Table 1). In the context of the dopamine system, which plays a major role in the effects of many drugs of abuse [*see* reviews by Koob *(32)*, Hyman *(11)*, and Fibiger et al. *(33)*], it is of interest that these Fras build up in time as part of the AP-1 complex that binds to the AP-1 site on the DNA as a result of chronic treatment with drugs of abuse *(34,35)*. It has been suggested *(36)* that "chronic Fras" mediate long-lasting changes in brain reward circuitries. A recent study with *FosB* mutant mice demonstrates that chronic Fras are products of the *FosB* gene, specifically the ΔfosB isoforms *(37)*.

Fifth Messengers and Beyond

In the classical messenger pathways as depicted in Table 1, fifth messengers are proteins expressed upon binding of fourth-messenger transcription factors, such as a heterodimer of Fos and Jun to the AP-1 site. A chain of

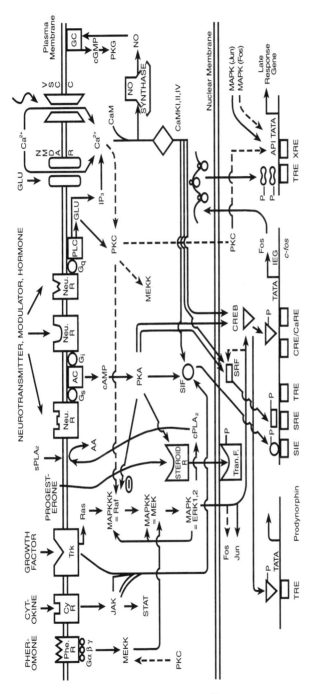

Fig. 1. Major signaling cascades and their interactions. Overview of signaling pathways leading from extracellular signals to the nucleus. Solid arrows denote direct stimulatory pathways, except where a minus sign is placed at the arrowhead (PKA to Raf). Broken arrows indicate indirect stimulatory pathways. AC, adenylate cyclase; Cy, cytokine; GC, guanylate cyclase; GLU, glutamate; Neu, neurotransmitter; Phe, pheromone; R, receptor; Tran. F, transcription factor; VSCC, voltage-sensitive Ca^{2+} channel; TRE, 12-O-tetradecanoylphorbol 13-acetate (TPA) response element in *c-fos* promoter. Other abbreviations, see last Subheading of text. The following sources were used in composing this figure: refs. *1,8,9,13,22,24,* and *38–48.*

9

events is possible with one transcription factor leading to another, and research is just beginning to unravel such pathways.

There are also pathways that do not use the classical second messengers as depicted in Table 1. One such example is the growth factor pathway (*see* Neurotrophic Factors and Pheromones and Fig. 1). Growth factor pathways contain many steps in which a kinase is itself the substrate of an other upstream kinase. For example, MAP kinase kinase kinase (MAPKKK = Raf) phosphorylates MAP kinase kinase (MAPKK = MEK), which phosphorylates MAPK (made up of two isoforms ERK-1 and -2, and an alternatively spliced form p40^{MAPK} *[8,39]*). If each step is counted as a messenger, with the growth factor being the first messenger, MAPK would be the fifth messenger, but this nomenclature is not commonly used. In addition, with the many interfaces occurring between various pathways (*see* Interfaces Between Signaling Cascades: Crosstalk), the assignment of second, third, and so on messenger would become a function of the pathway to which one is referring. Clearly, a circuitry approach (Fig. 1), although cumbersome, is more realistic than a linear approach (Table 1).

SIGNAL TRANSDUCTION CASCADES: VERTICAL APPROACH

cAMP/PKA Cascade

This signaling system is initiated by the action of neurotransmitters, neuromodulators, or hormones on G-protein-coupled receptors which stimulate or inhibit adenylate cyclase (Fig. 1). cAMP is required for the activity of PKA, which phosphorylates SIF, SRF, and CREB. These transcription factors in turn promote the expression of IEGs, and CREB–P also promotes the expression of prodynorphin. The example of the promoter region of the *c-fos* gene (Fig. 1) is a reminder of the fact that protein expression is the result of the integrated effects of multiple transcription factors. The *c-fos* promoter carries, in addition to the SIE and SRE, the bifunctional CaRE/CRE, which confers responsivity to Ca^{2+} and cAMP *(13,25)*. The signaling cascade that leads from cAMP to CREB has been implicated in the action of antidepressants (*see* Chapter 9) effects of opiates and cocaine (Chapter 12), and in learning and memory processes (Chapters 2 and 3). PKA is needed for the consolidation of short-term into long-term memory (*see* Chapters 2 and 3).

Ca^{2+}/Calmodulin-Dependent Protein Kinase Cascade

Ca^{2+} influx can occur through the cation channel of the NMDA receptor, a Ca^{2+} channel linked by a G-protein to a receptor, or a voltage-sensitive Ca^{2+} channel. In conjunction with CaM, Ca^{2+} stimulates CaM K which can

phosphorylate SIF, SRF, or CREB (Fig. 1). These are the same intermediates as we encountered earlier in the cAMP/PKA cascade, and this is a
reminder of how cell pathways can utilize common intermediates for multiple purposes. For example, cells with D_1 dopamine receptors will use the
Ca^{2+}/cAMP pathway, and cells with NMDA receptors will use the calmodulin kinase pathway. The type of proteins produced as a result will depend on
other transcription factors present in the cell. As we saw earlier, promoters
generally have response elements for more than one transcription factor and
the end result will reflect their combined effects. In addition, complex regulatory interactions can occur between transcription factors. For example,
protein products of the *fos* and *jun* families can form heterodimeric and
homodimeric complexes through their leucine zipper domains *(49)*. These
dimers, by binding to the AP-1 site, can either activate (c-Fos/c-Jun) or repress
(c-Fos/Jun-B) transcription. An additional complexity is that c-Jun can activate the *c-Jun* promoter, whereas Jun-B inhibits the *c-Jun* promoter. Thus,
an entire network of signals determines the array of IEGs in a particular cell
with likely different final outcomes depending on the cell type. However,
one can envision situations where it would be desirable for different stimuli
impacting the same cell to have the same end result. For example, the application of stress as well as growth factors can increase general protein synthesis, although even in this case two different kinases downstream from MAPK
appear to be involved (i.e., MAPK-activated protein kinase [MAPKAP-K]
1_a and MAPKAP-K2 *(40)*.

CaM KII is thought to play a role in synaptic strengthening needed for
proper spatial memory (*see* Chapter 2).

DAG/PKC Cascade

An additional effect of G-protein-coupled receptor activation is the stimulation of the PLC pathway (Fig. 1). As a result, DAG is produced; it is hydrophobic and stays associated with the membrane, where it in turn stimulates
PKC. In the absence of stimulation, PKC resides mostly in the cytosol. A
novel model is advanced in Chapter 4 for some forms of cytosolic PKC that
are "activation-prone." PKC can alter the phosphorylation state of many
proteins including several transcription factors (*see* Third Messengers and
Fourth Messengers). Examples of receptor-mediated effects of PKC are the
norepinephrine- or GABA (γ-aminobutyric acid)-induced decreases in Ca^{2+}
current in chick sensory neurons *(50)*. It is important to note that PKC is a
family of kinases, which include the conventional Ca^{2+}-dependent PKC-α,
-βI, -βII, and –γ, the newer Ca^{2+}-independent PKC-δ, -ε, -η, and -θ, and the
non-DAG/phorbol ester-activatable PKC-ξ, and -λ*(51,52)*.

Deficiencies in the DAG/PKC cascade, as well as in the cAMP/PKA cascade, have been implicated in the etiology of Alzheimer's disease (*see* Chapter 5). PKC activation has been shown to be involved in memory formation (Chapter 3). In addition, the PKC pathway is thought to play a prominent role in regulating inactivation of monoamine transporters, for example, by PKC-mediated phosphorylation of the dopamine transporter (Chapter 14) .

NO/PKG Cascade

This cascade involves the stimulatory effect of Ca^{2+}/CaM on NO synthase, the activation of guanylate cyclase by NO, and the increase in cGMP-dependent protein kinase G (PKG) (Fig. 1). Because NO is a gas, it can act on guanyate cyclase in the same cell where it is formed, or in neighboring cells following diffusion. There are various types of NO synthase, of which type I occurs in the brain. In the substantia nigra, the NO/PKG signaling pathway has been shown to regulate the state of phosphorylation of DARP-32, a protein modulated by dopamine through the cAMP/PKA pathway *(53)*. A recent and much publicized application of part of the NO/PKG signaling pathway is the treatment of erectile dysfunction with Viagra (sildenafil citrate), which inhibits phosphodiesterase type 5 *(54)*. This phosphodiesterase degrades cGMP in the corpus cavernosum of the penis, and its blockade with sildenafil increases the effect of local levels of NO released during sexual stimulation, leading to increased levels of cGMP, which is known for its smooth-muscle relaxing activity *(55)*; this causes increased blood flow to the corpus cavernosum. There is also recent evidence suggestive for the involvement of a central effect of NO in the control of penile erection in animal studies *(56)* (*see also* Chapter 6), but it is not known whether this contributes to the effect of sildenafil in humans. The inappropriate sexual behavior of neuronal NO synthase (NOS) null mice is discussed in Chapter 6. Multiple actions of NO in cell function and neurological disorders are also described in Chapters 6 and 8. In addition, the NO system is implicated in various aspects of learning (Chapter 3).

AA Cascade

The major pathway usually described for the formation of AA involves glutamate acting on NMDA receptors that can generate Ca^{2+} influx in situations of excitotoxicity (*see* ref. *57*). Intracellular Ca^{2+} activates PKC, which, in turn, stimulates cytosolic (c) PLA_2, releasing AA from membrane phospholipids (*see* Fig. 1). There is also a secreted (s) form of PLA_2, which acts from the outside under conditions where millimolar concentrations of Ca^{2+} are present, as is the case for the extracellular environment. In contrast. $cPLA_2$ is active at cytosolic micromolar levels of Ca^{2+} *(45)*. Released AA can alter transporter function in the brain, primarily acting in an inhibitory fashion *(16,18–21)*. AA also affects the release of dopamine *(58)*. Eicosanoids (i.e., metabolites of AA) have been reported to affect neuronal S-K$^+$ *(59)* and M-K$^+$ *(60)* currents and various activities of transmitters and peptides *(41)*. Finally, anandamide, the endogenous ligand for cannabinoid receptors, is arachidonylethanolamine, which can be synthesized from AA and ethanolamine [see the recent review by Axelrod and Felder *(61)*].

Neurotrophic Factors and Pheromones: MAP Kinase Cascade, JAK/STAT Pathway, and MEKK/MEK Circuit

Neurotrophic factors include the family of neurotrophins consisting of nerve growth factor (NGF), brain-derived neurotrophic factor (BDNF), and neurotrophins-3 and -4 (NT3 and NT4). In addition, the family of cytokines includes ciliary neurotrophic factor (CNTF), leukemia inhibitory factor (LIF), and oncostatin M. The neurotrophins act through growth-factor-receptor tyrosine kinase (Trk), which activates Ras. Ras is a small G-protein that enhances the activity of Raf, a cytoplasmic serine/threonine protein kinase that phosphorylates MAP kinase kinase (MAPKK, or MEK) (*see* Fig. 1). This, in turn, phosphorylates MAPK (ERK1,2), which activates a variety of substrate proteins such as the transcription factors Jun or CREB directly, and Fos or SRF indirectly *(1,8,13)*. Binding of the cytokine class of compounds activates the cytoplasmic protein kinase Janus kinase (JAK),which phosphorylates the Signal transducers and activators of transcription (STAT) family of proteins regulating expression of various proteins, including SIF (Fig. 1). Transcription factors such as SIF can be phosphorylated in the cytoplasm by a combination of two JAK-type kinases or one JAK-type kinase plus one Trk. These proteins can then form complexes with other regulatory proteins, including STAT, allowing translocation to the nucleus for transcription regulation *(13)*. Pheromones act on G-protein-coupled serpentine receptors resulting in MEKK, which phosphorylates MEK (Fig. 1). As pointed out by Bratigan *(38)*, *dephosphorylation* should not be ignored in all of these pathways. Protein phosphatase type-1 (PP1) and type-2A (PP2A), especially the latter, play a major role in these reversed steps.

Brain-derived neurotrophic factor (BDNF) has been implicated in depression and the action of antidepressants (*see* Chapters 9 and 10). The possibility of BDNF or other growth factors being involved in the beneficial action of transcranial magnetic sitimulation, a novel potential antidepressant treatment, is presented in Chapter 10. The involvement of several Ras signaling pathways in mediating cell death versus survival is discussed in Chapter 7.

INTERFACES BETWEEN SIGNALING CASCADES: CROSSTALK

There are many points of interaction between the different signaling pathways. Cascades can branch out because certain kinases act on more than one substrate. Alternatively, pathways can converge because different kinases act on the same substrate, or produce transcription factors that act on response elements in the same promoter. In addition, transcription factors can interact with each other or feedback on their own production, causing an intricate web for regulation of gene expression. In the following, attention is devoted to some of the major points of pathway divergence, pathway convergence, and transcription networking.

Pathway Divergence

Ca^{2+} appears to enhance action of PKC. In addition, Ca^{2+}/CaM stimulates NO synthase as well as CaM KI, II, IV (*see* Fig. 1). These characteristics in the Ca^{2+}/CaM cascade branch out into the DAG/PKC cascade and the NO/PKG cascade. In assessing a common role for Ca^{2+} in different cascades, it may be important where local Ca^{2+} changes occur. For example, influx of Ca^{2+} through the cation channel of the NMDA receptor may not equate with that through the voltage-sensitive Ca^{2+} channel. Indeed, Ca^{2+} influx through the cation channel of the NMDA receptor in cultured neurons induces *c-fos* through the SRE, whereas Ca^{2+} influx through the voltage-sensitive Ca^{2+} channel induces *c-fos* by a mechanism not involving the SRE (*see* ref. *13*). It is likely that the subcellular distribution of Ca^{2+}-activated proteins plays an important role in the final activation pattern. In this context, it is important to note that optical measurements of Ca^{2+} signals often show local changes in discrete regions of cells *(50)*.

CaM K phosphorylates CREB, SRF, and SIF (Fig. 1), thereby crossing over from the calmodulin cascade into the DAG/PKC, MAP kinase, and JAK/STAT pathways.

PKA phosphorylates CREB, SRF, SIF, and intracellular steroid receptors (Fig. 1). It induces the *Jun-B* gene and can also, directly or indirectly, inhibit *c-Jun (23)*. Furthermore, PKA has an inhibitory influence on the pathway that leads from Raf to MAPK, most likely by phosporylation of Ser 43 in Raf-1 (one of the isoforms of Raf) or Ser429 or Ser446 in B-Raf, which inhibits the binding of Raf to GTP *(1)*. The latter mechanism may not apply to all cell types and is still being debated. All together, PKA is a crossover point for the cAMP/PKA pathway, the MAP kinase cascade, and the calmodulin pathway.

PKC stimulates the pathway that leads from Raf to MAPK, indirectly activating $cPLA_2$ *(8,13,45,47)* (Fig. 1). The stimulation by PKC may occur through the newly isolated PKC-activated tyrosine kinase PYK-2, which increases MAP kinase activity in PC 12 cells *(1)*. PKC also activates MEKK by a mechanism not completely understood. MEKK can replace the yeast MAP3K, BCK1, which is placed after PKC in the cell-wall signaling cascade *(see* ref. *8)*. Furthermore, PKC can phosphorylate BCK1 in vitro *(see* ref. *62)*. PKC in the nucleus affects the phosphorylation state of a number of transcription factors, and it induces the genes *c-fos, c-Jun, Zif/268 (25)*, and *Jun-B (23)*. Taken together, PKC affects the MAP kinase cascade, the MEKK/ MEK circuit, the AA cascade, and other cascades that share transcription factors under the influence of PKC.

Recently, PKG has been proposed to be part of a neuroprotective pathway triggered by a metabolite of β-amyloid precursor protein (βAPP), sAPP. sAPP formation prevents the release of amyloid β peptide (Aβ) because a cleavage occurs between aminoacids 16 and 17 of Aβ. The latter protein forms senile plaques in Alzheimer's disease *(see* Chapter 5). sAPP acts through a putative receptor linked to a G-protein, which stimulates guanylate cyclase. cGMP, in turn, stimulates PKG, which could activate a protein phosphatase resulting in dephosphorylation of K^+ channels producing hyperpolarization. The latter counteracts the depolarizing action of glutamate during excitotoxic sequelae *(63,64)*. PKG may also activate the transcription factor NFκB, regulating expression of several neuroprotective gene products. NO may be involved in this process, as inhibition of NO synthase prevents NFκB activation *(see* ref. *65)*. This interaction may link this arm of the signaling pathway to the NO/PKG pathway (only the latter is shown in Fig. 1). In terms of neuronal cell death, reactive NO intermediates could be thought to counteract the potential neuroprotective effects of NO via PKG modification of K^+ channels and/or NFκB production.

MAPK can have the upstream kinases MAPKK and Raf as substrates *(8)* (Fig. 1). It can phosphorylate $cPLA_2$, CREB, and Jun, and, indirectly, Fos or SRF *(1,8,13,45)*. Thus, MAPK could trigger endpoints of the MAP kinase cascade, AA pathway, cAMP/PKA cascade, and calmodulin pathway.

As detailed in Chapter 6, NO, can, in addition to stimulating guanylate cyclase, also activate the Ras/MAPK pathway involving NMDA receptors. This effect of NO probably occurs through direct activation of Ras by redox modulation of Cys-118 in Ras (Chapter 6). In addition, NO stimulates cyclooxygenase leading to enhanced prostaglandin production (Chapter 6). Examples of Ras signaling pathway convergencies can be found in Chapter 7.

Pathway Convergence

One JAK-type kinase in combination with one Trk and STAT can result in phosphorylation of SIF. These interactions converge the JAK/STAT and MAP kinase pathways (Fig. 1).

$cPLA_2$ is the substrate for MAPK activated by PKC *(45,47)*, thereby linking the AA and DAG/PKC pathways (Fig. 1). An example of convergence of pathways through the $cPLA_2$–MAPK connection is the activation of the MAP kinase cascade, which, in turn, activates PKA in newborn human arterial smooth-muscle cells *(1)*. The growth factor signal platelet-derived growth factor (PDGF) stimulates MAPK, which, through $cPLA_2$, forms AA. The resulting prostaglandin E_2 (AA metabolite) acts on its receptor to stimulate adenylate cyclase leading to cAMP activation of PKA.

CREB can be phosphorylated by MAPK, PKA, and CaM K (Fig. 1). Thus, CREB is an end product shared by the MAP kinase cascade (indirectly affected by PKC), the cAMP/PKA pathway, and the calmodulin cascade. A CREB kinase has also been postulated *(66,67)* and recent evidence supports its existence. For example, in murine embryonic palate mesenchymal cells, transforming growth factor-beta (TGF-beta) induces CREB Ser133 phosphorylation and does not involve ERK1 or -2, PKA, or CaM K *(68)*. In other systems, MAPK activates a kinase that, in turn, phosphorylates Ser133 of CREB. This CREB kinase has been characterized as reactive phosphatase-treated S6 kinase (RSK) 2 *(69)* (which also phosphorylates SRF; *see* the next paragraph) in K562 and PC12 cells and as MAPKAP-K2 *(70)* in SK-N-MC cells. RSK2 belongs to the RSK family of MAPK-activated protein kinases. The RSK family is also called MAPKAP-K1, not to be confused with the enzyme MAPKAP-K2, another MAPK-activated protein kinase.

SRF can be directly phosphorylated by PKA and CaM K, and indirectly by MAPK, thereby linking the cAMP/PKA, calmodulin, and MAP kinase cascades (Fig. 1). The indirect activation of SRF by MAPK occurs through the intermediary kinase, RSK2 *(13)*.

SIF can be phosphorylated by PKA, CaM K, and a JAK/Trk/STAT or JAK/JAK/STAT complex (Fig. 1). Thus, SIF is an end product shared by the cAMP/PKA, calmodulin, and JAK/STAT cascades.

The production of Fos is stimulated by the binding of the phosphorylated transcription factors CREB, SRF, or SIF to the promoter region of *c-fos* (Fig. 1). Thus, expression of Fos can be activated by many different signaling pathways that contain these transcription factors upstream from Fos.

All NOS isoforms contain consensus phosphorylation sites, and neuronal NOS activity is decreased by activation of PKA, PKC, PKG, and CaM K (*see* Chapter 6).

Transcription Networking

It is likely that changes in gene expression underlie long-term changes in brain function as a result of learning and memory and through adaptive responses to drugs and other external agents. Changes in gene expression are thought to occur in two phases. First, IEGs are induced usually within 15 min after the stimulus and for no longer than 30–60 min. Second, late-response genes (LEGs) are induced on a slower time-scale and are dependent on the action of IEG proteins (*see* Chapter 10 for examples in amygdala kindling phenomena). IEGs respond to various second messenger systems (for *c-fos, see* Fig. 1) and the Ras-activated pathway (*see also* Chapter 7). Promoters of transcription factor genes typically respond to more than one type of signaling pathway, and transcription factors from different families combine to either induce or repress transcription of various proteins, including IEGs. Thus, most likely a network of signals activates an array of IEGs that together with other messenger systems affect transcription of several proteins capable of altering gene expression *(49)*. Future work will further characterize these networks in the brain, extending our knowledge as described in this book existing at this point in time.

ACKNOWLEDGMENTS

I would like to thank the National Institute on Drug Abuse (DA 08379) for supporting our experiments in which we study the involvement of first and second messengers in the action of cocaine, and J. C. Milbrandt for discussions regarding this manuscript.

NOMENCLATURE

Aβ	Amyloid β peptide
βAPP	β-amyloid precursor protein
sAPPα	Secreted α form of amyloid precursor protein
BDNF	Brain-derived neurotrophic factor
CaM KI, II, IV	Ca^{2+}/calmodulin-dependent protein kinase I, II, IV
CaM	Calmodulin
CNTF	Ciliary neurotrophic factor
CREB	cAMP response element binding protein
CRE	cAMP response element
DAG	Diacylglycerol
EGF	Epidermal growth factor
ERK	Extracellular signal regulated kinase
Fra	fos-Related antigen
GABA	γ-aminobutyric acid
IEG	Immediate early gene
IRBP	Inverted repeat element binding protein
IRE	Inverted repeat element

JAK	Janus kinase
LIF	Leukemia inhibitory factor
LSD	Lysergic acid diethylamine
MAP	Mitogen-activated protein
MAPK	MAP kinase
MAPKAP-K	MAPK-activated protein kinase
MEK	MAP kinase kinase
NMDA	N-methyl-D-aspartic acid
NT3	Neurotrophin-3
NT4	Neurotrophin-4
NO	Nitric oxide
PDGF	Platelet-derived growth factor
PIP_2	Phosphatidylinositol 4,5-bisphosphate
PKA	cAMP-dependent protein kinase A
PKC	Protein kinase C
PKG	(cGMP-dependent) Protein kinase G
PLA_2	Phospholipase A_2 (c- or s-form for cytosolic or secreted)
PLC	Phospholipase C
Raf	A cytoplasmic serine/threonine protein kinase
Ras	Small G-protein
RSK	Reactive phosphatase-treated S6 kinase = MAPKAP-K1
SIE	*sis*-Inducable factor response element
SIF	*sis*-Inducable factor
SRE	Serum response factor response element
SRF	Serum response factor
STAT	Signal transducers and activators of transcription
TGF-beta	Transforming growth factor-beta

REFERENCES

1. Graves, L.M., Bornfeldt, K., and Krebs, E. G. (1997) Historical perspectives and new insights involving the MAP kinase cascades, in *Signal Transduction in Health and Disease* (Corbin, J. D. and Francis, S. H., eds.), Lippincott-Raven, Philadelphia, pp. 49–62.
2. Gudermann, T., Kalkbrenner, F., Dippel, E., Laugwitz, K.-L., and Schultz, G. (1997) Specificity and complexity of receptor–G-protein interaction, in *Signal Transduction in Health and Disease* (Corbin, J. D. and Francis, S. H., eds.), Lippincott-Raven, Philadelphia, pp. 253–262.
3. Strittmatter, S. M., Valenzuela, D., Kennedy, T. E., Neer, E. J., and Fishman, M. C. (1990) G0 is a major growth cone protein subject to regulation by GAP-43. *Nature* **344,** 836–841.
4. Dym, M., Lamsam-Casalotti, S., Jia, M. C., Kleinman, H. K., and Papadopoulos, V. (1991) Basement membrane increases G-protein levels and follicle-stimulating

hormone responsiveness of Sertoli cell adenylyl cyclase activity. *Endocrinology* **128,** 1167–1176.

5. Cantley, L. C. and Songyang, Z. (1997) Specificity in protein-tyrosine kinase signaling, in *Signal Transduction in Health and Disease* (Corbin, J. D. and Francis, S. H., eds.), Lippincott-Raven, Philadelphia, pp. 41–48.

6. Zhao, A. Z., Yan, C., Sonnenburg, W. K., and Beavo, J. A. (1997) Recent advances in the study of Ca2+/CaM-activated phosphodiesterases, in *Signal Transduction in Health and Disease* (Corbin, J. D. and Francis, S. H., eds.), Lippincott-Raven, Philadelphia, pp. 237–251.

7. Sasaki, T., Shirataki, H., Nakanishi, H., and Takai, Y. (1997) Rab3A-Rabphilin-3A system in neurotransmitter release, in *Signal Transduction in Health and Disease* (Corbin, J. D. and Francis, S. H., eds.), Lippincott-Raven, Philadelphia, pp. 279–294.

8. Seger, R. and Krebs, E. G. (1995) The MAPK signaling cascade. *FASEB J.* **9,** 726–735.

9. Nestler, E. J., Berhow, M. T., and Brodkin, E. S. (1996) Molecular mechanisms of drug addiction: adaptations in signal transduction pathways. *Mol. Psychiatry* **1,** 190–199.

10. Cole, R. L., Konradi, C., Douglass, J., and Hyman, S. E. (1995) Neuronal adaptation to amphetamine and dopamine: molecular mechanisms of prodynorphin gene regulation in rat striatum. *Neuron* **14,** 813–823.

11. Hyman, S. E. (1996) Addiction to cocaine and amphetamine. *Neuron* **16,** 901–904.

12. Nestler, E. J., Hope, B. T., and Widnell, K. L. (1993) Drug addiction: a model for the molecular basis of neural plasticity. *Neuron* **11,** 995–1006.

13. Rogue, P. and Malviya, A. N. (1994) Regulation of signalling pathways to the nucleus by dopaminergic receptors. *Cell Signal.* **6,** 725–733.

14. Widnell, K. L., Self, D. W., Lane, S. B., Russell, D. S., Vaidya, V. A., Miserendino, M. J., Rubin, C. S., Duman, R. S., and Nestler, E. J. (1996) Regulation of CREB expression: in vivo evidence for a functional role in morphine action in the nucleus accumbens. *J. Pharmacol. Exp. Ther.* **276,** 306–315.

15. Monaco, L., Lamas, M., Tamai, K., Lalli, E., Zazopoulos, E., Penna, L., Nantel, F., Foulkes, N. S., Mazzucchelli, C., and Sassone-Corsi, P. (1997) Coupling transcription to signaling pathways, in *Signal Transduction in Health and Disease* (Corbin, J. D. and Francis, S. H., eds.), Lippincott-Raven, Philadelphia, pp. 63–74.

16. Trotti, D., Volterra, A., Lehre, K. P., Rossi, D., Gjesdal, O., Racagni, G., and Danbolt, N. C. (1995) Arachidonic acid inhibits a purified and reconstituted glutamate transporter directly from the water phase and not via the phospholipid membrane. *J. Biol. Chem.* **270,** 9890–9895.

17. Volterra, A., Trotti, D., Cassutti, P., Tromba, C., Salvaggio, A., Melcangi, R. C., and Racagni, G. (1992) High sensitivity of glutamate uptake to extracellular free arachidonic acid levels in rat cortical synaptosomes and astrocytes. *J. Neurochem.* **59,** 600–606.

18. Zerangue, N., Arriza, J. L., Amara, S. G., and Kavanaugh, M. P. (1995) Differential modulation of human glutamate transporter subtypes by arachidonic acid. *J. Biol. Chem.* **270,** 6433–6435.

19. Zafra, F., Alcantara, R., Gomeza, J., Aragon, C., and Gimenez, C. (1990) Arachidonic acid inhibits glycine transport in cultured glial cells. *Biochem. J.* **271,** 237–242.

20. Yu, A. C., Chan, P. H., and Fishman, R. A. (1986) Effects of arachidonic acid on glutamate and gamma-aminobutyric acid uptake in primary cultures of rat cerebral cortical astrocytes and neurons. *J. Neurochem.* **47,** 1181–1189.
21. Zhang, L. and Reith, M. E. A. (1996) Regulation of the functional activity of the human dopamine transporter by the arachidonic acid pathway. *Eur. J. Pharmacol.* **315,** 345–354.
22. Yokokura, H., Terada, O., Naito, Y., Sugita, R., and Hidaka, H. (1997) Cascade activation of the calmodulin kinase family. *Signal Transduction in Health and Disease* (Corbin, J. D. and Francis, S. H., eds.) Lippincott-Raven, Philadelphia, pp. 151–157.
23. de Groot, R. P., Auwerx, J., Karperien, M., Staels, B., and Kruijer, W. (1991) Activation of junB by PKC and PKA signal transduction through a novel cis-acting element. *Nucleic Acids Res.* **19,** 775–781.
24. Genova, L., Berke, J., and Hyman, S. E. (1997) Molecular adaptations to psychostimulants in striatal neurons: toward a pathophysiology of addiction. *Neurobiol. Dis.* **4,** 239–246.
25. Cochran, B. H. (1993) Regulation of immediate early gene expression, in *Activation of Immediate Early Genes by Drugs of Abuse* (Grzanna, R. and Brown, R. M., eds.), US Department of Health and Human Services, Rockville, MD, pp. 3–24.
26. Hope, B., Kosofsky, B., Hyman, S. E., and Nestler, E. J. (1992) Regulation of immediate early gene expression and AP-1 binding in the rat nucleus accumbens by chronic cocaine. *Proc. Natl. Acad. Sci. USA* **89,** 5764–5768.
27. Nestler, E. J. (1994) Molecular neurobiology of drug addiction. *Neuropsychopharmacology* **11,** 77–87.
28. Nestler, E. J. and Aghajanian, G. K. (1997) Molecular and cellular basis of addiction. *Science* **278,** 58–63.
29. Wang, J. Q. and McGinty, J. F. (1995) Differential effects of D1 and D2 dopamine receptor antagonists on acute amphetamine- or methamphetamine-induced up-regulation of zif/268 mRNA expression in rat forebrain. *J. Neurochem.* **65,** 2706–2715.
30. Wang, J. Q., Smith, A. J., and McGinty, J. F. (1995) A single injection of amphetamine or methamphetamine induces dynamic alterations in c-fos, zif/268 and preprodynorphin messenger RNA expression in rat forebrain. *Neuroscience* **68,** 83–95.
31. Moratalla, R., Vickers, E. A., Robertson, H. A., Cochran, B. H., and Graybiel, A. M. (1993) Coordinate expression of c-fos and jun B is induced in the rat striatum by cocaine. *J. Neurosci.* **13,** 423–433.
32. Koob, G. F. (1992) Drugs of abuse: anatomy, pharmacology and function of reward pathways. *Trends Pharmacol. Sci.* **13,** 177–184.
33. Fibiger, H. C., Phillips, A. G., and Brown, E. E. (1992) The neurobiology of cocaine-induced reinforcement, in *Ciba Found. Symp.* **166,** 96-111; discussion 111-124.
34. Hope, B. T., Nye, H. E., Kelz, M. B., Self, D. W., Iadarola, M. J., Nakabeppu, Y., Duman, R. S., and Nestler, E. J. (1994) Induction of a long-lasting AP-1 complex composed of altered Fos- like proteins in brain by chronic cocaine and other chronic treatments. *Neuron* **13,** 1235–1244.
35. Hyman, S. E. and Nestler, E. J. (1996) Initiation and adaptation: a paradigm for understanding psychotropic drug action. *Am. J. Psychiatry* **153,** 151–162.
36. Nye, H. E., Hope, B. T., Kelz, M. B., Iadarola, M., and Nestler, E. J. (1995) Pharmacological studies of the regulation of chronic FOS-related antigen induction by

cocaine in the striatum and nucleus accumbens. *J. Pharmacol. Exp. Ther.* **275,** 1671–1680.

37. Hiroi, N., Brown, J. R., Haile, C. N., Ye, H., Greenberg, M. E., and Nestler, E. J. (1997) FosB mutant mice: loss of chronic cocaine induction of Fos-related proteins and heightened sensitivity to cocaine's psychomotor and rewarding effects. *Proc. Natl. Acad. Sci. USA* **94,** 10,397–10,402.

38. Brautigan, D. L. (1997) Phosphatases as partners in signaling networks, in *Signal Transduction in Health and Disease* (Corbin, J. D. and Francis, S. H., eds.), Lippincott-Raven, Philadelphia, pp. 113–124.

39. Cobb, M. H., Xu, S., Cheng, M., Ebert, D., Robbins, D., Goldsmith, E., and Robinson, M. (1996) Structural analysis of the MAP kinase ERK2 and studies of MAP kinase regulatory pathways, in *Intracellular Signal Transduction* (Hidaka, H. and Nairn, A. C., eds.), Academic, New York, pp. 49–65.

40. Cohen, P. (1996) Dissection of protein kinase cascades that mediate cellular response to cytokines and cellular stress, in *Intracellular Signal Transduction* (Hidaka, H. and Nairn, A. C., eds.), Academic, New York, pp. 15–27.

41. Cooper, J. R., Bloom, F. E., and Roth, R. H. (1996a) Modulation of synaptic transmission, in *The Biochemical Basis of Neuropharmacology* (Cooper, J. R., Bloom, F. E., and Roth, R. H., eds.), Oxford University Press, Oxford, pp. 103–125.

42. Cooper, J. R., Bloom, F. E., and Roth, R. H. (1996b) Receptors, in *The Biochemical Basis of Neuropharmacology* (Cooper, J. R., Bloom, F. E., and Roth, R. H., eds.), Oxford University Press, Oxford, pp. 82–102.

43. Dennis, E. A. (1997) The growing phospholipase A2 superfamily of signal transduction enzymes. *Trends Biochem. Sci.* **22,** 1–2.

44. Feldman, R. S., Meyer, J. S., and Quenzer, L. F. (1997) Synaptic structure and function, in *Principles of Neuropsychopharmacology* (Feldman, R. S., Meyer, J. S., and Quenzer, L. F., eds.), Sinauer Associates, Inc., Sunderland, MA, pp. 185–232.

45. Lin, L. L., Wartmann, M., Lin, A. Y., Knopf, J. L., Seth, A., and Davis, R. J. (1993) cPLA2 is phosphorylated and activated by MAP kinase. *Cell* **72,** 269–278.

46. Luttrell, L. M., Van Biesen, T., Hawes, B. E., Koch, W. J., Krueger, K. M., Touhara, K., and Lefkowitz, R. J. (1997) G-Protein-coupled receptors and their regulation. Activation of the MAP kinase signaling pathway by G-protein-coupled receptors, in *Signal Transduction in Health and Disease* (Corbin, J. D. and Francis, S. H., eds.), Lippincott-Raven, Philadelphia, pp. 263–277.

47. Naor, Z., Shacham, S., Harris, D., Seger, R., and Reiss, N. (1995) Signal transduction of the gonadotropin releasing hormone (GnRH) receptor: cross-talk of calcium, protein kinase C (PKC), and arachidonic acid. *Cell. Mol. Neurobiol.* **15,** 527–544.

48. Tonks, N. K. (1996) Protein tyrosine phosphatases and the control of cellular signaling responses, in *Intracellular Signal Transduction* (Hidaka, H. and Nairn, A. C., eds.), Academic, New York, pp. 91–119.

49. Robertson, H. A. (1993) Immediate early gene activation and long-term changes in neural function: a possible role in addiction? in *Activation of Immediate Early Genes by Drugs of Abuse* (Grzanna, R. and Brown, R. M., eds.), NIDA Research Monograph 125, US Department of Health and Human Services, Rockville, MD, pp. 54–71.

50. Nicholls, J. G., Martin, A. R., and Wallace, B. G. (eds.) (1992) *From Neuron to Brain.* Sinauer Associates, Inc., Sunderland, MA, pp. 237–268.

51. Dekker, L. V. and Parker, P. J. (1994) Protein kinase C—a question of specificity. *Trends Biochem. Sci.* **19,** 73–77.
52. Wilkinson, S. E. and Hallam, T. J. (1994) Protein kinase C: is its pivotal role in cellular activation over-stated? *Trends Pharmacol. Sci.* **15,** 53–57.
53. Tsou, K., Snyder, G. L., and Greengard, P. (1993) Nitric oxide/cGMP pathway stimulates phosphorylation of DARPP-32, a dopamine- and cAMP-regulated phosphoprotein, in the substantia nigra. *Proc. Natl. Acad. Sci. USA* **90,** 3462–3465.
54. Stief, C. G., Uckert, S., Becker, A. J., Truss, M. C., and Jonas, U. (1998) The effect of the specific phosphodiesterase (PDE) inhibitors on human and rabbit cavernous tissue in vitro and in vivo. *J. Urol.* **159,** 1390–1393.
55. Katzung, B. G. and Chatterjee, K. (1995) Introduction, in *Basic and Clinical Pharmacology* (Katzung, B. G., ed.), Appleton & Lange, Norwalk, CT, pp. 171–187.
56. Melis, M. R. and Argiolas, A. (1997) Role of central nitric oxide in the control of penile erection and yawning. *Prog. Neuropsychopharmacol. Biol. Psychiatry* **21,** 899–922.
57. Volterra, A., Trotti, D., and Racagni, G. (1994) Glutamate uptake is inhibited by arachidonic acid and oxygen radicals via two distinct and additive mechanisms. *Mol. Pharmacol.* **46,** 986–992.
58. L'hirondel, M., Cheramy, A., Godeheu, G., and Glowinski, J. (1995) Effects of arachidonic acid on dopamine synthesis, spontaneous release, and uptake in striatal synaptosomes from the rat. *J. Neurochem.* **64,** 1406–1409.
59. Buttner, N., Siegelbaum, S. A., and Volterra, A. (1989) Direct modulation of Aplysia S-K$^+$ channels by a 12-lipoxygenase metabolite of arachidonic acid. *Nature* **342,** 553–555.
60. Schweitzer, P., Madamba, S., and Siggins, G. R. (1990) Arachidonic acid metabolites as mediators of somatostatin- induced increase of neuronal M-current. *Nature* **346,** 464–467.
61. Axelrod, J. and Felder, C. C. (1998) Cannabinoid receptors and their endogenous agonist, anandamide. *Neurochem. Res.* **23,** 575–581.
62. Cano, E. and Mahadevan, L. C. (1995) Parallel signal processing among mammalian MAPKs. *Trends Biochem. Sci.* **20,** 117–122.
63. Furukawa, K. (1998) Signaling by beta-amyloid precursor protein, in *Neuroprotective Signal Transduction* (Mattson, M. P., ed.), Humana, Totowa, NJ, pp. 197–220.
64. Mattson, M. P. (1998) Neuroprotective strategies based on targeting of postsynaptic signaling events, in *Neuroprotective Signal Transduction* (Mattson, M. P., ed.), Humana, Totowa, NJ, pp. 301–335.
65. Martin, D., Miller, G., and Fischer, N. (1998) Neuroprotective strategies based on interleukin signaling, in *Neuroprotective Signal Transduction* (Mattson, M. P., ed.), Humana, Totowa, NJ, pp. 185–196.
66. Segal, R. A. and Greenberg, M. E. (1996) Intracellular signaling pathways activated by neurotrophic factors. *Annu. Rev. Neurosci.* **19,** 463–489.
67. Springer, J. E. and Kitzman, P. H. (1998) Neuroprotective strategies involving the neurotrophins and their signaling pathways, in *Neuroprotective Signal Transduction* (Mattson, M. P., ed.), Humana, Totowa, NJ, pp. 1–21.
68. Potchinsky, M. B., Weston, W. M., Lloyd, M. R., and Greene, R. M. (1997) TGF-beta signaling in murine embryonic palate cells involves phosphorylation of the CREB transcription factor. *Exp. Cell Res.* **231,** 96–103.

69. Xing, J., Ginty, D. D., and Greenberg, M. E. (1996) Coupling of the RAS-MAPK pathway to gene activation by RSK2, a growth factor-regulated CREB kinase. *Science* **273,** 959–963.
70. Tan, Y., Rouse, J., Zhang, A., Cariati, S., Cohen, P., and Comb, M. J. (1996) FGF and stress regulate CREB and ATF-1 via a pathway involving p38 MAP kinase and MAPKAP kinase-2. *EMBO J.* **15,** 4629–4642.
71. Chen, J., Nye, H. E., Kelz, M. B., Hiroi, N., Nakabeppu, Y., Hope, B. T., and Nestler, E. J. (1995) Regulation of delta FosB and FosB-like proteins by electroconvulsive seizure and cocaine treatments. *Mol. Pharmacol.* **48,** 880–889.

Part II

Memory

2

Cellular and Molecular Mechanisms of Learning and Memory

Matthew Lattal and Ted Abel

The nature of the cellular basis of learning and memory remains an often-discussed, but elusive problem in neurobiology. A popular model for the physiological mechanisms underlying learning and memory postulates that memories are stored by alterations in the strength of neuronal connections within the appropriate neural circuitry. Thus, an understanding of the cellular and molecular basis of synaptic plasticity will expand our knowledge of the molecular basis of learning and memory.

The view that learning was the result of altered synaptic weights was first proposed by Ramon y Cajal in 1911 and formalized by Donald O. Hebb. In 1949, Hebb proposed his "learning rule," which suggested that alterations in the strength of synapses would occur between two neurons when those neurons were active simultaneously (1). Hebb's original postulate focused on the need for synaptic activity to lead to the generation of action potentials in the postsynaptic neuron, although more recent work has extended this to include local depolarization at the synapse.

One problem with testing this hypothesis is that it has been difficult to record directly the activity of single synapses in a behaving animal. Thus, the challenge in the field has been to relate changes in synaptic efficacy to specific behavioral instances of associative learning. In this chapter, we will review the relationship among synaptic plasticity, learning, and memory. We will examine the extent to which various current models of neuronal plasticity provide potential bases for memory storage and we will explore some of the signal transduction pathways that are critically important for long-term memory storage. We will focus on two systems—the gill and siphon withdrawal reflex of the invertebrate *Aplysia californica* and the mammalian hippocampus—and discuss the abilities of models of synaptic plasticity and learning to account for a range of genetic, pharmacological, and behavioral data.

From: *Cerebral Signal Transduction: From First to Fourth Messengers*
Edited by: M. E. A. Reith © Humana Press Inc., Totowa, NJ

The simpler model system provided by *Aplysia* has made possible the development of a cellular analog of Pavlovian conditioning, a form of associative learning, as well as habituation and sensitization, nonassociative forms of learning. In particular, studies in *Aplysia* have revealed some of the important cellular and molecular mechanisms underlying synaptic plasticity. As we will see, there have been significant advances, but it has remained difficult to relate these changes in synaptic strength to the behavior of the intact animal.

Studies of the mammalian hippocampus and its role in declarative memory have been undertaken at a variety of levels ranging from electrophysiological to pharmacological to genetic. Two major characteristics of the hippocampus —the ability of hippocampal neurons to undergo long-term potentiation (LTP), a persistent increase in synaptic strength resulting from repetitive electrical stimulation, and the existence of place cells, neurons that are active when an animal is located in a particular position in space—have been critically important in developing ideas about how the mammalian hippocampus functions in the acquisition and consolidation of spatial memories. To understand the role of synaptic plasticity in behavior, researchers have turned to genetically modified mice in an attempt to integrate information gained at the molecular and cellular levels with physiological and behavioral studies. The analysis of these mice allows researchers to test whether a particular gene product is important for LTP and provides a useful bridge between molecules and synaptic plasticity on the one hand and systems of neurons and behavior on the other. In this way, understanding the signal transduction mechanisms that underlie synaptic plasticity, combined with the use of powerful genetic technologies, has provided us with insights into the molecular basis of learning and memory.

SYNAPTIC PLASTICITY IN *APLYSIA*

Although much modern research focuses on synaptic plasticity in mice, a large portion of our current knowledge of signal transduction pathways involved in the learning process comes from the study of the opisthobranch *Aplysia*. This organism has been an attractive model for the cellular analysis of learning for several reasons. One is that the stimulus inputs and behavioral outputs are relatively simple, yet complex enough for the systematic investigation of the mechanisms of conditioning. A second is that the neurons in *Aplysia* are large and easily identifiable from organism to organism. This large size allows recording to occur from individual neurons, and the consistency across organisms allows the neuronal circuitry to be traced more effectively than in mammalian systems. Although learning systems in *Aplysia* often are simplified to show one or two sensory neurons and a motor neuron, in actuality there may be hundreds of neurons involved in the learning process *(2)*. Even so, the ability to isolate individual neurons and synapses in culture and in the organism itself has made the study of *Aplysia* particularly fruitful for understanding the cellular mechanisms of learning. A third reason for the study of *Aplysia* is that behavioral conditioning experiments have demonstrated important similarities between conditioning in *Aplysia* and conditioning in vertebrates *(3–5)*, although many questions remain about the extent of this generality. Because certain behavioral phenomena are common to both *Aplysia* and higher organisms, the study of *Aplysia* is predicated on the assumption that the cellular bases of these learning mechanisms can be extrapolated to synaptic plasticity in organisms with more complex nervous systems. Just as the use of model systems has revolutionized our understanding of the molecular processes underlying development, this reductionist approach to learning, pioneered by Eric Kandel and his colleagues, has been particularly fruitful in identifying molecular mechanisms of synaptic plasticity. Many discussions of *Aplysia* focus on the cellular and molecular mechanisms that form the basis of the effects of serotonin on the sensory–motor neuron coculture system. We will take a broader perspective, examining the possible roles that these mechanisms may play in behavioral processes.

Nonassociative Learning in Aplysia

Two behavioral processes that have been studied in great detail in *Aplysia* are habituation and sensitization. Research on habituation and sensitization in *Aplysia* has exploited the defensive response that occurs when an *Aplysia* is stimulated in certain ways. The components of this response are shown in Fig. 1. Shock delivered to the tail causes the tail to withdraw into the organism; shock delivered to the siphon causes the siphon and gill to withdraw. After repeated mild stimulation of either the tail or the siphon, habituation occurs, causing the withdraw response to attenuate *(7)*. A different form of learning, sensitization, occurs when, after a single shock to the tail, the gill withdrawal response produced by subsequent mild siphon stimulation is greater than if the animal did not receive the shock *(8)*. Both habituation and sensitization have received much attention at the cellular level because they are paralleled at the cellular level by specific changes in synaptic strength. Habituation is paralleled at the synaptic level by synaptic depression—a decrease in the amplitude of excitatory postsynaptic potentials (EPSPs; *9*) —and sensitization is paralleled by facilitation—an increase in the amplitude of EPSPs *(10)*.

In sensitization, shock to the tail increases the likelihood that a subsequent stimulation of another area, such as the siphon, will result in an increased

A

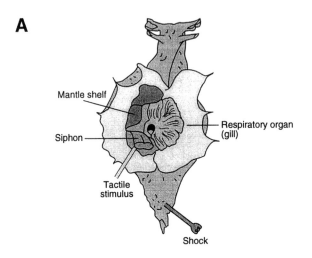

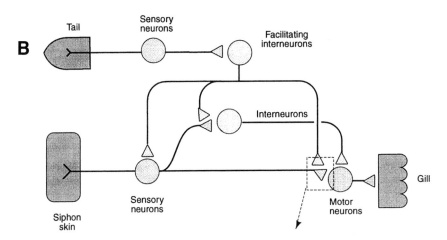

Fig. 1. (A) A dorsal view of *Aplysia*, showing the parts of the body stimulated in studies of habituation, sensitization, and Pavlovian conditioning. An electric shock to the tail causes the siphon to retract and the gill to withdraw. Stimulation of the siphon with a tactile stimulus causes very little response, but after this stimulation has been paired with tail shock, subsequent siphon stimulation causes a conditioned gill withdrawal response through Pavlovian conditioning. **(B)** Pathways involved in sensitization, habituation, and Pavlovian conditioning in the reduced *Aplysia* preparation. Stimulation of the tail by electric shock causes tail sensory neurons to excite facilitating interneurons, some of which are serotonergic. These interneurons synapse on sensory neurons from the siphon. Additional interneurons synapse on the motor neurons that control the gill withdrawal response. (Adapted from ref. *6*.)

gill withdrawal response. Thus, the stimulation of the tail lowers the threshold necessary for a stimulus to elicit a response. At a cellular level, sensitization is thought to involve several steps which are diagrammed in Fig. 2 *(11)*. Stimulation of the tail causes sensory neurons to excite interneurons, which, in turn, facilitate the release of serotonin from sensory neurons on which these interneurons synapse. Serotonin released by some of these facilitatory interneurons causes an increase in the level of cAMP in the sensory neurons, which activates cAMP-dependent protein kinase A (PKA), which phosphorylates a variety of targets, including K^+ channels or proteins closely associated with them. Protein kinase C (PKC) also is activated in response to serotonin, and it may play a more important role with prolonged exposure to serotonin *(11)*. The closure of K^+ channels by PKA or PKC prevents K^+ ions from escaping the cell to repolarize the cell membrane, meaning that the action potential produced by the depolarization of the neuron is broadened. As a result of broadened action potentials, more Ca^{2+} enters the presynaptic neuron, thus increasing the amount of neurotransmitter released. Increased transmitter release from the presynaptic neuron results in an increase in the amplitude of the EPSP in the postsynaptic neuron. This short-term facilitation is transient, lasting just minutes.

A longer-lasting form of sensitization occurs when stronger stimuli are used, or when weaker stimuli are applied repeatedly. This long-term sensitization results in long-term facilitation of synapses between sensory and motor neurons. Long-term facilitation differs from short-term facilitation in several key ways. First, long-term facilitation requires protein synthesis in the presynaptic neuron, whereas short-term facilitation does not *(12)*. Second, PKA, although transiently active in short-term facilitation, is persistently active and translocates to the cell nucleus of the presynaptic neuron during long-term facilitation *(13)*. Third, the cyclic AMP response element binding protein 1 (CREB1) is then activated in the cell nucleus, resulting in the gene transcription necessary for long-term facilitation. Recent findings suggest that another form of CREB, CREB2, may repress activation of CREB1 so that long-term facilitation does not occur as a result of mild serotonin stimulation. This repression may be released only in the presence of sufficient second-messenger activity needed for long-term facilitation to occur *(14)*.

These cellular parallels of habituation and sensitization were of tremendous historical importance because they mapped, for the first time, behavioral learning phenomena onto a cellular process. Other forms of learning, such as Pavlovian and operant conditioning, have been more difficult to model on a cellular level because of the more complex nature of those learning processes. Because the experimenter has more control over the subject's learn-

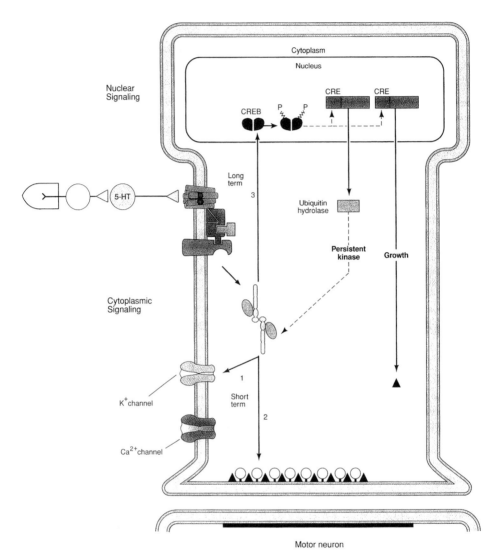

Fig. 2. Molecular mechanisms that underlie sensitization in *Aplysia*. Serotonin (5-HT) binds to G protein-coupled receptors, initiating a cascade of intracellular events. G-protein-coupled receptors stimulate adenylyl cyclase to synthesize cAMP, which activates PKA. During short-term facilitation, the effects of PKA include closing K^+ channels (pathway 1) and increasing transmitter release (pathway 2). Long-term effects result when PKA translocates into the nucleus and activates the transcriptional factor CREB. CREB induces the expression of effector genes which encode a variety of proteins. One class, ubiquitin hydrolases, leads to downregulation of the regulatory subunit of PKA, resulting in persistent PKA activation. The synthesis of another set of proteins (▲) results in growth of new synaptic connections. (Adapted from ref. 6.)

ing experience in Pavlovian paradigms compared to operant paradigms, in which a reinforcer is contingent upon an animal's response, Pavlovian paradigms have been more useful for modeling associative learning on a cellular level, although some attempts are being made at determining the pathways involved in operant conditioning *(15)*.

Associative Learning in Aplysia

Pavlovian conditioning involves the learning of a relation between a neutral stimulus, the conditioned stimulus (CS), which on its own does not elicit a response, and an unconditioned stimulus (US), which on its own elicits an unconditioned response (UR). After CS–US pairings, the CS comes to elicit a response on its own, the conditioned response (CR). The associative properties of Pavlovian conditioning—that a previously neutral stimulus can elicit a response as a result of that stimulus being paired with a biologically relevant stimulus—make it a particularly attractive behavioral model for the study of synaptic plasticity, which involves the strengthening of synapses after a learning experience. Because of the obvious parallels between the effects of Pavlovian conditioning on behavior and of coincident synaptic activity on synaptic plasticity, the characterization of Pavlovian conditioning on the cellular level has been a primary focus of research with *Aplysia (16,17)*.

Initial demonstrations of conditioning in *Aplysia* were performed in the intact animal *(18,19)*. These experiments demonstrated Pavlovian conditioning at the behavioral level by pairing weak tactile stimulation of the siphon or mantle (the CS) with a strong electric shock to the tail (the US), which elicited a UR in the form of gill withdrawal. After repeated CS–US pairings (i.e., stimulation of the siphon or mantle followed repeatedly by the presentation of tail shock), siphon stimulation on its own elicited the conditioned gill withdrawal response. Conditioning, like sensitization and habituation, is paralleled on the cellular level by an enhancement of EPSPs in motor neurons. After conditioning, a weak stimulus that did not elicit EPSPs on its own comes to elicit EPSPs as a result of its being paired with an EPSP-evoking US.

Analogs of conditioning at a cellular level have relied on the reduced preparation in which the central nervous system of the *Aplysia* is removed from the body *(20)*. The tail remains connected to the central nervous system so that it can be stimulated with electric shock. This reduced preparation circuit is diagrammed in Fig. 1B. In this preparation, the same stimulus inputs that occur in the intact animal can be delivered and intracellular recordings can be made from various sensory and motor neurons to determine the neuronal organization of the various pathways involved in conditioning. Specifically, tail shock can be paired with stimulation of the siphon sensory neurons and recordings can be made from synapses between sensory interneurons and motor neurons. In an even more reduced preparation, analogs of conditioning have been studied in cell culture, where treatment with serotonin substitutes for the tail shock US and spike activity in sensory neurons substitutes for the siphon stimulation CS *(21)*. Both the reduced preparation and cell-culture studies have been used in concert with behavioral studies to develop cellular models of Pavlovian conditioning in *Aplysia*.

Cellular models of Pavlovian conditioning incorporate some of the same pathways utilized in cellular models of sensitization. The US tail shock sensori-motor pathway has been defined quite precisely in the reduced *Aplysia* preparation through some of the aforementioned studies of sensitization to tail shock. In Pavlovian conditioning, however, repeated stimulation of another sensory neuron (the CS; e.g., the siphon sensory neuron) coincident with tail shock leads to increased excitatory postsynaptic potentials in the CS–motor neuron synapse. One proposed mode of action is diagrammed in Fig. 3. The exact molecular mechanism underlying this result is not yet clear, although it is thought that the activation of the CS pathway results in Ca^{2+} influx through voltage-gated Ca^{2+} channels, which enhances transmitter release through the activation of adenylyl cyclase. The activation of the US pathway triggers interneurons to release serotonin, which binds to G-protein-coupled receptors in the CS interneurons, resulting in a potentiation of adenylyl cyclase. This coincident activation of adenylyl cyclase by G-protein-coupled receptors and Ca^{2+} leads to both short- and long-term effects. In the short term, it results in an even greater enhancement of transmitter release from the CS interneuron and increases the activation of cAMP and PKA in the presynaptic neuron. The increase in PKA activity leads to an increase in gene transcription in the presynaptic neuron, which appears to be necessary for long-term memory storage *(11)*. Recent studies have outlined some additional mechanisms in the postsynaptic neuron that may also contribute to the learning underlying Pavlovian conditioning *(17,22)*. The evidence for pre- and post-synaptic mechanisms will be reviewed in later sections.

If coincident activation of adenylyl cyclase in the presynaptic neuron is important, then, as in behavioral studies with vertebrates, the precise temporal relation between the CS and the US should determine the extent of conditioning. One of the cornerstones of behavioral research on Pavlovian conditioning is that the CS must precede the US for learning to occur, although the optimal delay between the CS and US varies with different preparations *(23)*. Clark et al. *(24)* demonstrated that forward pairing was critical for EPSP enhancement after conditioning in the reduced *Aplysia* preparation.

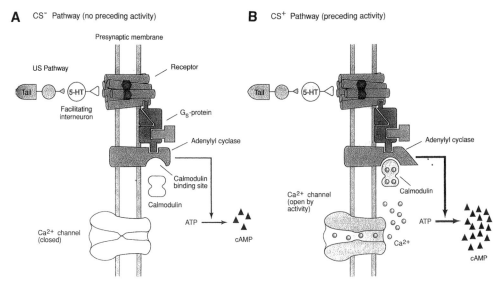

Fig. 3. Molecular mechanisms involved in activity-dependent presynaptic facilitation. US alone trials are shown in **A**. In these trials, as in sensitization, serotonin binds to G-protein-coupled receptors, which act on adenylyl cyclase. This results in a mild increase in cAMP levels. (**B**) shows results from trials in which CS sensory neuron stimulation precedes US stimulation. CS stimulation causes an influx of Ca^{2+}, which binds to calmodulin, which in turn activates adenylyl cyclase. The coincident activation of adenylyl cyclase, caused by Ca^{2+}/calmodulin as a result of the CS and G-protein-coupled receptor binding caused by the US, results in an increased level of cAMP. (Adapted from ref. *6*.)

Additionally, they demonstrated that enhancement was reduced with long CS–US intervals. It is thought that CS–US interval sensitivity occurs because the optimal window for facilitation of adenylyl cyclase in the CS sensory neurons is quite small and depends on simultaneous activation from the US pathway. For example, when the CS precedes the US by more than the optimal interval, the Ca^{2+} that enters the presynaptic neuron may have dissipated by the time the serotonin from the US pathway initiates G-protein-coupled receptor potentiation of adenylyl cyclase. Thus, with long CS–US intervals, no conditioning will occur. Similarly, backward conditioning, in which the US precedes the CS, also fails to produce learning. This may occur because the activation of adenylyl cyclase by G-protein-coupled receptors activated by the US pathway may be less persistent than that produced by Ca^{2+}/calmodulin, causing the priming of adenylyl cyclase to dissipate by the time the Ca^{2+} signal from the CS arrives *(25)*.

Although this analysis of temporal delay in forward CS–US pairings works quite nicely in *Aplysia*, further assumptions would need to be made to extend the analysis to other Pavlovian preparations, in which the optimal conditioning interval is much longer (e.g., flavor-aversion learning *[26]*; auto-shaping *[27]*). Another complication that cellular models should address is that *Aplysia* can form associations between a context, which is always present and thus not localizable in time, and a US *(3,28)*. More work is needed to determine how such constantly present stimuli enter into excitatory associations with the US.

Despite a large body of research examining conditioning at the cellular level, a clear understanding of the contribution of the presynaptic and post-synaptic mechanisms that drive learning remains elusive. One view of the synaptic changes required for learning holds that serotonin released from the US interneuron acts on the CS interneuron to ultimately activate PKA and CREB, causing the gene transcription necessary for long-term memory. A different view holds that postsynaptic mechanisms similar to those found in the mammalian hippocampus also are essential for long-term memory storage. Recent evidence suggests that both presynaptic and postsynaptic mechanisms are important for associative learning *(21)*.

Presynaptic Mechanisms

Much of our current understanding of presynaptic mechanisms involved in conditioning comes not only from studies of Pavlovian conditioning but also from cellular studies that attempt to model sensitization in cell culture. Although these cellular studies allow for rigorous control over the cellular processes involved in conditioning, little is known about the relevance of such studies to the behavioral phenomena they attempt to explain. Nonetheless, studies of synaptic plasticity in culture have led to several results that may help reveal synaptic mechanisms involved in Pavlovian conditioning.

The initial cellular model for Pavlovian conditioning in *Aplysia* relied on activity-dependent presynaptic facilitation (ADPF) of the sensory neuron *(20)*. The early evidence for ADPF came from studies in which the postsynaptic neuron was hyperpolarized before the analog of conditioning began. Hawkins and colleagues *(16,20)* showed that long-term facilitation occurred, even when the postsynaptic neuron was hyperpolarized and thus was ostensibly unable to fire. The presynaptic mechanism responsible appeared to be a broadened action potential that occurred following cellular CS–US pairings but that did not occur when the CS and US were explicitly unpaired. However, in a voltage-clamped cell-culture preparation, Klein *(29)* has shown that broadening of the presynaptic action potential may not contribute significantly to the postsynaptic response.

Recent experiments have demonstrated the involvement of CREB-mediated gene transcription and PKA activity in the presynaptic but not the postsynaptic neuron during long-term facilitation. These studies have focused on posttetanic potentiation and serotonin presentation in cultured *Aplysia* neurons. PKA inhibitors reduce pairing-specific facilitation when injected into the presynaptic neuron, but not when injected into the postsynaptic neuron, suggesting that PKA is not required in the postsynaptic neuron *(21)*. Martin et al. *(30)* found that long-term facilitation is specific to single axonal branches and that this facilitation relies on CREB-mediated transcription and growth of new synaptic connections exclusively at branches treated with serotonin. Interestingly, they found that presynaptic axons that were severed from their cell bodies maintained their ability to synthesize proteins in response to serotonin, suggesting that local presynaptic mechanisms were at work. These results demonstrate the importance of local synaptic action in long-term facilitation, but one needs to be cautious in extrapolating the findings of Martin et al. *(30)* to Pavlovian conditioning because their experiments examined the effects of repeated presentations of serotonin in culture. Similar demonstrations of the necessity for CREB-mediated presynaptic activity have yet to be performed in Pavlovian paradigms in the reduced *Aplysia* preparation. Such experiments are important, given demonstrations that conditioning in *Aplysia* is response-specific—a CS elicits different CRs depending on the US with which it is paired. A synapse-specific cellular mechanism would allow such response specificity to develop because differential local activity at the level of the synapse would enable a single neuron to be involved in multiple conditioning processes.

Postsynaptic Mechanisms

Although there is strong evidence that presynaptic mechanisms underlie learning in *Aplysia*, such evidence does not preclude postsynaptic mechanisms. The search for postsynaptic mechanisms underlying Pavlovian conditioning was motivated by Hebb's postulate and the findings that Hebbian mechanisms appear to be involved in synaptic plasticity in the mammalian brain. Such findings—including those showing that postsynaptic antagonists, such as the N-methyl-D-aspartate (NMDA) receptor antagonists CPP and APV, block LTP and certain forms of learning—have provided the impetus for recent research to examine the role of postsynaptic mechanisms underlying conditioning in *Aplysia*.

The initial searches for Hebbian mechanisms mediating Pavlovian conditioning in *Aplysia* failed to find evidence for postsynaptic mechanisms *(16, 20)*. One reason for this failure might be that these experiments prevented the postsynaptic neuron from firing by hyperpolarizing it. Hyperpolarization, while preventing the postsynaptic neuron from firing, does not necessarily prevent depolarization at the synapse, meaning that synaptic modifications may occur regardless of whether the neuron fires *(2)*.

Recent research has attempted to gain tighter control over the postsynaptic response by blocking receptors at the synapse. Murphy and Glanzman *(17)* have demonstrated that the cellular analog of Pavlovian conditioning is blocked when training occurs in the presence of APV. Although they offered convincing evidence that postsynaptic mechanisms are involved in this reduced preparation analog of conditioning, APV did not completely eliminate synaptic enhancement. Additionally, the synaptic enhancement produced by unpaired CS/US presentations was similar to that produced by paired CS/US presentations when assessed 15 min following stimulation, suggesting that short-term enhancement was not dependent on the contiguous relation between the CS and US. This suggests that short-term sensitization, the increase in responding to a CS resulting simply from the presentation of a US, may not be affected by NMDA receptor antagonists. However, at 60 min, only the group that received CS–US pairings showed synaptic enhancement above basal levels. The training produced by CS–US pairings therefore led to a longer-lasting change in synaptic strength and this strength was decreased, although not eliminated, by APV, demonstrating the importance of postsynaptic NMDA receptors in conditioning.

In another test of the idea that postsynaptic mechanisms play an important role in the cellular processes mediating learning, Murphy and Glanzman *(22)* injected BAPTA, a Ca^{2+} chelator, into the postsynaptic motor neuron. As can be seen in Fig. 4, groups that received CS–US pairings showed an enhanced postsynaptic response at both 15 and 60 min following training. That group which received the same CS–US pairings in the presence of BAPTA showed a similar postsynaptic response to the group that received the CS stimulation only. These results again implicate NMDA receptors, because they allow Ca^{2+} into the postsynaptic neuron. One needs to be cautious, however, in attributing these results exclusively to associative mechanisms because control groups that received either random or unpaired CS–US relations were not included. Thus, it is impossible to determine the extent to which nonassociative mechanisms contributed to the results. Indeed, in their APV experiments, Murphy and Glanzman *(17)* showed that unpaired presentations of the CS and US enhanced EPSPs, suggesting that nonassociative mechanisms may contribute to the enhancement observed with CS–US pairings. Nevertheless, these experiments by Murphy and Glanzman provide important initial evidence that postsynaptic mechanisms contribute to Pavlovian conditioning. Unfortunately, there still is no evidence that

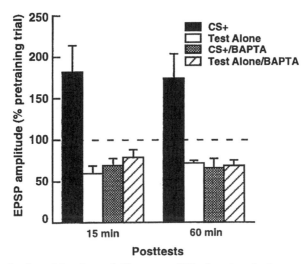

Fig. 4. Results from Murphy and Glanzman *(22)*, showing the importance of post-synaptic mechanisms in Pavlovian conditioning in the reduced *Aplysia* preparation. These data were collected at 15 and 60 min after one of four treatments. During training, the CS+/group received 12 action potentials in the sensory neuron (the CS) followed 500 ms later by the delivery of a 1-s tail nerve shock. Group CS+/BAPTA was treated identically to group CS+, except that CS–US pairings occurred in the presence of BAPTA, a Ca^{2+} chelator, in the motor neuron. Group Test Alone received only CS stimulation during training. Group Test Alone/BAPTA received CS stimulation in the presence of BAPTA in the motor neuron. (Adapted from ref. *22*.)

speaks to the issue of whether these postsynaptic mechanisms are involved in conditioning at the behavioral level.

Thus, there is evidence from different experiments that both presynaptic and postsynaptic mechanisms contribute to learning. Such mechanisms have been demonstrated recently in a single set of experiments in cell coculture *(21)*. Using serotonin stimulation as a substitute for tail shock, they found that presynaptic injection of EGTA, a Ca^{2+} chelator, or PKI6-22, a peptide inhibitor of PKA, reduced pairing-specific facilitation. Injection of PKI6-22 into the postsynaptic cell had no effect on pairing-specific facilitation, whereas postsynaptic injection of BAPTA, another Ca^{2+} chelator, reduced pairing-specific facilitation. Thus, they demonstrated that presynaptic and postsynaptic mechanisms contribute to long-term pairing-specific facilitation and that they may interact in a synergistic way, because interfering with either almost completely eliminated pairing-specific facilitation. Bao et al. *(21)* proposed that learning in *Aplysia* involves a hybrid mechanism consisting of presynaptic and postsynaptic elements. One possible mechanism relies on Ca^{2+} elevation in the postsynaptic neuron, causing a retrograde messenger to be sent to the presynaptic neuron that enhances transmitter release through interaction with Ca^{2+} or cAMP.

A general finding from studies of EPSPs in cell culture and the reduced preparation is that repeated weak stimulation leads to synaptic depression—EPSPs decrease in magnitude, presumably because the postsynaptic neuron habituates to the stimulation. This suggests that in the absence of the US, or serotonin, the postsynaptic response will decrease. One thus has to be cautious in attributing increased EPSPs after CS–US pairings to an associative mechanism because dishabituation or sensitization may be contributing to the results. Indeed, Murphy and Glanzman *(17)* clearly showed not only that the unpaired CS/US presentations prevent the habituation obtained in CS alone procedures, but also that the unpaired presentations actually increased EPSPs above basal levels, suggesting that sensitization caused by the simple presentation of the US may contribute to the response to the CS.

Another complication that results from explicitly unpaired CS/US presentations is that the organism may learn that the CS signals the absence of the US. Behaviorally, the organism might inhibit its response in the presence of the CS as a result of these explicitly unpaired presentations because the organism learns that the CS is a signal for the absence of shock. Thus, the sensitization that has been observed when the CS and US are explicitly unpaired may be incomplete.

More experiments clearly are needed for a full understanding of the role of postsynaptic mechanisms in Pavlovian conditioning, but these recent studies make valuable progress in showing the necessity of postsynaptic mechanisms. Importantly, these findings demonstrate similarities between the synaptic mechanisms involved in conditioning in *Aplysia* and those involved in long-term potentiation in the hippocampus. Determining the ways in which presynaptic and postsynaptic neurons work synergistically to cause synaptic facilitation will be an important step in formalizing a cellular theory of learning. Additionally, more work is necessary to determine the extent to which the synaptic mechanisms and signaling molecules discovered in *Aplysia* are relevant to the behavior of the whole organism. One model system in which the relations between signal transduction and the behavior of the organism have been more firmly established is the rodent hippocampus. The ability to generate genetically modified mice has enabled researchers to make strong connections between signaling molecules, synaptic plasticity, and a range of behaviors. We now turn to the analysis of molecular and cellular mechanisms involved in hippocampus-based learning and LTP.

SYNAPTIC PLASTICITY
IN THE MAMMALIAN HIPPOCAMPUS

In humans, the medial temporal lobe system, including the hippocampal formation, is critically important for declarative memory—the conscious recollection of memories for people, places, and things *(31)*. Beginning with studies of the patient H.M. and continuing with more recent analyses of patients with lesions restricted to the hippocampus proper, neuropsychologists have found that lesions to the hippocampus result in both anterograde and retrograde amnesia for facts and events while sparing procedural memory and motor skills.

In rodents, spatial and contextual learning are particularly well documented, and these forms of learning are sensitive to lesions of the hippocampal formation *(32)*. Two physiological properties of the rodent hippocampus are potential cellular mechanisms underlying memory storage. First, synapses within the hippocampus undergo long-term potentiation (LTP), a form of synaptic plasticity that is thought to be involved in at least some aspects of spatial memory *(33)*. Second, the hippocampus contains a cellular representation of space in the form of place cells that fire action potentials only when the animal is in a certain spatial location *(34)*. Because much research has attempted to determine the relationship between LTP and learning, we will first review the evidence linking these processes and then discuss the signal transduction pathways that are important for long-lasting forms of synaptic plasticity and long-term memory storage.

Synaptic plasticity, the change in the strength of synaptic connections in the brain, is thought to underlie memory storage and the acquisition of learned behaviors. One intensely studied form of synaptic plasticity is LTP, a persistent, activity-dependent form of synaptic enhancement that can be induced by brief, high-frequency stimulation of hippocampal neurons *(33)*. LTP, first described in detail by Bliss and Lomo in 1973, can be measured in hippocampal slices or in awake, behaving animals, where it can last for several weeks. The duration of LTP makes it an attractive model for certain types of long-term memory in the mammalian brain. In addition, LTP has other properties, including *associativity*, by which LTP induction at one synapse may be regulated by other inputs, *cooperativity*, which refers to the observation that a greater stimulus intensity will produce greater LTP, and *pathway specificity*, which refers to the observation that only synapses active at the time of LTP induction will be potentiated. These elements of LTP make it an ideal mechanism for memory storage from a computational perspective. On a molecular level, these properties derive, in large part, from the properties of a specific type of postsynaptic receptor for glutamate, the NMDA receptor, which serves as a molecular coincidence detector.

For the past 25 years, LTP has been studied as a potential cellular model of memory storage—a representative of the types of synaptic plasticity that may occur naturally during learning. There are four elements that comprise the basis of the hypothesis that spatial information is stored as activity-dependent alterations in synaptic weights in the hippocampal formation: *enhancement*, the idea that increased synaptic strength should accompany learning; *saturation*, the proposal that learning impairments should be observed after the saturation of LTP in hippocampal circuits; *blockade*, the hypothesis that spatial learning should be disrupted after the blockade of hippocampal LTP; and *erasure*, the idea that erasure of LTP should result in forgetting.

We will explore the experimental approaches that have been followed to investigate the relationship between LTP and learning. In particular, if LTP and learning recruit the same underlying physiological and cellular mechanisms, then modifying the ability of hippocampal synapses to undergo LTP should alter learning. In turn, learning should modify synaptic strength.

LTP experiments have focused particularly on the three major pathways in the hippocampal trisynaptic circuit: the perforant pathway between the entorhinal cortex and dentate gyrus granule cells, the mossy fiber pathway between dentate gyrus granule cells and CA3 pyramidal cells, and the Schaffer collateral pathway between CA3 and CA1 pyramidal cells (Fig. 5A). For many of these experiments, investigators have studied synaptic transmission in the perforant path, where recordings can be easily made in vivo in awake, behaving rodents.

Experience-Dependent Changes in Synaptic Strength

If synaptic plasticity underlies learning, then an increase in synaptic strength would be expected to accompany behavioral training. Early experiments to test this idea observed an increase in the size of the population spike and fEPSP in the dentate gyrus after the exposure of rats to a spatially complex environment *(35)*. As investigators attempted to extend this observation to hippocampus-dependent tasks such as the Morris water maze, they found, paradoxically, that spatial training in the Morris water maze resulted in a decrease in the size of the fEPSP as measured in the dentate gyrus. In exploring this observation, Moser et al. *(36)* found that striking changes in brain temperature occurred during different tasks. An increase in hippocampal temperature, as observed during active exploration or treadmill running, was correlated with an increase in fEPSP slope. During training in the Morris water maze, in which the water temperature is typically cooler than body temperature, the brain temperature drops and the fEPSP slope decreases.

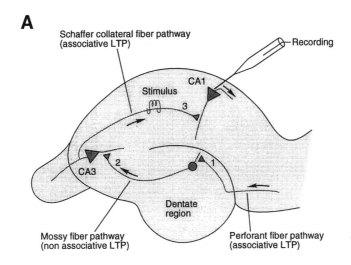

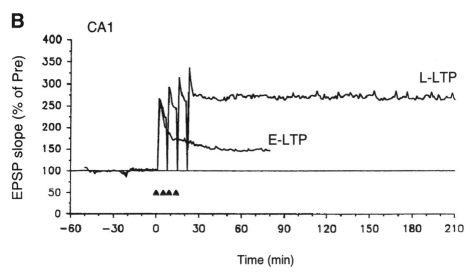

Fig. 5. (A) The major areas and pathways in the hippocampus. Input from the entorhinal cortex is relayed to the dentate gyrus via the perforant pathway. The mossy fiber pathway relays the signal from the dentate gyrus to area CA3. The Schaffer collateral pathway relays the signal from area CA3 to area CA1. In the example shown, LTP is induced by the stimulating electrode in the Schaffer collateral pathway and is recorded by the recording electrode in area CA1. **(B)** Different phases of LTP: E-LTP occurs after one tetanus (one filled triangle); L-LTP occurs after four tetani (four filled triangles). E-LTP lasts about 1 h, but L-LTP can last for up to 8 h in hippocampal slices. In contrast to E-LTP, L-LTP requires translation, transcription, and protein synthesis. (Adapted from ref. *6*.)

Further, changes in synaptic strength may also occur as a result of the stress that accompanies behavioral training *(37)*. Thus, the differentiation of learning-related changes in synaptic strength as measured in vivo from the large nonspecific changes that are produced during the performance of a task has proven difficult, especially for hippocampus-dependent tasks. If only a small number of synapses increase in strength during learning, or if some synapses are potentiated while others are depressed, it may be difficult to accurately measure changes in synaptic strength that result from learning. Indeed, after the temperature component of the field potential changes that occur during training is subtracted, only a short-lasting (15–20 min) increase in the fEPSP and population spike is found to occur during learning *(38)*. These experiments have focused on the idea that learning alters baseline-evoked synaptic transmission, but an alternative possibility, which has been the focus of only a few studies, is that learning alters synaptic plasticity, modulating the ability of synapses to undergo LTP *(39)*.

Although the demonstration of LTP-like enhancements in synaptic strength in the hippocampus in the behaving animal during learning has been elusive, studies of cued fear conditioning, a form of Pavlovian conditioning in which an animal learns to fear a previously neutral tone as a result of its association with an aversive stimulus such as footshock, have revealed that learning is indeed associated with an increase in synaptic strength in the appropriate neural circuit *(40,41)*. Information about the CS (tone) and the US (footshock) converge in the lateral nucleus of the amygdala. LTP, when induced by electrical stimulation in the connections between the auditory thalamus and the lateral nucleus, increases the response of neurons in the lateral nucleus to auditory stimulation. Thus, responses to natural stimuli in this system can be modulated by LTP *(42)*.

By monitoring extracellular potentials in the lateral nucleus in response to tones, Rogan et al. *(41)* demonstrated that cellular responses to the CS increased during the period of paired presentation of the CS and US as the animal acquired the fear response to the tone. Importantly, extinction of the fear response produced by the nonreinforced presentation of the tone caused the auditory-evoked potential to return to baseline. It is striking that the increased synaptic response to the CS parallels learning, but it has not been demonstrated that this increase is specific to the CS, nor is it clear exactly where in the auditory processing circuitry the increase occurs. The control used in these experiments was the explicitly unpaired presentation of the tone and shock, resulting in a decreased auditory-evoked potential following training, perhaps because the animal learns that the shock is not signaled by the tone as a result of this training procedure. Thus, unpaired training is not a "neutral" control protocol and may involve some learning. A better control might utilize a random pairing protocol in which the CS provides no information about the US *(43)*. Further, it remains to be explored whether these behavioral changes in synaptic strength are, like fear conditioning, dependent on NMDA receptor activation.

Behavioral Effects of LTP Saturation

If learning is the result of changes in the same sets of synapses as those modified by LTP, then the induction of LTP by tetanization—brief trains of high-frequency stimulation in presynaptic neurons—would be predicted to alter learning. In physiological studies, the induction of LTP blocks or occludes further potentiation, suggesting that tetanization of synaptic circuits within the hippocampus would impair learning. Many computational models assume that information is stored in the pattern of synaptic weights rather than in the absolute strength of synaptic connections *(44)*. By this view, then, the uniform saturation of synapses would block further learning by equalizing synaptic weights.

In the first saturation study published, McNaughton et al. *(45)* implanted electrodes bilaterally to stimulate the perforant pathway. Animals that received bilateral tetanic stimulation had deficits in the Barnes circular platform maze, a spatial task in which animals must learn the location of an escape hole on a circular maze. These observations were extended to the Morris water maze, in which tetanized rats exhibited a deficit in their ability to learn the location of a hidden platform, a task dependent on an intact hippocampus *(46)*. These initial experiments provided powerful support for the idea that interfering with plasticity in specific hippocampal pathways could disrupt learning, but several attempts to replicate these observations were unsuccessful *(47)*. Although these failures to replicate the saturation experiments have called into question the idea that synaptic plasticity is used during the learning of spatial tasks, several alternative explanations have been developed to explain why LTP saturation does not, in some instances, block learning. LTP induction techniques, for example, may not activate all of the fibers required for spatial learning. Indeed, tetanization at a single stimulation site fails to saturate the entire perforant path *(48)* and lesion experiments have revealed that learning can be supported by just a small fraction of the hippocampus *(49)*.

A recent experiment by Moser et al. *(50)* has revisited this question of the impact of LTP saturation on spatial learning in an ingenious way, and their results suggest that saturation does indeed impair spatial learning in the Morris water maze. To strengthen the experimental design and to increase the percentage of the synapses that were saturated, they made three modifications. First, based on their previous observation that unilateral hippocampal lesions

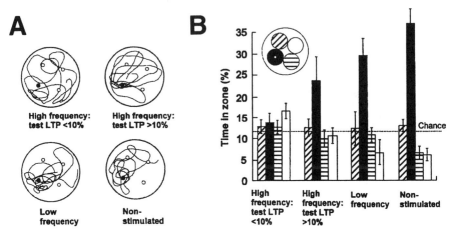

Fig. 6. Effects of saturation of LTP on behavior. Rats received either high-frequency tetanization, low frequency tetanization, or no tetanization. Rats that received high-frequency tetanization were divided into two subgroups: those that showed less than 10% LTP (saturated) on the test and those that showed greater than 10% LTP (nonsaturated) on the test. (**A**) Sample paths from Morris water maze probe trials, in which the hidden platform was removed. (**B**) Time spent in different zones in the pool. Filled bars indicate the target zone (the zone in which the platform was located during training). Error bars indicate standard errors of the mean and the dotted line indicates chance level. (Adapted from ref. *50*.)

do not impair performance in the water maze, they reduced the volume of functional hippocampal tissue by unilaterally lesioning the hippocampus. Second, to saturate a greater percentage of perforant path synapses, two bipolar stimulating electrodes were placed on each side of the angular bundle, the tract that carries the perforant path fibers into the hippocampus. Using these electrodes, multiple "cross-bundle" episodes of tetanic stimulation can be applied, thus saturating LTP in the maximal number of synapses. Finally, they monitored potentiation by implanting a third stimulating electrode.

After saturating LTP with five episodes of cross-bundle tetanization, animals were trained in the spatial version of the Morris water maze. Moser et al. *(50)* found a wide range of spatial learning in the rats following tetanization. Unlike previous studies, however, they determined the extent of saturation by measuring the amount of residual LTP using a "naive" test electrode and thus were able to correlate performance in the water maze with levels of residual LTP. Strikingly, they found that impairments in performance correlated with lower levels of residual LTP. Thus, animals in which LTP was saturated were poor learners, whereas animals in which LTP was unsaturated were good learners, in support of the hypothesis that synaptic plasticity underlies learning (Fig. 6). This positive result showing that LTP saturation impairs spatial learning, however, might be due to nonspecific effects of tetanization on basal synaptic transmission and thus does not prove that synaptic plasticity underlies learning. Thus, it will be particularly important to determine if the group with little residual LTP regains spatial learning ability as the level of saturation decays over time and the ability to induce LTP is restored. This reversibility might help address some concerns about non-specific side effects of high-frequency tetanization. Further, it will be interesting to determine the overall relationship between residual LTP (percent saturation) and spatial learning to see if they are correlated across a range of impairments.

Pharmacological and Genetic Blockage of LTP

Some of the strongest evidence linking LTP and learning is derived from experimental approaches using pharmacological or genetic manipulations that modulate LTP. By determining the effects of these manipulations on learning and memory, investigators have been able to correlate alterations in synaptic plasticity with behavioral impairments in hippocampal function. With the development of techniques such as targeted gene ablation and transgenesis, it has became clear that mice offer a superb genetic system for determining the role of individual gene products in synaptic plasticity and memory storage. The analysis of genetically modified mice allows researchers to test whether a particular gene product is important for LTP, and the use of this genetic approach to study neuronal physiology and behavior has drawn attention to the correlation between memory storage and hippocampal LTP.

As mentioned earlier, LTP exhibits synapse specificity and associativity because it occurs only when presynaptic activity is paired with postsynaptic depolarization. On a molecular level, these characteristics can be explained by the fact that the NMDA subtype of glutamate receptor is both ligand gated and voltage sensitive. Importantly, many forms of hippocampal LTP share a dependence on NMDA receptor function with many forms of spatial memory.

To explore the role of NMDA-receptor-dependent synaptic plasticity in spatial learning, Morris and co-workers *(51,52)* infused APV into the cerebral ventricles of rats and examined their performance in the Morris water maze. APV treatment resulted in longer latencies to find the hidden platform during training and little spatial specificity of search in a probe trial, during which the platform was removed from the pool. Although these experiments provide strong evidence in support of the link between LTP and learning, several caveats have emerged since these studies were published.

First, several forms of synaptic plasticity—such as long-term depression (LTD), a long-lasting decrease in EPSPs—depend on NMDA receptor function *(53)*, so it is difficult to make direct connections between any one form of synaptic plasticity and spatial learning. Second, APV may be affecting processes in brain regions other than the hippocampus when administered in this way. These processes may include modulation of sensory input, modifying anxiety, and altering motor abilities *(39)*. Third, the relationship between NMDA receptor function and spatial learning has become complicated by the observation that spatial learning is NMDA-receptor independent if animals are first pretrained in the water maze in a different environment *(54,55)*.

The strongest correlation between LTP and spatial memory comes from the study of mice in which the R1 subunit of the NMDA receptor was deleted in a regionally restricted fashion, only in hippocampal area CA1 *(53)*. This study is particularly important because it underscores the power of molecular approaches to study learning and memory. Conventional knockouts of the gene encoding the R1 subunit of the NMDA receptor were lethal, so Tsien and co-workers turned to a conditional knockout approach using Cre recombinase. They used the calcium/calmodulin-dependent protein kinase IIα (CaMKIIα) promoter to express Cre recombinase postnatally in neurons within the forebrain. To selectively delete the NMDA R1 gene, they inserted lox P sites, which are recognized by Cre recombinase, into the NMDA R1 locus. Using this approach, they achieved both temporal and regional restriction, knocking out NMDA receptor function only in hippocampal area CA1, a result not possible with pharmacological approaches. Thus, this genetic approach may overcome some of the difficulties encountered in the above-described pharmacological studies.

These mutant mice lacking NMDA receptor function only in hippocampal area CA1 have impaired Schaffer collateral LTP and deficits in spatial memory, providing evidence supporting a selectively important role for hippocampal area CA1, as suggested earlier by the study of the patient R.B. by Squire and colleagues *(56)*. However, these mutant mice also have impaired LTD *(53)*; thus, the nature of the synaptic plasticity deficit underlying their behavioral abnormality is unclear. In addition, it will be important to determine the effect of spatial and nonspatial pretraining on the impairments observed in these mutants in the spatial version of the water maze. It also will be important to explore whether these NMDA R1 transgenic mice exhibit deficits in nonspatial forms of hippocampus-dependent learning such as contextual fear conditioning *(57,58)* and olfactory-based tasks such as social transmission of food preferences *(59,60)*. Deficits in these nonspatial tasks would implicate NMDA receptors in area CA1 in a variety of learning tasks.

Protein Kinase A, Long-Term Memory, and the Late Phase of LTP

Like the study of mice lacking the R1 subunit of the NMDA receptor only in hippocampal area CA1, the study of other genetically modified mice has focused on the early, transient phase of LTP (E-LTP) in area CA1 that lasts about 1 h. These studies have shown that genetic manipulation of any one of several kinases interferes with not only E-LTP but also short-term memory *(61,62)*. The study of amnesiac patients and experimental animals has revealed, however, that the role of the hippocampus in memory storage extends from weeks to months *(63,64)*, suggesting that longer lasting forms of hippocampal synaptic plasticity may be required.

Long-term potentiation in the CA1 region of hippocampal slices, like many other forms of synaptic plasticity and memory, has distinct temporal phases *(65)*, as shown in Fig. 5B. In contrast to E-LTP, the late phase of LTP (L-LTP) lasts for up to 8 h in hippocampal slices *(66)* and for days in the intact animal *(67)*. Long-term memory storage, in contrast to short-term memory storage, is sensitive to disruption by inhibitors of protein synthesis *(68)*, and L-LTP in the CA1 region of hippocampal slices, unlike E-LTP, shares with long-term memory a requirement for translation and transcription *(66,69–71)*.

Although extensive information is available about E-LTP and its relationship to behavior, less is known about the behavioral role of L-LTP. Pharmacological experiments have suggested that PKA plays a critical role in L-LTP *(66,70)*. One of the nuclear targets of PKA is CREB *(72)*, and CRE-mediated gene expression is induced in response to stimuli that generate L-LTP *(73)*. Behavioral studies of mice lacking the α and Δ isoforms of CREB have suggested that this transcription factor plays a role in long-term memory storage, but the relationship between these memory deficits and L-LTP is unclear, because a deficit in LTP is observed during E-LTP following a single stimulus train *(74,75)*. Moreover, because CREB is a multifunctional transcription factor that can be activated by second-messenger systems other than PKA, including CaM kinases and the MAP kinase pathway *(76)*, these data on CREB knockout mice do not define a role for PKA in long-term memory.

To explore the role of PKA in long-lasting forms of synaptic plasticity and behavioral memory, Abel et al. *(77)* used transgenic techniques to reduce PKA activity in a specific subset of neurons within the mouse forebrain by using the CaMKIIα promoter to drive expression of R(AB), a dominant negative form of the regulatory subunit of PKA. R(AB) carries mutations in both cAMP binding sites and acts as a dominant inhibitor of both types of PKA catalytic subunits *(78–80)*. The transgenic approach is more spatially and temporally restricted than is the conventional gene knockout approach, thereby allowing for a more direct correlation between a behavioral deficit and synaptic physiology in the adult brain. Transgenic techniques are particularly powerful for the study of signaling molecules, such as PKA, that are encoded by multiple genes. Appropriately designed dominant-negative mutants can inhibit multiple related gene products simultaneously, an effect that cannot be obtained through conventional single-gene knockouts *(77)*. Further, appropriately designed constitutively active mutants can be used to activate an endogenous signaling pathway.

R(AB) transgenic mice have reduced hippocampal PKA activity, as well as impairments in L-LTP induced by repeated tetanization (four 100-Hz trains, 1-s duration, spaced 5 min apart) of Schaffer collateral pathway slices *(77)*. E-LTP induced by one or two stimulus trains is unchanged in the R(AB) transgenics, suggesting that L-LTP, unlike E-LTP, requires PKA and recruits distinct signaling pathways immediately following tetanization (Fig. 7).

For the behavioral analysis of hippocampal function, R(AB) transgenic animals have been tested in the hidden platform version of the Morris water maze task *(81)*. Transgenic animals improved during training, indicating that they learned the task, but when tested for memory in a probe trial, transgenic animals exhibited spatial memory deficits *(77)*. The Morris maze task requires repeated training over several days and does not, therefore, provide the temporal resolution necessary to distinguish between different phases of memory storage. Because L-LTP and long-term memory share a requirement for protein synthesis, one might predict that the R(AB) transgenics, which have a L-LTP deficit, would have normal short-term but defective long-term memory.

To define more precisely the time course of the memory deficit in R(AB) transgenics, contextual and cued fear-conditioning tasks, in which learning can be accomplished by a single training trial, have been used *(57,58)*. The R(AB) transgenics exhibited normal short-term memory, consistent with normal E-LTP, but deficient long-term memory for contextual fear conditioning *(77)*, a task that is sensitive to hippocampal lesions *(57,58,82,83)*. The time-course of the memory deficit of R(AB) transgenics in contextual fear conditioning parallels that of wild-type animals treated with the protein synthesis inhibitor anisomycin (Fig. 8). By contrast, the long-term memory for cued conditioning, a task sensitive to amygdala lesions but insensitive to hippocampal lesions, is not disrupted in R(AB) mice (Fig. 8). Importantly, R(AB) transgenic mice also showed normal long-term memory in the conditioned taste-aversion task *(77)*, a task which is sensitive to amygdala lesions *(84)*.

One concern about using genetically modified mice is that any observed behavioral effects may be the result of developmental effects of the transgene rather than a direct, acute effect of the transgene on memory storage in

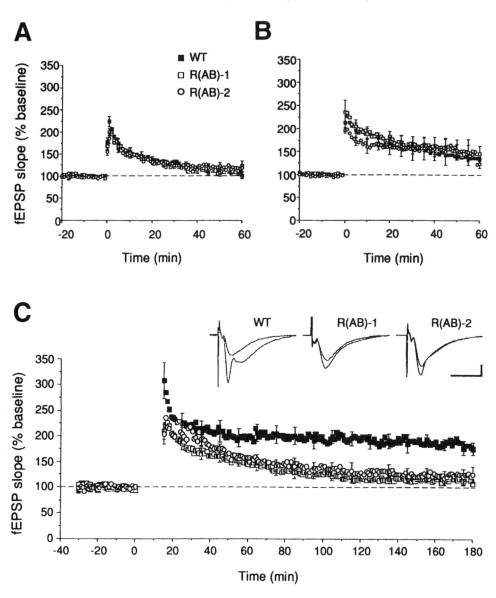

Fig. 7. LTP deficits in R(AB) transgenic mice. LTP was reduced in two different lines of transgenic mice (R[AB]-1 and R[AB]-2) following four 100-Hz trains (1-s d, 5 min apart) of stimulation. After about 2 h posttetanus, potentiation in the R(AB) mice returned to near-baseline levels but remained robust in wild-type mice. Sample fEPSP traces also are shown. They were recorded in area CA1 in wild-type, R(AB)-1, and R(AB)-2 slices 15 min before and 180 min after the four tetanic trains. Each super-imposed pair of sweeps was measured from a single slice. Scale bars: 2 mV, 10 ms. (Adapted from ref. *77*.)

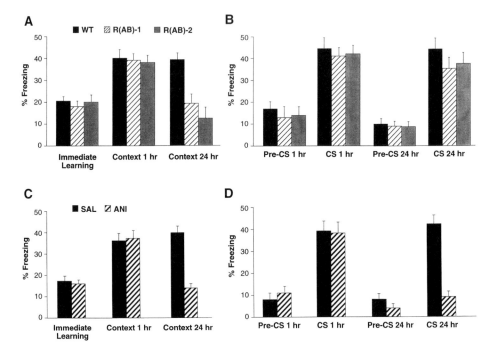

Fig. 8. Fear conditioning in R(AB) transgenics and anisomycin-injected wild-type mice. All mice received one CS–US pairing and then were tested immediately, 1 h, and 24 h after training. (**A**) R(AB) mice showed a deficit in long-term but not short-term memory for contextual fear conditioning. (**B**) R(AB) mice showed no deficits in cued fear conditioning. (**C**) Anisomycin disrupted long-term but not short-term memory for contextual fear conditioning. (**D**) Anisomycin disrupted long-term but not short-term memory for cued fear conditioning. (Adapted from ref. *77*.)

the adult. To provide a complementary way to study the role of PKA in memory storage and to address concerns about potential developmental effects of the R(AB) transgene, a pharmacological approach was taken using Rp-cAMPS, a membrane-permeant, phosphodiesterase-resistant inhibitor of PKA *(85)*. Intraventricular injection of Rp-cAMPS selectively affects long-term memory for contextual fear conditioning with a time-course similar to that seen in R(AB) transgenic animals or in wild-type mice after the administration of anisomycin *(86)*. The long-term memory deficits that occur in R(AB) transgenic mice and in wild-type mice after pharmacological inhibition of PKA demonstrate that the PKA pathway plays a crucial role in the hippocampus in initiating the molecular events leading to the consolidation of short-term memory into protein synthesis-dependent long-term memory in mammals. The molecular events involved in the short- and long-term synaptic plasticity that are thought to underlie memory are diagrammed in Fig. 9.

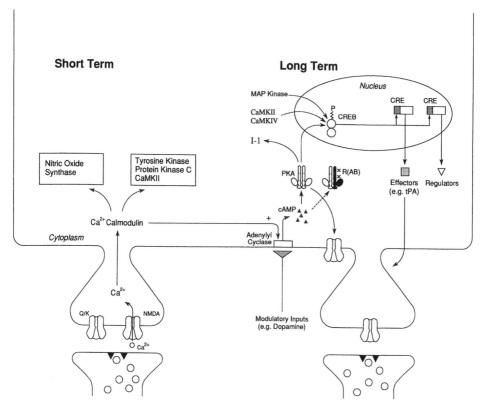

Fig. 9. Molecular schematic of LTP. In the absence of activity, the NMDA receptor is blocked by Mg^{2+}. This Mg^{2+} is expelled when non-NMDA (Q/K) receptors open and depolarize the membrane. Coincident binding of L-glutamate to the NMDA receptor allows Ca^{2+} influx, which activates several kinases, including tyrosine kinase, protein kinase C, and CaMKII. The activation of these kinases may be required for E-LTP, which corresponds to short-term memory. L-LTP, which corresponds to long-term memory, requires PKA activation and protein synthesis. When Ca^{2+}/calmodulin binds to adenylyl cyclase, cAMP levels rise, causing an activation of PKA, which phosphorylates ion channels, protein phosphatase inhibitor-1 (I-1), and nuclear targets such as CREB. CREB then activates effector genes encoding proteins that are necessary for alterations in synaptic strength and perhaps synaptic growth, such as tissue plasminogen activator (tPA). (Adapted from ref. *11*.)

These studies underscore the crucial interplay between pharmacological and genetic studies that are greatly expanding our knowledge of the molecular basis of synaptic plasticity, learning, and memory. By determining the signal transduction pathways critically important for long-lasting forms of synaptic plasticity, the sophisticated tools of mouse genetics can then be used to modify this signal transduction pathway in vivo. These functional experiments provide a rigorous test of the role of gene products, such as PKA, in learning and memory.

Hippocampal Place Cells

Basic Properties

The research described in the previous sections suggests that LTP may play a critical role in the synaptic plasticity underlying memory storage. Deficits in hippocampal LTP often are accompanied by deficits in spatial learning and in learning about the identities of contexts. This implies that one function of LTP in the hippocampus may be to mediate configural representations of multiple environmental stimuli *(87)*. In addition to undergoing LTP, another characteristic of neurons in the hippocampus is that some of them are place cells that respond selectively to particular locations in the environment *(88)*. The discovery of these place cells was an important step in placing physiological reality on theories of cognitive mapping, which was thought to be necessary to navigate the environment *(88,89)*. Because these cells seemed to be located almost exclusively in the hippocampus, their discovery was an important modern development in driving theories of hippocampal function. The recent emphasis on research with genetically modified mice has allowed strong links to be made among the molecular characteristics of place cells, synaptic plasticity, and behavior. The correlation between place cell function, synaptic plasticity, and spatial learning has generated specific hypotheses about the role of LTP in spatial learning, ranging from being important for the establishment of place fields to modifying synaptic strength both among place cells themselves and between place cells and other brain regions.

The majority of place cells in the brain are pyramidal cells found in areas CA1 and CA3 of the hippocampus. This does not mean that all pyramidal cells in the hippocampus are place cells; 70% to over 90% of pyramidal cells are place cells, and approximately half of them act as place cells in a given environment *(90)*; nor does it mean that there are no place cells outside of the hippocampus proper *(91)*. Additionally, the other major class of neurons in the hippocampus, theta cells, also code some spatial information *(92)*. Nevertheless, most of the work on place cells has focused on the pyramidal cells in areas CA1 and CA3, not only because these cells clearly function to a large degree as place cells but also because these areas of the hippocampus have been shown to be important in learning.

The most fundamental property of place cells is their place-specificity— a given cell fires only when the organism is in a certain location in the environment, although a small portion of place cells fire in more than one location in the same environment *(93)*. Place-specificity is observed after a single exposure to an environment *(88)*, and once established, place-specificity is stable for at least several months *(94)*. That it is established during the first exposure suggests that the initial formation of place fields is an unconditioned response that occurs to the environment. These place fields are dependent on cues in the environment; when environmental cues shift, place fields shift accordingly *(34)*. Learning may occur with prolonged or multiple exposure as the organism processes more and more about the stimulus environment. Similarly, although rewards are not necessary for the formation of place fields, an animal's ability to remember the spatial location of a reward may depend on associations between place fields and reward centers in the brain.

In contrast to simple sensory mapping, such as that which occurs in retinotopic mapping, place cell mapping does not occur topographically; adjacent cells do not represent adjacent points in the environment. Indeed, there is little correlation between the location of place fields and the location of place cells in the hippocampus; two physically adjacent place cells may fire at opposite points in the environment and two place cells located far apart within the hippocampus may fire at similar locations in the environment *(34)*. Additionally, a given place cell may fire differently depending on the environment, resulting in the same place cells potentially being involved in distinct cognitive maps in different environments. The system properties that lead to the unique configural representations required for each environment remain unknown, although connectionist models have been developed to provide a theoretical basis for such representations *(44,95)*.

Perhaps the most interesting and fundamental issue that has yet to be resolved about place cells is how these cells contribute to learning. Many place cell experiments have recorded from rats exploring different environments, in which there is neither a reward to be found nor an aversive situation to escape. Place fields form in these environments independent of the organism's having learned anything about the biological significance of a given environment. These experiments speak to the existence of place cells but are quiet on the issue of the involvement of place fields in tasks that require learning about spatially distributed stimuli. It is clear from maze experiments, in which animals must learn the relation between distal stimuli and a reward, that place fields are involved in locating rewards. O'Keefe and Speakman *(96)* recorded place fields from rats searching in a four-arm maze for a food pellet. They trained rats to locate a food reward by learning the relation between environmental cues and the reward. Some of their most interesting results occurred on trials in which the cues were removed. When the rat made an incorrect choice in the absence of cues, the place cells fired as if the rat had made the right choice. This suggests that place fields may be used to navigate through space to find a reward and that the firing of place fields is controlled not only by environmental cues but also by the expectations of the organism. Determining the mechanisms that incorporate place cells into such goal-directed behavior will be an important step in forming an integrative theory of place cell function.

Hippocampal Place Cells and LTP:
Pharmacological and Genetic Approaches

The pharmacological and genetic approaches aimed at studying the relation of LTP to spatial memory, which have been outlined in previous sections, also have been applied to study the relation between place cells, synaptic plasticity, and memory storage. Recent studies have demonstrated correlations among place cell function, LTP, and spatial learning. These experiments have shown a direct correlation between synaptic functioning and spatial learning—LTP deficits often are accompanied by spatial learning deficits in the Morris water maze and alterations in place cell properties *(53,90,97–99)*. These correlations suggest that LTP likely is involved in place cell function somewhere along the path leading to spatial memory. The precise role of LTP in this process remains unknown, but there are many possibilities. For example, LTP may be involved in the formation, maintenance, stability, or spatial selectivity of place cells. Additionally, LTP may play a role either in establishing connections between place cells themselves or in establishing connections between place cells and other brain regions, such as those involved in learning behavioral responses or those involved in processing rewards.

In parallel to deficits found in spatial learning experiments, there is pharmacological evidence that a blockade of NMDA receptors interferes with place cell function. Place field stability is disrupted after intraperitoneal injection of the NMDA receptor antagonist CPP *(98)*, suggesting a link between the ability of the mice to maintain stable place fields and their ability to retain spatial information. Although Kentros et al. *(98)* found effects of NMDA receptor antagonists on place field stability, they also found that CPP did not block previously formed spatial maps, nor did it block remapping in a new environment, suggesting that NMDA-mediated LTP may not be involved in the recall of previously formed maps or in the initial establishment of new place fields. However, these newly established fields are not retained, suggesting that LTP could be involved in the retention of these fields, at least in adult animals given an acute blockade of NMDA receptor function.

Experiments on genetically modified mice lacking NMDA R1 receptors in hippocampal area CA1 also have found spatial learning and place field deficits. As discussed in previous sections, these mutant mice perform poorly in the Morris water maze *(53)* and have decreased LTD and LTP. They also have both uncorrelated place cell firing and larger place field sizes *(99)*. These results differ somewhat from the pharmacological experiments studying CPP-treated mice. Kentros et al. *(98)* found no effect of CPP on field size, whereas McHugh et al. *(99)* found enlarged field sizes in the NMDA R1 knockout mice, suggesting that the place cells in the knockout mice were less spatially selective than were those in wild-type mice. One possible explanation for this difference is that the deficits in spatial selectivity in NMDA R1 mice occurred because place cells become spatially selective during development. The NMDA R1 deletion occurred by approx 19 d after birth, which is when mice begin to explore their environments and thus when place fields might begin to form. That they failed to form spatially selective place fields may reflect the idea that spatial selectivity forms during development. The acute CPP treatment given by Kentros et al. *(98)* to adult mice would not have affected such development.

The parallels between NMDA receptor function and place cell function are striking and strengthen the argument that LTP is involved at some level in place cell mechanisms. A similar connection has been made between synaptic plasticity and place cell function in mice overexpressing a calcium-independent form of CaMKII [CaMKII Asp 286 *(90)*]. These mice have deficits in the Barnes circular platform maze. They show normal LTP when stimulation occurs at 100 Hz, but they show deficits in LTP when stimulation occurs at 5–10 Hz *(62,100)*. This 5- to 10-Hz range of stimulation is similar to naturally occurring oscillations caused by theta cells, which are active in response to locomotion and also may encode some spatial information *(92)*. Rotenberg et al. *(90)* found that CaMKII Asp 286 transgenic mice had fewer place cells, and those place cells that did develop had larger fields, lower firing rates, and less stability than wild-type place fields. They hypothesized that the inability to strengthen synaptic connections at low frequencies resulted in the observed place cell deficits. These experiments establish a strong correlation between LTP deficits and place cell deficits, although they do not establish LTP as a causal mechanism for any particular place cell function.

There is additional evidence that place cell deficits are correlated with long-term memory deficits. R(AB) mice, which show a selective impairment in L-LTP, also have impairments in place cell function *(101)*. Rotenberg et al. *(101)* found that R(AB) mice had normal place cell firing rates and well-formed place fields, but these place fields were unstable over long periods of time. However, with repeated exposure to the same environment, the place fields became more stable, suggesting that other signal transduction pathways may be recruited to counter the place field deficit.

In addition to maintaining place fields, LTP may contribute to spatial learning and the formation of cognitive maps by strengthening synaptic connections between those neurons that fire at the same time and therefore have overlapping place fields *(44)*. Place fields that do not overlap will not fire contiguously and the strength of the synapse between them therefore will not change. Thus, the spatial distance between two fields may be represented by the strength of the synaptic connection between the neurons that give rise to those fields. What remains to be seen is what happens to synaptic strength between the synapses of two place cells that fire contiguously in one environment when the organism is placed in an environment where only one of those cells fires. The number of potential synapses in areas CA1 and CA3 alone is large enough to have unique connections in multiple environments, but any overlap of synapses in more than one environment may lead to degradation of the synaptic strength as a result of that synapse being differentially activated in different environments. Although some ideas have been put forward about mechanisms to deal with this issue *(44)*, it is not clear how the problem will be solved.

More is being learned about the properties of place cells and their importance for certain forms of synaptic plasticity and learning, but several issues remain unresolved. One concerns whether place cells are involved simply in forming a map of space or whether they play a more active role in navigational processes. The hippocampus has been shown to be important not only for solving explicit spatial problems but also for learning about simple associations between a given context and a shock *(57,58,63)*. It may be that place cells are important only for learning about the features of an environment, which would be necessary to solve context-based problems. Once a contextual representation is formed, place cells may work in conjunction with other systems to produce spatial navigation through an environment. One other type of cell that may contribute to a navigational system is the head-direction (HD) cell. Whereas the firing of place cells depends only on the position of the organism in space, HD cells respond only to head direction, independent of the organism's position in space. Unlike place cells, which may go silent in different environments, HD cells fire consistently in different environments, but their preferred directions may change *(102)*. Although the preferred directions of a given set of HD cells may change in different environments, the angle between the preferred direction of any given two cells has been observed to be constant across environments *(102)*. One result of such coordinated activation may be that the strength between two synapses reflects orientation angles in the same manner as synaptic strength could reflect distances between place cell fields. This orientation, in conjunction with specific location coded by place cells, may form the basis of a system necessary to orient and navigate in an environment.

The recent work on genetically modified mice has allowed a connection to be made among place fields, LTP, and behavior. The recent advances in the study of place cells opens the floodgates to a host of important questions that, prior to the development of genetic techniques, may have been unanswerable. For example, assuming that LTP contributes to the formation and maintenance of cognitive maps, one would like to know how this map is read. Is it read in the hippocampus, where it is produced, or is the map read in other brain regions important to instigating behavior, but that do not have place cells, such as the prefrontal cortex *(103,104)*? Place fields do not seem to be modulated by the location of food reward in the environment *(96)*, suggesting that place cells themselves do not code any information about the location of biologically significant stimuli. How, then, does an animal learn the location of food rewards in mazes? Where is this aspect of a cognitive map stored, and how might synaptic plasticity contribute to the linking of the location of reward stimuli to the brain areas responsible for processing rewards? These are important questions, and as more is learned about the respective contributions of synaptic plasticity and place cells to cognitive mapping, the answers to these questions will become clearer. The development of transgenic techniques to drive gene expression in restricted brain regions in an inducible fashion may be especially helpful for solving these unresolved issues.

FUTURE DIRECTIONS

We have reviewed some electrophysiological, biochemical, and genetic studies linking changes in synaptic strength to learning and memory. These studies have provided strong evidence that synaptic plasticity plays an important role in learning and memory. Nevertheless, it remains to be demonstrated conclusively that synaptic plasticity is the cellular mechanism underlying learning and memory. With this in mind, we now turn to a discussion of some unanswered questions in synaptic plasticity research. The continued integration of more sophisticated behavioral studies with more molecular and genetic approaches will be needed to resolve many of these questions.

Aplysia

In *Aplysia*, the argument in favor of a role for both postsynaptic LTP-like changes and presynaptic changes in neuronal excitability appears strong, especially for the cellular model of Pavlovian conditioning. Recent evidence gathered in the reduced preparation makes a case for the involvement of postsynaptic mechanisms in learning *(17,22)*, but the generality of these findings to the intact animal remains unclear. Indeed, the focus on the pre-synaptic and postsynaptic cellular mechanisms underlying learning has brought us back to where we started—the analysis of the behavior of the organism.

Contiguity, Synaptic Plasticity, and Learning

Behavioral demonstrations of such important phenomena as blocking *(3)*, conditional discrimination *(28)*, and second-order conditioning *(5)* in *Aplysia* suggest that the repertoire of behavioral learning processes in *Aplysia* may be similar to those in vertebrates. There are, however, many other avenues to explore to further examine the extent of this similarity. One critical area of research that needs to be explored more thoroughly in *Aplysia* is the role that CS–US contiguity plays in establishing and maintaining learning. The past 30 yr of research and theory in the behavioral analysis of animal learn-ing has been guided by several demonstrations showing that simple con-tiguity is neither sufficient nor necessary for learning to occur *(105–107)*. The implications of such results have not been explored thoroughly at the cellular and molecular levels. Indeed, models of synaptic plasticity such as long-term facilitation and long-term potentiation rely on contiguity as the determinant of changes in synaptic strength. However, given that contiguity often fails to engender learning at the behavioral level, do these cellular models have any relevance to behavior?

One way in which associative theories of learning have been able to save the notion of contiguity is to assume that contiguity between a CS and a US will in fact lead to learning, unless other cues present on a conditioning trial, such as the conditioning context, are already strongly associated with that US *(108)*. This competition among cues for conditioning has generated impor-tant insights about the learning process and has allowed contiguity-based theories such as the Rescorla–Wagner model to explain the situations in which contiguity fails to produce learning. Exploring the assumptions of successful contiguity-based models like the Rescorla–Wagner model may prove fruitful when developing theories of synaptic plasticity and when designing experiments at the cellular level.

The Structure of Learning

An exciting area of behavioral research that is now beginning to be inves-tigated at the cellular level involves determining the associative structure of the learning that underlies conditioning *(5,109)*. Second-order conditioning has been a powerful tool for revealing this associative structure *(110)*. In a second-order conditioning paradigm, a CS (A) is paired with a US for several trials and then a novel CS (X) is paired with the original CS (A) for several trials. A test of X alone often reveals a conditioned response to X, despite X's never having been paired with a US. Why does the animal respond to X? There are two possibilities. One is that X has entered into a direct association with the response (stimulus–response, S–R, learning); the other is that when X is presented, the animal recalls a representation of A (stimulus–stimulus, S–S, learning), which, in turn, recalls a representation of the US. If it can be shown in *Aplysia* that responding to X is mediated through A, cellular theories of learning would have to be modified because they have focused almost exclusively on the sensory–motor synapse. A finding of S–S learning in *Aplysia* would suggest that synapses between sensory interneurons may be of critical importance in mediating behavior *(5)*. Determining what role, if any, LTP might play in determining whether S–S or S–R learning occurs may help shed light on an important behavioral phenomenon from a cellular perspective.

Hippocampal LTP

Its ability to integrate information at the cellular, molecular, and behavioral levels makes the mouse hippocampus an ideal system to study because all the tools are in place to define the cellular mechanisms that underlie memory storage. What future directions in that field will be most fruitful for these studies?

Appropriate Protocols for Inducing LTP

Many fundamental questions remain about the appropriate protocols for inducing LTP. LTP often is studied in the hippocampus, but there are several areas within the hippocampus where LTP may be important for memory storage, including the Schaffer collateral, mossy fiber, and perforant pathways. The use of genetically modified mice in which LTP is selectively impaired at a subset of synapses within the hippocampal circuitry have recently underscored the important role of synaptic plasticity in hippocampal area CA1 *(53,111,112)*. Many of these studies, however, have examined LTP in hippocampal slices, and one might get different results based on whether LTP is recorded in slices or in vivo, as revealed by the recent studies of Thy-1 knockout mice *(112,113)*. Many questions will be engendered by the study of LTP in vivo, and there clearly is a need for more in vivo studies of LTP in mice, recording from a variety of synapses within the hippocampus. Given the differences observed between LTP in brain slices and LTP in living organisms, how can we be certain that mechanisms inferred from brain slices are the endogenous mechanisms controlling synaptic plasticity and memory storage?

On a behavioral level, also, there is a critical need to standardize the behavioral assays performed in different labs. Further, it is critical to ensure that behavioral studies are carried out and analyzed properly. In some Morris water maze experiments, for example, only the latency to find the platform during training is presented. To fully evaluate spatial learning and memory, however, other variables such as swim speed and the extent of thigmotaxis need to be analyzed. Importantly, performance on a probe trial, in which the platform is removed from the pool and the animal's swim path is recorded, needs to be determined. Because genetic background can have a dramatic effect on behavioral performance, this, too, needs to be standardized *(114)*. Unfortunately, even if behavioral tasks are standardized, different labs may generate different results based on subtle differences in equipment or training protocol *(121)*. It therefore also is important to use a variety of behavioral tasks that depend on similar underlying brain systems.

Effects of LTP Reversal

If LTP is an important cellular mechanism of memory storage, then the erasure of LTP following acquisition should lead to impaired performance. If one could "erase" LTP after learning has occurred, would this cause the animal to forget? Although this is an intriguing question, one of the reasons that it has not been addressed is that it is difficult to reverse LTP once it has been established. Most pharmacological treatments are active only when applied at or around the time of tetanus and have no effect when applied after LTP is established. One potential way of erasing LTP would be to genetically induce "depotentiation," an electrophysiological treatment (typically 5- to 10-Hz stimuli) that reduces LTP *(115)*. If the molecular reagents that would "erase" LTP could be identified from the study of depotentiation, then reversible gene expression systems, such as the tetracycline-regulated system *(116)*, could be used to activate expression of this LTP-erasing molecule just after learning a hippocampus-dependent task. Further, with the regional- and cell-type-specificity possible with genetically modified mice, LTP could be erased only in a subset of neurons within the hippocampus. If the erasure of LTP in a certain subset of neurons within the hippocampus impaired memory, then the link between synaptic plasticity and memory storage would be greatly strengthened.

Of course, discussion of the potential effects of erasing LTP begs the question of the effects of enhancing LTP. Would the enhancement of LTP enhance memory as a result of increasing synaptic efficiency and gene transcription, or would it impair memory perhaps due to saturation effects? Results that speak to this issue give an uncertain picture. Two recent articles have described genetically modified mice with enhanced LTP. PSD-95 mutant mice have enhanced LTP, but they exhibit spatial learning impairments *(117)*, whereas mice lacking nociceptin receptors exhibit enhanced LTP and improved spatial learning *(118)*.

Other Cellular Models of Memory Storage

One major criticism of the study of the cellular basis of memory storage is that the field has been dominated by the hypothesis that changes in synaptic strength mediate memory storage *(120)*. One of the reasons that LTP has been the focal point for so much research on synaptic plasticity is that there are few alternative cellular or systems models that might account for memory storage.

We have reviewed one of the systems properties of the hippocampus, the place cells that fire when an animal is located in a specific portion of the environment, and we have explored the relationship between these place cells and LTP. Studies examining this relationship have found that modulating LTP—either pharmacologically or genetically—leads to altered place cell properties. To distinguish between the role of synaptic plasticity and place cells in mediating memory storage, we need to expand the study of place cells in genetically modified mice, looking, for example, to see if place cell properties are normal in mice that behave normally but have impairments in LTP at a subset of synapses in the hippocampus *(111,112)*.

What other cellular mechanisms might be involved in memory storage? One possibility is that rather than altering the strength of existing synapses, learning may lead to the formation of new synapses or may activate previously "silent" synapses by the insertion or activation of AMPA-type glutamate receptors. Such mechanisms do not necessarily preclude the involvement of LTP—those synapses that are potentiated may act as a short-term marker for the formation of new synapses *(119)*. If this were the case, then erasing LTP after a critical period of time, as proposed earlier, would not affect memory because LTP may play a role only for a short time period after learning. Thus, although LTP may not be the cellular representation of the long-term store, it may be a critical step in establishing the growth necessary for long-term storage.

Another way in which memories might be stored at a cellular level is through changes in neuronal excitability, measured in terms of fEPSP-spike potentiation. In this model, neurons would be more or less able to fire action potentials, and this would be altered with learning. In a sense, this is reflected in (or forms the basis of) place cells because they are recordings of firing rates. The problem with this mechanism is that it would be hard to use in developing models for how neuronal firing could be regulated in an associative way.

Electrotonic changes in the way synaptic signals are integrated over distance in the dendrite may serve as another mechanism mediating long-term memory storage. In this way, the activity of individual synapses *per se* would not change, but the effective electrotonic distance of the synapses from the cell body and axon hillock would be altered. This would be accomplished by changing the active properties of dendrites, thus dramatically altering the ability of the neuron to respond to the activation of specific synapses. If this were restricted to specific branches within the dendritic tree, then this mechanism would mediate associative learning.

There clearly are many alternative cellular mechanisms that may drive memory storage. The introduction of a viable alternative to synaptic plasticity will be important not only for exploring alternatives to synaptic plasticity, but for establishing a theoretical basis for testing the limits of synaptic plasticity as a workable model for learning and memory.

Although many questions about the molecular and cellular mechanisms of learning and memory remain unanswered, the coordination of a variety of approaches—biochemical, genetic, pharmacological, electrophysiological, and behavioral—has generated a much richer understanding of the learning process than has ever been possible before. However, with this broad perspective comes the daunting task of assimilating these perspectives into meaningful explanations of learning and memory processes. Each of these approaches points to synaptic plasticity as the potential neurobiological building block for learning, and our study of the cellular and molecular mechanisms underlying learning and memory brings us back to the core problems that inspired us to investigate these issues in the first place—identifying the signal transduction pathways that underlie learning, the mechanisms by which learning modifies these pathways, and the way in which these modified pathways, in turn, influence memory and behavior.

REFERENCES

1. Hebb, D. O. (1949) *The Organization of Behavior.* Wiley, New York.
2. Glanzman, D. L. (1995) The cellular basis of classical conditioning in *Aplysia californica*—it's less simple than you think. *Trends Neurosci.* **18,** 30–36.
3. Colwill, R. M., Absher, R. A., and Roberts, M. L. (1988) Context-US learning in *Aplysia californica. J. Neurosci.* **8,** 4434–4439.
4. Hawkins, R. D., Carew, T. J., and Kandel, E. R. (1986) Effects of interstimulus interval and contingency on classical conditioning of the *Aplysia* siphon withdrawal reflex. *J. Neurosci.* **6,** 1695–1701.
5. Hawkins, R. D., Greene, W., and Kandel, E. R. (1998) Classical conditioning, differential conditioning, and second-order conditioning of the *Aplysia* gill-withdrawal reflex in a simplified mantle organ preparation. *Behav. Neurosci.* **112,** 636–645.

6. Kandel, E. R., Schwartz, J. H., and Jessell, T. M. (1995) *Essentials of Neural Science and Behavior*. Appleton & Lange, Stamford, CT.
7. Pinsker, H., Kupfermann, I., Castellucci, V., and Kandel, E. (1970) Habituation and dishabituation of the gill-withdrawal reflex in *Aplysia*. *Science* **167,** 1740–1742.
8. Carew, T. J., Castellucci, V. F., and Kandel, E. R. (1971) An analysis of dishabituation and sensitization of the gill-withdrawal reflex in *Aplysia*. *Int. J. Neurosci.* **2,** 79–98.
9. Castellucci, V., Pinsker, H., Kupfermann, I., and Kandel, E. R. (1970) Neuronal mechanisms of habituation and dishabituation of the gill-withdrawal reflex in *Aplysia*. *Science* **167,** 1745–1748.
10. Castellucci, V. and Kandel, E. R. (1976) Presynaptic facilitation as a mechanism for behavioral sensitization in *Aplysia*. *Science* **194,** 1176–1178.
11. Byrne, J. H. and Kandel, E. R. (1996) Presynaptic facilitation revisited: state and time dependence. *J. Neurosci.* **16,** 425–435.
12. Montarolo, P. G., Goelet, P., Castellucci, V. F., Morgan, J., Kandel, E. R., and Schacher, S. (1986) A critical period for macronuclear synthesis in long-term heterosynaptic facilitation in *Aplysia*. *Science* **234,** 1249–1254.
13. Dash, P., Hochner, B., and Kandel, E. R. (1990) Injection of the cAMP-responsive element into the nucleus of *Aplysia* sensory neurons blocks long-term facilitation. *Nature* **345,** 718–721.
14. Bartsch, D., Ghirardi, M., Skehel, P. A., Karl, K. A., Herder, S. P., Chen, M., Bailey, C. H., and Kandel, E. R. (1995) *Aplysia* CREB2 represses long-term facilitation: relief of repression converts transient facilitation into long-term functional and structural change. *Cell* **83,** 979–992.
15. Cook, D. G., Stopfer, M., and Carew, T. J. (1991) Identification of a reinforcement pathway necessary for operant conditioning of head waving in *Aplysia californica*. *Behav. Neural Biol.* **55,** 313–337.
16. Carew, T. J., Hawkins, R. D., Abrams, T. W., and Kandel, E. R. (1984) A test of Hebb's postulate at identified synapses which mediate classical conditioning in *Aplysia*. *J. Neurosci.* **4,** 1217–1224.
17. Murphy, G. G. and Glanzman, D. L. (1997) Mediation of classical conditioning in *Aplysia californica* by long-term potentiation of sensorimotor synapses. *Science* **278,** 467–471.
18. Carew, T. J., Walters, E. T., and Kandel, E. R. (1981) Classical conditioning in a simple withdrawal reflex in *Aplysia californica*. *J. Neurosci.* **1,** 1426–1437.
19. Walters, E. T., Carew, T. J., and Kandel, E. R. (1981) Associative learning in *Aplysia*: evidence for conditioned fear in an invertebrate. *Science* **211,** 504–506.
20. Hawkins, R. D., Abrams, T. W., Carew, T. J., and Kandel, E. R. (1983) A cellular mechanism of classical conditioning in *Aplysia*: activity-dependent amplification of presynaptic facilitation. *Science* **219,** 400–405.
21. Bao, J. X., Kandel, E. R., and Hawkins, R. D. (1998) Involvement of presynaptic and postsynaptic mechanisms in a cellular analog of classical conditioning at *Aplysia* sensory-motor neuron synapses in isolated cell culture. *J. Neurosci.* **18,** 458–466.
22. Murphy, G. G. and Glanzman, D. L. (1996) Enhancement of sensorimotor connections by conditioning-related stimulation in *Aplysia* depends upon postsynaptic Ca^{2+}. *Proc. Natl. Acad. Sci. USA* **93,** 9931–9936.

23. Rescorla, R. A. (1988) Behavioral studies of Pavlovian conditioning. *Annu. Rev. Neurosci.* **11,** 329–352.
24. Clark, G. A., Hawkins, R. D., and Kandel, E. R. (1994) Activity-dependent enhancement of presynaptic facilitation provides a cellular mechanism for the temporal specificity of classical conditioning in Aplysia. *Learning Memory* **1,** 243–258.
25. Abrams, T. W. and Kandel, E. R. (1988) Is contiguity detection in classical conditioning a system or a cellular property? Learning in *Aplysia* suggests a possible molecular site. *Trends Neurosci.* **11,** 128–135.
26. Schafe, G. E., Sollars, S. I., and Bernstein, I. L. (1995) The CS–US interval and taste aversion learning: a brief look. *Behav. Neurosci.* **109,** 799–802.
27. Gibbon, J., Baldock, M. D., Locurto, C., Gold, L., and Terrace, H. S. (1977) Trial and intertrial durations in autoshaping. *J. Exp. Psychol.: Animal Behav. Proc.* **3,** 264–284.
28. Colwill, R. M., Absher, R. A., and Roberts, M. L. (1988) Conditional discrimination learning in *Aplysia californica. J. Neurosci.* **8,** 4440–4444.
29. Klein, M. (1994) Synaptic augmentation by 5-HT at rested *Aplysia* sensorimotor synapses: independence of action potential prolongation. *Neuron* **13,** 159–166.
30. Martin, K. C., Casadio, A., Zhu, H. E. Y., Rose, J. C., Chen, M., Bailey, C. H., and Kandel, E. R. (1997) Synapse-specific, long-term facilitation of *Aplysia* sensory to motor synapses: a function for local protein synthesis in memory storage. *Cell* **91,** 927–938.
31. Squire, L. R. (1992) Memory and the hippocampus: a synthesis from findings with rats, monkeys and humans. *Psychol. Rev.* **99,** 195–231.
32. Cohen, N. J. and Eichenbaum, H. (1993) *Memory, Amnesia, and the Hippocampal System.* MIT Press, Cambridge, MA.
33. Bliss, T. V. and Collingridge, G. L. (1993) A synaptic model of memory: long-term potentiation in the hippocampus. *Nature* **361,** 31–39.
34. Muller, R. (1996) A quarter of a century of place cells. *Neuron* **17,** 813–822.
35. Sharp, P. E., McNaughton, B. L., and Barnes, C. A. (1985) Enhancement of hippocampal field potentials in rats exposed to a novel, complex environment. *Brain Res.* **339,** 361–365.
36. Moser, E., Mathiesen, I., and Andersen, P. (1993) Association between brain temperature and dentate field potentials in exploring and swimming rats. *Science* **259,** 1324–1326.
37. Shors, T. J., Seib, T. B., Levine, S., and Thompson, R. F. (1989) Inescapable versus escapable shock modulates long-term potentiation in the rat hippocampus. *Science* **244,** 224–226.
38. Moser, E. I. (1995) Learning-related changes in hippocampal field potentials. *Behav. Brain Res.* **71,** 11–18.
39. Jeffery, K. J. (1997) LTP and spatial learning—where to next? *Hippocampus* **7,** 95–110.
40. McKernan, M. G. and Shinnick-Gallagher, P. (1997) Fear conditioning induces a lasting potentiation of synaptic currents in vitro. *Nature* **390,** 607–611.
41. Rogan, M. T., Staubli, U. V., and LeDoux, J. E. (1997) Fear conditioning induces associative long-term potentiation in the amygdala. *Nature* **390,** 604–607.
42. Rogan, M. T. and LeDoux, J. E. (1995) LTP is accompanied by commensurate enhancement of auditory-evoked responses in a fear conditioning circuit. *Neuron* **15,** 127–136.

43. Rescorla, R. A. (1967) Pavlovian conditioning and its proper control procedures. *Psychol. Rev.* **74,** 71–80.
44. Muller, R. U., Stead, M., and Pach, J. (1996) The hippocampus as a cognitive graph. *J. Gen. Physiol.* **107,** 663–694.
45. McNaughton, B. L., Barnes, C. A., Rao, G., Baldwin, J., and Rasmussen, M. (1986) Long-term enhancement of hippocampal synaptic transmission and the acquisition of spatial information. *J. Neurosci.* **6,** 563–571.
46. Castro, C. A., Silbert, L. H., McNaughton, B. L., and Barnes, C. A. (1989) Recovery of spatial learning deficits after decay of electrically induced synaptic enhancement in the hippocampus. *Nature* **342,** 545–548.
47. Bliss, T. V. and Richter-Levin, G. (1993) Spatial learning and the saturation of long-term potentiation. *Hippocampus* **3,** 123–125.
48. Barnes, C. A., Jung, M. W., McNaughton, B. L., Korol, D. L., Andreasson, K., and Worley, P. F. (1994) LTP saturation and spatial learning disruption: effects of task variables and saturation levels. *J. Neurosci.* **14,** 5793–5806.
49. Moser, M. B., Moser, E. I., Forrest, E., Andersen, P., and Morris, R. G. (1995) Spatial learning with a minislab in the dorsal hippocampus. *Proc. Natl. Acad. Sci. USA* **92,** 9697–9701.
50. Moser, E. I., Krobert, K. A., Moser, M. B., and Morris, R. G. (1998) Impaired spatial learning after saturation of long-term potentiation. *Science* **281,** 2038–2042.
51. Morris, R. G. M. (1989) Synaptic plasticity and learning: selective impairment of learning in rats and blockade of long-term potentiation *in vivo* by the *N*-methyl-D-aspartate receptor agonist, AP5. *J. Neurosci.* **9,** 3040–3057.
52. Morris, R. G. M., Andersen, E., Lynch, G., and Baudry, M. (1986) Selective impairment of learning and blockade of long-term potentiation by an *N*-methyl-D-aspartate receptor antagonist, AP5. *Nature* **319,** 774–776.
53. Tsien, J. Z., Huerta, P. T., and Tonegawa, S. (1996) The essential role of hippocampal CA1 NMDA receptor-dependent synaptic plasticity in spatial memory. *Cell* **87,** 1327–1338.
54. Bannerman, D. M., Good, M. A., Butcher, S. P., Ramsay, M., and Morris, R. G. (1995) Distinct components of spatial learning revealed by prior training and NMDA receptor blockade. *Nature* **378,** 182–186.
55. Saucier, D. and Cain, D. P. (1995) Spatial learning without NMDA receptor-dependent long-term potentiation. *Nature* **378,** 186–189.
56. Rempel-Clower, N. L., Zola, S. M., Squire, L. R., and Amaral, D. G. (1996) Three cases of enduring memory impairment after bilateral damage limited to the hippocampal formation. *J. Neurosci.* **16,** 5233–5255.
57. Kim, J. J., Rison, R. A., and Fanselow, M. S. (1993) Effects of amygdala, hippocampus and periaqueductal gray lesions on short- and long-term contextual fear. *Behav. Neurosci.* **107,** 1093–1098.
58. Phillips, R. G. and LeDoux, J. E. (1992) Differential contribution of amygdala and hippocampus to cued and contextual fear conditioning. *Behav. Neurosci.* **106,** 274–285.
59. Bunsey, M. and Eichenbaum, H. (1995) Selective damage to the hippocampal region blocks long-term retention of a natural and nonspatial stimulus-stimulus association. *Hippocampus* **5,** 546–556.
60. Winocur, G. (1990) Anterograde and retrograde amnesia in rats with dorsal hippocampal or dorsomedial thalamic lesions. *Behav. Brain Res.* **38,** 145–154.

61. Chen, C. and Tonegawa, S. (1997) Molecular genetic analysis of synaptic plasticity, activity-dependent neural development, learning, and memory in the mammalian brain. *Annu. Rev. Neurosci.* **20,** 157–184.

62. Mayford, M., Abel, T., and Kandel, E. R. (1995) Transgenic approaches to cognition. *Curr. Opin. Neurobiol.* **5,** 141–148.

63. Kim, J. J. and Fanselow, M. S. (1992) Modality-specific retrograde amnesia of fear. *Science* **256,** 675–677.

64. Squire, L. R. and Alvarez, P. (1995) Retrograde amnesia and memory consolidation: a neurobiological perspective. *Curr. Opin. Neurobiol.* **5,** 169–177.

65. Huang, Y.-Y., Nguyen, P. V., Abel, T., and Kandel, E. R. (1996) Long-lasting forms of synaptic potentiation in the mammalian hippocampus. *Learning Memory* **3,** 74–85.

66. Frey, U., Huang, Y.-Y., and Kandel, E. R. (1993) Effects of cAMP simulate a late stage of LTP in hippocampal CA1 neurons. *Science* **260,** 1661–1664.

67. Abraham, W. C., Mason, S. E., Demmer, J., Williams, J. M., Richardson, C. L., Tate, W. P., Lawlor, P. A., and Dragunow, M. (1993) Correlations between immediate early gene induction and the persistence of long-term potentiation. *Neuroscience* **56,** 717–727.

68. Davis, H. P. and Squire, L. R. (1984) Protein synthesis and memory: a review. *Psychol. Bull.* **96,** 518–559.

69. Frey, U., Krug, M., Reymann, K. G., and Matthies, H. (1988) Anisomycin, an inhibitor of protein synthesis, blocks late phase of LTP phenomena in the hippocampal CA1 region *in vitro. Brain Res.* **452,** 57–65.

70. Huang, Y.-Y. and Kandel, E. R. (1994) Recruitment of long-lasting and protein kinase A-dependent long-term potentiation in the CA1 region of hippocampus requires repeated tetanization. *Learning Memory* **1,** 74–82.

71. Nguyen, P. V., Abel, T., and Kandel, E. R. (1994) Requirement for a critical period of transcription for induction of a late phase of LTP. *Science* **265,** 1104–1107.

72. Montminy, M. (1997) Transcriptional regulation by cyclic AMP. *Annu. Rev. Biochem.* **66,** 807–822.

73. Impey, S., Mark, M., Villacres, E. C., Poser, S., Chavkin, C., and Storm, D. R. (1996) Induction of CRE-mediated gene expression by stimuli that generate long-lasting LTP in area CA1 of the hippocampus. *Neuron* **16,** 973–982.

74. Bourtchuladze, R., Frenguelli, B., Blendy, J., Cioffi, D., Schütz, G., and Silva, A. J. (1994) Deficient long-term memory in mice with a targeted mutation of the cAMP-responsive element-binding protein. *Cell* **79,** 59–68.

75. Kogan, J. H., Frankland, P. W., Blendy, J. A., Coblentz, J., Marowitz, Z., Schütz, G., and Silva, A. (1997) Spaced training induces normal long-term memory in CREB mutant mice. *Curr. Biol.* **7,** 1–11.

76. Bito, H., Deisseroth, K., and Tsien, R. W. (1997) Ca^{2+}-dependent regulation in neuronal gene expression. *Curr. Opin. Neurobiol.* **7,** 419–429.

77. Abel, T., Nguyen, P. V., Barad, M., Deuel, T. A. S., Kandel, E. R., and Bourtchouladze, R. (1997) Genetic demonstration of a role for PKA in the late phase of LTP and in hippocampus-based long-term memory. *Cell* **88,** 615–626.

78. Clegg, C., Correll, L. A., Cadd, G. G., and McKnight, G. S. (1987) Inhibition of intracellular cAMP-dependent protein kinase using mutant genes of the regulatory type I subunit. *J. Biol. Chem.* **262,** 13,111–13,119.

79. Ginty, D. D., Glowacka, D., DeFranco, C., and Wagner, J. A. (1991) Nerve growth factor-induced neuronal differentiation after dominant repression of both Type I and Type II cAMP-dependent protein kinase activities. *J. Biol. Chem.* **266,** 15,325–15,333.

80. Mellon, P. L., Clegg, C. L., Correll, L. A., and McKnight, G. S. (1989) Regulation of transcription by cyclic AMP-dependent protein kinase A. *Proc. Natl. Acad. Sci. USA* **86,** 4887–4891.

81. Morris, R. G. M., Garrud, P., Rawlins, J. N. P., and O'Keefe, J. (1982) Place navigation impaired in rats with hippocampal lesions. *Nature* **297,** 681–683.

82. Chen, C., Kim, J. J., Thompson, R. F., and Tonegawa, S. (1996) Hippocampal lesions impair contextual fear conditioning in two strains of mice. *Behav. Neurosci.* **110,** 1177–1180.

83. Logue, S. F., Paylor, R., and Wehner, J. M. (1997) Hippocampal lesions cause learning deficits in inbred mice in the Morris water maze and conditioned fear task. *Behav. Neurosci.* **111,** 104–113.

84. Yamamoto, T., Shimura, T., Sako, N., Yasoshima, Y., and Sakou, N. (1994) Neural substrates for conditioned taste aversion in the rat. *Behav. Brain Res.* **65,** 123–137.

85. Rothermel, J. D. and Parker-Botelho, L. H. (1988) A mechanistic and kinetic analysis of the interactions of the diastereoisomers of adenosine 3',5'-(cyclic)phosphorothioate with purified cyclic AMP-dependent protein kinase. *Biochem. J.* **251,** 757–762.

86. Bourtchouladze, R., Abel, T., Berman, N., Gordon, R., Lapidus, K., and Kandel, E. R. (1998) Different training procedures for contextual memory in mice can recruit either one or two critical periods for memory consolidation that require protein synthesis and PKA. *Learning Memory* **5,** 365–374.

87. Rudy, J. W. and Sutherland, R. J. (1995) Configural association theory and the hippocampal formation: an appraisal and reconfiguration. *Hippocampus* **5,** 375–389.

88. O'Keefe, J. and Nadel, L. (1978) *The Hippocampus as a Cognitive Map.* Clarendon, Oxford.

89. O'Keefe, J. and Dostrovsky, J. (1971) The hippocampus as a spatial map. Preliminary evidence from unit activity in the freely-moving rat. *Brain Res.* **34,** 171–175.

90. Rotenberg, A., Mayford, M., Hawkins, R. D., Kandel, E. R., and Muller, R. U. (1996) Mice expressing activated CaMKII lack low frequency LTP and do not form stable place cells in the CA1 region of the hippocampus. *Cell* **87,** 1351–1361.

91. Sharp, P. E. (1997) Subicular cells generate similar spatial firing patterns in two geometrically and visually distinctive environments: comparison with hippocampal place cells. *Behav. Brain Res.* **85,** 71–92.

92. Kubie, J. L., Muller, R. U., and Bostock, E. (1990) Spatial firing properties of hippocampal theta cells. *J. Neurosci.* **10,** 1110–1123.

93. Muller, R. U., Kubie, J. L., and Ranck, J. B., Jr. (1987) Spatial firing patterns of hippocampal complex-spike cells in a fixed environment. *J. Neurosci.* **7,** 1935–1950.

94. Thompson, L. T. and Best, P. J. (1990) Long-term stability of the place-field activity of single units recorded from the dorsal hippocampus of freely behaving rats. *Brain Res.* **509,** 299–308.

95. Samsonovich, A. and McNaughton, B. L. (1997) Path integration and cognitive mapping in a continuous attractor neural network model. *J. Neurosci.* **17,** 5900–5920.

96. O'Keefe, J. and Speakman, A. (1987) Single unit activity in the rat hippocampus during a spatial memory task. *Exp. Brain Res.* **68,** 1–27.

97. Cho, Y. H., Giese, K. P., Tanila, H., Silva, A. J., and Eichenbaum, H. (1998) Abnormal hippocampal spatial representations in alphaCaMKIIT286A and CREB αΔ-mice. *Science* **279,** 867–869.

98. Kentros, C., Hargreaves, E., Hawkins, R. D., Kandel, E. R., Shapiro, M., and Muller, R. U. (1998) Abolition of long-term stability of new hippocampal place cell maps by NMDA receptor blockade. *Science* **280,** 2121–2126.

99. McHugh, T. J., Blum, K. I., Tsien, J. Z., Tonegawa, S., and Wilson, M. A. (1996) Impaired hippocampal representation of space in CA1-specific NMDAR1 knock-out mice. *Cell* **87,** 1339–1349.

100. Bach, M. E., Hawkins, R. H., Osman, M., Kandel, E. R., and Mayford, M. (1995) Impairment of spatial but not contextual memory in CaMKII mutant mice with a selective loss of hippocampal LTP in the range of the θ frequency. *Cell* **81,** 905–915.

101. Rotenberg, A., Abel, T., Kandel, E. R., and Muller, R. U. (1997) The firing fields of place cells in mice with impaired late-phase LTP are crisp but move to unpredictable locations. *Soc. Neurosci. Abst.* **23,** 501.

102. Muller, R. U., Ranck, J. B., Jr., and Taube, J. S. (1996) Head direction cells: properties and functional significance. *Curr. Opin. Neurobiol.* **6,** 196–206.

103. Jung, M. W., Qin, Y., McNaughton, B. L., and Barnes, C. A. (1998) Firing characteristics of deep layer neurons in prefrontal cortex in rats performing spatial working memory tasks. *Cereb. Cortex* **8,** 437–450.

104. Poucet, B. (1997) Searching for spatial unit firing in the prelimbic area of the rat medial prefrontal cortex. *Behav. Brain Res.* **84,** 151–159.

105. Kamin, L. J. (1968) "Attention-like" processes in classical conditioning, in *Miami Symposium on the Prediction of Behavior: Aversive Stimulation* (Jones, M. R., ed.), University of Miami Press, Coral Gables, FL, pp. 9–33.

106. Rescorla, R. A. (1966) Predictability and number of pairings in Pavlovian fear conditioning. *Psychonom. Sci.* **4,** 383–384.

107. Wagner, A. R., Logan, F. A., Haberlandt, K., and Price, T. (1968) Stimulus selection in animal discrimination learning. *J. Exp. Psychol.* **76,** 177–186.

108. Rescorla, R. A. and Wagner, A. R. (1972) A theory of Pavlovian conditioning: variations in the effectiveness of reinforcement and nonreinforcement, in *Classical Conditioning II: Current Research and Theory* (Black, A. and Prokasky, W. F., eds.), Appleton-Century-Crofts, New York, pp. 64–99.

109. Colwill, R. M. (1996) Detecting associations in Pavlovian conditioning and instrumental learning in vertebrates and in invertebrates, in *Neuroethological Studies of Cognitive and Perceptual Processes* (Moss, C. F. and Shettleworth, S. J., eds.), Westview, Boulder, CO, pp. 31–62.

110. Rescorla, R. A. (1980) *Pavlovian Second-Order Conditioning: Studies in Associative Learning.* L. Erlbaum Associates, Hillsdale, NJ.

111. Huang, Y.-Y., Kandel, E. R., Varshavsky, L., Brandon, E. P., Qi, M., Idzerda, R. L., McKnight, G. S., and Bourtchouladze, R. (1995) A genetic test of the effect of mutations in PKA on mossy fiber LTP and its relation to spatial and contextual learning. *Cell* **83,** 1211–1222.

112. Nosten-Bertrand, M., Errington, M. L., Murphy, K. P., Tokugawa, Y., Barboni, E., Kozlova, E., Michalovich, D., Morris, R. G., Silver, J., Stewart, C. L., Bliss, T. V., and Morris, R. J. (1996) Normal spatial learning despite regional inhibition of LTP in mice lacking Thy-1. *Nature* **379,** 826–829.

113. Errington, M. L., Bliss, T. V., Morris, R. J., Laroche, S., and Davis, S. (1997) Long-term potentiation in awake mutant mice. *Nature* **387,** 666–667.

114. Silva, A. J., Simpson, E. M., Takahashi, J. S., et al. (1997) Mutant mice and neuroscience: recommendations concerning genetic background. Banbury Conference on genetic background in mice. *Neuron* **19,** 755–759.

115. Staubli, U. and Chun, D. (1996) Factors regulating the reversibility of long-term potentiation. *J. Neurosci.* **16,** 853–860.

116. Mayford, M., Mansuy, I. M., Muller, R. U., and Kandel, E. R. (1997) Memory and behavior: a second generation of genetically modified mice. *Curr. Biol.* **7,** R580–R589.

117. Migaud, M., Charlesworth, P., Dempster, M., Webster, L. C., Watabe, A. M., Makhinson, M., He, Y., Ramsay, M. F., Morris, R. G., Morrison, J. H., O'Dell, T. J., and Grant, S. G. (1998) Enhanced long-term potentiation and impaired learning in mice with mutant postsynaptic density-95 protein. *Nature* **396,** 433–439.

118. Manabe, T., Noda, Y., Mamiya, T., Katagiri, H., Houtani, T., Nishi, M., Noda, T., Takahashi, T., Sugimoto, T., Nabeshima, T., and Takeshima, H. (1998) Facilitation of long-term potentiation and memory in mice lacking nociceptin receptors. *Nature* **394,** 577–581.

119. Frey, U. and Morris, R. G. (1998) Synaptic tagging: implications for late maintenance of hippocampal long-term potentiation. *Trends Neurosci.* **21,** 181–188.

120. Shors, T. J. and Matzel, L. D. (1997) Long-term potentiation: what's learning got to do with it? *Behav. Brain Sci.* **20,** 597–655.

121. Bevins, R. A., McPhee, J. E., Rauhut, A. S., and Ayres, J. J. B. (1997) Converging evidence for one-trial context fear conditioning with an immediate shock: importance of shock potency. *J. Exp. Psychol.: Animal Behav. Proc.* **23,** 312–324.

3

Signal Transduction Pathways in Well-Defined Models of Learning and Memory

Drosophila *and Honeybee*

Uli Müller

Mechanisms underlying learning and memory have been extensively investigated in a variety of systems, from invertebrates such as *Aplysia (1,2)*, *Drosophila (3–5)*, and the honeybee *(6,7)* to various vertebrates *(8–10)*. The study of learning in the fruitfly *Drosophila melanogaster* and the honeybee *Apis mellifera,* the two most powerful insect model systems, added important information to our knowledge of mechanisms implicated in learning and memory. Whereas *Drosophila* provides the opportunity for a genetic dissection of learning using molecular techniques, the honeybee is a complex yet electrophysiologically, biochemically, and pharmacologically accessible system for dissecting learning mechanisms. Thus, the characterization of *Drosophila* learning mutants elucidated the molecular and biochemical machinery underlying learning *(3–5,11)*, whereas investigations in the honeybee complementary enabled the identification of cellular substrates, transmitters, and second messengers involved in learning *(6,7)* (Fig. 1). The findings from both insect model systems provide information at different levels of analysis that complement each other beautifully.

GENETIC ANALYSIS OF LEARNING AND MEMORY IN *DROSOPHILA*

In addition to paradigms for habituation and sensitization, appetitive and aversive olfactory conditioning *(12,13)*, as well as complex behaviors like courtship, have been demonstrated in *Drosophila (14)*. The aversive conditioning to odors *(15)*, however, is the paradigm most widely used to investigate the processes of learning and memory in the fruitfly. In a training session, the flies are exposed to two odors in sequence for 1 min each. During exposure of one of the odors, repeated electric shocks are applied to the legs of the flies via a copper grid. The conditioned odor avoidance was

From: *Cerebral Signal Transduction: From First to Fourth Messengers*
Edited by: M. E. A. Reith © Humana Press Inc., Totowa, NJ

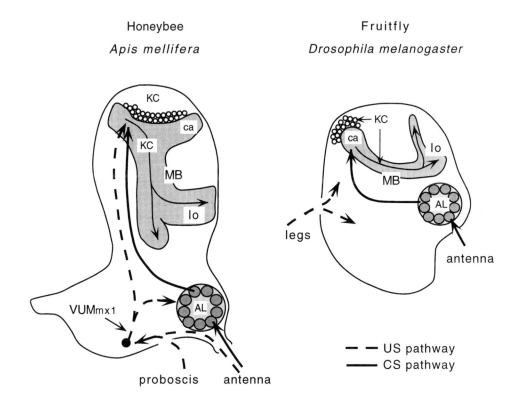

Fig. 1. Scheme of the signaling pathways involved in associative olfactory learning in the honeybee and *Drosophila*. In the honeybee, the neuronal circuits processing the odor stimulus (conditioned stimulus, CS) and the sucrose reward (unconditioned stimulus, US) in proboscis extension reflex conditioning are well described. CS pathway: Olfactory information from the antenna is processed in the antennal lobes (AL) which are connected to the calyces (ca) of the mushroom bodies (MB) and to the lateral protocerebrum. The MB, with their intrinsic Kenyon cells (KC), process input from different sensory modalities. US pathway: The information from chemo-sensory receptors for sucrose on the antenna and the proboscis is relayed to the VUMmx1 neuron. The VUMmx1 neuron aborizes in the AL, the calyces of the MB and the lateral protocerebrum and can substitute for the US. Thus, the CS and US pathways converge in the AL and the calyces of the MB, which are involved in different features of associative olfactory learning in the honeybee. The output path-ways that connect the lobes of the mushroom bodies (lo) to the motor circuits of PER are unknown. In *Drosophila,* the CS pathway of associative olfactory condi-tioning is similar to that in the honeybee, the circuitry of the US pathway is unclear. In aversive odor conditioning, the US information (electric shock) from the legs enters the brain via the ventral nerve cord. Neither the exact connections of the US pathway within the brain nor the output pathways are known.

tested in a T-maze, in which the flies are allowed to choose between the two odors. In this paradigm, only spaced training sessions (intersession intervals of 15 min) induced a long-term memory lasting days *(16)*.

The Central Role of the cAMP Cascade in Associative Olfactory Learning in **Drosophila**

Since the demonstration of associative learning in *Drosophila (12)*, genetic analysis has shown that the cAMP cascade plays a central role in associative olfactory learning in the fruitfly. A continuous line of evidence, starting from the early characterization of the behavioral *Drosophila* mutants *dunce* and *rutabaga*, which are deficient in enzymes regulating the cAMP level, to more recent investigations addressing the role of the cAMP-dependent protein kinase A (PKA) and the cAMP-responsive transcription, support the critical role of cAMP signal transduction in learning and memory *(5,11,17)* (Fig. 2). The role of the cAMP cascade in mechanisms of learning and memory has also been studied in other model systems, such as *Aplysia* and rodents. These studies emphasize the importance of the cAMP cascade, as cAMP/PKA-mediated processes are critically involved in neuronal plasticity in these models *(17,18)*.

Adenylate Cyclases and cAMP Phosphodiesterases in Drosophila *Learning*

The analysis of the early *Drosophila* mutants *dunce* and *rutabaga*, both displaying defects in associative olfactory learning, has shown that *dunce* is deficient in a cAMP-specific phosphodiesterase (PDE) *(19,20)* and *rutabaga* in a Ca^{2+}/calmodulin-dependent adenylate cyclase (AC) *(21,22)* (Fig. 2).

The *dunce* mutants, with reduced or missing cAMP PDE *(23,24)* have defects in various associative and nonassociative learning paradigms, but they also show pleiotropic effects, such as female sterility *(25)*. However, it has been demonstrated that induction of the *dunce* gene in adult flies can rescue the behavioral defect *(26)*. All mutant alleles of *dunce* show severe deficits in learning, independent of the enhanced cAMP level caused by disrupted PDE activity. Moreover, the normalization of the cAMP levels in the double mutant *dunce /rutabaga* does not rescue the deficits in learning *(22)*. This suggests that the defects in learning are more likely the result of a failure in the modulation of cAMP levels than of the general elevation of cAMP levels.

The *rutabaga* mutant with a loss of Ca^{2+}/calmodulin-dependent AC activity shows deficits in classical conditioning, operant conditioning, habituation, and distinct features of courtship behavior *(22,25)*. The mutation thus seems to affect different forms of learning. The primary sensory processes,

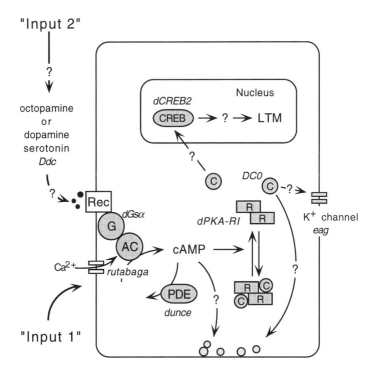

Fig. 2. Model of the cAMP cascade and its components involved in associative olfactory learning in *Drosophila*. Coincident activation of "Input 1" (presumably CS) and "Input 2" (presumably US) can lead to a synergistic stimulation of the adenylate cyclase (AC) and, thus, to elevated cAMP levels. Activation of cAMP-dependent protein kinase A (PKA) (R_2C_2) results in the phosphorylation of substrates (ion channels, etc.) and, thus, transient synaptic plasticity. Repeated elevations of cAMP levels by multiple training sessions can cause long-lasting modifications of synaptic connections, and thus, long-term memory (LTM) by inducing CREB-mediated gene expression. *Drosophila* learning mutants affecting the cAMP cascade are indicated beside the component. C, catalytic subunit of PKA; CREB, cAMP-response element binding protein; G, G-protein; PDE, cAMP-phosphodiesterase; R, regulatory subunit of PKA; Rec, receptor.

however, seem unaffected by a loss of *rutabaga* AC. Because the Ca^{2+}/calmodulin-dependent AC is synergistically activated by Ca^{2+}/calmodulin and receptor-mediated G-protein stimulation, this enzyme has been discussed as a molecular coincidence detector, thought to participate in learning *(27)*. Although such a function has not been demonstrated in *Drosophila*, studies on a similar adenylate cyclase in *Aplysia* reveal that AC stimulation is evidently a sequence-dependent interaction between calcium and transmitter stimuli *(28)*. In mammals, where a Ca^{2+}/calmodulin-dependent adenylate cyclase is located in the hippocampus and the neocortex, areas implicated in learning *(29)*, a similar scheme is feasible.

Further evidence of the important role of the *rutabaga* AC comes from the molecular characterization of the memory mutant *amnesiac (30)*, a neuropeptide encoding gene *(31)*. This peptide shows homologies to a mammalian adenylate cyclase-activating peptide (PACAP) that activates AC by a G-protein-coupled receptor *(32)*. In the larval neuromuscular junction of *Drosophila,* PACAP has been shown to coactivate the RAS/RAF pathway and the rutabaga AC *(33)*. These and several findings suggest that in mammals as in *Drosophila*, peptides and other neuromodulators are critically involved in G-protein-coupled neurotransmission in addition to neurotransmitters *(34)*. The central role of G-protein-mediated processes in *Drosophila* learning has been demonstrated for a G_α-subunit. Olfactory learning is totally disrupted by the expression of constitutively activated $G_{\alpha s}$ in the mushroom bodies *(35)*.

Protein Kinase A (PKA), a Major Target of cAMP

Although the exact targets of cAMP are unknown in *Drosophila*, several investigations support a major role of the PKA in *Drosophila* learning. PKAs are ubiquitous proteins and their highly conserved structure and properties have been demonstrated in *Drosophila melanogaster* and the honeybee, *Apis mellifera (36–39)*. The inactive enzyme is tetrameric (R_2C_2) with two regulatory (R) and two catalytic (C) subunits. Binding of cAMP to the regulatory subunits causes the dissociation of catalytic subunits. The free C-subunits are active and can phosphorylate various substrate proteins and, thus, modulate their function. After cAMP decay, the free C-subunits are inactivated by reassociation with the R-subunits. In mammals, four genes encode two classes of R-subunits classified as type I (RIα,β) and type II (RIIα,β) and another three genes encode the C-subunits (Cα, β, γ) *(40)*. Until now, one gene that encodes RI and several genes encoding the C-subunits have been cloned in *Drosophila (41,42)*. Biochemical studies failed to demonstrate the RI-subunit but could identify the as yet uncloned RII-subunit, the predominant form in adult *Drosophila* and honeybee *(36–38,43)*. Cloning of the RII-subunit, however, would be of considerable interest, because in mammals the RII-subunit serves as the instrument for targeting the catalytic action to specific cellular destinations via anchor proteins *(39,44)*. In *Drosophila,* the recently described A kinase anchor protein (AKAP) suggests a similar targeting of signals carried by cAMP for insects *(45)* as has been described for mammals.

Mutants of the *DCO* gene, which encodes a C-subunit, show reduced PKA activity depending on the severity of the allelic combination *(46,47)*. A reduction of PKA activity by 50% leads to very weak effects on learning and memory, as tested immediately after training. A reduction to 20% of the normal PKA activity causes a strong reduction in learning, whereas the effects on memory are very weak. Similar results were obtained in experiments using transgenic flies, which either express a PKA inhibitor *(48)* or disrupt expression of the gene encoding RI *(49)*.

Although these experiments do not address the function of the other C-subunits and the RII-subunit, the results clearly demonstrate the general role of PKA in *Drosophila* learning. Considering the complex expression patterns and the different function of PKA isoforms in mammals *(39,50)*, we also have to assume a similar complex function of cAMP-mediated processes in *Drosophila*.

The targets of the PKA or other protein kinases with respect to olfactory learning are unknown. However, the isolation of the *Drosophila* mutant *Su-var (3)* defective in both protein phosphatase 1 activity and olfactory learning *(51,52)* supports the central role of protein phosphorylation and dephosphorylation in mechanisms of learning.

The cAMP Response Element Binding Protein Is Essential for the Induction of Long-Term Memory

Recent work demonstrates that cAMP responsive transcription, mediated via the cAMP response element binding protein (CREB) plays a central role in the formation of long-term memory in invertebrates and mammals *(17,18)*. Members of the CREB family (CREB, CREM, and ATF-1) contain a leucine zipper domain that bind to cAMP-response element (CRE) sites *(53,54)*. PKA-mediated phosphorylation of CREB leads to the activation of the protein and, thus, induction of gene expression *(55)*. As in mammals, the *Drosophila* CREB (dCREB2) gene produces several isoforms, which act either as activators or repressors of transcription *(56)*. Experiments with transgenic flies have shown that overexpression of a CREB repressor isoform blocks long-term memory without affecting learning or short-term and middle-term memory *(16,57)*. Although the formation of long-term memory in *Drosophila* requires multiple training sessions, the induced overexpression of a CREB activator isoform leads to the formation of long-term memory after only a single training session *(58)*. This clearly demonstrates that CREB has a key function as a transcriptional switch in long-term memory formation and, moreover, it demonstrates that the balance between CREB activator and CREB repressor isoforms is critical for the induction of long-term memory in *Drosophila*. Similar findings in *Aplysia (59–61)* and rodents *(18,*

62) support the general role of CREB in long-lasting forms of synaptic plasticity and long-term memory formation. However, because CREB isoforms are phosphorylated by different kinases *(63,64)*, the contribution of the cAMP/PKA pathway to CREB-mediated long-term memory formation remains to be demonstrated.

Neurotransmitters Implicated in Drosophila *Learning*

In contrast to the impressive evidence for the central role of the cAMP/PKA cascade in *Drosophila* learning, little is known about the role of neurotransmitters that modulate the intracellular cAMP level. However, biogenic amines like dopamine, serotonin, and octopamine play a central role in learning and memory in both vertebrates and invertebrates *(65–70)*. In the dopa decarboxylase-defective *Drosophila* mutant *Ddc,* synthesis of serotonin and dopamine is disrupted and the mutant exhibits impaired learning *(70)*. Also, learning can be impaired by pharmacological manipulation of the octopamine system *(71)*. One approach for dissecting the function of neurotransmitters in learning is to clone receptors of biogenic amines.

The action of dopamine in mammals is mediated by distinct subclasses of G-protein-coupled receptors. Whereas members of the D1- and D5-like receptor subclasses stimulate adenylate cyclase and phoshatidylinositol-4,5-bisphosphate metabolism, members of D2-, D3-, and D4-like receptors activate potassium channels and inhibit adenylate cyclase activity *(72)*. D1-like receptors have been cloned and characterized in *Drosophila (73–76)* and the honeybee *(77)*. The D1 receptor subtypes and the cloned serotonin receptors *(78,79)* stimulate the adenylate cyclase activity in HEK cells and *Xenopus* oocytes. Their role with regard to the cAMP-mediated function in *Drosophila* learning, however, is unclear.

Octopamine is a biogenic amine that is found to play an important role in invertebrates, whereas its role in vertebrates is mysterious *(66,80)*. Pharmacological studies provided evidence that octopamine receptors are related to vertebrate α-adrenergic receptors and that different subclasses of octopamine receptors exist *(81,82)*. Meanwhile, a few octopamine receptors have been cloned and characterized with respect to their effects on second-messenger systems *(83–86)*. Although some octopamine receptors regulate levels of cAMP, the function of the receptors in *Drosophila* learning is unknown, as is the case for dopamine and serotonin. It is feasible, however, that octopamine, which plays a central role in olfactory conditioning in the honeybee *(6)*, may play a similar role in *Drosophila*. In the future, manipulation of a single class of receptors using genetic techniques available in *Drosophila* should allow studies of their function in learning.

The Drosophila Mushroom Bodies
as an Area of cAMP-Mediated Plasticity

The bilateral symmetric mushroom bodies are two prominent structures in the central brain of insects that have been implicated in higher control of distinct motor programs and learning and memory (3,87–89). In Drosophila, the somata of the mushroom-body intrinsic Kenyon cells (approx 2500) are located dorsally to the calyces (Fig. 1). The Kenyon cells arborize in the calyx neuropil, the main input region, and project via the pedunculi into the lobes of the mushroom bodies, the predominant output regions (90). The mushroom bodies receive input of different sensory modalities, which suggests an important function as centers of integration. The existence of structural defects in the mushroom bodies in developmental mutants (88) as well as the chemical ablation of mushroom bodies (91) impairs olfactory learning in Drosophila. The importance of the mushroom bodies for learning is supported by the fact that the learning mutants rutabaga, dunce, and DCO show a higher expression of the respective gene products in the mushroom bodies as compared to other brain areas (47,92,93). Moreover, the impairment of G-protein signaling in the mushroom bodies by expression of constitutively activated $G_{\alpha s}$-subunits eliminates associative olfactory learning in transgenic flies (35).

The dunce cAMP-specific PDE and the DCO catalytic subunit of PKA are evenly expressed in the somata, the dendrites, and the axons of the Kenyon cells (47,93). The rutabaga AC is elevated in the axons, which make up the peduncles and the axon terminals in the lobes, whereas the somata and the dendritic regions in the calyxes contain relatively little AC (92). This suggests that dunce PDE is involved in the regulation of cAMP levels throughout the mushroom bodies, whereas the rutabaga AC seems to modulate cAMP levels in the axons of the Kenyon cells via receptors.

Recently a dopamine receptor (DAMB) has been described that colocalizes with the rutabaga AC and mediates dopamine-induced cAMP formation (75). This makes DAMB a suitable initiator of the cAMP signaling implicated in learning and memory. However, until now no distinct role of dopamine or any other neurotransmitter has been demonstrated in Drosophila learning.

Immunohistological findings demonstrate that the regulatory subunit RII of PKAII, the predominat form of PKA in adult insects, is enriched in the mushroom bodies. This makes PKAII a most likely mediator of the cAMP-dependent processes in the Kenyon cells (43). Because the levels of PKAII immunostaining differ between subgroups of Kenyon cells, we have to assume a different contribution of Kenyon cells in cAMP-mediated processes. That the Kenyon cells are not a homogeneous cell population has also been demonstrated by other techniques (94,95), pointing to functional differences between subgroups of Kenyon cells that may also be relevant for olfactory learning. Although there is a good deal of evidence for the major role of cAMP-mediated signaling in the mushroom bodies (Fig. 2), the function of the individual components and their interaction on the cellular level is unknown.

Interestingly, defects in *Drosophila* mushroom bodies seem to affect only learning paradigms that use chemosensory cues, whereas paradigms using other sensory modalities show no defect in learning *(96)*. This finding not only supports the role of the mushroom bodies in olfactory learning but also suggests that different brain areas contribute to different aspects of learning, as has been already demonstrated in honeybees *(6,97,98)*.

Calcium-Dependent Protein Kinases and Their Role in Drosophila Learning

Calcium is the most common signal transduction molecule that controls many aspects of cellular function. Among the numerous Ca^{2+}-regulated processes, the Ca^{2+}-dependent protein phosphorylation mediated via Ca^{2+}/phospholipid-dependent protein kinases (PKC) and Ca^{2+}/calmodulin-dependent protein kinases (CaMII) play a major role in synaptic plasticity *(99,100)*. The role of PKC *(101,102)* and CaMII kinase *(103)* in *Drosophila* learning has been investigated using transgenic flies. Expressing inhibitors of either PKC or CaMII kinase in various paradigms provided evidence that these kinases are required for behavior in intact animals. Inhibition of the PKC activity impairs acquisition during courtship conditioning without affecting memory *(104)*. This suggests that PKC-blocked flies show an impaired retrieval of courtship behavior during the early phase of memory formation. The role of the PKC pathway with respect to olfactory learning, however, is unclear.

Blocking of CaMII kinase activity affects both nonassociative learning and memory in courtship conditioning *(105,106)*. A possible substrate for the CaMII kinase, and thus a target mediating the physiological changes during learning, is a K^+ channel subunit *(106)*. Flies with mutations in the K^+ channel gene ether-a-go-go *(eag)* show similar defects in courtship behaviors, as do flies with blocked CaMII kinase. CaMII kinase can phosphorylate *eag*, and inhibition of CaMII kinase has been shown to affect synaptic and developmental plasticity at the neuromuscular junction of *Drosophila* *(107,108)*. However, it remains unclear whether changes at neuromuscular junctions are the basis of behavioral plasticity in adult flies.

Implication of Other Signaling Pathways in Drosophila Olfactory Learning

The molecular analysis of *Drosophila* mutants that are defective in olfactory learning and that express the corresponding gene preferentially in the

mushroom bodies leads to the identification of further components important for learning. One of these factors is *leonardo,* which encodes a *Drosophila* homolog of the vertebrate 14-3-3 protein *(109–111).* The 14-3-3 proteins
are highly conserved and are found in animals and plants. Numerous functions have been ascribed to the 14-3-3 protein *(112),* and their association
with a diverse group of signaling proteins suggests a function in the coordination and modulation of various signaling cascades *(113,114).*

Leonardo, the *Drosophila* 14-3-3 protein, is preferentially expressed throughout the mushroom bodies *(110,111).* The reduction of *leonardo* in the mushroom bodies leads to a decrease in associative olfactory memory. Because
there is as yet no evidence for an interaction of 14-3-3 proteins with the
cAMP cascade, the findings suggest that *leonardo* interacts with another
signaling cascade. In the neuromuscular junction of *Drosophila* larvae, the
Leonardo protein is enriched in presynaptic boutons *(109).* Even though
leonardo mutants seem to be normal with regard to synaptogenesis, the basal
transmission and other properties of synaptic transmission are affected.
Although these findings suggest that the 14-3-3 protein is involved in the
regulation of the synaptic vesicle release, the concrete pathway with which
Leonardo interacts is unknown.

Recently, the cell-surface receptor α-integrin, which mediates cell adhesion and signal transduction has been shown to be involved in learning in
the *Drosophila* memory mutant *Volado* (*Vol*) with a deficient short-term
memory. The *Vol* gene encodes a new α-integrin *(115).* In heat-shock experiments, the transient conditional expression of *Vol* during training restored
the memory defects, suggesting a requirement for integrins during plastic
neuronal changes in the adult fly. Because of their functions in cell adhesion
and signal transduction, integrins may either act in the dynamic regulation
of synapse structure or in the modulation of signal transduction pathways
required for the formation of short-term memory.

LEARNING IN THE HONEYBEE:
BEHAVIOR, CIRCUITS, AND BIOCHEMICAL MACHINERY

The complex behavior of honeybees, including orientation, foraging, and
social communication, has attracted interest since the turn of the century
(116). Learning odor, color, and shape of flowers is an essential component
of the bee's behavior and reveals many characteristics of associative learning described in mammals *(6,7).* In particular, the classical conditioning of
the proboscis extension reflex (PER) of harnessed bees provides the opportunity to characterize the behavioral, cellular and molecular levels of learning and memory. In this paradigm, the PER elicited by the stimulation of an
antenna by sucrose solution is conditioned by pairing an odor stimulus (CS,

conditioned stimulus) with a sucrose reward (US, unconditioned stimulus). In the retrieval test, the PER is elicited in a high percentage of animals by application of the odor alone *(117,118)*. Interestingly, depending on the number of conditioning trials, different memories are induced. Whereas a single conditioning trial induces a medium-term memory that is sensitive to amnestic treatments, multiple-trial conditioning leads to an amnesia-insensitive long-term memory *(6,119,120)*.

Neuronal Basis of Olfactory Learning in the Honeybee

Neuronal Pathways Mediating CS and US

The brain areas and the circuits implicated in the processing of CS and US have been well described for the honeybee (Fig. 1). Olfactory information (CS) from the chemosensory receptors on the antennae project into the glomeruli of the antennal lobes. The glomeruli are sites of synaptic connections among sensory neurons, local interneurons, and projection neurons *(121,122)*. A considerable fraction of the local interneurons exhibits GABA-immunoreactivity *(123)*, suggesting an inhibitory function within the circuitry of the antennal lobe. The olfactory information is relayed via projection neurons to the calyces of the mushroom bodies and the lateral protocerebrum. The projection neurons from the antennal lobes terminate in the lip region of the calyces, whereas other regions of the calyces receive input from different sensory modalities. Because a portion of the olfactory projection neurons and the lip region show immunoreactivity to acetylcholinesterase (AChE) and ACh receptors *(124)*, the CS input into the mushroom bodies is partially cholinergic. Output neurons connect the two lobes of the mushroom bodies with its own input region, the lateral protocerebrum, the contralateral mushroom body, and several other brain regions *(89,125)*. The response properties of an identified neuron, the Pe1 neuron, which connects the peduncle with other areas of the brain, has been analyzed in the context of nonassociative and associative learning. The Pe1 neuron changes its properties specifically when the animal is sensitized or conditioned to an odor *(126)*. The output pathways that connect the mushroom bodies to the motor circuits responsible for proboscis extension are unknown.

Chemosensory receptors for sucrose at the antennae and the proboscis project to the subesophageal ganglion and terminate near motoric interneurons and motor neurons involved in proboscis extension *(127)*. A group of ventral unpaired median (VUM) neurons, which display octopamine-like immunoreactivity *(128)*, receives input from sucrose receptors. Intracellular recordings from one of these cells, the VUMmx1 neuron, demonstrated that VUMmx1 mediates the US in associative learning *(129)*. The VUMmx1 neuron ramifies in the antennal lobes, the mushroom bodies, and the lateral protocerebrum and, thus, converges with the CS pathway in three areas of the brain.

The implication of the antennal lobes and the mushroom bodies in associative learning has been demonstrated by local cooling *(119,120)*, functional calcium imaging *(130)* and by local injections of octopamine *(98)*.

Pharmacologically applied octopamine can substitute for the US function by pairing odor with local octopamine injections, either into the antennal lobes or the mushroom bodies. The results suggest that the antennal lobes and the mushroom bodies act partially independently with regard to associative learning and seem to contribute to different features of learning and memory formation. Thus, octopamine might be a transmitter mediating reinforcement in associative olfactory learning in the honeybee *(98)*.

The Transmitter of the US Pathway Modulates the cAMP Cascade

The general important role of biogenic amines in insect behavior *(65–67, 71)* suggests that these transmitters modulate second-messenger cascades in their target cells, some of which are important for learning and memory. In recent years, a couple of receptors for biogenic amines, such as serotonin, dopamine, and octopamine, have been pharmacologically characterized or cloned from different insect species. The majority of these receptor types are coupled to adenylate cyclases and thus modulate cAMP levels *(66,74,75, 77,79,81,84,131)*.

Biochemical techniques show that in the antennal lobes of the honeybee octopamine causes a transient activation of the cAMP-dependent protein kinase (PKA) *(132)*. Sucrose stimulation presenting the US induces a similar increase in PKA activity in the antennal lobes in vivo *(133)*. Odor stimulation or mechanical stimulation of the antennae have no effect on PKA activity. Within the antennal lobes, PKA is concentrated in the glomeruli, sites of synaptic connections between sensory neurons and interneurons. Although the sensory neurons exhibit PKA immunoreactivity, the major fraction of PKA in the glomeruli is the result of local interneurons *(43)*. This suggest that the US modulates the cAMP/PKA cascade in local interneurons by octopamine released from the VUMmx1 neuron in the antennal lobes (Fig. 3). Although serotonin and dopamine are also detected in the antennal lobes *(134)*, they do not seem to affect the level of PKA activity in the circuitry of the antennal lobes *(132)*.

In the mushroom bodies, another site of convergence of CS and US pathway in olfactory learning, the biogenic amines are also capable of stimulating the cAMP/PKA cascade *(43)*. Compared to serotonin and dopamine, octopamine shows a conspicuous concentration-dependent effect on PKA activity in cultured Kenyon cells. The different pharmacology of octopamine-binding sites in the mushroom bodies compared to other brain areas points to the expression of distinct subtypes of octopamine receptors in the Kenyon cells *(135)*. These subtypes of octopamine receptors are probably coupled to different second-messenger pathways *(81,85)*. This and the distinct levels of PKA expression in different subgroups of Kenyon cells *(43)* suggests a very complex neurotransmitter-mediated modulation of the cAMP cascade in the mushroom bodies.

Nicotinic receptors for ACh, the supposed transmitter of the CS pathway in the honeybee *(124)*, are very similar to vertebrate neuronal nicotinic ACh receptors *(136)*. Interestingly, application of ACh or other agents that increase intracellular Ca^{2+} levels *(137)* have no effect on PKA activity in the Kenyon cells. Although these experiments suggest a modulation of the cAMP/PKA cascade by octopamine, the supposed transmitter of the US pathway, future studies will have to demonstrate the modulatory function of these transmitters on second-messenger cascades in associative olfactory learning in intact animals.

Protein Kinase C in the Antennal Lobes of the Honeybee Is Involved in Associative Olfactory Learning

Although the central role of the cAMP cascade has been convincingly demonstrated in *Drosophila* olfactory learning *(3,11)* and the findings described in the above subheading also suggest a function of the cAMP/PKA cascade in processing the US in honeybee olfactory learning, very little is known about other second-messenger pathways. However, it has been demonstrated that the Ca^{2+}/phospholipid-dependent protein kinase (PKC) plays a role in the formation of a distinct associative olfactory memory in the honeybee *(97)*.

In contrast to the PKAII, the PKC is concentrated in local interneurons of the antennal lobes and in the mushroom body Kenyon cells. Associative learning induces changes in PKC activity in the antennal lobes (Fig. 3). Thus, a distinct function of PKC is revealed with regard to memory formation. Stimulation of the animals by the CS or by the US alone induces a transient activation of PKC in the antennal lobes. There is also no sequence-specific difference in PKC activation, as paired and unpaired application of CS and US induce comparable changes in PKC activity. The inhibition of the changes in PKC activity during the conditioning phase has no effect on the formation of olfactory memory. Thus, the PKC system in the antennal lobes does not seem to be implicated in processes of learning and memory formation but more likely seems to be involved in the processing of chemosensory information in general. The latter is in agreement with the localization of the PKC in the interneurons, which have an integrative function in chemosensory signal processing in the honeybee *(138)*. However, a learning-specific

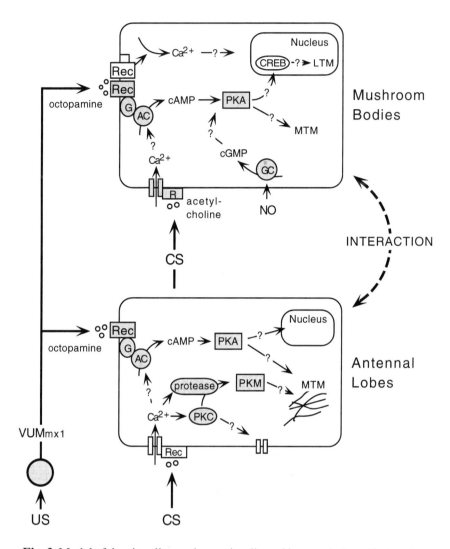

Fig. 3. Model of the signaling pathways implicated in associative olfactory learning in the honeybee. US stimulation of VUMm×1 releases octopamine in the antennal lobes and the mushroom bodies and thereby causes an activation of the cAMP/
PKA cascade. In the antennal lobes, the CS causes changes in Ca^{2+} levels and transiently activates the Ca^{2+}/phospholipid-dependent protein kinase C (PKC). The
coincident activation of a protease leads to a constitutive PKC (PKM) that contributes to medium-term memory (MTM) induced by multiple learning trials. In the
mushroom bodies, the CS leads to elevation of Ca^{2+} levels via acetylcholine receptors. In both neuropils, the synergistic activation of adenylate cyclase (AC) by Ca^{2+}
and octopamine can lead to an elevated PKA activity required for medium-term
memory. Activation of the nitric oxide (NO) system, most likely in the mushroom

modulation of PKC activity is detected after acquisition. Multiple-trial conditioning that induces an long-term memory leads to a long-lasting transient change in PKC activity *(97)*. This is in agreement with studies in vertebrates, where PKC seems to be implicated in memory formation but not in acquisition of memory *(139,140)*. A long-lasting activation of PKC has also been found in LTP *(141,142)*. Again, PKC is not involved in the initial induction, but it contributes to the expression of LTP at later times *(10)*.

In the antennal lobe of the honeybee, two different mechanisms contribute to the increase in PKC activity lasting from 1 h up to 3 d. Whereas the early phase of this increase in PKC activity in the range of hours is the result of a proteolytic mechanism, the late phase in the range of days depends on RNA and protein synthesis. Both mechanisms are independent of each other. Interestingly, only the inhibition of the proteolytic process has an immediate effect on an early phase of multiple-trial-induced memory. Thus, the PKC system in the antennal lobes seems to be required for an early phase of multiple-trial-induced memory but not for acquisition or late phases of memory (Fig. 3). This is in agreement with the finding that the antennal lobes and the mushroom bodies seem to contribute to different features of olfactory learning and memory formation *(98,119)*.

The late phase of PKC activation in the antennal lobes is probably only one of several parallel mechanisms that contribute to the formation of the late phase of long-term memory and occur in different brain areas. Because the mushroom bodies are very likely one of these areas, future investigations will have to analyze the interaction of processes occurring in the antennal lobes and the mushroom bodies.

Nitric Oxide System Is Implicated in Associative and Nonassociative Learning in the Honeybee

Nitric Oxide in the Nervous System of Invertebrates

Unlike classical transmitters in the nervous system, nitric oxide, which acts as a transmitter molecule, is not packaged in vesicles but, rather, diffuses

---

bodies, is required for the formation of a long-term memory (LTM) induced by multiple learning trials. The cGMP produced by the soluble guanylate cyclase (GC) is capable of interacting with the cAMP cascade, suggesting a synergistic activation of PKA by cAMP and cGMP and induction of CREB-mediated gene expression, as in *Drosophila*. The findings support the notion that different signaling pathways acting in parallel and in different neuropils are implicated in associative learning in the honeybee. CS, conditioned stimulus; CREB, cAMP-response element binding protein; G, G-protein; Rec, receptor; US, unconditioned stimulus.

from its site of production. It is, thus, independent of a complex release machinery *(143)*, but, like the release of classical neurotransmitters, its production by neuronal NO synthase (NOS) is a Ca^{2+}-dependent process *(144)*. Although its function as a signaling molecule has initially been demonstrated in the vertebrate nervous system, NO or NOS has, meanwhile, been detected in numerous invertebrates *(145–151)*. The NOS characterized in *Drosophila*, honeybee, locust, *Manduca*, and *Rhodnius prolixus* *(146,148,152–155)* exhibits properties similar to that described for the corresponding vertebrate enzyme *(143,156)*. This suggests that the diffusible signaling mediated by NO seems to be a common mechanism rather than an exception *(157,158)*.

The Soluble Guanylate Cyclase Is a Target of Nitric Oxide in the Insect Nervous System

The major target for NO found in neural tissues is the soluble form of guanylate cyclase (sGC) *(143,159)*. Activation of sGC is due to binding of NO to a heme group, which leads to formation of a nitrosyl–heme complex and consequent conformational change. NO-sensitive sGCs have been described in *Schistocerca* *(146,160,161)* and genes encoding for GC-subunits have been cloned in *Drosophila* and *Manduca sexta* *(153,162–164)*. To gain insight into the cellular organization of NO-producing cells and their potential targets, the sGC-containing cells, a histochemical approach was applied to the nervous system of various insects *(165)*. These investigations demonstrated that in the brain of the locust, the NO-releasing cells and the target cells are separated in most cases. In a few exceptions however, NOS and the NO-induced cGMP immunoreactivity is colocalized, suggesting that NO can also act as an intracellular messenger molecule in insects *(160,166,167)*.

Whereas cAMP exerts its physiological function mainly through the activation of cAMP-dependent protein kinase, cGMP modulates enzymes like cGMP-dependent protein kinases, cGMP-regulated cyclic nucleotide phosphodiesterases, and cyclic nucleotide-gated ion channels *(168,169)*. In *Drosophila* several cGMP-dependent protein kinases have been characterized *(170,171)*. Although natural polymorphism in behavior seems to be the result of cGMP-dependent protein kinases *(172)*, the role of the cGMP-dependent protein kinases in insect learning is unclear. Interestingly, the cAMP-dependent protein kinase type II from neuronal tissue of *Drosophila* and the honeybee is activated by cGMP at physiological levels *(36,38)*, suggesting a possible interaction of cGMP with the cAMP cascade. An interaction between cAMP- and cGMP-dependent pathways can also occur at the level of phosphodiesterases (PDE). Of the seven gene families of vertebrate PDE, currently the PDE 2 and PDE 3 families are known to be regulated by cGMP *(173,174)*.

Although three forms of PDE have been described in *Drosophila (175)*, there is, as yet, no evidence for cGMP-regulated forms of PDE in insects. The third class of cGMP-regulated enzymes are the cyclic nucleotide-gated channels. Members of this class of ion channels have been characterized in photoreceptor and olfactory cells in vertebrates *(176,177)* and *Drosophila (178)*. Although in vertebrates these channels are activated by NO and CO *(179,180)* such a modulation has, as yet, not been demonstrated in insects.

The NO/cGMP System Is Implicated
in Adaptive Chemosensory Signal Processing

It is conspicuous that in all insect species tested thus far, the antennal lobes, the primary center of chemosensory information processing, exhibit a high concentration of NOS activity *(158)*. High concentrations of NOS are also found in the olfactory bulb in mammals *(181)*, suggesting a conserved function of the NO system in processing olfactory signals at the site of the first synaptic integration. Because of the properties of the diffusible messenger NO and based on the glomerular structure within the olfactory system, Breer and Shepherd *(182)* suggest a functional role for the NO/cGMP system in synaptic integration of sensory signals within the olfactory bulb.

Recently it has been demonstrated that the NO/cGMP system in the antennal lobe of *Apis* is implicated in the processing of adaptive mechanisms during chemosensory neuronal processing *(183)*. Appetitive stimuli, like water or sucrose solution, applied to the antennae of a honeybee elicits the extension of the proboscis. Repetitive stimulation of one antenna, with intertrial intervals in the range of seconds, leads to side-specific disappearance of the proboscis extension response. This behavioral plasticity is restricted to the stimulated side and matches all characteristic parameters for habituation *(184)*.

Interneurons in the antennal lobe innervating different glomeruli have been implicated in integrative processes in chemosensory signal processing in the honeybee *(138)*. Interestingly, these interneurons contain the highest level of NOS in the antennal lobes *(144,160,183)* and are capable of releasing NO upon the increase of cytosolic Ca^{2+} or application of transmitters in cell culture *(144)*.

Inhibition of NOS in the antennal lobes specifically interferes with the habituation to repetitive chemosensory stimuli, but it does not affect the response to single stimuli *(183)*. The blocking of NOS in a single antennal lobe only affects the habituation of the corresponding side, suggesting a contribution of NO in integrative neuronal processing restricted to the circuitry of a single antennal lobe. The function of NO is specific for the integrative component, because parameters such as satiation level, stimulus strength, interstimulus interval, and duration of sensory stimuli are not impaired by inhibition of NOS. Local inhibition of the sGC, a major target of NO, in a single antennal lobe leads to effects identical to those observed with NOS inhibition. This strongly suggests that the physiological effects of NO are mediated by NO-sensitive GC and thus by changes in cGMP levels in the target neurons *(183)*. The target of cGMP with regard to chemosensory adaptation, however, is unclear.

Whereas the function of NO in habituation is most likely the result of the release of NO by local interneurons and, thus, a simultaneous modulation of synaptic transmission in many glomeruli, other aspects in chemosensory signaling may be modulated by the restricted release of NO from sensory neurons in single glomeruli *(158)*. The different connectivity of local interneurons and sensory neurons, together with the observation that NO produces distinct effects depending on the actual synaptic activity of the target site *(185,186)*, suggests that within the antennal lobe, NO may be implicated in totally different features of chemosensory signal processing.

Irrespective of this speculation, future analysis will lead to a further understanding of the different aspects of NO signaling in general and hopefully will reveal information whether the high concentration of NOS in primary chemosensory neuropils reflects an evolutionary conserved feature of NO signaling in these neuronal circuits.

NOS Is Implicated in the Formation
of a Distinct Long-Term Memory in the Honeybee

In vertebrates, NO has been implicated in various forms of learning *(187–191)* and in mechanisms of long-term depression (LTD) and long-term potentiation (LTP) of synaptic connections *(159,186,192)*. In the honeybee, a specific role of NO in the formation of a distinct form of olfactory long-term memory has also been demonstrated *(193)*.

In the honeybee, memories with different properties are formed, depending on the number of associative conditioning trials applied to the animals. Whereas a single conditioning trial leads to a medium-term memory that lasts for hours, multiple conditioning trials induce a long-term memory that lasts for days *(6,7)*.

It is remarkable that the highest levels of NOS are found in the antennal lobes and the mushroom bodies *(148)*, as both neuropils are known to participate in olfactory learning *(98)*. Whereas the antennal lobes, primary centers of olfactory processing, receive input from the sensory neurons of the antennae, the mushroom bodies process input from the antennal lobes in addition to input from other sensory modalities.

Systemic inhibition of NOS activity during associative olfactory learning impairs the formation of the long-term memory induced by multiple learning trials but has no effect on the formation of a single-trial-induced memory *(193)*. The memory induced by multiple-trial conditioning in the presence of NOS inhibitors shows properties similar to the single-trial-induced memory. NOS inhibition has no effect on nonassociative components, such as motivational factors, satiation, or sensitization which interfere with associative olfactory learning. Moreover, acquisition or retrieval is also not affected by blocking NOS.

The features of NO-mediated mechanisms in memory formation in honeybees reveal remarkable parallels to findings in vertebrates. Only the NOS inhibition during training impairs memory formation, whereas blocking NOS activity after training or during retention does not affect retention of memory *(187,189,194)*. Moreover, that NO is required has been demonstrated for distinct learning paradigms only, whereas learning paradigms using other parameters are unaffected by NOS blockers. This suggests that NO is implicated in very defined processes only, depending on special features, such as sensory modalities, parameters of stimuli application, and other yet unknown parameters *(185,195–197)*.

Similar results have been reported for the in vitro model of synaptic plasticity, the long-term potentiation (LTP). In LTP, NO produces distinct effects depending on parameters such as the stimuli applied and the synaptic activity of the target site *(185,186,198)*. However, the function of NO in LTP and its relation to behavioral plasticity is controversial *(195,196,19)*.

In the honeybee, two brain structures are potentially involved in NO-mediated memory formation, the antennal lobes, and the mushroom bodies. Both neuropils exhibit high NOS activity *(148)* and have been shown to be implicated in distinct aspects of olfactory learning in the honeybee *(6,97,98)*. Based on findings from the olfactory system in mammals however, the NO system in the antennal lobes are likely candidates. In mice and sheep, the NO system in the accessory olfactory system has been implicated in the formation of distinct olfactory memory. Whereas female mice form a memory of the pheromones of the mating male, sheep learn to recognize the odors of their lambs in the first hours after birth *(9,190,200)*. The formation of this memory is mediated by a reduced inhibitory transmission of gamma-aminobutyric acid (GABA) from the granule cells to the mitral cells. In both cases, NO has been demonstrated to mediate the formation of this memory. Whereas, in mice, the coincident activation of pheromonal inputs and exogenous administration of nitric oxide in the accessory olfactory system can induce a pheromone-specific olfactory memory without mating *(191)*, blocking of NOS activity prevents the formation of memory in sheep *(190)*. Although, in the latter, NO modulates the soluble guanylate cyclase, the target of the resulting changes in cGMP concentration is unknown.

The sGC is also a feasible target for NO concerning its role in memory formation in the honeybee *(183)*. Because the NO/cGMP system can interact with the cAMP cascade, it is feasible that NO mediates components of the cAMP cascade in the mushroom bodies. It is noticeable that in the honeybee mushroom body, the area that receives olfactory input shows a high NOS labeling compared to the areas that receive input from other sensory modalities *(148)*. In all insect species as yet tested, the mushroom-body intrinsic cells (Kenyon cells) exhibit no NOS staining. Evidence from the locust, where NO-sensitive sGC exists at least in a subset of Kenyon cells *(160)*, supports the idea that NO released from olfactory inputs interferes with cAMP-mediated signaling in the Kenyon cells.

Irrespective of the exact function of the NO system in synaptic plasticity, the parallel findings in vertebrates and invertebrates support the idea that the general features of NO-mediated signaling seems to be conserved from insects to mammals *(158)*. Thus, future investigations in accessible systems such as the honeybee are ideally suited to uncover the general features of NO-mediated signaling.

REFERENCES

1. Byrne, J. H. and Kandel, E. R. (1996) Presynaptic facilitation revisited: state and time dependence. *J. Neurosci.* **16**, 425–435.
2. Carew, T. J. (1996) Molecular enhancement of memory formation. *Neuron* **16**, 5–8.
3. Davis, R. L. (1993) Mushroom bodies and *Drosophila* learning. *Neuron* **11**, 1–14.
4. DeZazzo, J. and Tully, T. (1995) Dissection of memory formation: from behavioral pharmacology to molecular genetics. *Trends Neurosci.* **18**, 212–218.
5. Dubnau, J. and Tully, T. (1998) Gene discovery in *Drosophila*: new insights for learning and memory. *Annu. Rev. Neurosci.* **21**, 407–444.
6. Hammer, M. and Menzel, R. (1995) Learning and memory in the honeybee. *J. Neurosci.* **15**, 1617–1630.
7. Menzel, R. and Müller, U. (1996) Learning and memory in honeybees: from behavior to neural substrates. *Annu. Rev. Neurosci.* **19**, 379–404.
8. Bliss, T. V. P. and Collingridge, G. L. (1993) A synaptic model of memory: long-term potentiation in the hippocampus. *Nature* **361**, 31–39.
9. Brennan, P. A. and Keverne, E. B. (1997) Neural mechanisms of mammalian olfactory learning. *Prog. Neurobiol.* **51**, 457–481.
10. Roberson, E. D., English, J. D., and Sweatt, J. D. (1996) A biochemist's view of long-term potentiation. *Learning Memory* **3**, 1–24.
11. Davis, R. L. (1996) Physiology and biochemistry of *Drosophila* learning mutants. *Physiol. Rev.* **76**, 299–317.
12. Quinn, W. G., Harris, W. A., and Benzer, S. (1974) Conditioned behavior in *Drosophila melanogaster*. *Proc. Natl. Acad. Sci. USA* **71**, 708–712.
13. Tempel, B. L., Bonini, N., Dawson, D. R., and Quinn, W. G. (1983) Reward learning in normal and mutant *Drosophila*. *Proc. Natl. Acad. Sci. USA* **80**, 1482–1486.

14. Hall, J. (1994) The mating of a fly. *Science* **264**, 1702–1714.
15. Tully, T. and Quinn, W. G. (1985) Classical conditioning and retention in normal and mutant *Drosophila melanogaster. J. Comp. Physiol.* **157**, 263–277.
16. Tully, T., Preat, T., Boynton, S. C., and Del Vecchio, M. (1994) Genetic dissection of consolidated memory in *Drosophila. Cell* **79**, 35–47.
17. Yin, J. C. P. and Tully, T. (1996) CREB and the formation of long-term memory. *Curr. Opin. Neurobiol.* **6**, 264–268.
18. Abel, T., Martin, K. C., Bartsch, D., and Kandel, E. R. (1998) Memory suppressor genes: inhibitory constraints on the storage of long-term memory. *Science* **279**, 338–341.
19. Byers, D., Davis, R. L., and Kiger, J. A. J. (1981) Defect in cyclic AMP phosphodiesterase due to the *dunce* mutation of learning in *Drosophila melanogaster. Nature* **289**, 79–81.
20. Dudai, Y., Jan, Y. N., Byers, D., Quinn, W. G., and Benzer, S. (1976) *Dunce*, a mutant of *Drosophila* deficient in learning. *Proc. Natl. Acad. Sci. USA* **73,** 1684–1688.
21. Levin, L. R., Han, P. -L., Hwang, P. M., Feinstein, P. G., Davis, R. L., and Reed, R. R. (1992) The *Drosophila* learning and memory gene *rutabaga* encodes a Ca^{2+}/calmodulin-responsive adenylyl cyclase. *Cell* **68,** 479–489.
22. Livingstone, M. S., Sziber, P. P., and Quinn, W. G. (1984) Loss of calcium/calmodulin responsiveness in adenylate cyclase of *rutabaga*, a *Drosophila* learning mutant. *Cell* **37,** 205–215.
23. Chen, C. N., Denome, S., and Davis, R. L. (1986) Molecular analysis of cDNA clones and the corresponding genomic coding region of the *Drosophila dunce* locus, the structure gene for cAMP phosphodiesterase. *Proc. Natl. Acad. Sci. USA* **86**, 3599–3603.
24. Qui, Y., Chen, C. N., Malone, T., Richter, L., Beckendorf, S. K., and Davis. R. L. (1991) Characterization of the memory gene *dunce* of *Drosophila melanogaster. J. Mol. Biol.* **222,** 553–565.
25. Dudai, Y. (1988) Neurogenetic dissection of learning and short-term memory in *Drosophila. Annu. Rev. Neurosci.* **11,** 537–563.
26. Dauwalder, B. and Davis, R. L. (1995) Conditional rescue of the *dunce* learning/memory and female fertility defects with *Drosophila* or rat transgenes. *J. Neurosci.* **15**, 3490–3499.
27. Anholt, R. R. H. (1994) Signal integration in the nervous system: adenylate cyclases as molecular coincidence detectors. *Trends Neurosci.* **17**, 37–41.
28. Abrams, T. W., Yovell, Y., Onyike, C. U., Cohen, J. E., and Jarrard, H. E. (1998) Analysis of sequence-dependent interactions between transient calcium and transmitter stimuli in activating adenylyl cyclase in *Aplysia*: possible contribution to CS–US sequence requirement during conditioning. *Learning Memory* **4**, 496–509.
29. Xia, Z., Refsdal, C. D., Merchant, K. M., Dorsa, D. M., and Storm, D. R. (1991) Distribution of mRNA for the calmodulin-sensitive adenylate cyclase in rat brain: expression in areas associated with learning and memory. *Neuron* **6**, 431–443.
30. Quinn, W. G., Sziber, P. P., and Booker, R.. (1979) The *Drosophila* memory mutant *amnesiac. Nature* **277**, 212–214.
31. Feany, M. B. and Quinn, W. G. (1995) A neuropeptide gene defined by the *Drosophila* memory mutant amnesiac. *Science* **268**, 869–873.

32. Ogi, K., Kimura, C., Onda, H., Arimura, A., and Fujino, M. (1990) Molecular cloning and characterization of cDNA for the precursor of rat pituitary adenylate cyclase activating polypeptide (PACAP). *Biochem. Biophys. Res. Commun.* **173**, 1271–1279.

33. Zhong, Y. (1995) Mediation of PACAK-like neuropeptide transmission by coactivation of Ras/Raf and cAMP signal transduction pathways in *Drosophila. Nature* **375**, 588–592.

34. Guo, H. F., The, I., Hannan, F., Bernards, A., and Zhong, Y. (1997) Requirement of *Drosophila* NF1 for activation of adenylyl cyclase by PACAP38-like neuropeptides. *Science* **276**, 795–798.

35. Connolly, J. B., Roberts, I. J., Armstrong, D., Kaiser, K., Forte, M., Tully, T., and O'Kane, C. J. (1996) Associative learning disrupted by impaired Gs signaling in *Drosophila* mushroom bodies. *Science* **274**, 2104–2107.

36. Altfelder, K. and Müller, U. (1991) Cyclic nucleotide-dependent protein kinases in the neural tissue of the honeybee *Apis mellifera. Insect Biochem.* **21**, 487–494.

37. Foster, J. L., Guttman, J. J., Hall, L. M., and Rosen, O. M. (1984) *Drosophila* cAMP-dependent protein kinase *J. Biol. Chem.* **259**, 13,049–13,055.

38. Müller, U. and Spatz, H.-C. (1989) Ca^{2+}-dependent proteolytic modification of the cAMP-dependent protein kinase in *Drosophila* wild-type and *dunce* memory mutants. *J. Neurogenet.* **6**, 95–114.

39. Taylor, S. S., Buechler, J. A., and Yonemoto, W. (1990) cAMP-dependent protein kinase: framework for a diverse family of regulatory enzymes. *Annu. Rev. Biochem.* **59**, 971–1005.

40. Francis S. H. and Corbin J. D. (1994) Structure and function of cyclic nucleotide-dependent protein kinases. *Annu. Rev. Physiol.* **56**, 237–272.

41. Foster, J. L., Higgins, G. C., and Jackson, F. R. (1988) Cloning, sequence, and expression of the *Drosophila* cAMP-dependent protein kinase catalytic subunit gene. *J. Biol. Chem.* **263**, 1676–1681.

42. Kalderon, D. and Rubin, G. M. (1988) Isolation and characterization of *Drosophila* cAMP-dependent protein kinase genes. *Genes Dev.* **2**,1539–1556.

43. Müller, U. (1997) Neuronal cAMP-dependent protein kinase type II is concentrated in mushroom bodies of *Drosophila melanogaster* and the honeybee *Apis mellifera. J. Neurobiol.* **33**, 33–44.

44. Rubin, C. S. (1994) A kinase anchor proteins and the intracellular targeting of signals carried by cyclic AMP. *Biochim. Biophys. Acta* **1224**, 467–479.

45. Han, J. D., Baker, N. E., and Rubin, C. S. (1997) Molecular characterization of a novel A kinase anchor protein from *Drosophila melanogaster. J. Biol. Chem.* **272**, 26,611–26,619.

46. Li, W., Tully, T., and Kalderon, D. (1996) Effects of a conditional *Drosophila* PKA mutant on learning and memory. *Learning Memory* **2**, 320–333.

47. Skoulakis, E. M. C., Kalderon, D., and Davis, R. L. (1993) Preferential expression in mushroom bodies of the catalytic subunit of protein kinase A and its role in learning and memory. *Neuron* **11**, 197–208.

48. Drain, P., Folkers, E., and Quinn, W. G. (1991) cAMP-dependent protein kinase and the disruption of learning in transgenic flies. *Neuron* **6**, 71–82.

49. Goodwin, S. F., Del Vecchio, M., Velinzon, K., Hogel, C., Russell, S. R. H., Tully, T., and Kaiser, K. (1997) Defective learning in mutants of the *Drosophila* gene for a regulatory subunit of cAMP-dependent protein kinase. *J Neurosci.* **17**, 8817–8827.

50. Brandon, E. P., Idzerda, R. L., and McKnight, G. S. (1997) PKA isoforms, neural pathways, and behavior: making the connection. *Curr. Opin. Neurobiol.* **7**, 397–403.

51. Asztalos, Z., Von Wegerer, J., Wustmann, G., Dombradi, V., Gausz, J., Spatz, H.-C., and Friedrich, P. (1993) Protein phosphatase 1-deficient mutant *Drosophila* is affected in habituation and associative learning. *J. Neurosci.* **13**, 924–930.

52. Dombradi, V., Gausz, J., and Cohen, P. T. W. (1991) The structure and function of protein phosphatase 1 (87B) gene in *Drosophila* phosphatase mutants. *Adv. Protein Phosphatases* **6**, 75–76.

53. DeGroot, R. P. and Sassone-Corsi, P. (1993) Hormonal control of gene expression: multiplicity and versatility of cyclic adenosine 3',5'-monophosphate-responsive nuclear regulators. *Mol. Endocrinol.* **7**, 145–153.

54. Habener, J. F., Miller, C. P., and Vallejo, M. (1995) cAMP-dependent regulation of gene transcription by cAMP response element-binding protein and cAMP response element modulator. *Vitam. Horm.* **51**, 1–57.

55. Hagiwara, M., Brindle, P., Harootunian, A., Armstrong, R., Rivier, J., Vale, W., Tsien, R., and Montminy, M. R. (1993) Coupling of hormonal stimulation and transcription via the cyclic AMP-responsive factor CREB is rate limited by nuclear entry of protein kinase A. *Mol. Cell. Biol.* **13**, 4852–4859.

56. Yin, J. C. P., Wallach, J. S., Wilder, E. L., Klingensmith, J., Dang, D., Perrimon, N., Zhou, H., Tully, T., and Quinn, W. G. (1995) A *Drosophila* CREB/CREM homolog encodes multiple isoforms, including a cyclic AMP-dependent protein kinase-responsive transcriptional activator and antagonist. *Mol. Cell. Biol.* **15**, 5123–5130.

57. Yin, J. C. P., Wallach, J. S., Del Vecchio, M., Wilder, E. L., Zhou, H., Quinn, W. G., and Tully, T. (1994) Induction of a dominant negative CREB transgene specifically blocks long-term memory in *Drosophila*. *Cell* **79**, 49–58.

58. Yin, J. C. P., Del Vecchio, M., Zhou, H., and Tully, T. (1995) CREB as a memory modulator: induced expression of a dCREB2 activator isoform enhances long-term memory in *Drosophila*. *Cell* **81**, 107–115.

59. Bartsch, D., Ghirardi, M., Skehel, P. A., Karl, K. A., Herder, S. P., Chen, M., Bailey, C. H., and Kandel, E. R. (1995) *Aplysia* CREB2 represses long-term facilitation: relief of repression converts transient facilitation into long-term functional and structural change. *Cell* **83**, 979–992.

60. Bartsch, D., Casadio, A., Karl, K. A., Serodio, P., and Kandel, E. R. (1998) *CREB1* encodes a nuclear activator, a repressor, and a cytoplasmic modulator that form a regulatory unit critical for long-term facilitation. *Cell* **95**, 211–223.

61. Dash, P. K., Hochner, B., and Kandel, E. R. (1990) Injection of the cAMP-responsive element into the nucleus of *Aplysia* sensory neurons blocks long-term facilitation. *Nature* **345**, 718–721.

62. Bourtchuladze, R., Frenguelli, B., Blendy, J., Cioffi, D., Schutz, G., and Silva, A. J. (1994) Deficient long-term memory in mice with a targeted mutation of the cAMP-responsive element-binding protein. *Cell* **79**, 59–68.

63. Bito, H., Deisseroth, K., and Tsien, R. W. (1997) Ca^{2+}-dependent regulation in neuronal gene expression. *Curr. Opin. Neurobiol.* **7**, 419–429.

64. Martin, K. C., Michael, D., Rose, J. C., Barad, M., Casadio, A., Zhu, H., and Kandel, E. R. (1997) MAP kinase translocates into the nucleus of the presynaptic cell and is required for long-term facilitation in *Aplysia*. *Neuron* **18**, 899–912.

65. Bicker, G. and Menzel, R. (1989) Chemical codes for the control of behavior in arthropods. *Nature* **337**, 33–39.
66. Evans, P. D. (1980) Biogenic amines in the insect nervous system. *Adv. Insect Physiol.* **15**, 317–473.
67. Mercer, A. and Menzel, R. (1982) The effects of biogenic amines on conditioned and unconditioned responses to olfactory stimuli in the honeybee, *Apis mellifera*. *J. Comp. Physiol.* **145**, 363–368.
68. Sawaguchi, T. and Goldman-Rakic, P. S. (1994) The role of D1-dopamine receptor in working memory: local injections of dopamine antagonists into the prefrontal cortex of rhesus monkeys performing an oculomotor delayed-response task. *J. Neurophysiol.* **71**, 515–528.
69. Stark, H. and Scheich, H. (1997) Dopaminergic and serotonergic neurotransmission systems are differentially involved in auditory cortex learning: a long-term microdialysis study of metabolites. *J. Neurochem.* **68**, 691–697.
70. Tempel, B. L., Livingstone, M. S., and Quinn, W. G. (1984) Mutations in the dopa decarboxylase gene affect learning in *Drosophila*. *Proc. Natl. Acad. Sci. USA* **81**, 3577–3581.
71. Dudai, Y., Buxbaum, J., Corfas, G., and Ofraim, M. (1987) Formamidines interact with *Drosophila* octopamine receptors, alter the flies' behavior and reduce their learning ability. *J. Comp. Physiol. A* **161**, 739–746.
72. Gingrich, J. A. and Caron, M. G. (1993) Recent advances in the molecular biology of dopamine receptors. *Annu. Rev. Neurosci.* **16**, 299–321.
73. Feng, G. P., Hannan, F., Reale, V., Hon, Y. Y., Kousky, C. T., Evans, P. D., and Hall, L. M. (1996) Cloning and functional characterization of a novel dopamine receptor from *Drosophila melanogaster*. *J. Neurosci.* **16**, 3925–3933.
74. Gotzes, F., Balfanz, S., and Baumann, A. (1994) Primary structure and functional characterization of a *Drosophila* dopamine receptor with high homology to human D1/5 receptors. *Recept. Chan.* **2**, 131–141.
75. Han, K. A., Millar, N. S., Grotewiel, M. S., and Davis, R. L. (1996) DAMB, a novel dopamine receptor expressed specifically in *Drosophila* mushroom bodies. *Neuron* **16**, 1127–1135.
76. Reale, V., Hannan, F., Hall, L. M., and Evans, P. D. (1997) Agonist-specific coupling of a cloned *Drosophila melanogaster* D1-like dopamine receptor to multiple second messenger pathways by synthetic agonists. *J. Neurosci.* **17**, 6545–6553.
77. Blenau, W., Erber, J., and Baumann, A. (1998) Characterization of a dopamine D1 receptor from *Apis mellifera*: cloning, functional expression, pharmacology, and mRNA localization in the brain. *J. Neurochem.* **70**, 15–23.
78. Obosi, L. A., Schuette, D. G., Europe-Finner, G. N., Beadle, D. J., Hen, R., King, L. A., and Bermudez, I. (1996) Functional characterisation of the *Drosophila* 5-HTdro1 and 5-HTdro2B serotonin receptors in insect cells: activation of a Galphas-like protein by 5-HTdro1 but lack of coupling to inhibitory G-proteins by 5-HTdro2B. *FEBS Lett.* **381**, 233–236.
79. Witz, P., Amlaiky, N., Plassat, J.-L., Maroteaux, L., Borrelli, E., and Hen, R. (1990) Cloning and characterization of a *Drosophila* serotonin receptor that activates adenylate cyclase. *Proc. Natl. Acad. Sci. USA* **87**, 8940–8944.
80. David, J. C. and Coulon, J.-F. (1985) Octopamine in invertebrates and vertebrates. A review. *Prog. Neurobiol.* **24**, 141–185.

81. Evans, P. D. and Robb, S. (1993) Octopamine receptor subtypes and their modes of action. *Neurochem. Res.* **18**, 869–874.

82. Venter, J. C., DiPorzio, U., Robinson, D. A., Shreeve, S. M., Lai, J., Kerlavage, A. R., Fracek, S. P., Jr., Lentes, K.-U., and Fraser, C. M. (1988) Evolution of neurotransmitter receptor systems. *Prog. Neurobiol.* **30**, 105–169.

83. Arakawa, S., Gocayne, J. D., McCombie, W. R., Urquhart, D. A., Hall, L. M., Fraser, C. M., and Venter, J. C. (1990) Cloning, localization, and permanent expression of a *Drosophila* octopamine receptor. *Neuron* **4**, 343–354.

84. Han, K. A., Millar, N. S., and Davis, R. L. (1998) A novel octopamine receptor with preferential expression in *Drosophila* mushroom bodies. *J. Neurosci.* **18**, 3650–3658.

85. Robb, S., Cheek, T. R., Hannan, F. L., Hall, L. M., Midgley, J. M., and Evans, P. D. (1994) Agonist-specific coupling of a cloned *Drosophila* octopamine/tyramine receptor to multiple second messenger systems. *EMBO J.* **13**, 1325–1330.

86. Saudou, F., Amlaiky, N., Plassat, J.-L., Borrelli, E., and Hen, R. (1990) Cloning and characterization of a *Drosophila* tyramine receptor. *EMBO J.* **9**, 3611–3617.

87. Heisenberg, M. (1998) What do the mushroom bodies do for the insect brain? An introduction. *Learning Memory* **5**, 1–10.

88. Heisenberg, M., Borst, A., Wagner, S., and Byers, D. (1985) *Drosophila* mushroom body mutants are deficient in olfactory learning. *J. Neurogenet.* **2**, 1–30

89. Menzel, R., Durst, C., Erber, J., Eichmüller, S., Hammer, M., Hildebrandt, H., Mauelshagen, J., Müller, U., Rosenboom, H., Rybak, J., Schäfer, S., and Scheidler, A. (1994) The mushroom bodies in the honeybee: from molecules to behaviour, in *Neural Basis of Behavioural Adaptations*, Fortschritte der Zoologie 39, (Schildberger, K. and Elsner, N., eds.), Gustav Fischer, Stuttgart, pp. 81–102.

90. Schürmann, F. W. (1987) The architecture of the mushroom bodies and related neuropiles in the insect brain, in *Arthropod Brain: Its Evolution, Development, Structure, and Functions* (Gupta, A. P., ed.), Wiley, New York, pp. 231–264.

91. deBelle, J. S. and Heisenberg, M. (1994) Associative odor learning in *Drosophila* abolished by chemical ablation of mushroom bodies. *Science* **263**, 692–695.

92. Han, P.-L., Levin, L. R., Reed, R. R., and Davis, R. L. (1992) Preferential expression of the *Drosophila rutabaga* gene in mushroom bodies, neural centers for learning in insects. *Neuron* **9**, 619–627.

93. Nighorn, A., Healy, M. J., and Davis, R. L. (1991) The cyclic AMP phosphodiesterase encoded by the *Drosophila dunce* gene is concentrated in the mushroom body neuropil. *Neuron* **6**, 455–467.

94. Crittenden, J. R., Skoulakis, E. M. C., Han, K.-A., Kalderon, D., and Davis, R. L. (1998) Tripartite mushroom body architecture revealed by antigenic markers. *Learning Memory* **5**, 38–51.

95. Yang, M. Y., Armstrong, J. D., Vilinsky, I., Strausfeld, N. J., and Kaiser, K. (1995) Subdivision of the *Drosophila* mushroom bodies by enhancer trap expression patterns. *Neuron* **15**, 45–54.

96. Wolf, R., Wittig, T., Wustmann, G. Eyding, D., and Heisenberg, M. (1998) *Drosophila* mushroom bodies are dispensable for visual, tactile, and motor learning. *Learning Memory* **5**, 166- 178.

97. Grünbaum, L. and Müller, U. (1998) Induction of a specific olfactory memory leads to a long-lasting activation of protein kinase C in the antennal lobe of the honeybee. *J. Neurosci.* **18**, 4384–4392.

98. Hammer, M. and Menzel, R. (1998) Multiple sites of associative odor learning as revealed by local brain microinjections of octopamine in honeybees. *Learning Memory* **5**, 146–156.
99. Braun, A. P. and Schuman, H. (1995) The multifunctional calcium/calmodulin dependent protein kinase: From form to function. *Annu. Rev. Physiol.* **57**, 417–445.
100. Tanaka, C. and Nishizuka, Y. (1994) The protein kinase C family for neuronal signaling. *Annu. Rev. Neurosci.* **17**, 551–567.
101. Rosenthal, A., Rhee, L., Yadegari, R., Paro, R., Ullrich, A., and Goeddel, D. V. (1987) Structure and nucleotide sequence of a *Drosophila* protein kinase C gene. *EMBO J.* **2**, 433–441.
102. Schaeffer, E., Smith, D., Mardon, G., Quinn, W., and Zuker, C. (1989) Isolation and characterization of two new *Drosophila* protein kinase C genes, including one specifically expressed in photoreceptor cells. *Cell* **3**, 403–412.
103. Cho, K.-O., Wall, J. B., Pugh, P. C., Ito, M., Mueller, S. A., and Kennedy, M. B. (1991) The a subunit of type II Ca^{2+}/calmodulin-dependent protein kinase is highly conserved in *Drosophila*. *Neuron* **7**, 439–450.
104. Kane, N. S., Robichon, A., Dickinson, J. A., and Greenspan, R. J. (1997) Learning without performance in PKC-deficient *Drosophila*. *Neuron* **18**, 307–314.
105. Griffith, L. C., Verselis, L. M., Aitken, K. M., Kyriacou, C. P., Danho, W., and Greenspan, R. J. (1993) Inhibition of calcium/calmodulin-dependent protein kinase in *Drosophila* disrupts behavioral plasticity. *Neuron* **10**, 501–509.
106. Griffith, L. C., Wang, J., Zhong, Y., Wu, C.-F., and Greenspan, R. J. (1994) Calcium/calmodulin-dependent protein kinase II and potassium channel subunit Eag similarly affect plasticity in *Drosophila*. *Proc. Natl. Acad. Sci. USA* **91**, 10,044–10,048.
107. Jia, X.-X., Gorczyca, M., and Budnik, V. (1993) Ultrastructure of neuromuscular junctions in *Drosophila*: comparison of wild type and mutants with increased excitability. *J. Neurobiol.* **24**, 1025–1044.
108. Wang, J., Renger, J.J., Griffith, L.C., Greenspan, R.J., and Wu, C.-F. (1994) Concomitant alterations of physiological and developmental plasticity in *Drosophila* CaM kinase II-inhibited synapses. *Neuron* **13**, 1373–1384.
109. Broadie, K., Rushton, E., Skoulakis, E. M. C., and Davis, R. L. (1997) Leonardo, a *Drosophila* 14-3-3 protein involved in learning, regulates presynaptic function. *Neuron* **19**, 391–402.
110. Skoulakis, E. M. C. and Davis, R. L. (1996) Olfactory learning deficits in mutants for leonardo, a *Drosophila* gene encoding a 14-3-3 protein. *Neuron* **17**, 931–944.
111. Skoulakis, E. M. C. and Davis, R. L. (1998) 14-3-3 proteins in neuronal development and function. *Mol. Neurobiol.* **16**, 269–284.
112. Aitken, A. (1995) 14-3-3 proteins on the MAP. *Trends Biochem. Sci.* **20**, 95–97.
113. Morrison, D. (1994) 14-3-3: modulators of signaling proteins? *Science* **266**, 56–57.
114. Muslin, A. J., Tanner, J. W., Allen, P. M., and Shaw, A. S. (1996) Interaction of 14-3-3 with signaling proteins is mediated by the recognition of phosphoserine. *Cell* **84**, 889–897.
115. Grotewiel, M. S., Beck, C. D. O., Wu, K. H., Zhu, X. R., and Davis, R. L. (1998) Integrin-mediated short-term memory in *Drosophila*. *Nature* **391**, 455–460.
116. von Frisch, K. (ed.). (1967) *The Dance Language and Orientation of Bees.* Harvard University Press, Cambridge, MA.

117. Menzel, R. (1985) Learning in honey bees in an ecological and behavioral context, in *Experimental Behavioral Ecology* (Hölldobler, B. and Lindauer, M., eds.), Fischer, Stuttgart, pp. 55–74.
118. Menzel, R. (1990) Learning, memory, and "cognition" in honey bees, in *Neurobiology of Comparative Cognition* (Kesner, R. P. and Olten, D. S. eds.), Erlbaum, Hillsdale, NJ, pp. 237–292.
119. Erber, J., Masuhr, T., and Menzel, R. (1980) Localization of short-term memory in the brain of the bee, *Apis mellifera. Physiol. Entomol.* **5**, 343–358.
120. Menzel, R., Erber, J., and Masuhr, T. (1974) Learning and memory in the honeybee, in *Experimental analysis of Insect Behaviour* (Browne, L. B., ed.), Springer-Verlag, Berlin, pp. 195–217.
121. Homberg, U., Christensen, T. A., and Hildebrandt, J. G. (1989) Structure and function of the deutocerebrum in insects. *Annu. Rev. Entomol.* **34**, 477–501.
122. Stocker, R. F. (1994) The organisation of the chemosensory system in *Drosophila melanogaster* a review. *Cell Tissue Res.* **275**, 3–26.
123. Schäfer, S. and Bicker, G. (1986) Distribution of GABA-like immunoreactivity in the brain of the honeybee. *J. Comp. Neurol.* **246**, 287–300.
124. Kreissl, S. and Bicker, G. (1989) Histochemistry of acetylcholinesterase and immunocytochemistry of an acetylcholine receptor-like antigen in the brain of the honeybee. *J. Comp. Neurol.* **286**, 71–84.
125. Rybak, J. and Menzel, R. (1993) Anatomy of the mushroom bodies in the honey bee brain: the neuronal connections of the alpha-lobe. *J. Comp. Neurol.* **344**, 444–465.
126. Mauelshagen, J. (1993) Neural correlates of olfactory learning in an identified neuron in the honey bee brain. *J. Neurophysiol.* **69**, 609–625.
127. Rehder, V. (1989) Sensory pathways and motoneurons of the proboscis reflex in the suboesophageal ganglion of the honeybee. *J. Comp. Neurol.* **279**, 499–513.
128. Kreissl, S., Eichmüller, S., Bicker, G., Rapus, J., and Eckert, M. (1994) Octopamine-like immunoreactivity in the brain and suboesophageal ganglion of the honeybee. *J. Comp. Neurol.* **348**, 583–595.
129. Hammer, M. (1993) An identified neuron mediates the unconditioned stimulus in associative olfactory learning in honeybees. *Nature* **366**, 59–63.
130. Faber, T., Joerges, J., and Menzel, R. (1999) Associative learning modifies neural representations of odors in the insect brain. *Nature Neurosci.* **2**, 74–78.
131. Blenau, W. and Erber, J. (1998) Behavioural pharmacology of dopamine, serotonin and putative aminergic ligands in the mushroom bodies of the honeybee (*Apis mellifera*). *Behav. Brain Res.* **96**, 115–124.
132. Hildebrandt, H. and Müller, U. (1995) Octopamine mediates rapid stimulation of PKA in the antennal lobe of honeybees. *J. Neurobiol.* **27,** 44–50.
133. Hildebrandt, H. and Müller, U. (1995) PKA activity in the antennal lobe of honeybees is regulated by chemosensory stimulation *in vivo. Brain Res.* **679**, 281–288.
134. Mercer, A., Mobbs, P. G., Evans, P. D., and Davenport, A. (1983) Biogenic amines in the brain of the honey bee, *Apis mellifera. Cell Tissue Res.* **234**, 655–677.
135. Erber, J., Kloppenburg, P., and Scheidler, A. (1993) Neuromodulation by serotonin and octopamine in the honeybee: behaviour, neuroanatomy and electrophysiology. *Experientia* **49,** 1073–1083.

136. Goldberg, F., Grünewald, B., Rosenboom, H., and Menzel, R. (1999) Nicotinic acetylcholine currents of cultured Kenyon cells from the mushroom bodies of the honey bee *Apis mellifera*. *J. Physiol.* **514,** 759–768.

137. Bicker, G. and Kreissl, S. (1994) Calcium imaging reveals nicotinic acetylcholine receptors on cultured mushroom body neurons. *J. Neurophysiol.* **71,** 808–810.

138. Fonta, C., Sun, X. J., and Masson, C. (1991) Cellular analysis of odour integration in the honeybee antennal lobe, in *The Behaviour and Physiology of Bees* (Goodman, L. J. and Fisher, R. C., eds.), CAB International, London, pp. 227–241.

139. Burchuladze, R., Potter, J., and Rose, S. P. R. (1990) Memory formation in the chick depends on membrane-bound protein kinase C. *Brain Res.* **535,** 131–138.

140. Zhao, W. Q., Gibbs, M. E., Sedman, G. L., and Ng, K. T. (1994) Effect of PKC inhibitors and activators on memory. *Behav. Brain Res.* **60,** 151–160.

141. Colley, P. A., Sheu, F.-S., and Routtenberg, A. (1990) Inhibition of protein kinase C blocks two components of LTP persistence, leaving initial potentiation intact. *J. Neurosci.* **10,** 3353–3360.

142. Klann, E., Chen, S.-J., and Sweatt, J. D. (1993) Mechanism of protein kinase C activation during the induction and maintenance of long-term potentiation probed using a selective peptide substrate. *Proc. Natl. Acad. Sci. USA* **90,** 8337–8341.

143. Garthwaite, J. and Boulton, C. L. (1995) Nitric oxide signaling in the central nervous system. *Annu. Rev. Physiol.* **57,** 683–706.

144. Müller, U. and Bicker, G. (1994) Calcium-activated release of nitric oxide and cellular distribution of nitric oxide-synthesizing neurons in the nervous system of the locust. *J. Neurosci.* **14,** 7521–7528.

145. Elofsson, E., Carlber, M., Moroz, L., Nezlin, L., and Sakharov, D. (1993) Is nitric oxide (NO) produced by invertebrate neurones? *NeuroReport* **4,** 279–282.

146. Elphick, M. R., Green, I. C., and O'Shea, M. (1993) Nitric oxide synthesis and action in an invertebrate brain. *Brain Res.* **619,** 344–346.

147. Moroz, L. L. and Gillette, R. (1995) From polyplacophora to cephalopoda: a comparative analysis of nitric oxide signalling in *Mollusca. Acta Biol. Hung.* **46,** 169–182.

148. Müller, U. (1994) Ca^{2+}/calmodulin-dependent nitric oxide synthase in *Apis mellifera* and *Drosophila melanogaster. Eur. J. Neurosci.* **6,** 1362–1370.

149. Müller, U. and Buchner, E. (1993) Histochemical localization of NADPH-diaphorase in adult *Drosophila* brain: is nitric oxide a neuronal messenger also in insects? *Naturwissenschaften* **80,** 524–526.

150. Ribeiro, J. M. C. and Nussenzveig, R. H. (1993) Nitric oxide synthase activity from a hematophagous insect salivary gland. *FEBS Lett.* **330,** 165–168.

151. Salleo, A., Musci, G., Barra, P. F. A., and Calabrese, L. (1996) The discharge mechanism of acontial nematocytes involves the release of nitric oxide. *J. Exp. Biol.* **199,** 1261–1267.

152. Elphick, M. R., Rayne, R. C., Riveros-Moreno, V., Moncada, S., and O'Shea, M. (1995) Nitric oxide synthesis in locust olfactory interneurones. *J. Exp. Biol.* **198,** 821–829.

153. Nighorn, A., Gibson, N. J., Rivers, D. M., Hildebrand, J. G., and Morton, D. B. (1998) The nitric oxide–cGMP pathway may mediate communication between sensory afferents and projection neurons in the antennal lobe of *Manduca sexta. J. Neurosci.* **18,** 7244–7255.

154. Regulski, M. and Tully, T. (1995) Molecular and biochemical characterization of dNOS: a *Drosophila* Ca^{2+}/calmodulin-dependent nitric oxide synthase. *Proc. Natl. Acad. Sci. USA* **92**, 9072–9076.

155. Yuda, M., Hirai, M., Miura, K., Matsumura, H., Ando, K., and Chinzei, Y. (1996) cDNA cloning, expression and characterization of nitric-oxide synthase from the salivary glands of the blood-sucking insect *Rhodnius prolixus. Eur. J. Biochem.* **242**, 807–812.

156. Bredt, D. S. and Snyder, S. H. (1994) Nitric oxide: a physiologic messenger molecule. *Annu. Rev. Biochem.* **63**, 175–195.

157. Jacklet, J. W. (1997) Nitric oxide signaling in invertebrates. *Invert. Neurosci.* **3**, 1–14.

158. Müller, U. (1997) The nitric oxide system in insects. *Prog. Neurobiol.* **51**, 363–381.

159. Schuman, E. M. and Madison, D. V. (1994) Nitric oxide and synaptic function. *Annu. Rev. Neurosci.* **17**, 153–183.

160. Bicker, G., Schmachtenberg, O., and De Vente, J. (1996) The nitric oxide/cyclic GMP messenger system in olfactory pathways of the locust brain. *Eur. J. Neurosci.* **8**, 2635–2643.

161. Elphick, M. R. and Jones, I. W. (1998) Localization of soluble guanylyl cyclase α-subunit in identified insect neurons. *Brain Res.* **800**, 174–179.

162. Liu, W., Yoon, J., Burg, M., Chen, L., and Pak, W.L. (1995) Molecular characterization of two *Drosophila* guanylate cyclases in the nervous system. *J. Biol. Chem.* **270**, 12,418–12,427.

163. Shah, S. and Hyde, D. R. (1995) Two *Drosophila* genes that encode the α and β-subunits of the brain soluble guanylyl cyclases. *J. Biol. Chem.* **270**,15,368–15,376.

164. Yoshikawa, S., Miyamoto, I., Aruga, J., Furuichi, T., Okano, H., and Mikoshiba, K. (1993) Isolation of a *Drosophila* gene encoding a head-specific guanylyl cyclase. *J. Neurochem.* **60**, 1570–1573.

165. De Vente, J. and Steinbusch, H. W. M. (1992) On the stimulation of soluble and particulate guanylate cyclase in the rat brain and the involvement of nitric oxide as studied by cGMP immunocytochemistry. *Acta Histochem.* **92**, 13–38.

166. Bicker, G. and Schmachtenberg, O. (1997) Cytochemical evidence for nitric oxide cyclic GMP signal transmission in the visual system of the locust. *Eur. J. Neurosci.* **9**, 189–193.

167. Bicker, G., Schmachtenberg, O., and De Vente, J. (1997) Geometric considerations of nitric oxide cyclic GMP signalling in the glomerular neuropil of the locust antennal lobe. *Proc. R. Soc. Lond. (Biol.)* **264**, 1177–1181.

168. Goy, M. F. (1991) cGMP: the wayward child of the cyclic nucleotide family. *Trends Neurosci.* **14**, 293–299.

169. MacFarland, R. T. (1995) Molecular aspects of cyclic GMP signaling. *Zool. Sci.* **12**, 151–163.

170. Foster, J. L., Higgins, G. C., and Jackson, F. R. (1996) Biochemical properties and cellular localization of the *Drosophila* DG1 cGMP-dependent protein kinase. *J. Biol. Chem.* **271**, 23,322–23,328.

171. Kalderon, D. and Rubin, G. M. (1989) cGMP-dependent protein kinase genes in *Drosophila. J. Biol. Chem.* **264**, 10,738–10,748.

172. Osborne, K. A., Robichon, A., Burgess, E., Butland, S., Shaw, R. A., Coulthard, A., Pereira, H. S., Greenspan, R. J., and Sokolowski, M. B. (1997) Natural behavior

polymorphism due to a cGMP-dependent protein kinase of *Drosophila. Science*
277, 834–836.

173. Beavo, J. A. (1995) Cyclic nucleotide phosphodiesterases: functional implications
of multiple isoforms. *Physiol. Rev.* **75**, 725–748.

174. Takemoto, D. J., Gonzalez, K., Udovichenko, I., and Cunnick, J. (1993) Cyclic
GMP-regulated cyclic nucleotide phosphodiesterases. *Cell. Signal.* **5**, 549–553.

175. Kiger, J. A., Jr., Davis, R. L., Salz, H., Fletscher, T., and Bowling, M. (1981)
Genetic analysis of cyclic nucleotide phosphodiesterases in *Drosophila melanogaster. Adv. Cycl. Nucl. Res.* **14**, 273–288.

176. Kaupp, U. B. (1991) The cyclic nucleotide-gated channels of vertebrate photoreceptors and olfactory epithelium. *Trends Neurosci.* **14**, 150–157.

177. Yau, K.-W. (1994) Cyclic nucleotide gated channels, an expanding new family of
ion channels. *Proc. Natl. Acad. Sci. USA* **91**, 3481–3483.

178. Baumann, A., Frings, S., Godde, M., Seifert, R., and Kaupp, U. B. (1994) Primary
structure and functional expression of a *Drosophila* cyclic nucleotide-gated channel present in eyes and antennae. *EMBO J.* **13**, 5040–5050.

179. Ahmad, I., Leinders-Zufall, T., Kocsis, J. D., Shepherd, G. M., Zufall, F., and
Barnstable, C. J. (1994) Retinal ganglion cells express a cGMP-gated cation conductance activatable by nitric oxide donors. *Neuron* **12**, 155–165.

180. Leinders-Zufall, T., Shepherd, G. M., and Zufall, F. (1995) Regulation of cyclic
nucleotide-gated channels and membrane excitability in olfactory receptor cells
by carbon monoxide. *J. Neurophysiol.* **74**, 1498–1508.

181. Bredt, D. S., Glatt, C. E., Hwang, P. M., Fotuhi, M., Dawson, T. M., and Snyder,
S. H. (1991) Nitric oxide synthase protein and mRNA are discretely localized in
neuronal populations of the mammalian CNS together with NADPH diaphorase.
Neuron **7**, 615–624.

182. Breer, H. and Shepherd, G. M. (1993) Implications of the NO/cGMP system for
olfaction. *Trends Neurosci.* **16**, 5–9.

183. Müller, U. and Hildebrandt, H. (1995) The nitric oxide/cGMP system in the antennal lobe of *Apis mellifera* is implicated in integrative processing of chemosensory
stimuli. *Eur. J. Neurosci.* **7**, 2240–2248.

184. Braun, G. and Bicker, G. (1992) Habituation of an appetitive reflex in the honeybee. *J. Neurophysiol.* **67**, 588–598.

185. Haley, J. E., Malen, P. L., and Chapman, P. F. (1993) Nitric oxide synthase inhibitors block long-term potentiation induced by weak but not strong tetanic stimulation at physiological brain temperatures in rat hippocampal slices. *Neurosci. Lett.*
160, 85–88.

186. Zhuo, M., Kandel, E. R., and Hawkins, R. D. (1994) Nitric oxide and cGMP can
produce either synaptic depression or potentiation depending on the frequency of
presynaptic stimulation in the hippocampus. *NeuroReport* **5**, 1033–1036.

187. Böhme, G. A., Bon, C., Lemaire, M., Reibaud, M., Piot, O., Stutzmann, J.-M.,
Doble, A., and Blanchard, J.-C. (1993) Altered synaptic plasticity and memory
formation in nitric oxide synthase inhibitor-treated rats. *Proc. Natl. Acad. Sci.
USA* **90**, 9191–9194.

188. Chapman, P. F., Atkins, C. M., Allen, M. T, Haley, J. E., and Steinmetz, J. E.
(1992) Inhibition of nitric oxide synthesis impairs two different forms of learning.
NeuroReport **3**, 567–570.

189. Hölscher, C. and Rose, S. P. R. (1992) An inhibitor of nitric oxide synthesis prevents memory formation in the chick. *Neurosci. Lett.* **145**, 165–167.

190. Kendrick, K. M., Guevara-Guzman, R., Zorrilla, J., Hinton, M. R., Broad, K. D., Mimmack, M., and Ohkura, S. (1997) Formation of olfactory memories mediated by nitric oxide. *Nature* **388**, 670–674.

191. Okere, C. O., Kaba, H., and Higuchi, T. (1996) Formation of an olfactory recognition memory in mice: reassessment of the role of nitric oxide. *Neuroscience* **71**, 349–354.

192. Dawson, T. M. and Snyder, S. H. (1994) Gases as biological messengers: nitric oxide and carbon monoxide in the brain. *J. Neurosci.* **14**, 5147–5159.

193. Müller, U. (1996) Inhibition of nitric oxide synthase impairs a distinct form of long-term memory in the honeybee, *Apis mellifera. Neuron* **16**, 541–549.

194. Qiang, M., Chen, Y. C., Wang, R., Wu, F. M., and Qiao, J. T. (1997) Nitric oxide is involved in the formation of learning and memory in rats: studies using passive avoidance response and Morris water maze task. *Behav. Pharmacol.* **8**, 183–187.

195. Bannerman, D. M., Chapman, P. F., Kelly, P. A. T., Butcher, S. P., and Morris, R. G. M. (1994) Inhibition of nitric oxide synthase does not impair spatial learning. *J. Neurosci.* **14**, 7404–7414.

196. Bannerman, D. M., Chapman, P. F., Kelly, P. A. T., Butcher, S. P., and Morris, R. G. M. (1994) Inhibition of nitric oxide synthase does not prevent the induction of long-term potentiation *in vivo. J. Neurosci.* **14**, 7415–7425.

197. Williams, J. H., Li, Y.-G., Nayak, A., Errington, M. L., Murphy, K. P. S. J., and Bliss, T. V. P. (1993) The suppression of long-term potentiation in rat hippocampus by inhibitors of nitric oxide synthase is temperature and age dependent. *Neuron* **11**, 877–884.

198. Gribkoff, V. K. and Lum-Ragan, J. T. (1992) Evidence for nitric oxide synthase inhibitor-sensitive and insensitive hippocampal synaptic potentiation. *J. Neurophysiol.* **68**, 639–642.

199. Blokland, A., De Vente, J., Prickaerts, J., Honig, W., Markerink-van Ittersum, M., and Steinbusch, H. (1999) Local inhibition of hippocampal nitric oxide synthase does not impair place learning in the Morris water escape task in rats. *Eur. J. Neurosci.* **11**, 223–232.

200. Kendrick, K. M., Lévy, F., and Keverne, E. B. (1992) Changes in the sensory processing of olfactory signals induced by birth in sheep. *Science* **256**, 833–836.

4

Protein Kinase C Signaling in Learning and Memory

Eddy A. Van der Zee, Bas R. K. Douma, John F. Disterhoft, and Paul G. M. Luiten

BRIEF HISTORY OF THE DISCOVERY OF PROTEIN KINASE C

It is a generally accepted concept that highly selective changes in the strength of synaptic connections between neurons in the brain contribute to learning and memory. Storage of information in the brain appears to involve persistent, use-dependent alteration in the efficacy of synaptic transmission. Approximately a decade ago, the first experimental data in support of a putative role of protein kinase C (PKC) in synaptic plasticity and information storage were reported. PKC is a cellular second messenger involved in various neuronal signal transduction pathways through phosphorylation of specific substrate proteins by which neurons increase their excitability in response to external inputs (1,2). Protein kinases phosphorylate many cellular proteins, catalyzing the transfer of phosphate to certain amino acid residues within proteins. Phosphorylation can alter the folding of the protein and, hence, their function. The discovery of a novel, cyclic nucleotide-independent, protein kinase took place relatively recently in the late 1970s (for a historical review on PKC in learning and memory, see ref. 3 and references therein). In 1977, co-workers from the Nishizuka group at Kobe University in Japan first reported to have found a new type of kinase, and 2 yr thereafter, it was referred to as PKC. It was demonstrated by this group that the kinase could be enzymatically fully active in the presence of Ca^{2+} and the phospholipid phosphatidylserine. The kinase is activated in a reversible manner by attachment to membrane phospholipid in the presence of Ca^{2+}. Further analysis showed that a small amount of diacylglycerol (DAG; a minor component of the cellular lipids) significantly increases the affinity of this enzyme for Ca^{2+} and phospholipid. DAG (and inositoltrisphosphate) is produced by

From: *Cerebral Signal Transduction: From First to Fourth Messengers*
Edited by: M. E. A. Reith © Humana Press Inc., Totowa, NJ

the hydrolysis of phosphatidylinositol bisphosphate (PI turnover). Interestingly, DAG permitted activation of PKC at resting intracellular Ca^{2+} levels. PKC is usually present in an inactive form in the cytosol. As a result of the specific binding of PKC by DAG, which is transiently formed in the membrane, activation of PKC is accompanied by its translocation from the cytosol to the membrane. The duration and magnitude of the DAG signal determines the activation of PKC at the cellular membrane.

The direct link between receptor stimulation and PKC activation proved to be a critical step in the history of PKC because it indicated the implication of PKC in signal transduction between nerve cells and the translation of an extracellular message to an intracellular biochemical signal. PKC gained much interest after the mid-1980s, when two major findings (i.e., that it was the main target of PI turnover as well as the receptor for phorbol ester [a DAG analog]) became widely known. Thereafter, it was discovered that PKC comprises a family of closely related structures *(4,5)*. To date, at least 12 PKC isozymes have been identified and classified into 3 groups based on their structure and cofactor regulation: classical (cPKC), novel (nPKC), and atypical (aPKC) isoforms. The classical group consists of four calcium-dependent isoforms, PKCα, βI, βII, and γ, of which only the γ isoform (*see* Fig. 1 for a schematic representation of PKCγ) is specific to brain tissue *(1,6)*. The discovery of the diverse PKC isoforms contributed to our understanding of the mystery of how a great diversity of messages in cellular communication can be generated by only a limited number of different components. In brain, PKCs are widely distributed. Here, we will focus on PKC distribution in the hippocampus, a brain region intimately involved in learning and memory mediated (in part) through PKC-dependent pathways (for a review, *see* ref. *7*). It has been demonstrated that PKC subtypes have an isoform-specific cellular localization in various tissues and cell types, including the hippocampus *(7,8)*. Apparently, each isozyme of the PKC family has its own spatial organization and is present in the right intracellular compartment of the cell, presumably in association with its specific target substrate proteins.

WHY PKC IS THOUGHT TO BE CRUCIAL IN LEARNING AND MEMORY: OUTLASTING THE INITIAL STIMULUS

Activation of PKC by phorbol ester or intracellular injection of the purified enzyme induced a long-lasting enhancement of the excitability of neurons *(9)*. Before this report, there was no direct evidence for the involvement of PKC in the control of neuronal excitability. Baraban et al. *(10)* studied the effect of phorbol ester in rat hippocampal slices to further clarify the role of PKC in neuronal function. They found that phorbol ester blocked the late

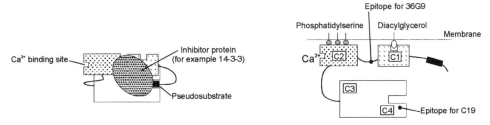

Fig. 1. Schematic representation of PKCγ. The epitope for 36G9 is located at the V2 (near C1) part of the regulatory domain, close to (but not identical to) the diacylglycerol (DAG) binding site (the essential amino acid residues are contained in the sequence comprised between positions 164 and 197 *[43,44]*), whereas C19 binds to the C-terminal portion of the catalytic domain (the small black dots indicate the epitopes). A 30-kDa PKC-inhibitory protein, 14-3-3 (the τ isoform of 14-3-3 inhibits PKCγ *[45]*), has putative interaction sites at the C1 part of the regulatory domain and at the substrate binding site of the catalytic domain of PKC *(45)*. 14-3-3 is attached to inactive, cytosolic PKCγ, blocking the binding sites of both antibodies. It should be noted, however, that endogenous inhibitor proteins other than 14-3-3 *(46)* might well be responsible for the epitope blocking effect; 14-3-3, however, seems at present the most likely candidate. (This figure is adapted from a previous version and is reprinted from *Prog. Neuro-Psychopharmacol. Biol. Psychiatry*, Vol. 21, E. A. Van der Zee, P. G. M. Luiten, and J. F. Disterhoft, Learning-induced alterations in hippocampal PKC-immunoreactivity: a review and hypothesis of its functional significance, pp. 531–572, 1997, with permission from Elsevier Science.)

hyperpolarization elicited by synaptic stimulation, and their results also suggested that PKC regulates membrane conductance, possibly through translocation of PKC. These studies clearly implicated PKC in the modulation of membrane function, and it became clear that PKC could be a key regulator of neuronal excitability.

Persistent changes in phosphorylation of PKC substrates that outlast an initial signal is most likely an important molecular event in information storage. Prolonged PKC activation, therefore, may serve as a critical step in the chain of biological events leading to memory formation. In the mid-1980s, the group of Routtenberg suggested that the liberation of free cis-fatty acid from membrane phospholipids by phospholipase A2 (PLA2) was an important mechanism for PKC activation in long-term potentiation (LTP). LTP is a cellular experimental model of activity-dependent enhancement of synaptic efficacy and cellular excitability widely studied as a cellular mechanism contributing to memory formation *(11)*. Activation of this pathway was thought to stabilize PKC in an activated state, thus contributing to the maintenance of the potentiated synaptic response *(12)*. A synergistic action of free cis-fatty acids and DAG for the activation of PKC was suggested by Seifert et al. *(13)*. In the early part of the 1990s, the group of Nishizuka emphasized that stimuli somewhat separated in time but acting through different types of phospholipases (phosphoinositide- and phosphatidylcholine-specific phospholipase C, and phospholipase D plus a phosphatase *[14]*) and on the same target could induce sustained elevation of DAG *(4,15)*. PKC activity could be maintained if both DAG and free cis-fatty acid are available at a relatively late phase after the initial signal occurred, which would be essential for long-term cellular responses such as necessary in learning and memory.

Persistent enhancement of cell responsiveness after removal of extracellular stimuli was suggested to be due to the continued association, or anchoring, of PKC to the membrane *(16)*. In this way, a "cellular memory" could be formed, where information has been stored or "remembered" by the cell in the form of constitutively active PKC. In a model of associative learning presented by Bank et al. *(17)*, such stable anchoring of PKC to the membrane and, hence, its prolonged activity was suggested to occur also by phosphorylation of PKC by calcium–calmodulin-dependent protein kinase. However, stable anchoring of PKC has been challenged by research on the dynamics of PKC translocation in the 1990s (*see* Translocation of PKCγ and Learning and Memory).

Outlasting the initial stimulus can also be achieved by a change in protein–protein interaction. The translocation-induced removal of an inhibitor protein (e.g., 14-3-3) of PKC after signal transduction has been suggested by us as a mechanism in the regulation of PKCγ *(7)*. In this way, activated PKCγ retains some of its activity after redistribution from the membrane to the cytosol (the "activation-prone" PKCγ, *see* Protein–Protein Interaction). In Learning-Induced Alterations, immunocytochemical and biochemical evidence will be presented for such a mechanism in relation to learning and memory.

EVIDENCE FOR A ROLE OF PKC
IN LEARNING AND MEMORY

The involvement of PKC in learning and memory processes has been demonstrated extensively in previous work *(17–30)*. Spatial learning accuracy correlated positively with hippocampal PKC activity, as determined by mouse strain comparison *(30)*, and changes in mouse hippocampal PKC activity following spatial learning were demonstrated by Noguès et al. *(21)*. A single intracerebroventricular injection of the phorbol ester phorbol-12,13-dibutyrate (PDBu) prior to training improved spatial learning performance in rats *(26)*, whereas intraperitoneal injection of a kinase inhibitor immediately following acquisition impaired memory consolidation in mice

(20). PKC activation appears to be a common mechanism among cognition-stimulating nootropic drugs *(19)*.

[³H]PDBu binding has been used as a method to detect changes in the PKC signal transduction pathway. Spatial discrimination learning in a water maze induced a significant reduction of [³H]PDBu binding in the CA3 region of the rat hippocampus *(25)*. Golski et al. *(18)* demonstrated that the changes in hippocampal [³H]PDBu binding only occurred in rats performing hippocampally dependent spatial discrimination, but not hippocampally independent-cued discrimination in the water maze. Spieler et al. *(31)*, however, found that it was the experience of an enriched environment by the repeated behavioral stimulation rather than the acquisition of maze learning that leads to enhanced [³H]PDBu binding in the rat hippocampus. Examples of rat hippocampal changes in binding of this radioligand induced by learning tasks other than spatial learning was shown by Bernabeu et al. *(32)*. They demonstrated that one trial inhibitory avoidance learning induced a short-term increase in [³H]PDBu binding in the CA1, CA2, CA3, and dentate gyrus subfields of the hippocampus.

Alterations in the (subcellular) distribution of [³H]PDBu were shown in the rabbit hippocampus during acquisition, consolidation, and retention of delay eye-blink conditioning, a form of hippocampally independent associative learning *(24,28)*. A clear increase in [³H]PDBu binding was seen in the CA3 stratum oriens, but not in the CA1, during eye-blink acquisition *(28)*, whereas an increase was found in the CA1 stratum pyramidale in conditioned rabbits 24 h after reaching a behavioral criterion *(24)*. This increase seemed to shift to the CA1 stratum radiatum and, most notably, the stratum oriens 48 h after the previous time point *(24)*. Because of the limited spatial resolution of the [³H]PDBu results, however, it is somewhat difficult to tell in these studies in which cellular compartments or in which hippocampal cell types (principal cells, interneurons, and/or glia cells) these changes take place.

LEARNING-INDUCED ALTERATIONS IN PKCγ IMMUNOLABELING

Analysis of the neural substrate that utilizes PKC-related signal transduction pathways—while engaged in learning and memory processes in rat, mouse, and rabbit hippocampus—through immunocytochemical procedures has provided us with a way to study the functional role and molecular cascade of PKC activation within the cells that constitute the active neural circuit. Antibodies are excellent probes to study the distribution of PKC *in situ* in the hippocampus with a high anatomical resolution (for a review of PKC immunoreactivity in hippocampus, *see* ref. 7).

A hole board has been used by us as a test apparatus *(33,34)* for spatial orientation to study alterations in PKC immunoreactivity (PKC-ir) induced by spatial learning. The hole board contains 16 equidistant holes in the floor plate, of which 4 are baited in a fixed pattern over time, and the subjects have to orient themselves within the hole board to obtain the food. Both rats *(35,36)* and mice *(37,38)* were tested in the hole board. A strong increase in PKCγ-ir was observed in pyramidal neurons and dentate granule cells of trained compared with naive rats after 17 training sessions of 2 trials every second day. A group tested for nine training sessions reached an intermediate degree of enhanced PKCγ-ir, suggesting that this increase was somehow related to the behavioral history and degree of acquisition of the animals *(35)*. Moreover, animals that were not able to improve their reference memory in the hole board because of experimental induction of seizures did not reveal such an increase in PKCγ-ir, which further suggested that the alterations in PKCγ-ir were learning-specific. In contrast to PKCγ, no clear changes were seen for PKCαβ in any group, using an antibody that did not discriminate among PKCα, β1, and β2. A pilot experiment revealed the long-term nature of the increase in PKCγ-ir, which may last up to 2–3 wk after the last training trial.

In the experiments using mice and rats *(36–38)* that were more specifically aimed at describing the learning-related aspects of the altered immunostaining, additional control groups beside the naive control group were added (habituated and pseudotrained groups, which were respectively exposed to the test apparatus only and pseudotrained with all holes baited). Like in trained rats, the trained mice revealed a strong increase in PKCγ-ir throughout the hippocampus (*see* Fig. 2), whereas the pyramidal and granule cells of the pseudotrained group were nearly indistinguishable from those of naive animals. The pyramidal and granule cells of the habituated animals, however, revealed somewhat intermediate values in PKCγ-ir. These results clearly showed that although the initial exposure to the new environment (the hole board) enhanced PKCγ-ir, the strongest increase in hippocampal PKCγ-ir was spatial-learning-specific.

A similar immunocytochemical approach was used in associative learning in rabbits. Trace eye-blink conditioning, in which a stimulus-free "trace" period intervened between the tone (conditioned stimulus; CS) and airpuff (unconditioned stimulus; US), was used as a learning task because it is a more complex (higher order) form of conditioning than delay eye-blink conditioning. Furthermore, it has been demonstrated to be dependent on the intact hippocampus for its successful acquisition and proper consolidation in the rabbit *(39–41,76)*, and the activity of hippocampal pyramidal cells increases during training *(42)*. Animals were trained up to 80% conditioned responses

(CRs) in a daily 80-trial session. Based on their rate of acquisition, fast and slow learners could be distinguished. The slow learners showed a slow increase in the daily percentage CRs and reached approximately 40% CRs at the last day of training, whereas all fast learners reached the criterion of 80% CRs in a training session. Optical densities (ODs) of the immunostaining for all four cPKC isoforms were measured in several subfields of the hippocampus. Large and significant ($p < 0.05$) increases in staining intensity in trace conditioned compared with naive and pseudoconditioned subjects were found for PKCγ only in the dentate gyrus, CA3, and CA1, but not in the subiculum. Neurons of the lateral hypothalamus, serving as a nonhippocampal control region, showed no learning-specific changes, demonstrating that the immunocytochemical PKCγ alterations are region-specific as well as subregion-specific. This was true with either the monoclonal antibody 36G9 or the polyclonal antibody C19, reactive with the regulatory and catalytic domain of PKCγ, respectively (*see* Fig. 1). A fluorescent double-labeling experiment was carried out to examine whether the changes for the two anti-PKCγ antibodies occurred in the same cells and same cellular compartments. This double-labeling experiment verified that both PKCγ antibodies recognized the same set of hippocampal neurons characterized by enhanced immunoreactivity. The increase in immunostaining found for each antibody separately in the DG, CA1, and CA3 regions therefore occurs in individual neurons of the same cell population.

The slow learners had intermediate OD values for PKCγ-ir between the pseudoconditioned and trace conditioned subjects but did not differ significantly from either group at any hippocampal subfield. However, the slow learners differed significantly from the fast learners ($p = 0.027$), but not from the pseudoconditioned rabbits ($p = 0.409$), when the DG, CA3, and CA1 were taken together. In addition to the neuronal increase in PKCγ-ir, the number of astrocytes immunopositive for PKCβ2 and γ (these isoforms seem to be colocalized in astrocytes, in contrast to PKCα and β1) was significantly reduced in the CA1 stratum radiatum in trace conditioned over naive and pseudoconditioned rabbits *(47)*. This reduction is most likely the result of a downregulation of the PKC isoforms rather than the actual loss of the astroglial cells. No changes were found for PKCα and β1, and no conditioning-specific changes were seen in the subiculum, suggesting that the coincidental alterations in neuronal and astroglial PKC are somehow related, learning-specific, and probably underlying learning-induced neural activation.

Interestingly, the level of PKCγ-ir in the CA1 and CA3 pyramidal cells and the total number of CRs across training sessions correlated positively when all trace conditioned animals (fast and slow learners) were taken together ($r^2 = 0.761$, $p = 0.0002$), whereas no such correlation was found for the dentate granule cells. This correlation suggests that each training trial in which a CR is produced also produces a further increase in PKCγ-ir.

As can be seen in Fig. 2, a considerable part of the hippocampal cells reveal increased levels of PKCγ-ir, but not all cells were affected by spatial learning. This was seen in the rabbit hippocampus after associative learning as well *(48)*. On average, three-quarters of the pyramidal cells were affected by trace conditioning, and similar numbers of cells change biophysically during trace eye-blink conditioning *(49)*. At first glance, it may seem that this number is rather high in relation to the functional demand on the hippocampus to perform the learning task. It should be noted, however, that PKCγ-mediated signal transduction at relatively few synapses per neuron could cause a dramatic increase in cellular staining throughout the neuron if the activated PKCγ migrates from the site of action (the synapse) to intracellular targets (*see* Protein–Protein Interaction and Fig. 3).

In conclusion, the learning-induced changes in PKCγ-ir in the hippocampus of rat, mouse, and rabbit are similar, although the degree of changes differs depending on the involvement of the hippocampus (delay versus trace eye-blink conditioning) and the degree of acquisition (fully trained and partly trained rats; fast- versus slow-learning rabbits). The results described clearly demonstrate that PKC is involved in the sequence of molecular events that underlie learning and memory and that only PKCγ shows a consistent and robust increase in immunoreactivity following hippocampally dependent spatial and associative learning. Bowers et al. *(51)* reported that C57BL mice, which are good spatial learners, have more membrane-bound PKCγ than DBA mice, which are poor spatial learners. In contrast, no differences were found for the α, β1, or β2 PKC isoforms between these strains. Translocation of PKCγ, but not PKCαβ, was observed following the induction of LTP in the rat DG, CA3, and CA1 *(52)*. The enhanced activity of PKC found during the maintenance phase of LTP *(53)* with an antibody that does not discriminate between the γ and β isoforms *(54)* could also be due solely to the contribution of (a putatively active form of) PKCγ. Moreover, an increase in γ (but not α, β1, β2, ε or ζ) PKC mRNA was found 2–24 h after LTP *(55)*. These results and our observations provide evidence for isoform-specific functions and strongly point to brain-specific PKCγ as the primary isoform involved in memory processes.

FUNCTION OF HIPPOCAMPAL PKCγ
IN LEARNING AND MEMORY

The ultimate function of PKCγ is to phosphorylate substrate proteins in order to create persistent cellular changes that outlast the initial signal (i.e., a sensory stimulus relevant to the learning task) that induced the activation

NAIVE   TRAINED

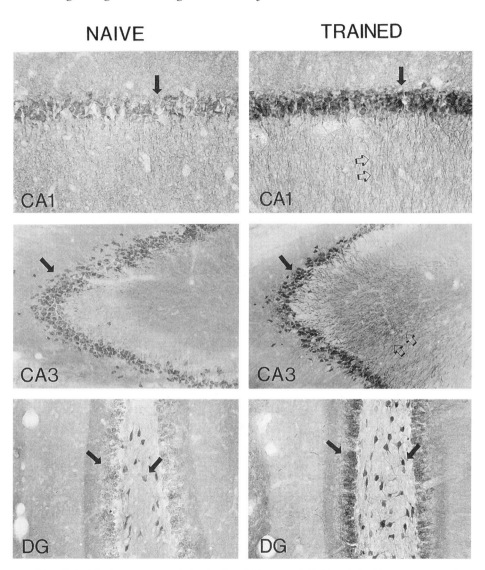

Fig. 2. PKCγ immunoreactivity in the different subfields of the hippocampus of naive (left panel) and hole-board trained (right panel) mice. PKCγ immunoreactivity is significantly higher in trained mice 24 h after a patial learning than in naive mice. The solid arrows point to immunoreactive neurons. The open arrows indicate the learning-induced increase of PKCγ immunoreactivity in dendrites. CA1 and CA3 = cornu ammonis 1 and 3; DG = dentate gyrus.

of PKCγ. One putative function of PKC in both spatial learning and associative learning is related to membrane conductance changes in neurons engaged in learning and memory. The postburst after-hyperpolarization (AHP) in hippo-

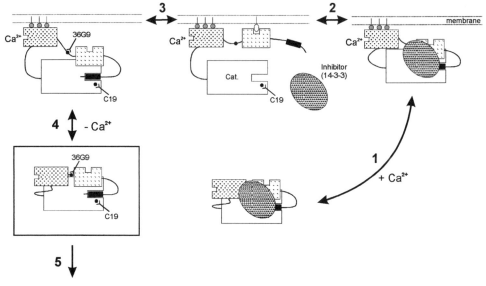

Fig. 3. PKCγ-regulation during learning. See Protein–Protein Interaction for explanation. The "activation-prone" form of PKCγ, depicted in the rectangular box, may subsequently bind to intracellular binding sites, whereas PKCγ witch the inhibitor attached to it cannot. (Part of this figure is adapted from a previous version and is reprinted from *Prog. Neuro-Psychopharmacol. Biol. Psychiatry*, Vol. 21, E. A. Van der Zee, P. G. M. Luiten, and J. F. Disterhoft, Learning-induced alterations in hippocampal PKC-immunoreactivity: a review and hypothesis of its functional significance, pp. 531–572, 1997, with permission from Elsevier Science.) The schematic representation of PKC and pathways are modified from ref. *50*.

campal CA1 neurons, generated by a Ca^{2+}-activated potassium conductance (*56–58*), is reduced after eye-blink conditioning (*49,59–61*). Phosphorylation of potassium channels through Ca^{2+}-dependent kinases seems crucial in the decrease of this conductance after conditioning (*62*). Intracellular injection of PKC into CA1 pyramidal neurons reduced the AHP and enhanced the excitability of the neurons (*63*), whereas phorbol ester activation of PKC in CA1 pyramidal cells also led to the reduction in the Ca^{2+}-dependent AHP (*10,64,65*). Thus, it is reasonable to expect reductions in the AHP in the same hippocampal pyramidal cell populations in which PKCγ is changed.

The 500-ms trace eye-blink conditioning requires the formation of a short-term "memory trace" of the CS (the tone) to bridge the interstimulus interval between the CS and US (the airpuff) in order to successfully form an association and correctly time the conditioned response. The hippocampus is involved in the association of temporal events, and timing of the eye-blink response is crucial; it is suggested that in trace eye-blink conditioning, the animal is likely to learn that "a tone followed by an empty interval means to blink" *(41)*. The observed positive correlation of the increase in PKCγ-ir and the number of correct behavioral CRs may suggest that alterations in PKCγ are related to the proper timing of the eye-blink response and that a certain amount of previously immunonegative PKCγ becomes available for binding of 36G9 and C19 in relation to each successfully timed eye-blink. This increase in PKCγ-ir could take place at one given set of synapses, or at more and more synapses that are recruited as training continues, or a combination of both. Hippocampal input corresponding to the CS and US pathways most likely stimulates receptors coupled to PKCγ. If the increase in PKCγ-ir is a consequence of activation, it may have catalyzed the phosphorylation of specific substrate proteins in the cytosol, membrane, or the postsynaptic density. Some key substrate proteins are Ca^{2+}-activated potassium channels, glutamate receptors, proteins involved in the translation of mRNA in spine ribosomes (to stimulate the production of synaptic proteins necessary for synaptic restructuring *[66,67]*), or nuclear proteins that alter gene expression. All these possible phosphorylation events may eventually result in enhanced synaptic strength (and the formation of perforated synapses, as, for example, postulated in a model of structural synaptic plasticity associated with LTP *[68]*), which seems pivotal in (temporary) memory storage necessary for timing the behavioral response. Recently, Geinisman and co-workers *(69)* demonstrated that trace eye-blink conditioning results in an enlargement of postsynaptic density areas in axospinous nonperforated synapses of pyramidal cells in the CA1 stratum radiatum. In contrast, the total number of synapses was not altered. These observations indicate that learning involves the remodeling of synapses rather than net synaptogenesis, and PKCγ might participate in the mechanism underlying these morphological changes.

TRANSLOCATION OF PKCγ
AND LEARNING AND MEMORY

Protein kinase C activity is primarily regulated by translocation of the protein form the cytosol to the membrane. Phorbol esters activate PKC, and phorbol ester experiments strongly suggest that the increase in PKCγ-ir is indeed related to active or previously activated PKCγ. PKC stimulation by PDBu enhances PKCγ-ir and mimicks the effect of learning. This was observed in both mouse and rabbit tissue. Mildly fixed brain sections containing the hippocampus of naive mice were treated with 1.0 m*M* PDBu prior to immunocytochemical staining. This treatment induced a clear increase in PKCγ-ir to comparable levels as that seen in mice trained in the hole board *(37)*. The difference, however, was that PKCγ-ir was increased in every cell that expressed PKCγ, whereas in trained animals, an increase was found with variation in cell number, type, and brain region. Similarly, rabbit hippocampal slices were used to study whether the conditioning-induced increase in PKCγ-ir could be experimentally mimicked by activating PKCs through the phorbol ester PDBu *(48)*. Incubation (20 min) of hippocampal slices obtained from naive rabbits in a medium containing 1.0 mM PDBu resulted in a clear increase in the immunoreactivity for PKCγ as compared to control slices similarly treated but with the omission of PDBu from the incubation medium. As in the mouse brain sections, all subregions of the hippocampal formation showed this increase in PKCγ-ir, including the subicular area that showed no such increase after conditioning. OD measures of the immunostaining in slices revealed a significant ($p < 0.05$) increase in the CA1 pyramidal cell bodies and associated dendrites, resembling the differences between naive and trace conditioned animals. Immunoblots of control and PDBu-treated slices revealed that there was no generation of PKM (the isolated, approx 30-kDa catalytic fragment of PKC), no increase in the total amount of PKCγ, and no significant translocation (a translocation from the cytosol to the pellet fraction of 4.3% was found). These findings were strikingly similar to those observed after trace eye-blink conditioning. In contrast to PKCγ, PKCα revealed translocation from the cytosol to the pellet fraction of 17.1% in the slice experiment after PDBu treatment, indicating that PKC can translocate under our experimental conditions.

Western blot analyses were performed to examine whether the training-induced increase in PKCγ-ir could be the result of the generation of the constitutively active catalytic fragment PKM of PKCγ, to an increase in the total amount of PKCγ, or to translocation. The immunoblots showed no evidence for any of these possibilities. No immunoreactive bands were seen at 30 or 40–45 kDa, which would be indicative of proteolytic activation of PKCγ; the total content of PKCγ was similar in naive, pseudoconditioned and trace conditioned rabbits; and no significant translocation of PKCγ was seen at the 24 h postconditioning time-point studied *(48)*.

It should be noted that the increase in PKCγ-ir induced by spatial and associative learning is not restricted to the hippocampus, but is also present in certain regions of the neocortex and thalamus. Moreover, other learning tasks such as passive shock avoidance induce PKCγ-ir changes in the neocortex, but not in the hippocampus *(70)*, and delay eye-blink conditioning, the independent type of associative learning, induced an increase in PKCγ-ir in the dendrites of hippocampal pyramidal cells, but not in the pyramidal cell bodies or in dentate gyrus granule cells *(71)*. These results further demonstrate that the type of learning is directly related to the pattern of altered PKCγ-ir in brain.

Lack of significant PKCγ translocation in the above-described studies for spatial learning (hole board) and associative learning (trace eye-blink conditioning) may be the result of the choice of the 24-h posttraining time point, although Bank et al. *(17)* reported a translocation of PKC activity from the cytosol to the membrane in delay eye-blink conditioning (which, however, was not reproduced in later studies of the same laboratory *[29]*). A nonsignificant increase of 4.7% in cytosolic localization of PKCγ was observed in trace conditioned over pseudoconditioned rabbits *(48)*. We should stress that PKC translocation, often found to be a relatively rapid and transient process *(50)*, could have occurred at earlier time-points in the learning or consolidation process. Membrane-associated PKCγ could have been redistributed back to the cytosol by the 24-h postconditioning time-point. To examine whether translocation of PKCγ does occur at earlier time-points than the 24-h posttraining time-point during spatial learning, a hole-board experiment with rats was performed that were tested after 4 or 11 d of training *(72)*. Their rate of learning as well as their reference memory differed significantly between the groups at these time-points. Subcellular localization of PKCγ was studied by Western blotting 10 min after their last training trial. These experiments revealed that PKCγ did translocate from the cytosol to a membrane-associated fraction in the early acquisition group (4 d of training), but not in the group of rats that was at two-thirds of the maximal reference-memory score usually found in the hole board. In other words, PKCγ seems to be predominantly involved in spatial learning through translocation during the period of novel information processing. This translocation was only seen from the cytosol to the membrane-associated fraction, but not to the membrane-incorporated fraction. This is in contrast to the suggestion that prolonged translocation of PKC from the soluble to an integral membrane protein form plays an important role in memory processes *(16,73)*. Rather, it is in line with the observation of rapid redistribution of membrane-associated PKC to the cytosol after the dissociation of applied phorbol esters in vivo *(74,75)*.

Interestingly, trace eye-blink conditioning specifically enhanced synaptic transmission between CA3 and CA1 immediately after training, whereas this enhancement was no longer present 24 h after training *(76,77)*. These results are in line with the interpretation that the hippocampus, probably through PKCγ-mediated pathways, is primarily involved in the early processing of information.

A PROTEIN–PROTEIN INTERACTION MODEL
FOR THE REGULATION OF PKCγ IMMUNOREACTIVITY

Several mechanisms could be responsible for the increase in the exposure of the epitopes and, hence, the increase in PKCγ-ir. These include a conformational change of PKCγ, a change in protein–protein interaction between PKCγ and another protein, or a change in the phosphorylation state near or at the epitope of 36G9 and C19 on the PKCγ protein. The immunoblot analyses, however, do not favor the last possibility. We have proposed a hypothetical model based on a protein–protein interaction between PKCγ and a binding protein, which may or may not include the induction of conformational changes in PKCγ to explain our results (Fig. 3; ref. 7). In short, the equivalent increase in PKCγ-ir observed with the antibodies 36G9 and C19 indicates that the access of both antibodies was affected by learning.

Hippocampal input related to spatial orientation or to the CS–US association in eye-blink conditioning stimulates PKC-coupled receptors in pyramidal neurons, which results in transient membrane translocation induced by the generation of Ca^{2+} (increasing the affinity for PKCγ for phosphatidylserine [PS] at the C2 domain; step 1 in Fig. 3) and DAG (step 2). Binding of DAG overcomes the binding of an inhibitor (e.g., 14-3-3) and fully activates PKCγ by releasing the pseudosubstrate (black rectangle) from the catalytic domain. The inhibitor (14-3-3) loses its affinity for PKCγ and binds to other acceptors. 36G9 is unable to bind to PKCγ when both DAG and PS are attached to it *(44)*, whereas C19 can bind to PKCγ in this state. However, DAG is only transiently present (step 3), and PKCγ is redistributed back to the cytosol at the 24-h postconditioning time-point (step 4), ready for reactivation (the so-called "activation-prone cytosolic form" of PKCγ depicted in the box). PKCγ is still free from the inhibitor (hence, the increased binding of both 36G9 and C19), and this may reflect a form of persistent activation of the enzyme. Because of the loss of the inhibitor, this form of PKCγ may bind to intracellular acceptors (step 5), such as the Golgi apparatus, endoplasmic reticulum, or nucleus.

Sunayashiki-Kusuzaki et al. *(29)* showed that associative learning potentiates subsequent PKC activation in the rabbit hippocampus, which may be related to this "activation-prone" condition of PKCγ. A biochemical modification of PKCγ (e.g., a result of a conformational change, or a change in phosphorylation state) and/or a modification of 14-3-3 (e.g., as a result of sequestration, a change in phosphorylation state, or a change in content over time) may prohibit rebinding of 14-3-3 to PKCγ, explaining the relatively long-term nature of the change in PKCγ-ir. The overall effect of learning is a shift toward binding of 36G9 and C19 to previously undetectable PKCγ.

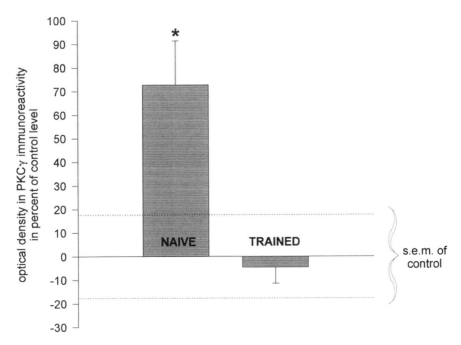

Fig. 4. Removal of an inhibitor putatively attached to PKCγ by pronase/pepsin treatment of mildly fixated hippocampal sections. In sections of naive rats, the treatment significantly increased PKCγ immunoreactivity, whereas it did not change PKCγ labeling in sections of trained rats.

A first set of data favors the model of the "activation-prone" PKCγ after learning. The inhibitor can be removed experimentally by degradation of the protein (e.g., induced by proteases). In this way, the epitope for 36G9 and C19 can be made accessible. We used a pronase/pepsine treatment applied to hippocampal sections of hole-board-trained and naive rats to induce limited protein degradation (0.0025% pronase followed by 0.1% pepsin, each for 25 min at room temperature). Our protein–protein interaction model, with a shift to more PKCγ without an inhibitor attached to it shortly after learning than in naive subjects, predicts a higher impact of the protease treatment on PKCγ-ir in sections of naive than of trained rats. Indeed, PKCγ-ir was increased significantly ($p < 0.05$) with 73% in CA1 pyramidal cells of naive rats, but slightly decreased PKCγ-ir with 8% in the hippocampus of trained rats (Fig. 4). This was found for both 36G9 and C19. These results demonstrate a difference in the PKCγ protein between naive and trained rats, which can best be explained by the loss of an inhibitor after training. Furthermore, it should be noted that the change in PKCγ of trained subjects is protein-specific; that is, similar protease experiments for other proteins coexpressed in PKCγ-positive hippocampal cells (such as heat-shock proteins) did not reveal any difference between naive and trained rats.

In summary, the immunocytochemical approach to decipher the role of PKCγ in learning and memory offers some new insights. The learning-induced changes in PKCγ-ir can be set in motion by several transmitter systems. In the hippocampus, especially the cholinergic and glutamatergic systems may act in tandem to regulate PKCγ *(7)*. PKCγ, therefore, can be viewed as the point of focus of postsynaptic signal transduction, playing a key role in learning and memory functions. The proposed protein–protein interaction model also addresses another issue: the intracellular direction (trafficking) of PKC toward its substrates. Because PKC has many substrates (cytoskeleton, endoplasmic reticulum, Golgi apparatus, nucleus) that are at a relatively far distance from the site of signal transduction (e.g., the dendritic spine), migration of the PKC molecule from this site to other intracellular targets while retaining, to a certain extent, its activational state seems highly relevant. The activation-prone form of PKCγ may rapidly interact with acceptors (inducing the release of the pseudosubstrate, and hence inducing the catalytic activity of the enzyme), which were unable to take place in the presence of an inhibitor. Future research on the function of PKCγ in learning and memory at the level of intracellular PKCγ pathways and protein–protein interactions with other interacting substrates may, therefore, eventually shed more light on the mechanisms underlying cognitive behavior.

ACKNOWLEDGMENTS

The authors thank Ina Everts and Susanne La Fleur for their contributions to the manuscript.

REFERENCES

1. Nishizuka, Y. (1986) Studies and perspectives of protein kinase C. *Science* **233**, 305–312.
2. Nishizuka, Y. (1988) The molecular heterogeneity of protein kinase C and its implications for cellular recognition. *Nature* **334**, 661–664.
3. Van der Zee, E. A. and Douma, B. R. K. (1997) Historical review of research on protein kinase C in learning and memory. *Prog. Neuro-Psychopharmacol. Biol. Psychiatry* **21**, 379–406.
4. Nishizuka, Y. (1995) Protein kinase C and lipid signaling for sustained cellular responses. *FASEB J.* **9,** 484–496.
5. Pears, C. (1995) Structure and function of the protein kinase C gene family. *J. Biosci.* **20**, 311–332.
6. Yoshida, Y., Huang, F. L., Nakabayashi, H., and Huang, K.-P. (1988) Tissue distribution and developmental expression of protein kinase C isozymes. *J. Biol. Chem.* **263**, 9868–9973.

7. Van der Zee, E. A., Luiten, P. G. M., and Disterhoft, J. F. (1997) Learning-induced alterations in hippocampal PKC-immunoreactivity: a review and hypothesis of its functional significance. *Prog. Neuro-Psychopharmacol. Biol. Psychiatry* **21**, 531–572.

8. Tanaka, C. and Saito, N. (1992) Localization of subspecies of protein kinase C in the mammalian central nervous system. *Neurochem. Int.* **21**, 499–512.

9. De Riemer, S. A., Strong, J. A., Albert, K. A., Greengard, P., and Kaczmarek, L. K. (1985) Enhancement of calcium current in Aplysia neurones by phorbol ester and protein kinase C. *Nature* **313**, 313–316.

10. Baraban, J. M., Snyder, S. H., and Alger, B. E. (1985) Protein kinase C regulates ionic conductance in hippocampal pyramidal neurons: electrophysiological effects of phorbol esters. *Proc. Natl. Acad. Sci. USA* **82**, 2538–2542.

11. Bliss, T. V. P. and Collingridge, G. L. (1993) A synaptic model of memory: long-term potentiation in the hippocampus. *Nature* **361**, 31–39.

12. Linden, D. J. and Routtenberg, A. (1989) The role of protein kinase C on long-term potentiation: a testable model. *Brain Res. Rev.* **14**, 279–296.

13. Seifert, R., Schächtele, C., and Schultz, G. (1987) Activation of protein kinase C by cis- and trans-octadecadienoic acids in intact human platelets and its potentiation by diacylglycerol. *Biochem. Biophys. Res. Commun.* **149**, 762–768.

14. Liscovitch, M. (1992) Crosstalk among multiple signal-activated phospholipases. *Trends Biochem. Sci.* **17**, 393–399.

15. Asaoka, Y., Nakamura, S., Yoshida, K., and Nishizuka, Y. (1992) Protein kinase C, calcium and phospholipid degradation. *Trends Biochem. Sci.* **17**, 414–417.

16. Alkon, D. L. and Rasmussen, H. (1988) A spatial–temporal model of cell activation. *Science* **239**, 998–1005.

17. Bank, B., Deweer, A., Kuzirian, A. M., Rasmussen, H., and Alkon, D. L. (1988) Classical conditioning induces long-term translocation of protein kinase C in rabbit hippocampal CA1 cells. *Proc. Natl. Acad. Sci. USA* **85**, 1988–1992.

18. Golski, S., Olds, J. L., Mishkin, M., Olton, D. S., and Alkon, D. L. (1995) Protein kinase C in the hippocampus is altered by spatial but not cued discriminations: a component task analysis. *Brain Res.* **676**, 53–62.

19. Lucchi, L., Pascale, A., Battaini, F., Govoni, S., and Trabucchi, M. (1993) Cognition stimulating drugs modulate protein kinase C activity in cerebral cortex and hippocampus of adult rats. *Life Sci.* **53**, 1821–1832.

20. Mathis, C., Lehmann, J., and Ungerer, A. (1992) The selective protein kinase C inhibitor, NPC 15437, induces specific deficits in memory retention in mice. *Eur. J. Pharmacol.* **220**, 107–110.

21. Noguès, X., Micheau, J., and Jaffard, R. (1994) Protein kinase C activity in the hippocampus following spatial learning tasks in mice. *Hippocampus* **4**, 71–78.

22. Noguès, X., Jaffard, R., and Micheau, J. (1996) Investigations on the role of hippocampal protein kinase C on memory processes: pharmacological approach. *Behav. Brain Res.* **75**, 139–146.

23. Olds, J. L. and Alkon, D. L. (1991) A role for protein kinase C in associative learning. *New Biol.* **3**, 27–35.

24. Olds, J. L., Anderson, M. L., McPhie, D. L., Staten, L. D., and Alkon, D. L. (1989) Imaging of memory-specific changes in the distribution of protein kinase C in the hippocampus. *Science* **245**, 866–869.

25. Olds, J. L., Golski, S., McPhie, D. L., Olton, D., Mishkin, M., and Alkon, D. L. (1990) Discrimination learning alters in the distribution of protein kinase C in the hippocampus of rats. *J. Neurosci.* **10**, 3707–3713.
26. Paylor, R., Rudy, J. W., and Wehner, J. M. (1991) Acute phorbol ester treatment improves spatial learning performance in rats. *Behav. Brain Res.* **45**, 189–193.
27. Paylor, R, Morrison, S. K., Rudy, J. W., Waltrip, L. T., and Wehner, J. M. (1992) Brief exposure to an enriched environment improves performance on the Morris water task and increases hippocampal cytosolic protein kinase C activity in young rats. *Behav. Brain Res.* **52**, 49–59.
28. Scharenberg, A. M., Olds, J. L., Schreurs, B. G., Craig, A. M., and Alkon, D. L. (1991) Protein kinase C redistribution within CA3 stratum oriens during acquisition of nictitating membrane conditioning in the rabbit. *Proc. Natl. Acad. Sci. USA* **88**, 6637–6641.
29. Sunayashiki-Kusuzaki, K., Lester, D. S., Schreurs, B. G., and Alkon, D. L. (1993) Associative learning potentiates protein kinase C activation in synaptosomes of the rabbit hippocampus. *Proc. Natl. Acad. Sci. USA* **90,** 4286–4289.
30. Wehner, J. M., Sleight, S., and Upchurch, M. (1990) Hippocampal protein kinase C activity is reduced in poor spatial learners. *Brain Res.* **523**, 181–187.
31. Spieler, K., Schoch, P., Martin, J. R., and Haefely, W. (1993) Environmental stimulation promotes changes in the distribution of phorbol ester receptors. *Pharmacol. Biochem. Behav.* **46**, 553–560.
32. Bernabeu, B., Izquierdo, I., Cammarota, M., Jerusalinsky, D., and Medina, J. H. (1995) Learning-specific, time-dependent increase in [^{3}H]phorbol dibutyrate binding to protein kinase C in selected regions of the rat brain. *Brain Res.* **685**, 163–168.
33. Oades, R. D. (1981) Types of memory or attention? Impairments after lesions of the hippocampus and limbic ventral tegmentum. *Brain Res. Bull.* **7**, 221–226.
34. Oades, R. D. and Isaacson, R. L. (1978) The development of food search behavior by rats: the effects of hippocampal damage and haloperidol. *Behav. Biol.* **24**, 327–337.
35. Beldhuis, H. J. A., Everts, H. G. J., Van der Zee, E. A., Luiten, P. G. M., and Bohus, B. (1992) Amygdala kindling-induced seizures selectively impair spatial memory. 1. Behavioral characteristics and effects on hippocampal neuronal protein kinase C isoforms. *Hippocampus* **2**, 397–410.
36. Douma, B. R. K., Van der Zee, E. A., and Luiten, P. G. M. (1997) Changes in hippocampal PKCγ, mAChR- and trkB-immunoreactivity after a spatial learning paradigm, in *Neurochemistry: Cellular, Molecular and Clinical Aspects* (Teelken, A. and Korf, J., eds.), Plenum, New York, pp 507–511.
37. Van der Zee, E. A., Compaan, J. C., De Boer, M., and Luiten, P. G. M. (1992) Changes in PKCγ immunoreactivity in mouse hippocampus induced by spatial learning. *J. Neurosci.* **12**, 4808–4815.
38. Van der Zee, E. A., Compaan, J. C., Bohus, B., and Luiten, P. G. M. (1995) Alterations in the immunoreactivity for muscarinic acetylcholine receptors and colocalized PKCγ in mouse hippocampus induced by spatial discrimination learning. *Hippocampus* **5**, 349–362.
39. Solomon, P. R., Vander Schaaf, E., Thompson, R. F., and Weisz, D. J. (1986) Hippocampus and trace conditioning of rabbit's classical conditioned nictitating membrane response. *Behav. Neurosci.* **100**, 729–744.

40. Moyer, J. R., Jr., Deyo, R. A., and Disterhoft, J. F. (1990) Hippocampectomy disrupts trace eye-blink conditioning in rabbits. *Behav. Neurosci.* **104**, 243–252.

41. Kim, J. J., Clark, R. E., and Thompson, R. F. (1995) Hippocampectomy impairs the memory of recently, but not remotely, acquired trace eyeblink conditioned responses. *Behav. Neurosci.* **109**, 195–203.

42. McEchron, M. D. and Disterhoft, J. F. (1997) Sequence of single neuron changes in CA1 hippocampus of rabbits during acquisition of trace eyeblink conditioned responses. *J. Neurophysiol.* **78**, 1030–1044.

43. Cazaubon, S., Marais, R. M., Parker, P., and Strosberg, A. D. (1989) Monoclonal antibodies to protein kinase Cγ. Functional relationship between epitopes and cofactor binding sites. *Eur. J. Biochem.* **182**, 401–406.

44. Cazaubon, S., Webster, C., Camoin, L., Strosberg, A. D., and Parker, P. (1990) Effector dependent conformational changes in protein kinase Cγ through epitope mapping with inhibitory monoclonal antibodies. *Eur. J. Biochem.* **194,** 799–804.

45. Robinson, K., Jones, D., Patel, Y., Martin, H., Madrazo, J., Martin, S., Howell, S., Elmore, M., Finnen, M. J., and Aitken, A. (1994) Mechanism of inhibition of protein kinase C by 14-3-3 isoforms. 14-3-3 isoforms do not have phospholipase A2 activity. *Biochem. J.* **299**, 853–861.

46. Melner, M. H. (1996) Physiological inhibitors of protein kinase C. *Biochem. Pharmacol.* **51**, 69–877.

47. Van der Zee, E. A., Kronforst-Collins, M. A., and Disterhoft, J. F. (1996) Associative learning down-regulates PKCβ2- and γ-immunoreactivity in astrocytes. *NeuroReport* **7**, 2753–2756.

48. Van der Zee, E. A., Kronforst-Collins, M. A., Maizels, E. T., Hunzicker-Dunn, M., and Disterhoft, J. F. (1997) γ Isoform-selective changes in PKC-immunoreactivity after trace eyeblink conditioning in the rabbit hippocampus. *Hippocampus* **7**, 271–285.

49. Moyer, J. R., Jr., Thompson, L. T., and Disterhoft, J. F. (1996) Trace eyeblink conditioning increases CA1 excitability in a transient and learning-specific manner. *J. Neurosci.* **16**, 5536–5546.

50. Newton, A. C. (1995) Protein kinase C: structure, function, and regulation. *J. Biol. Chem.* **270**, 28,495–28,498.

51. Bowers, B. J., Christensen, S. C., Pauley J. R., Paylor, R., Yuva, L., Dunbar, S. E., and Wehner, J. M. (1995) Protein and molecular characterization of hippocampal protein kinase C in C57BL/6 and DBA/2 mice. *J. Neurochem.* **64**, 2737–2746.

52. Angenstein, F., Riedel, G., Reymann, K. G., and Staak, S. (1994) Hippocampal long-term potentiation in vivo induces translocation of protein kinase Cγ. *NeuroReport* **5**, 381–384.

53. Klann, E., Chen, S.-J., and Sweatt, J. D. (1993) Mechanism of protein kinase C activation during the induction and maintenance of long-term potentiation probed using a selective peptide substrate. *Proc. Natl. Acad. Sci. USA* **90**, 8337–8341.

54. Huang, K.-P., Nakabayashi, H., and Huang, F. L. (1986) Isozymic forms of rat brain Ca^{2+}-activated and phospholipid-dependent protein kinase. *Proc. Natl. Acad. Sci. USA* **83**, 8535–8539.

55. Thomas, K. L., Laroche, S., Errington, M. L., Bliss, T. V. P., and Hunt, S. P. (1994) Spatial and temporal changes in signal transduction pathways during LTP. *Neuron* **13**, 737–745.

56. Hotson, J. R. and Prince, D. A. (1980) A calcium-activated hyperpolarization follows repetitive firing in hippocampla neurons. *J. Neurophysiol.* **43**, 409–419.

57. Lancaster, B. and Adams, P. R. (1986) Calcium-dependent current generating the afterhyperpolarization of hippocampal neurons. *J. Neurophysiol.* **55**, 1268–1282.

58. Storm, J. F. (1990) Potassium currents in hippocampal pyramidal cells. *Prog. Brain Res.* **83**, 161–187.

59. Disterhoft, J. F., Coulter, D. A., and Alkon, D. L. (1986) Conditioning-specific membrane changes of rabbit hippocampal neurons measured in vitro. *Proc. Natl. Acad. Sci. USA* **83**, 2733–2737.

60. Disterhoft, J. F., Golden, D. T., Read, H. L., Coulter, D. A., and Alkon, D. L. (1988) AHP reductions in rabbit hippocampal neurons during conditioning correlate with acquisition of the learned response. *Brain Res.* **462**, 118–125.

61. Thompson, L. T., Moyer, J. R., and Disterhoft, J. F. (1996) Transient changes in excitability of rabbit CA3 neurons with a time-course appropriate to support memory consolidation. *J. Neurophysiol.* **76**, 1836–1849.

62. Etcheberrigaray, R., Matzel, L. D. F., Lederhendler, I. I., and Alkon, D. L. (1992) Classical conditioning and protein kinase C activation regulate the same single potassium channel in Hermissenda crassicornis. *Proc. Natl. Acad. Sci. USA* **89**, 7184–7188.

63. Hu, G.-Y., Hvalby, O., Walaas, S. I., Albert, K. A., Skjeflo, P., Anderson, P., and Greengard, P. (1987) Protein kinase C injection into hippocampal pyramidal cells elicits features of long term potentiation. *Nature* **328**, 426–429.

64. Agopyan, N. and Agopyan, I. (1991) Effects of protein kinase C activators and inhibitors on membrane properties, synaptic responses, and cholinergic actions in CA1 subfield of rat hippocampus in situ and in vitro. *Synapse* **7**, 193–206.

65. Malenka, C., Madison, D. V., Andrade, R., and Nicoll, R. A. (1986) Phorbol esters mimic some cholinergic actions in hippocampal pyramidal neurons. *J. Neurosci.* **6**, 475–480.

66. Weiler, I. J. and Greenough, W. T. (1993) Metabotropic glutamate receptors trigger postsynaptic protein synthesis. *Proc. Natl. Acad. Sci. USA* **90**, 7168–7171.

67. Weiler, I. J., Hawrylak, N., and Greenough, W. T. (1995) Morphogenesis in memory formation: synaptic and cellular mechanisms. *Behav. Brain Res.* **66**, 1–6.

68. Geinisman, Y., De Toledo-Morrell, L., Morrell, F., Persina, I. S., and Beatty, M. A. (1996) Synapse restructuring associated with the maintenance phase of hippocampal long-term potentiation. *J. Comp. Neurol.* **368**, 413–423.

69. Geinisman, Y., Disterhoft, J. F., Gundersen, H. J. G., McEchron, M. D., Persina, I. S., Power, J. M., Van der Zee, E. A., and West, M. J. Structural substrate of hippocampus-dependent associative learning: remodeling of synapses, submitted.

70. Van der Zee, E. A., Douma, B. R. K., Bohus, B., and Luiten, P. G. M. (1994) Passive avoidance training induces enhanced levels of immunoreactivity for muscarinic acetylcholine receptor and coexpressed PKCγ and MAP-2 in rat cortical neurons. *Cereb. Cortex* **4**, 376–390.

71. Van der Zee, E. A., Palm, I. F., Kronforst, M. A., Maizels, E. T., Shanmugam, M., Hunzicker-Dunn, M., and Disterhoft, J. F. (1995) Trace and delay eyeblink conditioning induce alterations in the immunoreactivity for PKCγ in the rabbit hippocampus. *Soc. Neurosci. Abstr.* **21**, 1218.

72. Douma, B. R. K., Van der Zee, E. A., and Luiten, P. G. M. (1998) Translocation of protein kinase Cγ occurs during the early phase of acquisition of food rewarded spatial learning. *Behav. Neurosci.* **112**, 496–501.

73. Burgoyne, R. D. (1989) A role for membrane-inserted protein kinase C in cellular memory? *Trends Biochem. Sci.* **14,** 87–88.

74. Mosior, M. and Newton, A. C. (1995) Mechanism of interaction of protein kinase C with phorbol esters. Reversibility and nature of membrane association. *J. Biol. Chem.* **270**, 25,526–25,533.

75. Szallasi, Z., Smith, C. B., and Blumberg, P. M. (1994) Dissociation of phorbol esters leads to immediate redistribution to the cytosol of protein kinase C α and δ in mouse keratinocytes. *J. Biol. Chem.* **269**, 27,159–27,162.

76. Power, J. M., Thompson, L. T., Moyer, J. R., Jr., and Disterhoft, J. F. (1997) Enhanced synaptic transmission in CA1 hippocampus after eyeblink conditioning. *J. Neurophysiol.* **78**, 1184–1187.

77. Weiss, C., Kronforst-Collins, M. A., and Disterhoft, J. F. (1996) Activity of hippocampal pyramidal neurons during trace eyeblink conditioning. *Hippocampus* **6,** 192–209.

Part III

Neurodegeneration and Apoptosis

5

Neurotransmitter Receptor–G-Protein-Mediated Signal Transduction in Alzheimer's Disease

Richard F. Cowburn

INTRODUCTION

Alzheimer's disease is the most frequent cause of dementia in the elderly and is characterized clinically by a progressive loss of memory, intellect, and personality. The characteristic neuropathological hallmarks of Alzheimer's disease are the extracellular deposition of a 39–43 amino acid protein termed β-amyloid (or Aβ), in the cerebrovasculature *(1)* and cores of senile plaques *(2)*, as well as the formation of paired helical filaments (PHFs) that comprise intracellular neurofibrillary tangles (NFTs) *(3)*, neuropil threads, and senile plaque neurites. The principal component of PHFs is an abnormally hyperphosphorylated form of the microtubule-associated protein tau *(4)*. Other features of Alzheimer's disease pathology include neuronal and synaptic fallout that disrupts neurotransmission via the ascending cholinergic, noradrenergic, and serotonergic projections to the neocortex, as well as cortical excitatory amino acidergic pyramidal neurones.

It has become apparent in recent years that disrupted neurotransmission in Alzheimer's disease is unlikely to occur as a simple consequence of the disease process, but it may also contribute to the underlying mechanisms leading to senile plaque and NFT formation. This chapter provides a brief overview of the evidence from experimental and postmortem brain studies that have implicated altered neurotransmission via the receptor–G-protein-mediated phospholipase C (PLC)/protein kinase C (PKC) and adenylyl cyclase (AC)/protein kinase A (PKA) signaling pathways in Alzheimer's disease pathogenesis.

AMYLOID AND ALZHEIMER'S DISEASE

The Aβ peptide that comprises the cerebrovascular and senile plaque amyloid deposits is derived from a single transmembrane spanning glycoprotein, the amyloid precursor protein (APP), which is encoded by a gene on chromosome 21. Aβ is a natural metabolic product of APP generated by

From: *Cerebral Signal Transduction: From First to Fourth Messengers*
Edited by: M. E. A. Reith © Humana Press Inc., Totowa, NJ

the combined actions of two as yet unidentified proteases termed β-secretase, for that which cleaves APP just to the N-terminal of the Aβ sequence, and γ-secretase, which cleaves APP near the C-terminus of Aβ. APP can also be processed by an alternative nonamyloidogenic route that involves cleavage within the Aβ sequence by a so-called α-secretase. Metabolism of APP by α-secretase results in the release of a secreted amino terminal portion of APP (APPs) and precludes generation of full length Aβ (for a review, *see* ref. 5).

In recent years, the study of the basic pathogenic mechanisms underlying Alzheimer's disease has been dominated by the so-called "amyloid cascade hypothesis." In its simplest form, this hypothesis states that mismetabolism of APP to generate excessive Aβ deposition in the form of senile plaques provides the driving force for a cascade of events including NFT formation and neuronal degeneration, finally resulting in the clinical symptomatology of dementia that brings the patient to the physician *(6)*.

The neuronal degeneration seen in Alzheimer's disease may occur as a result of direct neurotoxicity of Aβ or the result of Aβ sensitizing neurones to other injurious stimuli *(7)*. Alternatively, neuronal degeneration in Alzheimer's disease may be caused by a decreased production of secreted APPs, there being evidence that this molecule plays a role in modulating neuronal excitability, synaptic plasticity, neurite outgrowth, synaptogenesis, and cell survival *(8)*.

The major support for the amyloid cascade hypothesis has come from the identification and understanding the effects of Alzheimer's disease causing mutations in rare autosomal dominant inherited forms of the disorder. Alzheimer's disease gene mutations have been identified in the APP gene as well as in the unrelated presenilin-1 (PS1) and presenilin-2 (PS2) genes located on chromosomes 14 and 1, respectively. Mutations in the APP gene produce different effects according to the mutation type, including an increased Aβ production, as seen with the so-called "Swedish" APP 670/671 mutation that occurs at the site for β-secretase processing of APP and an increased production of longer Aβ peptides (Aβ 1-42/43) that have an increased propensity to form fibrils and thus deposit in the brain, as seen with APP 717 mutations found at the C-terminus of Aβ. The pathogenic mechanisms of PS1 and PS2 gene mutations also involve an increased production of the longer (1-42/43) more fibrillogenic forms of Aβ. Because increased Aβ 1-42/43 production is seen both with APP and PS gene mutations, it has been suggested that this provides a common pathogenic step for early-onset familial Alzheimer's disease gene defects (reviewed in ref. 9).

Opponents of the amyloid cascade hypothesis claim that Aβ deposition occurs in the normal aged brain without evidence of dementia or neuronal damage and is therefore not a phenomenon specific for Alzheimer's disease. Also, the inability to establish a direct relationship (if it exists) between Aβ deposition and senile plaque development on the one hand and tau hyperphosphorylation and development of PHFs and NFTs on the other has led to the idea that these two pathologies may be coincidental or, as discussed later in this chapter, that they occur as a result of a third factor such as disturbed signal transduction *(10)*. This, together with studies showing that the number and distribution of NFTs rather than senile plaques correlate better with clinical symptomatology of dementia, has provided support for those wishing to study mechanisms involved in tau protein hyperphosphorylation and PHF formation *(11)*.

TAU PROTEIN AND ALZHEIMER'S DISEASE

The principle component of the PHFs, which make up NFTs, neuropil threads, and senile plaque neurites, is an abnormally hyperphosphorylated form of the microtubule-associated protein tau. Human brain tau occurs as six different isoforms that are normally located in the axons, where they bind to tubulin, thereby promoting microtubule assembly and stability. The microtubule stabilizing function of tau is greatly diminished by its hyperphosphorylation to PHF–tau, which binds poorly to tubulin. The concentration of PHF–tau has been shown to be increased in Alzheimer's disease cerebral cortex, whereas the concentration of normal tau is reduced, suggesting that the hyperphosphorylation of normal tau rather than an increased synthesis of tau protein is important in Alzheimer's disease *(12)*.

Both proline-directed and non-proline-dependent protein kinases are likely to be important for PHF–tau hyperphosphorylation *(10,13)*. The proline-directed protein kinases that have been studied with respect to PHF–tau phosphorylation include MAP kinase, cdc2 kinase, cdk2, cdk5, and glycogen synthase kinase-3 (GSK-3). Of these, GSK-3 has been shown to induce Alzheimer's disease-like phosphorylation of tau following transfection of cells with the enzyme, suggesting that GSK-3 might be crucially involved in PHF–tau hyperphosphorylation *(14)*. Moreover, GSK-3 has been recently shown to be preferentially localized in NFT-bearing neurones *(15)*. Less attention has been given to the non-proline-dependent protein kinases such as cAMP-dependent PKA, PKC, casein kinases I and II, and calcium/calmodulin-dependent kinase II (CaM kinase II), it having been shown that tau is, in general, a poor substrate for these enzymes.

In addition to overactive kinases, it has been suggested that hyperphosphorylation of PHF–tau could occur as a result of decreased protein phosphatase activity in Alzheimer's disease. Protein phosphatases PP-2A and PP-2B have been shown to revert tau that has been in vitro hyperphosphorylated by

CaM kinase II, PKA, MAP kinase, and cdc2 kinase to a normal-like state *(13)*. PP-2A and PP-2B can also catalyze the dephosphorylation of Alzheimer's disease PHF–tau at select sites *(16)* and PP-2B shows a selectively reduced expression in NFT-bearing neurones *(17)*. Furthermore, phosphatase enzyme activities are reported to be reduced in Alzheimer's disease brain *(18)*.

Although hyperphosphorylation of tau prevents its binding to microtubules and is also thought to occur prior to PHF assembly, it has been argued that tau hyperphosphorylation *per se* is neither necessary nor sufficient for PHF formation *(19)*. In this respect, Goedert and colleagues have provided convincing evidence that early interactions between tau and sulfated glycosaminoglycans can promote recombinant tau assembly into Alzheimer-like filaments, prevent tau binding to microtubules, and also stimulate tau protein phosphorylation *(19,20)*.

DISTURBED SIGNAL TRANSDUCTION AS A CONTRIBUTOR TO ALZHEIMER'S DISEASE PATHOLOGY

As mentioned previously, disease-causing mutations in the APP and PS genes account for only a small minority of all Alzheimer's disease cases. In the vast majority of sporadic cases, it remains unclear as to the mechanisms that cause the presumed APP mismetabolism leading to senile plaque formation, as well as tau protein hyperphosphorylation to produce PHFs and NFTs. One factor that may underly the hallmark pathologies of sporadic Alzheimer's disease is that of altered signal transduction via the neurotransmitter receptor–G-protein-mediated PLC/PKC and AC/PKA pathways. The following section provides a brief overview of the experimental evidence showing how these signaling pathways contribute to APP metabolism and tau protein phosphorylation. A scheme of these pathways, adapted from a recent review by Lovestone and Reynolds *(10)*, is given in Fig. 1.

That metabolism of APP could be influenced by activation of signal transduction pathways was first suggested by studies showing that phorbol ester stimulation of PKC markedly increases APPs production in a variety of transfected cell-line systems *(21,22)*. Concomitant with an increased APPs secretion, PKC activation was shown to lower $A\beta$ production *(23)* (although not in all cell systems *[24,25]*) suggesting that PKC-stimulated α-secretase APP processing can have direct consequences for $A\beta$ production.

Nonamyloidogenic α-secretase processing of APP can also be modulated upstream of PKC following activation of a number of neurotransmitter-receptor types linked to the PLC-stimulated phosphoinositide hydrolysis pathway, including the acetylcholine muscarinic M1 and M3 *(26–28)* and glutamate metabotropic *(29,30)* receptor subtypes. Muscarinic M1, glutamate metabotropate, as well as serotonin $5HT_{2a}$ and $5HT_{2c}$ receptors have also

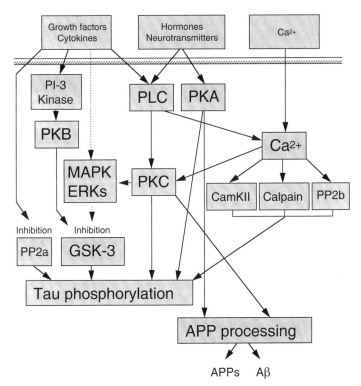

Fig. 1. Interplay between different signaling pathways implicated in APP metabolism and tau protein phosphorylation. An overview of how PLC/PKC and AC/PKA signaling influence APP metabolism and tau protein phosphorylation is given in the text. A description of the other pathways shown can be found in ref. *10*. (Redrawn and modified from ref. *10*.)

been demonstrated to modulate APPs production via phospholipase A2 (PLA2) activation *(31–33)* and there is also evidence that APP α-secretase processing can be regulated by altered calcium homeostasis *(34,35)*. Such studies, when taken together with evidence that the APPs' release is regulated by neuronal activity, have led to the notion that disturbed neurotransmission mediated by the above signaling pathways could limit α-secretase processing of APP to give an increased Aβ production *(36)*.

In an attempt to understand the mechanisms underlying phorbol–ester-stimulated APPs secretion, Xu and colleagues have provided evidence that PKC is involved in regulation of APP-containing secretory vesicle formation from the trans-Golgi network (TGN) *(37)*. Subsequent studies from this group have shown that cAMP-dependent PKA can converge with PKC at the level of TGN secretory vesicle formation to increase APPs production *(38)*. However, in apparent contrast, Efthimiopoulos and colleagues have provided evidence that increased intracellular cAMP inhibits both constitutive and phorbol–ester-stimulated APPs secretion in C6 glioma cells *(39)*. These discrepancies are likely to reflect differences in the experimental paradigms used to investigate PKA modulation of APP processing. The role of PKA in modulating neuronal processing of APP remains elusive. Evidence for a nonbeneficial effect of cAMP-mediated signaling on APP metabolism is also provided by a study showing that cAMP can regulate APP mRNA through transcriptional mechanisms *(40,41)* to give an accumulation of cell-associated APP holoprotein containing amyloidogenic Aβ peptides *(42)*.

In addition to the regulation of APP metabolism, both PKC and PKA have been implicated in tau protein hyperphosphorylation. Althought tau is a relatively poor substrate for these enzymes, it is thought that these non-proline-dependent protein kinases may be important for unmasking sites otherwise inaccessible to the proline-dependent kinases. In this respect, it has been shown that prior tau phosphorylation by PKA stimulates subsequent GSK-3 phosphorylation severalfold *(43,44)*. Phosphorylation of tau by PKA also inhibits its proteolytic breakdown by the calcium-activated protease, calpain *(45)*. Finally, as suggested by Lovestone and Reynolds, PKC may also influence tau protein phosphorylation by activating MAP kinases leading, in turn, to an inhibition of GSK-3 activity *(10)*.

Having provided a theoretical basis by which altered neurotransmitter receptor–G-protein-mediated PLC/PKC and AC/PKA signal transduction could exacerbate Alzheimer's disease senile plaque and NFT pathologies, I will now summarize the data showing how these signaling pathways are altered in Alzheimer's disease postmortem brain.

THE PHOSPHOINOSITIDE HYDROLYSIS
PATHWAY IN ALZHEIMER'S DISEASE

The phosphoinositide hydrolysis pathway (as shown in its simplest form in Fig. 2) has been shown to be disrupted at a number of levels in Alzheimer's disease brain. The first evidence suggesting that this pathway was impaired in Alzheimer's disease was provided by Stokes and Hawthorne, who reported large (30–50%) decreases in the levels of phosphatidylinositol, phosphatidylinositol 4-phosphate and phosphatidylinositol 4,5-bisphosphate in Alzheimer's disease anterior temporal cortex *(46)*. Subsequently, others reported that coupling of acetylcholine muscarinic M1 receptors, presumably to the G_q-protein that modulates PLC, was disrupted in Alzheimer's disease parietal *(47)* and frontal *(48)* cortices and thalamus *(49)*. More recently, Ladner and colleagues showed that acetylcholine muscarinic M1 receptor–G-protein coupling was disrupted in Alzheimer's disease superior frontal and pri-

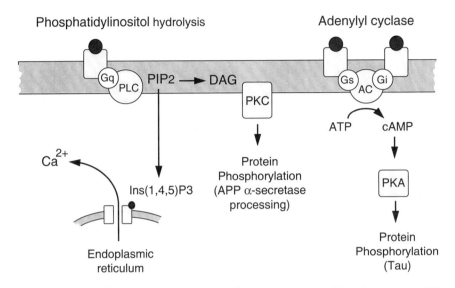

Fig. 2. A simplified scheme of the prinicipal components of the phosphoinositide hydrolysis and AC signal transduction pathways. The key components of the phosphoinositide hydrolysis pathway include the effector enzyme phospholipase C (PLC), the receptor activation of which is regulated by an intermediate heterotrimeric G-protein, termed Gq. Receptor and G-protein activation of phospholipase C results in the hydrolysis of phosphatidylinositol-4,5-bisphosphate (PIP$_2$) to give the second messengers diacylglycerol (DAG) and inositol(1,4,5)-trisphosphate (Ins[1,4,5]P$_3$). DAG, in turn, activates protein kinase C (PKC), whereas Ins(1,4,5)P$_3$ mediates intracellular calcium homeostasis. Ins(1,4,5)P$_3$ can be further phosphorylated to inositol(1,3,4,5)-tetrakisphosphate (Ins[1,3,4,5]P$_4$), which may act synergistically with Ins(1,4,5)P$_3$ to regulate intracellular calcium levels. Adenylyl cyclase (AC) activity can be either stimulated or inhibited, depending on the receptor type occupied. Receptor stimulation and inhibition of AC occurs via G$_s$- and G$_i$-proteins, respectively, and results in altered production of cyclic AMP (cAMP), an important intracellular second messenger for ion channel gating and protein phosphorylation via cAMP-dependent protein kinase A (PKA).

mary visual cortices, but not in dorsal striatum. Based on the differential pathology seen in these regions, these authors concluded that disrupted M1 receptor G-protein coupling occurs in those brain regions showing abundant Alzheimer's disease neuritic senile plaques rather than NFTs *(50)*. Data in support of a muscarinic receptor–G-protein uncoupling in Alzheimer's disease brain have also been provided by Cutler et al., who reported that the magnitude of carbachol-stimulated low K$_m$ GTPase activity in Alzheimer's disease basal ganglia, superior frontal gyrus, and hippocampus showed significant inverse relationships with disease duration *(51)*.

More direct evidence for an impaired regulation of the phosphoinositide hydrolysis pathway in Alzheimer's disease has been provided by studies looking at the regulation of phospholipase C (PLC) activity in Alzheimer's disease brain following stimulation by different neurotransmitter agonists and G-protein activators. Thus, Jope and colleagues reported that in Alzheimer's disease prefrontal cortex, the hydrolysis of exogenous [^{3}H]phosphatidylinositol was significantly reduced in response to a number of cholinergic agonists *(52–54)*, as well as following stimulation of glutamate metabotropate, histamine, and serotonin receptor types *(52)*. These authors also reported decreased [^{3}H]phosphatidylinositol hydrolysis in response to direct G-protein activation with GTPγS and aluminum fluoride *(52)*, which taken together with data from studies showing that G_q-protein α-subunit levels are relatively preserved in a number of Alzheimer's disease brain regions *(52,53, 55,56)*, suggests that disrupted neurotransmitter receptor–G_q-protein regulation of phosphoinositide hydrolysis in Alzheimer's disease brain likely results from an altered G_q-protein function.

A disregulated phosphoinositide hydrolysis has also been shown in Alzheimer's disease frontal cortex using physiological mixtures of exogenous phosphoinositides *(57)* and [^{3}H]phosphatidylinositol-bisphosphate *(58)* as enzyme substrates. However, it should be noted that others have failed to show this deficit *(59,60)*. A more extensive review of these works, together with a discussion of the experimental considerations of measuring phosphoinositide hydrolysis in postmortem brain, has been published recently *(61)*.

The most repeatedly demonstrated deficit in phosphoinositide hydrolysis signaling in Alzheimer's disease is that of an altered PKC regulation. The PKC deficit in Alzheimer's disease was first shown by Cole and colleagues, who reported significant reductions in both [^{3}H]phorbol-dibutyrate binding to PKC and PKC enzyme activities in particulate fractions of Alzheimer's disease frontal cortex *(62)*. The mechanism underlying the latter has been suggested to involve a defective sensitivity of the enzyme to its major cofactors phosphatidylserine, calcium, and diacylglycerol, rather than altered protein levels or abundance of different PKC isozymes *(63)*. However, data from other studies are more consistent with the idea that altered PKC activities in Alzheimer's disease brain reflect a loss of the protein. Thus, Masliah et al. showed decreased concentrations of the calcium-dependent PKCβII and PKCα isoforms in particulate fractions prepared from Alzheimer's disease hippocampus and mid-frontal cortex *(64)*. Similarly, Shimohama et al. reported significant reductions of PKCβ in membranous fractions from Alzheimer's disease temporal cortical tissue *(65)*. More recently, Matsushima et al. reported a significant reduction of PKCϵ in Alzheimer's disease temporal cortex that occurred in the absence of altered levels of two other calcium-independent isoforms, namely PKCδ and PKCζ *(66)*. In contrast, Wang et al. reported that PKCα and PKCβ were significantly increased in membrane fractions of Alzheimer's disease frontal cortex, whereas in soluble fractions from the same tissue, PKCβ levels were decreased *(67)*.

That PKC alterations in Alzheimer's disease represent an early phenomenon that has consequences for the development of hallmark pathologies of this disorder has been suggested by Masliah and colleagues. These workers showed that early or diffuse Aβ-containing plaques without PHF-containing neuritic components showed a strong immunostaining with anti-PKCβII antibodies that was associated with the membranes of cellular processes that extended into these structures *(68)*. This, together with data showing that both PKCβII levels and PKC-dependent phosphorylation of the endogenous substrate P86 were reduced in the neocortex of clinically nondemented individuals with cortical plaques, led these authors to conclude that there is a PKCβII aberration in Alzheimer's disease that precedes clinical deficits and that correlates better with neuritic plaque rather than NFT formation *(68)*.

In apparent contrast to the above, a recent study indicates that reduced PKC levels correlate well with the progression of Alzheimer's disease-related neurofibrillary changes. By performing [^{3}H]phorbol 12,13-dibutyrate autoradiography on sections of entorhinal cortex and anterior hippocampus that had been staged for disease pathology according to a protocol devised by Braak and Braak *(69)*, Kurumatani et al. showed that levels of total PKC in the entorhinal cortex, subiculum, CA1, CA2, CA3, and CA4 pyramidal layers, as well as dentate gyrus decline with the progressive accumulation of neurofibrillary changes, but not amyloid deposition in these regions *(70)*.

In addition to the above deficits, there is convincing evidence for a disruption of the phosphoinositide hydrolysis pathway at the level of the intracellular actions of the second messenger, inositol 1,4,5-trisphosphate (Ins[1,4,5]P$_3$). Young et al. and Garlind and colleagues have demonstrated significant reductions in the binding of [^{3}H]Ins(1,4,5)P$_3$ to receptor sites in homogenates of Alzheimer's disease parietal cortex, hippocampus, superior frontal, and superior temporal cortices, as well as cerebellum *(71,72)*. Subsequently, Haug and colleagues reported that reduced Ins(1,4,5)P$_3$ receptor protein levels in Alzheimer's disease temporal cortex, as determined by immunoblotting, correlated with a semiquantitative score for neuritic plaque and NFT accumulation *(73)*. More recently, Kurumatani and colleagues showed that loss of Ins(1,4,5)P$_3$ receptors in the entorhinal cortex, subiculum, and CA1 pyramidal layer of the anterior hippocampus correlated with the progression of both neurofibrillary changes and amyloid deposition *(70)*.

ADENYLYL CYCLASE IN ALZHEIMER'S DISEASE

In contrast to the multiple deficits seen in the phosphoinositide hydrolysis pathway, it appears that disrupted AC signaling (*see* Fig. 2 for the scheme) in Alzheimer's disease brain is more circumscribed in that it occurs primarily at the level of neurotransmitter receptor–G_s-protein–enzyme coupling. Somewhat analogous to the deficit in acetylcholine muscarinic M1 receptor–G_q-protein coupling described earlier, it has been shown using radioligand binding techniques that β_1-adrenoceptor coupling to presumed G_s-protein α-subunits is impaired in Alzheimer's disease temporal cortex *(74)*. Similarly, Wang and Friedman have reported a reduced ability of the β-adrenoceptor agonist isoprenaline to increase [^{35}S]GTPγS binding to $G_{s\alpha}$-subunits immunoprecipitated from Alzheimer's disease frontal cortex *(63)*.

More concrete evidence for an impaired G_s-protein function in Alzheimer's disease has come from studies that have assayed AC enzyme activities in the disorder. Using assay conditions favoring G_s-protein stimulation of the calcium/calmodulin-insensitive AC isoforms, it was reported that the G-protein activators, GTPγS, and aluminum fluoride gave markedly blunted stimulations of enzyme activity in membranes prepared from Alzheimer's disease hippocampus, frontal, temporal and occipital cortices, angular gyrus, and cerebellum, as compared to controls. In the same studies, it was found that forskolin stimulations of enzyme catalytic activity were not significantly different between Alzheimer's disease and control groups *(75–77)*. These studies have been interpreted as indicating a specific lesion at the level of G_s-protein–AC interactions. As such, they are in general agreement with those published by Ohm and colleagues on AC regulation in Alzheimer's disease hippocampus *(78,79)*. In contrast, Ross and colleagues failed to show an impaired G_s-protein–AC interaction in the Alzheimer's disease neocortex. This discrepancy has been suggested to reflect differences in assay conditions for an enzyme with many different isoforms and regulators *(80)*.

Of the above-described studies, it should be noted that Ohm and colleagues, in addition to finding an impaired G_s-protein regulation of AC, also showed reduced forskolin-stimulated enzyme activities in Alzheimer's disease hippocampus *(78)*. This suggestive loss of AC enzyme units in the disorder has been corroborated by more recent immunoblotting studies using isoform-specific antibodies. Thus, Yamamoto and colleagues have shown that in Alzheimer's disease parietal cortex, there is a significant loss of the AC type I and II, but not IV or V/VI, isoforms *(81,82)*. Moreover, loss of the calcium/calmodulin-sensitive type I isoform was shown to correlate with reduced basal and forskolin-stimulated enzyme activities measured in the presence, but not absence, of calcium and calmodulin *(82)*. These findings raise the possibility that any loss of AC enzyme units in Alzheimer's disease brain is restricted to calcium/calmodulin-sensitive forms of the enzyme.

The G_s-protein–AC disregulation seen in Alzheimer's disease brain does not appear to result from gross changes in total G_s-protein α-subunit levels *(77,80,83)*. However, in some brain regions, such as the hippocampus and angular gyrus, subtle changes in the number of high- and low-molecular-weight $G_{s\alpha}$ isoforms may be important *(77)*.

That altered AC activity may be related to the progression of Alzheimer's disease pathology has been recently suggested by Ohm and colleagues, who showed that in the mid-portion of the hippocampus there is a reduced forskolin-stimulated enzyme activity that occurs prior to the development of severe neurofibrillary changes *(84)*.

The notion that disrupted receptor G-protein regulation of AC occurs selectively at the level of the stimulatory G_s-protein input to the enzyme has received support from a number of studies showing that the inhibitory coupling of receptors such as the noradrenaline α_2 *(85)*, serotonin 5-HT_{1A} *(86)*, adenosine A_1 *(79)*, somatostatin *(87)*, and κ1 opiate *(88)* types to presumed G_i and G_o proteins appears relatively intact in Alzheimer's disease neocortex. In comparison, recent studies have shown that in Alzheimer's disease superior temporal cortex, the number of high-affinity, G-protein-coupled acetylcholine muscarinic M2 receptor sites appears even to be increased *(89,90)*. In contrast, the integrity of muscarinic M2 receptor G-protein coupling has been shown to be compromised in Alzheimer's disease hippocampus, a region that typically shows severe senile plaque and NFT pathology *(89)*.

Also, in contrast to the G_s-protein–AC dysfunction seen in Alzheimer's disease, it has been reported that inhibition of the enzyme following either direct G_i-protein activation *(91)* or stimulation of either somatostatin or adenosine A_1 receptor types *(79)* is relatively preserved in a number of Alzheimer's disease brain regions.

Unlike the well-documented PKC deficit described for the phosphoinositide hydrolysis pathway, it appears that AC signaling deficits in Alzheimer's disease brain are not accompanied by widespread changes in PKA levels and activity. Meier-Ruge et al. were the first to show comparable PKA activities in crude tissue homogenates of Alzheimer's disease and control temporal cortex *(92)*, whereas Bonkale et al. showed that particulate and soluble PKA activities were unaltered in Alzheimer's disease superior temporal cortex and cerebellum *(75)*. In a recent autoradiographic study, Bonkale et al. showed that [^{3}H]cAMP binding to soluble and particulate forms of PKA did not show any changes in the dentate gyrus and CA1–CA4 subfields of the hippocampus with staging for either Alzheimer's disease neurofibrillary changes or amyloid deposition. In contrast, apparent levels of soluble PKA in the entorhinal cortex were found to be decreased with the progression of both of these pathologies *(93)*. In an immunohistochemical study, Licameli et al. reported similar staining intensities for the regulatory (RIIβ) subunits of type II PKA in Alzheimer's disease and control hippocampal pyramidal neurones and in nonpyramidal neurones of the amygdala and putamen. In contrast, RIIβ immunoreactivity was substantially decreased in the superior temporal and occipital cortices, but not in the frontal cortex of Alzheimer's disease cases, compared to controls *(94)*.

DISTURBED SIGNAL TRANSDUCTION
AS A CONSEQUENCE OF THE DISEASE PROCESS

From the above discussion, it is clear that both the receptor–G-protein, mediated PLC/PKC and AC/PKA pathways are severely disrupted in Alzheimer's disease brain. So far, disruptions to these pathways have been presented from the point of view of contributing to Alzheimer's disease senile plaque and NFT formation by enhancing Aβ production and tau protein hyperphosphorylation, respectively. The alternative possibility worth considering is that impaired signal transduction occurs as a consequence of Alzheimer's disease pathology (e.g., as a result of the effects of Aβ). In this respect, it has been shown that Aβ exerts multiple effects on cellular calcium homeostasis, either by interactions with existing calcium channels or by *de novo* channel formation (for a review, *see* ref. *95*). Studies in intact cells have shown that physiological levels of Aβ can stimulate PKC translocation and activity *(96)* and also enhance phosphoinositide hydrolysis by amplifying depolarization-induced calcium responses *(97)*. An effect of Aβ on phosphoinositide hydrolysis signaling is also suggested from a study showing that the peptide can enhance $[^3H]Ins(1,4,5)P_3$ and $[^3H]Ins(1,3,4,5)P_4$ binding to their respective receptor sites *(98)*. A preliminary report has also provided evidence that Aβ can stimulate GTPase and AC activities in rat hippocampal and cortical membrane preparations *(99)*.

It is feasible that prolonged Aβ activations of the above systems could result in their downregulation to give the pattern of dysfunction seen in Alzheimer's disease brain. In this context, it is worth noting that Kelly and colleagues have shown that exposure of rodent cortical neurones to subtoxic levels of Aβ leads to an attenuation of carbachol-induced GTPase activity and phosphoinositide hydrolysis by a mechanism thought to involve free-radical-induced uncoupling of muscarinic receptors *(100)*.

Apart from the effects of Aβ, it is feasible that signal transduction disturbances in Alzheimer's disease occur also as a consequence of a disrupted cytoskeleton. The work of Rasenick and colleagues has shown that tubulin can regulate AC activity through an interaction with the G-protein α-sub-

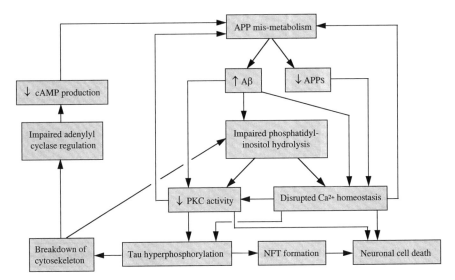

Fig. 3. The pivotal role of altered signal transduction in Alzheimer's disease pathogenesis.

units, $G_{\alpha s}$ and $G_{\alpha i1}$ *(101,102)*. In this scenario, tubulin activates these G-proteins by transferring GTP and stabilizing the active nucleotide-bound $G\alpha$ conformation. Tubulin has been shown to regulate PLC activity by a similar mechanism thought to involve GTP transfer from tubulin to $G_{\alpha q}$ *(101)*. Interestingly, Khatoon and colleagues have shown that tau can promote GTP binding to tubulin and that tubulin prepared from Alzheimer's disease brain has less bound GTP compared to control brain tubulin *(103)*. Taken together, the above studies provide strong suggestive evidence that impaired G-protein modulation of AC and PLC activities in Alzheimer's disease brain may result from decreased GTP transfer from tubulin to G-protein α-subunits that occurs as a consequence of the lowered binding of PHF–tau to tubulin.

CONCLUSIONS

This chapter has provided an overview of current knowledge as to the integrity of receptor–G-protein-mediated PLC/PKC and AC/PKA signaling in Alzheimer's disease brain. Given the involvement of these signaling pathways in APP metabolism and tau protein phosphorylation, together with evidence of the detrimental effects of $A\beta$ peptides on certain components of these pathways, a number of workers have suggested scenarios whereby signal transduction abnormalities play a pivotal role in exacerbating the progressive neurodegeneration seen in Alzheimer's disease *(10,54,104)*. One such scheme that attempts to incorporate some of the ideas discussed in this chapter is shown in Fig. 3. Although speculative, this scheme provides a basis by which disturbed neurotransmission via receptor–G-protein-mediated signaling pathways could contribute to APP mismetabolism and tau protein hyperphosphorylation. The subsequent increase in Aβ production and breakdown of the cytoskeleton would, in turn, undermine the integrity of these signaling pathways thereby further compromising neuronal viabil-ity. Such a scheme with disturbed signal transduction playing a pivotal role in exacerbating Alzheimer's disease neurodegeneration can be easily elaborated to include other proposed pathogenic mechanisms, including the effects of oxidative stress *(104)*, altered membrane composition and fluidity *(105)*, and differential effects of apolipoprotein E isoforms *(106)*.

Finally, recent studies have shown that Alzheimer's disease causing APP 717 and PS gene mutations can cause apoptotic cell death by activating a G_o-protein-mediated signaling pathway (reviewed in ref. *107*). This, together with data showing that APP and PS mutations exert effects on cellular calcium homeostasis *(108)* and regulation of AC activity *(109)*, raises the intriguing possibility that abnormal signal transduction may also play a role in the rare genetic forms of Alzheimer's disease.

ACKNOWLEDGMENTS

The author would like to thank the following organizations for funding research in laboratory: the Swedish Medical Research Council (K97-19X-12244-01A), the European Union Biomed-2 concerted action (contract number bmh4-ct-96-0162). the Swedish Alzheimer's disease fund, the Karolinska Institute research fund, Loo and Hans Ostermans Fund, Stiftelsen för Gamla Tjänarinnor, and the Sigurd and Elsa Goljes, Gun and Bertil Stohnes, and Åke Wibergs foundations.

REFERENCES

1. Glenner, G. G. and Wong, C. W. (1984) Alzheimer's disease: initial report of the purification and characterization of a novel cerebrovascular amyloid protein. *Biochem. Biophys. Res. Commun.* **120,** 885–890.
2. Masters, C. L., Simms, G., Weinman, N. A., Multhaup, G., McDonald, B. L., and Beyreuther, K. (1985) Amyloid plaque core protein in Alzheimer disease and Down syndrome. *Proc. Natl. Acad. Sci. USA* **82,** 4245–4249.
3. Kidd, M. (1963) Paired helical filaments in electron microscopy of Alzheimer's Disease. *Nature* **197,** 192–193.
4. Grundke-Iqbal, I., Iqbal, K., Tung, Y. C., Quinlan, M., Wisniewski, H. M., and Binder, L. I. (1986) Abnormal phosphorylation of the microtubule-associated protein tau (tau) in Alzheimer cytoskeletal pathology. *Proc. Natl. Acad. Sci. USA* **83,** 4913–4917.
5. Selkoe, D. J. (1996) Cell biology of the beta-amyloid precursor protein and the genetics of Alzheimer's disease. *Cold Spring Harb. Symp. Quant. Biol.* **61,** 587–596.

6. Hardy, J. A. and Higgins, G. A. (1992) Alzheimer's disease: the amyloid cascade hypothesis. *Science* **256,** 184–185.

7. Iversen, L. L., Mortishire-Smith, R. J., Pollack, S. J., and Shearman, M. S. (1995) The toxicity in vitro of beta-amyloid protein. *Biochem. J.* **311,** 1–16.

8. Mattson, M. P. (1997) Cellular actions of beta-amyloid precursor protein and its soluble and fibrillogenic derivatives. *Physiol. Rev.* **77,** 1081–1132.

9. Hardy, J. (1997) Amyloid, the presenilins and Alzheimer's disease. *Trends Neurosci.* **20,** 154–159.

10. Lovestone, S. and Reynolds, C. H. (1997) The phosphorylation of tau: a critical stage in neurodevelopment and neurodegenerative processes. *Neuroscience* **78,** 309–324.

11. Neve, R. L. and Robakis, N. K. (1998) Alzheimer's disease: a re-examination of the amyloid hypothesis. *Trends Neurosci.* **21,** 15–19.

12. Blennow, K. and Cowburn, R. F. (1996) The neurochemistry of Alzheimer's disease. *Acta Neurol. Scand.* **168**(Suppl), 77–86.

13. Goedert, M. (1996) Tau protein and the neurofibrillary pathology of Alzheimer's disease. *Ann. NY Acad. Sci.* **777,** 121–131.

14. Lovestone, S., Hartley, C. L., Pearce, J., and Anderton, B. H. (1996) Phosphorylation of tau by glycogen synthase kinase-3 beta in intact mammalian cells: the effects on the organization and stability of microtubules. *Neuroscience* **73,** 1145–1157.

15. Pei, J. J., Tanaka, T., Tung, Y. C., Braak, E., Iqbal, K., and Grundke-Iqbal, I. (1997) Distribution, levels, and activity of glycogen synthase kinase-3 in the Alzheimer disease brain. *J. Neuropathol. Exp. Neurol.* **56,** 70–78.

16. Gong, C.-X., Grundke-Iqbal, I., and Iqbal, K (1994) Dephosphorylation of Alzheimer's disease abnormally phosphorylated tau by protein phosphatase-2A. *Neuroscience* **61,** 765–772.

17. Pei, J. J., Sersen, E., Iqbal, K., and Grundke-Iqbal, I. (1994) Expression of protein phosphatases (PP-1, PP-2A, PP-2B and PTP-1B) and protein kinases (MAP kinase and P34cdc2) in the hippocampus of patients with Alzheimer disease and normal aged individuals. *Brain Res.* **655,** 70–76.

18. Gong, C.-X., Shaikh, S., Wang, J.-Z., Zaidi, T., Grundke-Iqbal, I., and Iqbal, K. (1995) Phosphatase activity toward abnormally phosphorylated τ: decrease in Alzheimer disease brain. *J. Neurochem.* **65,** 732–738.

19. Goedert, M., Jakes, R., Spillantini, M. G., Hasegawa, M., Smith, M. J., and Crowther, R. A. (1996) Assembly of microtubule-associated protein tau into Alzheimer-like filaments induced by sulphated glycosaminoglycans. *Nature* **383,** 550–553.

20. Hasegawa, M., Crowther, R. A., Jakes, R., and Goedert, M. (1997) Alzheimer-like changes in microtubule-associated protein Tau induced by sulfated glycosaminoglycans. Inhibition of microtubule binding, stimulation of phosphorylation, and filament assembly depend on the degree of sulfation. *J. Biol. Chem.* **272,** 33,118–33,124.

21. Caporaso, G. L., Gandy, S. E., Buxbaum, J. D., Ramabhadran, T. V., and Greengard, P. (1992) Protein phosphorylation regulates secretion of Alzheimer beta/A4 amyloid precursor protein. *Proc. Natl. Acad. Sci. USA* **89,** 3055–3059.

22. Checler, F. J. (1995) Processing of the beta-amyloid precursor protein and its regulation in Alzheimer's disease. *J. Neurochem.* **65,** 1431–1444.

23. Hung, A. Y, Haass, C., Nitsch, R. M., Qiu, W. Q., Citron, M., Wurtman, R. J., Growdon, J. H., and Selkoe, D. J. (1993) Activation of protein kinase C inhibits cellular production of the amyloid beta-protein. *J. Biol. Chem.* **268,** 22,959–22,962.

24. Dyrks, T., Mönning, U., Beyreuther, K., and Turner, J. (1994) Amyloid precursor protein secretion and beta A4 amyloid generation are not mutually exclusive. *FEBS Lett.* **349,** 210–214.

25. LeBlanc, A. C., Koutroumanis, M., and Goodyer, C. G. (1998) Protein kinase C activation increases release of secreted amyloid precursor protein without decreasing Abeta production in human primary neuron cultures. *J. Neurosci.* **18,** 2907–2913.

26. Buxbaum, J. D., Oishi, M., Chen, H. I., Pinkas-Kramarski, R., Jaffe, E. A., Gandy, S. E., and Greengard, P. (1992) Cholinergic agonists and interleukin 1 regulate processing and secretion of the Alzheimer beta/A4 amyloid protein precursor. *Proc. Natl. Acad. Sci. USA* **89,** 10,075–10,078.

27. Nitsch, R. M., Slack, B. E., Wurtman, R. J., and Growdon, J. H. (1992) Release of Alzheimer amyloid precursor derivatives stimulated by activation of muscarinic acetylcholine receptors. *Science* **258,** 304–307.

28. Wolf, B. A., Wertkin, A. M., Jolly, Y. C., Yasuda, R. P., Wolfe, B. B., Konrad, R. J., Manning, D., Ravi, S., Williamson, J. R., and Lee, V. M. (1995) Muscarinic regulation of Alzheimer's disease amyloid precursor protein secretion and amyloid beta-protein production in human neuronal NT2N cells. *J. Biol. Chem.* **270,** 4916–4922.

29. Jolly-Tornetta, C., Gao, Z. Y., Lee, V. M., and Wolf, B. A. (1998) Regulation of amyloid precursor protein secretion by glutamate receptors in human Ntera 2 neurons. *J. Biol. Chem.* **273,** 14,015–14,021.

30. Lee, R. K., Wurtman, R. J., Cox, A. J., and Nitsch, R. M. (1995) Amyloid precursor protein processing is stimulated by metabotropic glutamate receptors. *Proc. Natl. Acad. Sci. USA* **92,** 8083–8087.

31. Emmerling, M. R., Moore, C. J., Doyle, P. D., Carroll, R. T., and Davis, R. E. (1993) Phospholipase A2 activation influences the processing and secretion of the amyloid precursor protein. *Biochem. Biophys. Res. Commun.* **197,** 292–297.

32. Nitsch, R. M., Deng, M., Growdon, J. H., and Wurtman, R. J. (1996) Serotonin 5-HT2a and 5-HT2c receptors stimulate amyloid precursor protein ectodomain secretion. *J. Biol. Chem.* **271,** 4188–4194.

33. Nitsch, R. M., Deng, A., Wurtman, R. J., and Growdon, J. H. (1997) Metabotropic glutamate receptor subtype mGluR1alpha stimulates the secretion of the amyloid beta-protein precursor ectodomain. *J. Neurochem.* **69,** 704–712.

34. Buxbaum, J. D., Ruefli, A. A., Parker, C. A., Cypess, A. M., and Greengard, P. (1994) Calcium regulates processing of the Alzheimer amyloid protein precursor in a protein kinase C-independent manner. *Proc. Natl. Acad. Sci. USA* **91,** 4489–4493.

35. Querfurth, H. W. and Selkoe, D. J. (1994) Calcium ionophore increases amyloid beta peptide production by cultured cells. *Biochemistry* **33,** 4550–4561.

36. Nitsch, R. M. and Growdon, J. H. (1994) Role of neurotransmission in the regulation of amyloid beta-protein precursor processing. *Biochem. Pharmacol.* **47,** 1275–1284.

37. Xu, H., Greengard, P., and Gandy, S. (1995) Regulated formation of Golgi secretory vesicles containing Alzheimer beta-amyloid precursor protein. *J. Biol. Chem.* **270,** 23,243–23,245.

38. Xu, H., Sweeney, D., Greengard, P., and Gandy, S. (1996) Metabolism of Alzheimer beta-amyloid precursor protein: regulation by protein kinase A in intact cells and in a cell-free system. *Proc. Natl. Acad. Sci. USA* **93**, 4081–4084.

39. Efthimiopoulos, S., Punj, S., Manolopoulos, V., Pangalos, M., Wang, G. P., Refolo, L. M., and Robakis, N. K. (1996) Intracellular cyclic AMP inhibits constitutive and phorbol ester-stimulated secretory cleavage of amyloid precursor protein. *J. Neurochem.* **67**, 872–875.

40. Bourbonnière, M., Shekarabi, M., and Nalbantoglu, J. (1997) Enhanced expression of amyloid precursor protein in response to dibutyryl cyclic AMP is not mediated by the transcription factor AP-2. *J. Neurochem.* **68**, 909–916.

41. Shekarabi, M., Bourbonnière, M., Dagenais, A., and Nalbantoglu, J. (1997) Transcriptional regulation of amyloid precursor protein during dibutyryl cyclic AMP-induced differentiation of NG108-15 cells. *J. Neurochem.* **68**, 970–978.

42. Lee, R. K., Araki, W., and Wurtman, R. J. (1997) Stimulation of amyloid precursor protein synthesis by adrenergic receptors coupled to cAMP formation. *Proc. Natl. Acad. Sci. USA* **94**, 5422–5426.

43. Blanchard, B. J., Raghunandan, R., and Roder, H. M. (1994) Hyperphosphorylation of human tau by brain kinase PK40^{erk} beyond phosphorylation by cAMP-dependent PKA: relation to Alzheimer's disease. *Biochem. Biophys. Res. Commun.* **200**, 187–194.

44. Singh, T. J., Zaidi, T., Grundke-Iqbal, I., Iqbal, K. (1995) Modulation of GSK-3-catalyzed phosphorylation of microtubule-associated protein tau by non-proline-dependent protein kinases. *FEBS Lett.* **358**, 4–8.

45. Litersky, J. M. and Johnson, G. V. W. (1992) Phosphorylation by cAMP-dependent protein kinase inhibits the degradation of tau by calpain. *J. Biol. Chem.* **267**, 1563–1568.

46. Stokes, C. E. and Hawthorne, J. N. (1987) Reduced phosphoinositide concentrations in anterior temporal cortex of Alzheimer-diseased brains. *J. Neurochem.* **48**, 1018–1021.

47. Smith, C. J., Perry, E. K., Perry, R. H., Fairbairn, A. F., and Birdsall, N. J. M. (1987) Guanine nucleotide modulation of muscarininc cholinergic receptor binding in postmortem human brain—a preliminary study in Alzheimer's disease. *Neurosci. Lett.* **82**, 227–232.

48. Flynn, D. D., Weinstein, D. A., and Mash, D. C. (1991) Loss of high-affinity agonist binding to M1 muscarinic receptors in Alzheimer's disease: implications for the failure of cholinergic replacement therapies. *Ann. Neurol.* **29**, 256–262.

49. Warpman, U., Alafuzoff, I., and Nordberg, A. (1993) Coupling of muscarinic receptors to GTP proteins in postmortem human brain—alterations in Alzheimer's disease. *Neurosci. Lett.* **150**, 39–43.

50. Ladner, C. J., Celesia, C. G., Magnuson, D. J., and Lee, J. M. (1995) Regional alterations in M1 muscarinic receptor-G-protein coupling in Alzheimer's disease. *J. Neuropathol. Exp. Neurol.* **54**, 783–789.

51. Cutler, R., Joseph, J. A., Yamagami, K., Villalobos-Molina, R., and Roth, G. S. (1995) Area specific alterations in muscarinic stimulated low K_m GTPase activity in aging and Alzheimer's disease: implications for altered signal transduction. *Brain Res.* **664**, 54–60.

52. Greenwood, A. F., Powers, R. T. E., and Jope, R. S. (1995) Phosphoinositide hydrolysis, Gq alpha, phosholipase C, and protein kinase C in post mortem human

brain: effects of post mortem interval, subject age, and Alzheimer's disease. *Neuroscience* **69,** 125–138.

53. Jope, R. S., Song, L., Li, X., and Powers, R. (1994) Impaired phosphoinositide hydrolysis in Alzheimer's disease brain. *Neurobiol. Aging* **15,** 221–226.

54. Jope, R. S., Song, L., and Powers, R. E. (1997) Cholinergic activation of phosphoinositide signaling is impaired in Alzheimer's disease brain. *Neurobiol. Aging* **18,** 111–120.

55. Li, X., Greenwood, A. F., Powers, R., and Jope, R. S. (1996) Effects of postmortem interval, age, and Alzheimer's disease on G-proteins in human brain. *Neurobiol. Aging* **17,** 115–122.

56. Shanahan, C., Deasy, M., Ravid, R., and O'Neill, C. (1995) Comparative immunoblot analysis of the guanine nucleotide binding protein G_{qa} in control and Alzheimer's disease brains. *Biochem. Soc. Trans.* **23,** 363S.

57. Crews, F. T., Kurian, P., and Freund, G. (1994) Cholinergic and serotonergic stimulation of phosphoinositide hydrolysis is decreased in Alzheimer's disease. *Life Sci.* **55,** 1992–2002.

58. Ferraro-DiLeo, G. and Flynn, D. D. (1993) Diminished muscarinic receptor-stimulated [^{3}H]-PIP$_2$ hydrolysis in Alzheimer's disease. *Life Sci.* **53,** 439–444.

59. Alder, J. T., Chessell, I. P., and Bowen, D. M. (1995) A neurochemical approach for studying response to acetylcholine in Alzheimer's disease. *Neurochem. Res.* **20,** 769–771.

60. Wallace, M. A. and Claro, E. (1993) Transmembrane signalling through phospholipase C in human cortical membranes. *Neurochem. Res.* **18,** 139–145.

61. Jope, R. S. (1996) Cholinergic muscarinic receptor signaling by the phosphoinositide signal transduction system in Alzheimer's disease. *Alzheimer's Dis. Rev.* **1,** 2–14.

62. Cole, G., Dobkins, K., Hansen, L. A., Terry, R. D., and Saitoh, T. (1988) Decreased levels of protein kinase C in Alzheimer brain. *Brain Res.* **452,** 165–174.

63. Wang, H.-Y. and Friedmann, E. (1994) Receptor-mediated activation of G-proteins is reduced in postmortem brains from Alzheimer's disease patients. *Neurosci. Lett.* **173,** 37–39.

64. Masliah, E., Cole, G. M., Shimohama, S., Hansen, L. A., DeTeresa, R., Terry, R. D., and Saitoh, T. (1990) Differential involvement of protein kinase C isozymes in Alzheimer's disease. *J. Neurosci.* **10,** 2113–2124.

65. Shimohama, S., Narita, M., Matsushima, H., Kimura, J., Kameyama, M., Hagiwara, M., Hidaka, H., and Taniguchi, T. (1993) Assessment of protein kinase C isozymes by two-site enzyme immunoassay in human brains and changes in Alzheimer's disease. *Neurology* **43,** 1407–1413.

66. Matsushima, H., Shimohama, S., Chachin, M., Taniguchi, T., and Kimura, J. (1996) Ca^{2+}-dependent and Ca^{2+}-independent protein kinase C changes in the brains of patients with Alzheimer's disease. *J. Neurochem.* **67,** 317–323.

67. Wang, H.-Y., Pisano, M. R., and Friedman, E. (1994) Attenuation of protein kinase C activity and translocation in Alzheimer's disease brain. *Neurobiol. Aging* **15,** 293–298.

68. Masliah, E., Cole, G. M., Hansen, L. A., Mallory, M., Allbright, T., Terry, R. D., and Saitoh, T. (1991) Protein kinase C alteration is an early biochemical marker in Alzheimer's disease. *J. Neurosci.* **11,** 2759–2767.

69. Braak, H. and Braak, E. (1991) Neuropathological stageing of Alzheimer-related changes. *Acta Neuropathol. (Berlin)* **82**, 239–259.
70. Kurumatani, T., Fastbom, J., Bonkale, W. L., Bogdanovic, N., Winblad, B., Ohm, T. G., and Cowburn, R. F. (1998) Loss of inositol 1,4,5-trisphosphate receptor sites and decreased PKC levels correlate with staging of Alzheimer's disease neurofibrillary pathology. *Brain Res.* 796, 209–221.
71. Garlind, A., Cowburn, R. F., Forsell, C., Ravid, R., Winblad, B., and Fowler, C. J. (1995) Diminished [^{3}H]inositol(1,4,5)P$_3$ but not [^{3}H]inositol(1,3,4,5)P$_4$ binding in Alzheimer's disease brain. *Brain Res.* **681**, 160–166.
72. Young, L. T., Kish, S. J., Li, P. P., and Warsh, J. J. (1988) Decreased brain [^{3}H]inositol 1,4,5-trisphosphate binding in Alzheimer's disease. *Neurosci. Lett.* **94**, 198–202.
73. Haug, L.-S., Østvold, A. C., Cowburn, R. F., Garlind, A., Winblad, B., and Walaas, S. I. (1996) Decreased inositol(1,4,5)trisphosphate receptor levels in Alzheimer's disease cerebral cortex: selectivity of changes and correlation to pathological severity. *Neurodegeneration* **5**, 169–176.
74. Cowburn, R. F., Vestling, M., Fowler, C. J., Ravid, R., Winblad, B., and O'Neill, C. (1993) Disrupted β1-adrenoceptor–G-protein coupling in the temporal cortex of patients with Alzheimer's disease. *Neurosci. Lett.* **155**, 163–166.
75. Bonkale, W., Fastbom, J., Wiehager, B., Ravid, R., Winblad, B., and Cowburn, R. F. (1996) Impaired G-protein stimulated adenylyl cyclase activity in Alzheimer's disease brain is not accompanied by reduced cyclic-amp dependent protein kinase activity. *Brain Res.* **737**, 155–161.
76. Cowburn, R. F., O'Neill, C., Ravid, R., Alafuzoff, I., Winblad, B., and Fowler, C. J., (1992) Adenylyl cyclase activity in post-mortem human brain: evidence of altered G protein mediation in Alzheimer's disease. *J. Neurochem.* **58**, 1409–1419.
77. O'Neill, C., Fowler, C. J., Wiehager, B., Ravid, R., Winblad, B., and Cowburn, R. F. (1994) Region selective alterations in G-protein subunit levels in the Alzheimer's disease brain. *Brain Res.* **636**, 193–201.
78. Ohm, T. G., Bohl, J., and Lemmer, B. (1991) Reduced basal and stimulated (iso-prenaline, Gpp[NH]p, forskolin) adenylate cyclase activity in Alzheimer's disease correlated with histopathological changes. *Brain Res.* **540**, 229–236.
79. Schnecko, A., Witte, K. Bolh, J., Ohm, T., and Lemmer, B. (1994) Adenylyl cyclase activity in Alzheimer's disease brain: stimulatory and inhibitory signal transduction pathways are differentially affected. *Neurosci. Lett.* **644**, 291–296.
80. Ross, B. M., McLaughlin, M., Roberts, M., Milligan, G., McCulloch, J., and Knowler, J. T. (1993) Alterations in the activity of adenylyl cyclase and high affinity GTPase in Alzheimer's disease. *Brain Res.* **622**, 35–42.
81. Yamamoto, M., Ozawa, H., Saito, T., Frolich, L., Riederer, P., and Takahata, N. (1996) Reduced immunoreactivity of adenylyl cyclase in dementia of the Alzheimer type. *NeuroReport* **7**, 2965–2970.
82. Yamamoto, M., Ozawa, H., Saito, T., Hatta, S., Riederer, P., and Takahata, N. (1997) Ca2+/CaM-sensitive adenylyl cyclase activity is decreased in the Alzheimer's brain: possible relation to type I adenylyl cyclase. *J. Neural. Transm.* **104**, 721–732.
83. McLaughlin, M., Ross, B. M., Milligan, G., McCulloch, J., and Knowler, J. T. (1991) Robustness of G proteins in Alzheimer's disease: an immunoblot study. *J. Neurochem.* **57**, 9–14.

84. Ohm, T. G., Schmitt, M., Bohl, J., and Lemmer, B. (1997) Decrease in adenylate cyclase activity antecedes neurofibrillary tangle formation. *Neurobiol. Aging* **18**, 275–279.
85. O'Neill, C., Fowler, C. J., Wiehager, B., Cowburn, R. F., Alafuzoff, I., and Winblad, B. (1991) Coupling of human brain cerebral cortical alpha 2-adrenoceptors to GTP-binding proteins in Alzheimer's disease. *Brain Res.* **563**, 39–43.
86. O'Neill, C., Cowburn, R. F., Wiehager, B., Alafuzoff, I., Winblad, B., and Fowler, C. J. (1991) Preservation of 5-hydroxytryptamine1A receptor–G protein interactions in the cerebral cortex of patients with Alzheimer's disease. *Neurosci. Lett.* **133**, 15–19.
87. Bergström, L., Garlind, A., Nilsson, L., Alafuzoff, I., Fowler, C. J., Winblad, B., and Cowburn, R. F. (1991) Regional distribution of somatostatin receptor binding and modulation of adenylyl cyclase activity in Alzheimer's disease brain. *J. Neurol. Sci.* **105**, 225–233.
88. Garlind, A., Cowburn, R. F., Wiehager, B., Ravid, R., Winblad, B., and Fowler, C. J. (1995) Preservation of kappa 1 opioid receptor recognition site density and regulation by G-proteins in the temporal cortex of patients with Alzheimer's disease. *Neurosci. Lett.* **185**, 131–134.
89. Cowburn, R. F., Wiehager, B., Ravid, R., and Winblad, B. (1996) Acetylcholine muscarinic M2 receptor stimulated [^{35}S]GTPγS binding shows regional selective changes in Alzheimer's disease postmortem brain. *Neurodegeneration* **5**, 19–26.
90. Hérnandez-Hérnandez, A., Adem, A., Ravid, R., and Cowburn, R. F. (1995) Preservation of acetylcholine muscarinic M2 receptor G-protein interactions in the neocortex of patients with Alzheimer's disease. *Neurosci. Lett.* **186**, 57–60.
91. Cowburn, R. F., O'Neill, C., Ravid, R., Winblad B., and Fowler, C. J. (1992) Preservation of Gi-protein inhibited adenylyl cyclase activity in the brains of patients with Alzheimer's disease. *Neurosci. Lett.* **141**, 16–20.
92. Meier-Ruge, W., Iwangoff, P., and Reichlmeier, K. (1984) Neurochemical enzyme changes in Alzheimer's and Pick's disease. *Arch. Gerontol. Geriatr.* **3**, 161–165.
93. Bonkale, W. L, Cowburn, R. F, Bogdanovic, N., Ohm, T. G., and Fastbom, J. (1998) Reduced [^{3}H]cAMP binding in brain staged for Alzheimer's disease neurofibrillary changes and amyloid deposits. *Neurobiol. Aging* **19(4S)**, S52.
94. Licameli, V., Mattiace, L. A., Erlichman, J., Davies, P., Dickson, D., and Shafit-Zagardo, B. (1992) Regional localization of the regulatory subunit (RII beta) of the type II cAMP-dependent protein kinase in human brain. *Brain Res.* **578**, 61–68.
95. Fraser, S. P., Suh,, Y. H., and Djamgoz, M. B. (1997) Ionic effects of the Alzheimer's disease beta-amyloid precursor protein and its metabolic fragments. *Trends Neurosci.* **20**, 67–72.
96. Luo, Y, Hawver, D. B., Iwasaki, K., Sunderland, T., Roth, G. S., and Wolozin, B. (1997) Physiological levels of beta-amyloid peptide stimulate protein kinase C in PC12 cells. *Brain Res.* **769**, 287–295.
97. Hartmann, H., Eckert, A., Crews, F. T., and Müller, W. E. (1996) β-amyloid amplifies PLC activity and Ca^{2+} signalling in fully differentiated brain cells of mice. *Amyloid. Int. J. Exp. Clin. Invest.* **3**, 234–241.
98. Cowburn, R. F., Wiehager, B., and Sundström, E. (1995) Beta-amyloid peptides enhance binding of the calcium mobilising second messengers, inositol(1,4,5)trisphosphate and inositol-(1,3,4,5)tetrakisphosphate to their receptor sites in rat cortical membranes. *Neurosci. Lett.* **191**, 31–34.

99. Langel, Ü, Soomets, U., Mahlapuu, R., Karelson, E., Zilmer, M., and Zorko, M. (1998) Regulation of GTPase and adenylate cyclase by amyloid-γ peptides. *Neurobiol. Aging* **19**(4S), S48.

100. Kelly, J. F., Furukawa, K., Barger, S. W., Rengen, M. R., Mark, R. J., Blanc, E. M., Roth, G. S., and Mattson, M .P. (1996) Amyloid beta-peptide disrupts carbachol-induced muscarinic cholinergic signal transduction in cortical neurons. *Proc. Natl. Acad. Sci. USA* **93**, 6753–6758.

101. Popova, J. S., Garrison, J. C., Rhee, S. G., and Rasenick, M. M. (1997) Tubulin, Gq, and phosphatidylinositol 4,5-bisphosphate interact to regulate phospholipase Cbeta1 signaling. *J. Biol. Chem.* **272**, 6760–6765.

102. Roychowdhury, S. and Rasenick, M. M. (1994) Tubulin–G protein association stabilizes GTP binding and activates GTPase: cytoskeletal participation in neuronal signal transduction. *Biochemistry* **33**, 9800–9805.

103. Khatoon, S., Grundke-Iqbal, I., and Iqbal, K. (1995) Guanosine triphosphate binding to beta-subunit of tubulin in Alzheimer's disease brain: role of microtubule-associated protein tau. *J. Neurochem.* **64**, 777–787.

104. Fowler, C. J. (1997) The role of the phosphoinositide signalling system in the pathogenesis of sporadic Alzheimer's disease: a hypothesis. *Brain Res. Rev.* **25**, 373–380.

105. Roth, G. S., Joseph, J. A., and Mason, R. P. (1995) Membrane alterations as causes of impaired signal transduction in Alzheimer's disease and aging. *Trends Neurosci.* **18**, 203–206.

106. De Sarno, P. and Jope, R. S. (1998) Phosphoinositide hydrolysis activated by muscarinic or glutamatergic, but not adrenergic, receptors is impaired in ApoE-deficient mice and by hydrogen peroxide and peroxynitrite. *Exp. Neurol.* **152**, 123–128.

107. Neve, R. L. (1996) Mixed signals in Alzheimer's disease. *Trends Neurosci.* **19**, 371–372.

108. Gibson, G. E., Vestling, M., Zhang, H., Szolosi, S., Alkon, D., Lannfelt, L., Gandy, S., and Cowburn, R. F. (1997) Abnormalities in Alzheimer's disease fibroblasts bearing the APP670/671 mutation. *Neurobiol. Aging* **18**, 573–580.

109. Vestling, M., Adem, A., Racchi, M., Gibson, G. E., Lannfelt, L., Cowburn, R. F. (1997) Differential regulation of adenylyl cyclase in fibroblasts from sporadic and familial Alzheimer's disease cases with PS1 and APP mutations. *NeuroReport* **8**, 2031–2035.

6

The NO Signaling Pathway in the Brain
Neural Injury, Neurological Disorders, and Aggression

Masayuki Sasaki,
Valina L. Dawson, and Ted M. Dawson

Nitric oxide (NO) is a widespread and multifunctional biological messenger molecule. It mediates vasodilation of blood vessels, host defense against infectious agents and tumors, and neurotransmission of the central and peripheral nervous systems *(1–3)*. The discovery of NO as a messenger molecule in the nervous system also revised conventional concepts of neurotransmitters. Compared with the traditional neuronal messenger molecules, NO has a variety of distinguished features. NO is probably the smallest and most versatile bioactive molecule identified, it diffuses freely across membranes, it is not stored in synaptic vesicles, and it is not released by exocytosis upon membrane depolarization. NO seems to be terminated primarily by reactions with its targets. In the nervous system, NO may play a role not only in physiologic neuronal functions, such as neurotransmitter release, neural development, regeneration, synaptic plasticity, and regulation of gene expression, but also in pathological conditions in which deregulated excessive production of NO leads to neural injury. Furthermore, rapid progress is now being made in understanding the regulation of NOS activity and the cellular and molecular targets of NO under physiologic and pathologic conditions. Some of the newly revealed roles for NO in the nervous system include regulation or control of neuronal morphogenesis, short-term or long-term synaptic plasticity, regulation of gene expression, and modification of sexual and aggressive behavior. Excess formation of NO plays a role in neural injury in several kinds of neurologic insults, which has promoted the development of selective NOS inhibitors for the treatment of neurologic disorders.

NO SYNTHASE ISOFORMS
AND REGULATION OF NO GENERATION

Nitric oxide is formed by the enzymatic conversion of the guanidino nitrogen of L-arginine by NO synthase (NOS). There are three NOS isoforms–neuronal

From: *Cerebral Signal Transduction: From First to Fourth Messengers*
Edited by: M. E. A. Reith © Humana Press Inc., Totowa, NJ

NOS (nNOS; type I), inducible NOS (iNOS; type II), and endothelial NOS (eNOS; type III) *(4–6)*—and all three isoforms are expressed in the central nervous system (CNS). NO can also be generated nonenzymatically by the direct reduction of nitrite to NO under acidic and highly reduced conditions that may occur under ischemic conditions *(7,8)*. nNOS and eNOS are calcium/calmodulin-dependent and, under most conditions, are constitutively expressed. However, recent studies indicate that both enzymes can be induced under conditions of cell stress, injury, and differentiation *(9)*. iNOS is calcium independent and its expression is primarily regulated at the level of transcription. Interestingly, in the absence of the substrate, L-arginine, NOS is able to produce superoxide anion ($O_2^{\bullet-}$) and hydrogen peroxide *(10,11)*. NOS generation of $O_2^{\bullet-}$ and NO in arginine-depleted cells can lead to peroxynitrite-mediated cellular injury *(12)*.

Because NO diffuses freely and cannot be stored, the only way to functionally control NO levels is to regulate the level and activity of the NOS enzymes. Previous studies have shown that the NO-synthesizing enzymes have multiple regulatory sites, including a calmodulin binding site and phosphorylation recognition sequences. All NOS isoforms have been shown to possess consensus phosphorylation sites. It appears likely that phosphorylation by protein kinase A (PKA), protein kinase C (PKC), cyclic GMP-dependent protein kinase (PKG), and calcium/calmodulin protein kinase (CaMK) decrease nNOS catalytic activity *(13,14)*. The calcium/calmodulin-dependent protein phosphatase, calcineurin, dephosphorylates nNOS to increase its catalytic activity *(15)*. This may be the mechanism by which the immunosuppressant and calcineurin inhibitor, FK506 (Tacrolimus), protects cultured cortical neurons against glutamate neurotoxicity mediated by nNOS activation *(15)*. Intracellular availability of various cofactors for NOS considerably influences production *(3)*. Recent studies identified a novel protein, PIN, which interacts with nNOS to inhibit its activity *(16)*. PIN is thought to destabilize the nNOS dimer, a conformation necessary for activity. Control of iNOS activity primarily occurs at the level of gene transcription under a complex mechanisms of regulation *(17,18)*. iNOS expression is induced in response to cytokines, lipopolysaccharide (LPS), or other agents *(17)*. In the nervous system, NO derived from nNOS in glial cells negatively regulates iNOS expression through inhibition of nuclear factor-kappa of B-cells (NF-κB).

The increase in nNOS levels under several physiologic and pathologic conditions suggests that control of nNOS activity may be also regulated through the transcription of the nNOS gene. To understand the mechanisms of transcription we characterized the 5'-flanking regions of mouse nNOS gene and identified multiple promoters that appear to control nNOS tran-

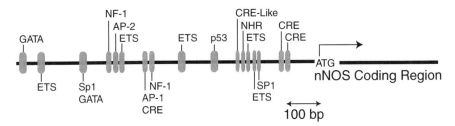

Fig. 1. Potential cis-regulatory elements in mouse nNOS exon 2 promoter.

scription in a temporal and tissue/region-specific manner *(19)*. The putative promoter sequences identified lack a typical TATA box and are rich in guanine and cytosine. This is consistent with the presence of multiple transcription initiation sites. Comparison of the mouse and human 5'-flanking regions revealed 5 regions of high homology (>90%), implicating these regions as important regulatory components of nNOS transcription. Promoter–reporter gene constructs indicate that these promoters drive expression of the reporter genes in mouse neuroblastoma, Neuro 2A, cells with varying degrees of efficiency. We found that the major promoter for nNOS transcription is contained within exon 2. Deletion and mutational analysis of the promoter–reporter constructs indicate that basal and cAMP-stimulated nNOS transcription is tightly regulated by two canonical CRE regulatory sequences contained within exon 2 (Fig. 1). The importance of CRE-binding proteins in nNOS transcription is supported by the examination of a series of CRE-binding protein mutants. Through these studies, a member of the CRE-binding (CREB) protein family of transcription factors was found to be necessary and sufficient to mediate basal and cAMP-dependent nNOS expression. Thus, alternative splicing and multiple promoters are employed to regulate the diversity of nNOS expression, and nNOS gene expression is regulated primarily by a transcription mechanism depending on a CREB protein family member *(20)*.

After transcription, nNOS may also be regulated by changes in its intracellular localization. Electron microscopic immunohistochemistry has localized nNOS to the Golgi apparatus, endoplasmic reticulum, spines, dendritic shafts, and axon terminals *(21)*. The presence of a PSD-95/Discs-Large/ZO-1 (PDZ) domain in nNOS *(22)* and its role in the membrane association of nNOS in skeletal muscle *(23)* may provide insight into the physiological roles of nNOS. PDZ domain interactions may mediate binding of nNOS to synaptic junctions through interactions with postsynaptic density-95 protein (PSD-95) and PSD-93 *(24)*. Skeletal-muscle membranes from patients with Duschene muscular dystrophy that lack dystrophin also lack nNOS *(23)*.

REACTION OF NO IN BIOLOGICAL SYSTEM

Nitric oxide markedly differs from traditional neuronal messenger molecules in that NO is a highly reactive free radical with a short half-life (3–6 s). Once produced, NO is either rapidly oxidized, reduced, or complexed to various biomolecules. The fate of NO depends on the local microenvironment in which it is generated. In addition to the free-radical form (NO$^\bullet$), there are oxidized and reduced forms of NO: NO$^+$ (nitrosonium ion) and NO$^-$ (nitroxide ion), respectively. The presence of these various states of oxidation of NO may explain why NO has contrasting functions in some situations *(25)*. The free-radical form (NO$^\bullet$) is probably the primary form produced by NOS under physiological conditions.

As defined first for its functional target, NO binds to the heme moiety of guanylyl cyclase (GC) to evoke a conformational change that activates the enzyme resulting in cGMP formation. NO activates another heme-containing enzyme, cyclooxygenase, to increases prostaglandin production *(26)*. NO also reacts with nonheme iron contained with iron sulfur clusters in numerous enzymes. These reactions readily occur after macrophage activation. In contrast with the reversible reaction of NO with heme, the reaction of NO with iron sulfur clusters results in the dissolution of the cluster *(27)*. Because the major enzymes within the oxidative respiratory chain in the mitochondria contain iron sulfur clusters at their active sites, this interaction is thought to be the primary mechanism by which NO may inhibit cellular ATP production, resulting in cytotoxicity. In the presence of NO, cytosolic aconitase functions as the iron-responsive element-binding protein; in its absence, it functions as a cytosolic aconitase *(28,29)*. NO disrupts aconitase activity and exposes its RNA-binding site, permitting the iron-responsive element-binding protein to bind to the iron-responsive element *(30)*.

In biological systems, the most permissive interaction of NO is the reaction with the $O_2^{\bullet-}$ to produce highly toxic radical, peroxynitrite ($ONOO^-$). This is such a potent oxidant that it seems to mediate most of NO neurotoxicity. Peroxynitrite may further decompose to hydroxyl and nitrogen dioxide radicals in physiological solution, which are also highly reactive and biologically destructive *(31)*. The reaction of NO with $O_2^{\bullet-}$ to produce $ONOO^-$ is so rapid, at a speed more than three times faster than the catalytic decomposition of $O_2^{\bullet-}$ by superoxide dismutase (SOD), that it influences the fate of $O_2^{\bullet-}$, another virulent reactive oxygen species (ROS) *(32)*. Peroxynitrite can readily react with sulfhydryls and with zinc thiolate moieties, directly nitrate and hydroxylate aromatic amino acid residues, and it can also oxidize lipids, proteins, and DNA. Peroxynitrite, not NO, seems to oxidize iron sulfur clusters *(33)*.

Nitrosylation appears to be a potentially important regulatory system through which NO invokes its many physiological functions. Protein-associated targets of NO include heme, cysteine (Cys), and tyrosine (Tyr) residues. At physiological pH, Cys residues are efficiently nitrosylated, and nitration of Tyr occur at a much slower rate *(34)*. In general, thiols are assumed to be one of the major targets for NO. Nitrosothiols with biological relevance have been isolated and characterized, including *S*-nitrosoglutathione and the nitrosothiol of serum albumin *(35)*. Another potential physiological example of nitrosylation occurs at the NMDA receptor. This may be one of the molecular mechanisms through which NO physiologically modulates glutamatergic neurotransmission *(25)*. However, other studies suggest that NO inhibits *N*-methyl-D-aspartate (NMDA)-induced currents through an interaction with cations rather than through the redox modulatory site of the NMDA receptor channel *(36)*. NO reacts with an active site of cysteine within glyceraldehyde-3-phosphate dehydrogenase (GAPDH), resulting in the direct binding of NAD to cysteine and inhibition of GAPDH catalytic activity and depression of glycolysis *(37,38)*. NO also reacts with cysteine residue of hemoglobin, forming an *S*-nitrosyl hemoglobin *(39)*. *S*-Nitrosyl hemoglobin is an alternative source of NO during arterial–venous transit and appears to control blood pressure, which may facilitate efficient oxygen delivery to tissues.

NO-MEDIATED NEURAL FUNCTION

Autonomic Nerve System

In the peripheral nervous system, NO functions as a neurotransmitter. The most prominent evidence come from studies on the autonomic nerve system, including the digestive system. nNOS is selectively localized in the myenteric plexus and NOS inhibitors selectively block nonadrenergic- and noncholinergic (NANC)-mediated relaxation of the gastrointestinal tract *(2)*, indicating that NO is the NANC neurotransmitter *(40)*. This is further supported by studies using mice with targeted disruption of the nNOS gene *(41)*. These mice are normal in most respects despite their lack of NOS catalytic activity in the brain and loss of NOS immunostaining in the central and peripheral nervous systems. No morphological abnormalities were observed in the brain or in most peripheral tissue at the gross and microscopic level. However, these mice have the markedly enlarged stomachs and histological examination reveals hypertrophy of the inner circular muscle layer. NO is thought to mediate relaxation of the pyloric sphincter in response to a food bolus. The distended stomach and hypertrophy of the inner circular muscle layer seems to be a compensatory reaction against the dysfunction of the pyloric sphincter to relax in response to a food bolus. The pathology resembles hypertrophic pyloric stenosis, in which the lack of NO may play a role because a recent study shows an absence of NADPH diaphorase staining in the myenteric neurons in human male infants with pyloric stenosis *(42)*. Furthermore, the nNOS gene is a susceptibility locus for the infantile pyloric stenosis *(43)*. In the circular smooth muscle of gastric fundus, NO acts as an inhibitory neurotransmitter that mediates the slow component of inhibitory junction potentials *(44)*.

Neuronal NOS is highly expressed within the nerves of the urogenital system, including the pelvic plexus, the cavernosal nerve, and plexus. It is also present in the adventitia of the deep cavernosal arteries and sinusoids in the periphery of the corpus cavernosa. Both eNOS and nNOS may contribute to penile erections as mice lacking either isoform are capable of reproduction. nNOS knockout mice still have NO-dependent erections because NOS inhibitors block penile erections in the nNOS knockout mice *(45)*. eNOS is expressed at high levels in the endothelium of the penile vasculature and sinusoidal endothelium within the corpora cavernosa *(45)*. NO functions as the NANC neurotransmitter that mediates penile erections because penile erections elicited by electrical stimulation of the cavernous nerve are blocked by low doses of NOS inhibitors in intact rats and NOS inhibitors block NANC-stimulated relaxation of isolated cavernosal strips *(46,47)*. nNOS null mice also display hypertrophied urinary bladders and loss of neurally mediated relaxation of urethral and bladder muscle, providing a model for idiopathic voiding disorder in humans *(48)*.

Nitric oxide may play a role in modulating behavior. Inhibition of NOS attenuates alcohol consumption in alcohol-preferring rats and prevents the acquisition of tolerance to the ataxic effects of alcohol *(49)*. NO may be involved in light phase shifts of circadian rhythm in that NOS inhibitors block light-induced resetting of behavioral rhythms *(50)*.

Aggressive Behavior

The most prominent feature of nNOS null mice is that male mice exhibit inappropriate sexual behavior, mounting females indiscriminately regardless of whether the females are in estrus or not. Furthermore, when male mice are housed together, they are markedly aggressive and do not respond to appropriate submissive postures by wild-type males nor do they elicit submissive postures in response to attack by another mouse *(51)*. These behaviors may be a selective consequence of the loss of nNOS, because nNOS null mice have no detectable abnormalities in brain structure and have normal synaptic plasticity in the hippocampus and cerebellum *(52,53)*. Furthermore, plasma testosterone levels, which could account for both aggression and sexual behavior, do not differ between wild-type and nNOS null mice. In newt, brain NOS activity increases transiently during the males' courtship behavior, suggesting that a role for NO in reproductive behavior may be conserved throughout vertebrate evolution *(54)*. Because aggressive behavior was virtually abolished in nNOS –/– mice by castration, abnormal development of the neural circuitry regulating aggression (i.e., not requiring testosterone) is unlikely to be responsible for the increased aggressive behavior seen in the knockout mice. The lack of nNOS might cause some abnormalities in neuronal development that is undetectable in these studies, resulting in aggressive behavior in adult animals. This possibility does not seem likely, as only male, but not female mice have abnormalities in behavior. Futhermore, selective nNOS inhibitors given to adult male mice causes aggressive behavior *(55)*.

Synaptic Plasticity

Long-term potentiation (LTP) in the hippocampus and long-term depression (LTD) in the cerebellum are two forms of synaptic plasticity and widely accepted as models for studying the molecular basis of learning and memory. Induction of LTP in the CA1 region of hippocampal slices usually requires calcium influx through postsynaptic NMDA receptor channels. Enhancement of glutamate release presynaptically is thought to be the primary mechanism of LTP. NO has been proposed as a candidate for the retrograde messenger from the postsynaptic neuron to the presynaptic neuron. This notion is support by several findings. In hippocampal neurons, NO produces an increase in the frequency of spontaneous miniature excitatory postsynaptic potentials. Direct application of NO to CA1 neurons mimics LTP. Furthermore, NOS inhibitors including hemoglobin block the establishment of LTP *(56–58)*. Intracellular application of a NO scavenger or an ultraviolet (UV)-sensitive NO donor also reinforced the role of NO as the retrograde messenger by blocking or enhancing LTP, respectively *(59)*. Similar evidence exists for a role of NO in LTD *(60)*. Several studies suggests that cGMP, to which most of the function of NO are attributed, is an important intracellular messenger that is required for the establishment of LTP or LTD *(61,62)*. Hippocampal CA1 neuron express nNOS at negligible levels, but these neurons express significant levels of eNOS *(63)*. eNOS plays a primary role in NO-mediated LTP in wild-type mice, as adenovirus-mediated disruption of eNOS activity in wild-type hippocampal slices eliminates NO-dependent LTP *(64)*. In spite of these findings, LTP in nNOS null or eNOS null mice appears to be intact. Interestingly LTP in nNOS –/– or eNOS –/– mice was further blocked by NOS inhibition *(53,65)*. To clarify the role of

NO in LTP, Kandel and colleagues recently generated double-knockout mice lacking both nNOS and eNOS gene. They found that hippocampal LTP in the stratum radiatum of CA1, which contains high levels of eNOS, was markedly reduced *(65)*. However, in the stratum oriens of hippocampal CA1, which contains low levels of eNOS, LTP was intact. Thus, there is both NO-dependent and NO-independent LTP.

Neurotransmitter Release

Findings that NOS inhibitors block NMDA receptor-mediated neuro-transmitter release from the cerebral cortex or striatal synaptosomes suggest that NO plays a role in regulation of NMDA receptor-mediated neurotrans-mitter release *(66,67)*. NO is involved in the glutamate- or norepinephrine-stimulated and GABA-inhibited release of the hypothalamic luteinizing hormone-releasing hormone (LHRH) *(68)*. In synaptosomes, neurotransmit-ter release evoked by stimulation of NMDA receptors is blocked by NOS inhibitors, whereas release by potassium depolarization is unaffected. Pre-sumably, potassium depolarizes all terminals so that the effects of NO are diminished as a result of the limited population of NOS-containing terminals.

Nitric oxide may act on both calcium-sensitive and calcium-insensitive pools of synaptic vesicles *(69)*. NO influences neurotransmitter release by activating the cGMP–PKG pathway, resulting in augmention of the phos-phorylation of synaptic vesicle proteins, or it may occur through activation of cyclic GMP-dependent cation channels *(70)*. Recent studies indicate that NO may facilitate vesicle docking events by promoting VAMP/SNAP-25–syntaxin 1a complex formation and by inhibiting binding of n-sec1 to syntaxin 1a *(71)*. This effect of NO may be mediated by nitrosylation of sulfhydryl groups of protein(s) involved.

Differentiation and Morphogenesis

Nitric oxide may play a key role in nervous system morphogenesis and synaptic plasticity. nNOS is transiently expressed in the cerebral cortical plate neurons from embryonic days E15–E19 of rats *(72)*. At this stage, cor-tical neuroblasts exit from the final cell-division cycle to start differentia-tion toward mature neurons. This intense labeling of the cortical plate neurons and their processes decreases rapidly and vanishes by the 15th postnatal day. Embryonic sensory ganglia are also nNOS positive and this decreases to less than 1% of the cells at birth. Thus, the transient expression of NOS may be a reflection of physiological processes that are important for neu-ronal development. This notion is consistent with observations in PC12 cells. Nerve growth factor (NGF) triggers the differentiation of PC12 cells, which is accompanied by the marked induction of nNOS *(66,73)*. During neuronal differentiation, NO may trigger a switch to growth arrest during differentiation of neuronal cells; thus, NOS serves as a growth-arrest gene initiating the switch to cytostasis during differentiation. This is further supported by the finding that in *Drosophila*, NOS is highly expressed in developing imaginal disk. Inhibition of NOS at the stage of larvae causes organ hypertrophy, and ectopic NOS expression in larvae has the opposite effects *(74)*.

Maturation of adult spinal cord motor neurons may involve NO *(75)*. NO may also play a role in the development of proper patterns of connections in the retinotectal system, as NOS expression peaks at the time when refinement of the initial pattern of connections is occurring *(76)*. Neurons of the developing olfactory epithelium during migration and the establishment of primary synapses in the olfactory bulb of rats express nNOS during embryonic development *(77)*. Olfactory nNOS expression rapidly declines after birth and is undetectable by postnatal day 7. These obsevations reinforce the notion that nNOS induction is a manifestation of cellular processes essential for neuronal cell differentiation. In addition, nNOS expression is rapidly induced in regenerating olfactory receptor neurons after bulbectomy and is particularly enriched in their outgrowing axons. Thus, NO may play a role in activity-dependent establishment of connections in both developing and regenerating olfactory neurons.

Ras–MAPK Pathway

The Ras–mitogen-activated protein kinase (MAPK) pathway is widely accepted as a major intracellular signaling cascade that induces long-term changes in gene expression. How NO can influence neuronal development is unknown, but one intriguing hypothesis has come from the finding that NO activates this important pathway *(78,79)*. Stimulation of NMDA receptors in cultured cortical neurons activates the Ras–ERK pathway via calcium-dependent activation of nNOS and NO generation *(79)*. NMDA-stimulated phosphorylation of CREB, a downstream effector of ERK, is also NO dependent *(79)*. Stimulation of the NMDA receptor directly activates Ras through activation of nNOS and NO formation. NMDA-induced Ras activation is an nNOS-dependent process because the NOS inhibitor, L-NAME, blocked Ras activation, which was reversed by an excess substrate, L-Arg. Furthermore, even though Ras is functionally intact in nNOS –/– mice, Ras failed to be activated by NMDA stimulation in the primary cortical cultures derived from nNOS –/– mice. Because NMDA-induced calcium currents occurs at the same degree from primary cultures of wild-type mice and nNOS –/– mice, alterations in intracellular calcium signaling are not responsible for the differences in Ras activation. Most of NO's physiologic effects in the nervous system are attributable to activation of GC and increases in intracellular cGMP levels. However, the potent and selective GC inhibitor, ODQ, and the cell permeable cGMP analog, 8Br-cGMP, have no effect on Ras activation.

Multiple pathways including Src, Ras–GRF, and PYK have been shown to activate Ras in a calcium-dependent manner in several cell lines; however, in neurons, NMDA stimulation directly activates Ras through NO. Ras appears to have a modulatory site highly sensitive to the cellular redox state. Ras has a critical cysteine at Cys-118, which fits with the consensus sequence for a redox modulation site sensitive to NO *(80)*. Site-directed mutagenesis of Cys-118 to Ser-118 eliminates NO's ability to activate Ras, but this mutant Ras retains the ability to be stimulated by growth factors *(81)*. Thus, NO appears to directly activate Ras by redox modulation of Cys-118 following NMDA receptor stimulation.

The finding that NMDA stimulation activates Ras through NO is significant in that this may explain the mechanism of prior observations that NMDA receptor stimulation results in the tyrosine phosphorylation of Erk. This phosphorylation was shown to occur through changes in cytoplasmic calcium, but the molecular mechanisms of these events have been poorly understood *(82)*. We recently showed the blockade of NMDA-mediated Erk phosphorylation by the NOS inhibitor, L-NAME, and the reversal by an excess substrate, L-Arg, as well as the failure of NMDA to induce tyrosine phosphorylation of Erk in nNOS –/– neuronal cultures. Because the Ras–ERK kinase pathway is widely accepted as one of the major pathways of neuronal activity-dependent long-term changes in nervous system through gene expression *(82)*, our findings give rise to the possibility that NO may be a key mediator linking activity to gene expression and long-term plasticity. NO donors activate all three types of MAPK (ERK, JNK/SAPK, p38 MAPK) with different extent and time-courses in T-cell leukemic cell lines *(78)*. ERK and JNK/p38 MAPK pathways are implicated in neuronal survival and death, respectively. Because the MAPK signaling pathway plays a central role in growth factor response (ERK) or stress response (JNK, p38 MAPK) in the nervous system, NO–MAPK signaling may underlie NO's role in neuronal survival, differentiation, and apoptotic cell death during development (Fig. 2).

NO IN NEUROLOGICAL DISORDERS

When deregulated generation of NO reaches an excess level, it induces a neurotoxic cascade *(83)*. Overactivation of NMDA receptors by excess glutamate mediates cell death in focal cerebral ischemia *(84)*. Involvement

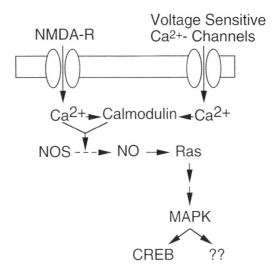

Fig. 2. Schematic diagram showing the role of NO in coupling NMDA receptor-mediated and voltage-sensitive calcium influx to Ras and downstream effectors. MAPK: mitogen-activated protein kinase. [Reproduced with permission from Dawson, T. M., Gonzalez-Zulueta, M., Kusel, J., and Dawson, V. L. (1998) Nitric oxide: diverse actions in the central and peripheral nervous systems. *Neuroscientist* **4,** 96–112.]

of NO in glutamate neurotoxicity was first demonstrated in in vitro experiments using primary cortical cultures *(85)*. NMDA applied only for a short (5 min) period of time is able to elicit cell death in cortical cultures assessed 24 h later. This type of neuronal cell death has been called delayed neurotoxicity, in which irreversible processes are set in motion by the 5-min application of NMDA *(86)*. This type of neuronal cell death is exquisitely dependent on calcium. Furthermore, calcium influx via NMDA receptors elicits more potent toxicity than other modes of calcium entry. This type of toxicity involves NO, as treatment with NOS inhibitors, removal of L-arginine, or reduced hemoglobin which scavenges NO, blocks this form of toxicity *(85)*. The nNOS null mice provide further evidence of the role of NO in NMDA neurotoxicity. These mice are relatively resistant to NMDA neurotoxicity as well as combined oxygen and glucose deprivation *(87)*. Furthermore, recent studies indicate that the conditions under which neurons are cultured can profoundly influence the expression level of nNOS and subsequent involvement of NO in excitotoxicity *(88)*. The expression of nNOS in primary cultures at levels equivalent to in vivo expression (1–2% of the neuronal population) is critical for the assessment of NO-mediated neurotoxicity. In primary cultures, the expression of nNOS is dependent on both the method of culture and the age of the cultures *(88,89)*. Neurotrophins, despite their general role in attenuating excitotoxic neuronal injury, were recently shown to increase the number of nNOS neurons in cortical culture grown on glial feeder layers and render neurons more sensitive to NMDA *(88)*. This finding is consistent with a previous report by Choi and colleagues on the enhancement of excitotoxic neuronal injury by neurotrophins under certain conditions *(90)*. However, when neurons are grown on a polyornithine matrix, neurotrophins failed to enhance the expression of nNOS and are neuroprotective, consistent with the neuroprotective role of neurotrophins in neurotoxicity studies in intact animals. Glutamate neurotoxicity also contributes in some degree to the pathogenesis of neurodegenerative diseases such as Huntington's disease and Alzheimer's disease *(91)*, implicating NO in these disorders.

Neuronal NOS neurons have garnered much attention from a neuropathological point of view because they are identical to NADPH diaphorase positive neurons. They have been shown to be resistant to a variety of neurodegenerating conditions, including NMDA neurotoxicity, Huntington's disease, Alzheimer's disease, and vascular stroke *(92–94)*. These observations give rise to the possibility that nNOS neurons might have protective factors that enable them to tolerate the toxic effects of NO. To clarify this enigmatic phenomenon, we established a NO-resistant PC12 cell line. Using sequential analysis of gene expression (SAGE), we determined that manganese SOD (MnSOD) is one of the neuroprotective factors against NO-induced neurotoxicity. nNOS neurons in the cortex are enriched in MnSOD and this enzyme may prevent the local formation of toxic peroxynitrite, rendering nNOS neurons resistant to NO's toxic actions *(95)*. Overexpression of MnSOD by recombinant adenovirus in cultured cortical neurons markedly increased resistance to NMDA-mediated cell death *(95)*. Furthermore, application of antisense MnSOD oligos to knock-down MnSOD expression makes nNOS neurons more susceptible to NO and NMDA toxicity (Fig. 3).

Nitric oxide presumably kills neurons via peroxynitrite ($ONOO^-$) *(96)*. In vitro experiments indicate that NMDA-mediated neurotoxicity occurs through $ONOO^-$. In high doses, exposure to $ONOO^-$ results in necrosis, on the other hand, in low doses it elicits apoptosis *(97)*. Peroxynitrite causes lipid peroxidation that changes the fluidity of membranes and impairs a variety of intramitochondrial and extramitochondrial membrane transport systems, including ATPase, glucose transporters, and mitochondrial ion transport systems *(98)*.

In brain ischemia, deregulated release of glutamate overstimulates NMDA receptors. This is thought to be the primary mechanism that leads to the delayed neuronal cell death after the onset of ischemia *(86)*. NO produced

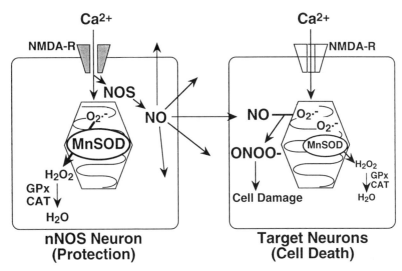

Fig. 3. Schematic diagram of the postulated mechanism of nNOS neuron survival. NO is formed on sustained glutamate stimulation of NMDA receptors. nNOS neurons function as "killer neurons" and are enriched in MnSOD, which allows them to effectively scavenge NMDA-induced mitochondrial production of $O_2^{\bullet-}$. NO freely diffuses to the adjacent target neuron, where it combines with the $O_2^{\bullet-}$ to form the peroxynitrite anion ($ONOO^-$), which is an extremely potent oxidant, setting in motion cell death programs. CAT: catalase; GPx: gluthathione peroxidase. [Reproduced with permission from Dawson, T. M., Gonzalez-Zulueta, M., Kusel, J., and Dawson, V. L. (1998) Nitric oxide: diverse actions in the central and peripheral nervous systems. *Neuroscientist* **4,** 96–112.]

excessively through the NMDA receptor overstimulation mediates the neuronal cell death because selective nNOS inhibitors block ischemic damage following middle cerebral artery (MCA) ligation in several animal models *(99)*. This notion is further supported by the studies employing nNOS null mice. These mice have reduced infarct volumes compared to age-matched wild-type controls following permanent MCA occlusion *(100)*. Inhibition of eNOS may account for the deleterious effects of nonselective NOS inhibitors. MCA occlusion of eNOS null mice leads to larger infarcts than in wild-type mice *(101)*. This observation is associated with a reduction in cerebral blood flow in the ischemic penumbra region. The selective nNOS inhibitor 7-NI, which has a minimal effect on cerebral blood flow, is neuroprotective in animal models of focal ischemia *(102)*. Furthermore, when the nNOS null mice are treated with the nonselective NOS inhibitor nitro-L-arginine methyl-ester, there are adverse effects on cerebral blood flow and stroke volume is increased. Conversely, when eNOS null mice are treated with NOS inhibitors during MCA occlusion, infarct volume is reduced *(100)*. Thus, neuronally derived NO plays an important role in mediating neuronal cell death, and endothelial-derived NO plays an important protective role by regulating and maintaining proper cerebral blood flow *(101–103)*. In the late stage of cerebral ischemia (approx 6 h), postischemic inflammation induces iNOS expression and sustained generation of large amounts of NO leads to additional delayed neural injury *(104)*. Supporting this idea is the observation that iNOS null mice have smaller infarcts after cerebral ischemia *(105)*.

Nitric oxide may mediate other forms of neurotoxicity. NOS inhibitors protect against CNS oxygen toxicity *(106)*. Selective nNOS inhibitors protect against the dopaminergic neurotoxin (MPTP), which is a model of Parkinson's disease *(107,108)*. Further evidence of a role for neuronally derived NO in Parkinson's disease is the recent observation that mice lacking the nNOS gene *(109)* and 7-nitroindazole-treated baboon *(110)* are markedly resistant to MPTP neurotoxicity.

Nitric oxide may also play a role in the pathogenesis of AIDS dementia. NO derived from both neuronal and inducible NOS may contribute to AIDS dementia. The HIV coat protein, gp120, causes cell death in primary cortical cultures, in part, via the activation of nNOS *(111)*. Additionally, the coat proteins gp160, gp120, and gp41 induce iNOS in microglial and astrocyte cultures *(112)*. The role of iNOS in AIDS dementia has been established by recent studies that show that iNOS is markedly induced in the brains of patients suffering from severe AIDS dementia *(113)*.

Of the multiple mechanisms of the toxic effects of NO or peroxynitrite (ONOO⁻), the best established candidate for the target of NO and peroxynitrite is a nuclear enzyme, poly(ADP-ribose) polymerase [PARP; also called poly(ADP-ribose) synthetase or NAD^+ protein(ADP-ribosyl)transferase]. PARP is highly enriched in the nucleus and is involved in the DNA repair process. PARP senses DNA damage, including DNA nicks that are generated from oxidization by radicals like NO. Once activated, it adds up to several hundreds of ADP-ribose units to nuclear proteins such as histone and PARP itself *(114)*. The physiologic significance of ADP-ribosylation still remains unknown, but during the process of ADP-ribose attachment, 1 mol of NAD is consumed and four free-energy equivalents of ATP are consumed to regenerate NAD. Thus, activation of PARP can rapidly deplete cellular energy stores and lead to cell death. PARP inhibitors have been shown to be neuroprotective against NO-induced neurotoxicity in several model systems, including cortical cell cultures *(114)*, cerebellar granule cell cultures *(32)*, and hippocampal slices *(115)*. Targeted disruption of the PARP gene provided compelling evidence for involvement of PARP in NO-mediated toxicity *(116)*. Islet cells of mutant mice lacking PARP do not show DNA damage-

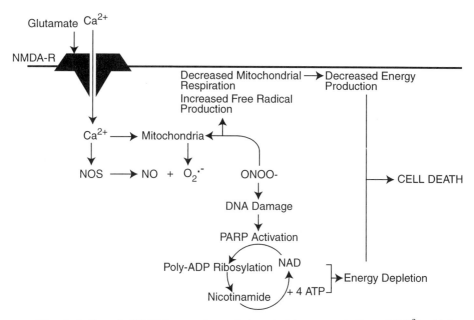

Fig. 4. Acting via NMDA receptors, glutamate triggers an influx of CA^{2+} which induces the production of NO from nNOS neurons. Mitochondrial activation results in an increase in superoxide anion ($O_2{}^{-}$) production. NO is a diffusible molecule that combines with superoxide to form peroxynitrite ($OONO^-$). Peroxynitrite damages mitochondrial enzymes, decreasing mitochondrial respiration and the production of ATP, and peroxynitrite damages MnSOD, further increasing the production of superoxide anion. Peroxynitrite has limited diffusion across membranes and can leave the mitochondria and enter the nucleus, where it damages DNA. Nicks and fragments of DNA activate PARP. Massive activation of PARP leads to ADP-ribosylation and depletion of NAD. ATP is depleted in an effort to resynthesize NAD. In the setting of impaired energy generation resulting from mitochondrial dysfunction, this loss of NAD and ATP leads to cell death.

induced NAD depletion and are more resistant to NO toxicity. Furthermore, PARP null mice are dramatically resistant to focal cerebral ischemia, and cortical cultures from PARP null mice are completely resistant to the NMDA toxicity and oxygen glucose depreviation *(117)* (Fig. 4).

Another potentially important target of NO-mediated neurotoxicity is mitochondrial respiration. NO binds to mitochondrial complex I and II and cis-aconitase, leading to inhibition of oxidative phosphorylation and gly-colysis. NO also reversibly inhibits mitochondrial respiration by competing with oxygen at cytochrome oxidase *(118)* presumably by nitrosylation of tyrosine *(119)*. Inactivation of MnSOD by peroxynitrite ($ONOO^-$) would lead to more $O_2{}^{\bullet-}$ generation, therefore more peroxynitrite formation. Mitochondrial respiratory decline would result in more free-radical formation and membrane depolarization as a result of the failure of Na^+, K^+-ATPase. Glutamate receptor channels and voltage-dependent sodium channels also contribute to membrane depolarization and further glutamate release *(120)*. This would promote more calcium influx through the NMDA receptor and sustained elevation of intracellular calcium and NOS activation, which could lead to a self-propagating destructive cycle that involves mitochondrial dysfunction, increases in $O_2^{\bullet-}$ and peroxynitrite, ultimately leading to disturbance of cellular redox state and depletion of energy stores.

CONCLUSION

Nitric oxide has revolutionized our conceptions about neurotransmitters. Since its discovery as EDRF, its role in physiologic and pathologic processes are becoming more diverse and widespread. Understanding the molecular mechanisms that control NO's disposition, and production can lead to a better understanding of physiologic processes as well as novel therapeutics for pathologic conditions.

ACKNOWLEDGMENTS

One of the authors (V. L. D.) is supported by US PHS Grant NS 33142 and the American Heart Association and is a NARSAD Staglin Music Festival Investigator. T. M. D. is an Established Investigator of the American Heart Association and is supported by US PHS Grants NS 37090 and NS33277, and the Paul Beeson Faculty Scholar Award in Aging Research. Under an agreement between the Johns Hopkins University and Guilford Pharmaceuticals, TMD and VLD are entitled to a share of sales royalty received by the University from Guilford. TMD and the University also own Guilford stock, and the University stock is subject to certain restrictions under University policy. The terms of this arrangement are being managed by the University in accordance with its conflict of interest policies.

REFERENCES

1. Dawson, T. M. and Snyder, S. H. (1994) Gases as biological messengers: nitric oxide and carbon monoxide in the brain. *J. Neurosci.* **14,** 5147–5159.
2. Moncada, S. and Higgs, A. (1993) The L-arginine-nitric oxide pathway. *N. Engl. J. Med.* **329,** 2002–2012.
3. Nathan, C. and Xie, Q. W. (1994) Nitric oxide synthases: roles, tolls, and controls. *Cell* **78,** 915–918.
4. Bredt, D. S., Hwang, P. M., Glatt, C. E., Lowenstein, C., Reed, R. R., and Snyder, S. H. (1991) Cloned and expressed nitric oxide synthase structurally resembles cytochrome P-450 reductase. *Nature* **351,** 714–718.

5. Sessa, W. C., Harrison, J. K., Barber, C. M., Zeng, D., Durieux, M. E., D'Angelo, D. D., Lynch, K. R., and Peach, M. J. (1992) Molecular cloning and expression of a cDNA encoding endothelial cell nitric oxide synthase. *J. Biol. Chem.* **267,** 15,274–15,276.

6. Xie, Q. W., Cho, H. J., Calaycay, J., Mumford, R. A., Swiderek, K. M., Lee, T. D., Ding, A., Troso, T., and Nathan, C. (1992) Cloning and characterization of inducible nitric oxide synthase from mouse macrophages. *Science* **256,** 225–228.

7. Meyer, D. J. (1995) Enzymatic/non-enzymatic formation of nitric oxide [Letter; comment]. *Nat. Med.* **1,** 1103–1104.

8. Zweier, J. L., Wang, P., Samouilov, A., and Kuppusamy, P. (1995) Enzyme-independent formation of nitric oxide in biological tissues. *Nat. Med.* **1,** 804–809.

9. Forstermann, U., Boissel, J. P., and Kleinert, H. (1998) Expressional control of the "constitutive" isoforms of nitric oxide synthase (NOS I and NOS III). *FASEB J.* **12,** 773–790.

10. Heinzel, B., John, M., Klatt, P., Bohme, E., and Mayer, B. (1992) $Ca2^+$/calmodulin-dependent formation of hydrogen peroxide by brain nitric oxide synthase. *Biochem. J.* **281,** 627–630.

11. Pou, S., Pou, W. S., Bredt, D. S., Snyder, S. H., and Rosen, G. M. (1992) Generation of superoxide by purified brain nitric oxide synthase. *J. Biol. Chem.* **267,** 24,173–24,176.

12. Xia, Y., Dawson, V. L., Dawson, T. M., Snyder, S. H., and Zweier, J. L. (1996) Nitric oxide synthase generates superoxide and nitric oxide in arginine-depleted cells leading to peroxynitrite-mediated cellular injury. *Proc. Natl. Acad. Sci. USA* **93,** 6770–6774.

13. Bredt, D. S., Ferris, C. D., and Snyder, S. H. (1992) Nitric oxide synthase regulatory sites. Phosphorylation by cyclic AMP-dependent protein kinase, protein kinase C, and calcium/calmodulin protein kinase; identification of flavin and calmodulin binding sites. *J. Biol. Chem.* **267,** 10,976–10,981.

14. Dinerman, J. L., Steiner, J. P., Dawson, T. M., Dawson, V., and Snyder, S. H. (1994) Cyclic nucleotide dependent phosphorylation of neuronal nitric oxide synthase inhibits catalytic activity. *Neuropharmacology* **33,** 1245–1251.

15. Dawson, T. M., Steiner, J. P., Dawson, V. L., Dinerman, J. L., Uhl, G. R., and Snyder, S. H. (1993) Immunosuppressant FK506 enhances phosphorylation of nitric oxide synthase and protects against glutamate neurotoxicity. *Proc. Natl. Acad. Sci. USA* **90,** 9808–9812.

16. Jaffrey, S. R. and Snyder, S. H. (1996) PIN: an associated protein inhibitor of neuronal nitric oxide synthase. *Science* **274,** 774–777.

17. Togashi, H., Sasaki, M., Frohman, E., Taira, E., Ratan, R. R., Dawson, T. M., and Dawson, V. L. (1997) Neuronal (type I) nitric oxide synthase regulates nuclear factor kappaB activity and immunologic (type II) nitric oxide synthase expression. *Proc. Natl. Acad. Sci. USA* **94,** 2676–2680.

18. Xie, Q. W. and Nathan, C. (1993) Promoter of the mouse gene encoding calcium-independent nitric oxide synthase confers inducibility by interferon-gamma and bacterial lipopolysaccharide. *Trans. Assoc. Am. Physicians* **106,** 1–12.

19. Sasaki, M., Huang, H., Wei, X., Dillman, J. F., Dawson, V. L., and Dawson, T. M. (1997) Transcriptional regulation of nNOS during neuronal development and plasticity is controled by multiple promoters. *Soc. Neurosci. Abstr.* **23,** 2225.

20. Sasaki, M., Huang, H., Ahn, S., Ginty, D. D., Dawson, V. L., and Dawson, T. M. (1998) Expression of neuronal nitric oxide synthase (nNOS) is regulated through calcium/cyclic AMP response element. *Soc. Neurosci. Abst.* **24,** 1247.

21. Darius, S., Wolf, G., Huang, P. L., and Fishman, M. C. (1995) Localization of NADPH-diaphorase/nitric oxide synthase in the rat retina: an electron microscopic study. *Brain Res.* **690,** 231–235.

22. Ponting, C. P. and Phillips, C. (1995) DHR domains in syntrophins, neuronal NO synthases and other intracellular proteins [Letter]. *Trends Biochem. Sci.* **20,** 102–103.

23. Brenman, J. E., Chao, D. S., Xia, H., Aldape, K., and Bredt, D. S. (1995) Nitric oxide synthase complexed with dystrophin and absent from skeletal muscle sarco-lemma in Duchenne muscular dystrophy. *Cell* **82,** 743–752.

24. Brenman, J. E., Chao, D. S., Gee, S. H., McGee, A. W., Craven, S. E., Santillano, D. R., Wu, Z., Huang, F., Xia, H., Peters, M. F., Froehner, S. C., and Bredt, D. S. (1996) Interaction of nitric oxide synthase with the postsynaptic density protein PSD-95 and alpha1-syntrophin mediated by PDZ domains. *Cell* **84,** 757–767.

25. Lipton, S. A., Choi, Y. B., Pan, Z. H., Lei, S. Z., Chen, H. S., Sucher, N. J., Loscalzo, J., Singel, D. J., and Stamler, J. S. (1993) A redox-based mechanism for the neuroprotective and neurodestructive effects of nitric oxide and related nitroso-compounds. *Nature* **364,** 626–632.

26. Salvemini, D., Misko, T. P., Masferrer, J. L., Seibert, K., Currie, M. G., and Needleman, P. (1993) Nitric oxide activates cyclooxygenase enzymes. *Proc. Natl. Acad. Sci. USA* **90,** 7240–7244.

27. Henry, Y., Lepoivre, M., Drapier, J. C., Ducrocq, C., Boucher, J. L., and Guissani, A. (1993) EPR characterization of molecular targets for NO in mammalian cells and organelles. *FASEB J.* **7,** 1124–1134.

28. Drapier, J. C., Hirling, H., Wietzerbin, J., Kaldy, P., and Kuhn, L. C. (1993) Bio-synthesis of nitric oxide activates iron regulatory factor in macrophages. *EMBO J.* **12,** 3643–3649.

29. Weiss, G., Goossen, B., Doppler, W., Fuchs, D., Pantopoulos, K., Werner-Felmayer, G., Wachter, H., and Hentze, M. W. (1993) Translational regulation via iron-respon-sive elements by the nitric oxide/NO-synthase pathway. *EMBO J.* **12,** 3651–3657.

30. Jaffrey, S. R., Cohen, N. A., Rouault, T. A., Klausner, R. D., and Snyder, S. H. (1994) The iron-responsive element binding protein: a target for synaptic actions of nitric oxide. *Proc. Natl. Acad. Sci. USA* **91,** 12,994–12,998.

31. Moro, M. A., Darley-Usmar, V. M., Goodwin, D. A., Read, N. G., Zamora-Pino, R., Feelisch, M., Radomski, M. W., and Moncada, S. (1994) Paradoxical fate and biological action of peroxynitrite on human platelets. *Proc. Natl. Acad. Sci. USA* **91,** 6702–6706.

32. Klug, D., Rabani, J., and Fridovich, I. (1972) A direct demonstration of the cata-lytic action of superoxide dismutase through the use of pulse radiolysis. *J. Biol. Chem.* **247,** 4839–4842.

33. Castro, L., Rodriguez, M., and Radi, R. (1994) Aconitase is readily inactivated by peroxynitrite, but not by its precursor, nitric oxide. *J. Biol. Chem.* **269,** 29,409–29,415.

34. Mirza, U. A., Chait, B. T., and Lander, H. M. (1995) Monitoring reactions of nitric oxide with peptides and proteins by electrospray ionization-mass spectrom-etry. *J. Biol. Chem.* **270,** 17,185–17,188.

35. Stamler, J. S., Simon, D. I., Osborne, J. A., Mullins, M. E., Jaraki, O., Michel, T., Singel, D. J., and Loscalzo, J. (1992) *S*-Nitrosylation of proteins with nitric oxide:

synthesis and characterization of biologically active compounds. *Proc. Natl. Acad. Sci. USA* **89**, 444–448.

36. Fagni, L., Olivier, M., Lafon-Cazal, M., and Bockaert, J. (1995) Involvement of divalent ions in the nitric oxide-induced blockade of *N*-methyl-D-aspartate receptors in cerebellar granule cells. *Mol. Pharmacol.* **47**, 1239–1247.

37. Dimmeler, S., Lottspeich, F., and Brune, B. (1992) Nitric oxide causes ADP-ribosylation and inhibition of glyceraldehyde-3-phosphate dehydrogenase. *J. Biol. Chem.* **267**, 16,771–16,774.

38. Zhang, J. and Snyder, S. H. (1992) Nitric oxide stimulates auto-ADP-ribosylation of glyceraldehyde-3-phosphate dehydrogenase. *Proc. Natl. Acad. Sci. USA* **89**, 9382–9385.

39. Jia, L., Bonaventura, J., and Stamler, J. S. (1996) *S*-Nitrosohaemoglobin: a dynamic activity of blood involved in vascular control. *Nature* **380**, 221–226.

40. Bredt, D. S., Hwang, P. M., and Snyder, S. H. (1990) Localization of nitric oxide synthase indicating a neural role for nitric oxide. *Nature* **347**, 768–770.

41. Huang, P. L., Dawson, T. M., Bredt, D. S., Snyder, S. H., and Fishman, M. C. (1993) Targeted disruption of the neuronal nitric oxide synthase gene. *Cell* **75**, 1273–1286.

42. Vanderwinden, J. M., Mailleux, P., Schiffmann, S. N., Vanderhaeghen, J. J., and De Laet, M. H. (1992) Nitric oxide synthase activity in infantile hypertrophic pyloric stenosis. *N. Engl. J. Med.* **327**, 511–515.

43. Chung, E., Curtis, D., Chen, G., Marsden, P. A., Twells, R., Xu, W., and Gardiner, M. (1996) Genetic evidence for the neuronal nitric oxide synthase gene (NOS1) as a susceptibility locus for infantile pyloric stenosis. *Am. J. Hum. Genet.* **58**, 363–370.

44. Mashimo, H., He, X. D., Huang, P. L., Fishman, M. C., and Goyal, R. K. (1996) Neuronal constitutive nitric oxide synthase is involved in murine enteric inhibitory neurotransmission. *J. Clin. Invest.* **98**, 8–13.

45. Burnett, A. L., Nelson, R. J., Calvin, D. C., Liu, J.-X., Demas, G. E., Klein, S. L., Kriegsfeld, L. J., Dawson, V. L., Dawson, T. M., and Snyder, S. H. (1996) Nitric oxide dependent penile erection in mice lacking neuronal nitric oxide synthase. *Mol. Med.* **2**, 288–296.

46. Burnett, A. L., Lowenstein, C. J., Bredt, D. S., Chang, T. S., and Snyder, S. H. (1992) Nitric oxide: a physiologic mediator of penile erection. *Science* **257**, 401–403.

47. Rajfer, J., Aronson, W. J., Bush, P. A., Dorey, F. J., and Ignarro, L. J. (1992) Nitric oxide as a mediator of relaxation of the corpus cavernosum in response to nonadrenergic, noncholinergic neurotransmission. *N. Engl. J. Med.* **326**, 90–94.

48. Burnett, A. L., Calvin, D. C., Chamness, S. L., Liu, J. X., Nelson, R. J., Klein, S. L., Dawson, V. L., Dawson, T. M., and Snyder, S. H. (1997) Urinary bladder-urethral sphincter dysfunction in mice with targeted disruption of neuronal nitric oxide synthase models idiopathic voiding disorders in humans. *Nat. Med.* **3**, 571–574.

49. Rezvani, A. H., Grady, D. R., Peek, A. E., and Pucilowski, O. (1995) Inhibition of nitric oxide synthesis attenuates alcohol consumption in two strains of alcohol-preferring rats. *Pharmacol. Biochem. Behav.* **50**, 265–270.

50. Ding, J. M., Chen, D., Weber, E. T., Faiman, L. E., Rea, M. A., and Gillette, M. U. (1994) Resetting the biological clock: mediation of nocturnal circadian shifts by glutamate and NO. *Science* **266**, 1713–1717.

51. Nelson, R. J., Demas, G. E., Huang, P. L., Fishman, M. C., Dawson, V. L., Dawson, T. M., and Snyder, S. H. (1995) Behavioural abnormalities in male mice lacking neuronal nitric oxide synthase. *Nature* **378,** 383–386.

52. Linden, D. J., Dawson, T. M., and Dawson, V. L. (1995) An evaluation of the nitric oxide/cGMP/cGMP-dependent protein kinase cascade in the induction of cerebellar long-term depression in culture. *J. Neurosci.* **15,** 5098–5105.

53. O'Dell, T. J., Huang, P. L., Dawson, T. M., Dinerman, J. L., Snyder, S. H., Kandel, E. R., and Fishman, M. C. (1994) Endothelial NOS and the blockade of LTP by NOS inhibitors in mice lacking neuronal NOS. *Science* **265,** 542–546.

54. Zerani, M. and Gobbetti, A. (1996) NO sexual behaviour in newts. *Nature* **382,** 31.

55. Demas, G. E., Eliasson, M. J., Dawson, T. M., Dawson, V. L., Kriegsfeld, L. J., Nelson, R. J., and Snyder, S. H. (1997) Inhibition of neuronal nitric oxide synthase increases aggressive behavior in mice. *Mol. Med.* **3,** 610–616.

56. Bohme, G. A., Bon, C., Stutzmann, J. M., Doble, A., and Blanchard, J. C. (1991) Possible involvement of nitric oxide in long-term potentiation. *Eur. J. Pharmacol.* **199,** 379–381.

57. O'Dell, T. J., Hawkins, R. D., Kandel, E. R., and Arancio, O. (1991) Tests of the roles of two diffusible substances in long-term potentiation: evidence for nitric oxide as a possible early retrograde messenger. *Proc. Natl. Acad. Sci. USA* **88,** 11,285–11,289.

58. Schuman, E. M. and Madison, D. V. (1991) A requirement for the intercellular messenger nitric oxide in long-term potentiation. *Science* **254,** 1503–1506.

59. Arancio, O., Kiebler, M., Lee, C. J., Lev-Ram, V., Tsien, R. Y., Kandel, E. R., and Hawkins, R. D. (1996) Nitric oxide acts directly in the presynaptic neuron to produce long-term potentiation in cultured hippocampal neurons. *Cell* **87,** 1025–1035.

60. Daniel, H., Hemart, N., Jaillard, D., and Crepel, F. (1993) Long-term depression requires nitric oxide and guanosine 3':5' cyclic monophosphate production in rat cerebellar Purkinje cells. *Eur. J. Neurosci.* **5,** 1079–1082.

61. Hawkins, R. D., Zhuo, M., and Arancio, O. (1994) Nitric oxide and carbon monoxide as possible retrograde messengers in hippocampal long-term potentiation. *J. Neurobiol.* **25,** 652–665.

62. Lev-Ram, V., Jiang, T., Wood, J., Lawrence, D. S., and Tsien, R. Y. (1997) Synergies and coincidence requirements between NO, cGMP, and Ca2+ in the induction of cerebellar long-term depression. *Neuron* **18,** 1025–1038.

63. Dinerman, J. L., Dawson, T. M., Schell, M. J., Snowman, A., and Snyder, S. H. (1994) Endothelial nitric oxide synthase localized to hippocampal pyramidal cells: implications for synaptic plasticity. *Proc. Natl. Acad. Sci. USA* **91,** 4214–4218.

64. Kantor, D. B., Lanzrein, M., Stary, S. J., Sandoval, G. M., Smith, W. B., Sullivan, B. M., Davidson, N., and Schuman, E. M. (1996) A role for endothelial NO synthase in LTP revealed by adenovirus-mediated inhibition and rescue. *Science* **274,** 1744–1748.

65. Son, H., Hawkins, R. D., Martin, K., Kiebler, M., Huang, P. L., Fishman, M. C., and Kandel, E. R. (1996) Long-term potentiation is reduced in mice that are doubly mutant in endothelial and neuronal nitric oxide synthase. *Cell* **87,** 1015–1023.

66. Hirsch, D. B., Steiner, J. P., Dawson, T. M., Mammen, A., and Snyder, S. H. (1993) Neurotransmitter release regulated by nitric oxide in PC-12 cells and brain synaptosomes. *Curr. Biol.* **3,** 749–754.

67. Muller, U. (1996) Inhibition of nitric oxide synthase impairs a distinct form of long-term memory in the honeybee, *Apis mellifera*. *Neuron* **16**, 541–549.
68. Seilicovich, A., Duvilanski, B. H., Pisera, D., Theas, S., Gimeno, M., Rettori, V., and McCann, S. M. (1995) Nitric oxide inhibits hypothalamic luteinizing hormone-releasing hormone release by releasing gamma-aminobutyric acid. *Proc. Natl. Acad. Sci. USA* **92**, 3421–3424.
69. Meffert, M. K., Premack, B. A., and Schulman, H. (1994) Nitric oxide stimulates Ca(2+)-independent synaptic vesicle release. *Neuron* **12**, 1235–1244.
70. Ahmad, I., Leinders-Zufall, T., Kocsis, J. D., Shepherd, G. M., Zufall, F., and Barnstable, C. J. (1994) Retinal ganglion cells express a cGMP-gated cation conductance activatable by nitric oxide donors. *Neuron* **12**, 155–165.
71. Meffert, M. K., Calakos, N. C., Scheller, R. H., and Schulman, H. (1996) Nitric oxide modulates synaptic vesicle docking fusion reactions. *Neuron* **16**, 1229–1236.
72. Bredt, D. S. and Snyder, S. H. (1994) Transient nitric oxide synthase neurons in embryonic cerebral cortical plate, sensory ganglia, and olfactory epithelium. *Neuron* **13**, 301–313.
73. Peunova, N. and Enikolopov, G. (1995) Nitric oxide triggers a switch to growth arrest during differentiation of neuronal cells. *Nature* **375**, 68–73.
74. Kuzin, B., Roberts, I., Peunova, N., and Enikolopov, G. (1996) Nitric oxide regulates cell proliferation during Drosophila development. *Cell* **87**, 639–649.
75. Kalb, R. G. and Agostini, J. (1993) Molecular evidence for nitric oxide-mediated motor neuron development. *Neuroscience* **57**, 1–8.
76. Williams, C. V., Nordquist, D., and McLoon, S. C. (1994) Correlation of nitric oxide synthase expression with changing patterns of axonal projections in the developing visual system. *J. Neurosci.* **14**, 1746–1755.
77. Roskams, A. J., Bredt, D. S., Dawson, T. M., and Ronnett, G. V. (1994) Nitric oxide mediates the formation of synaptic connections in developing and regenerating olfactory receptor neurons. *Neuron* **13**, 289–299.
78. Lander, H. M., Jacovina, A. T., Davis, R. J., and Tauras, J. M. (1996) Differential activation of mitogen-activated protein kinases by nitric oxide-related species. *J. Biol. Chem.* **271**, 19,705–19,709.
79. Yun, H. Y., Gonzalez-Zulueta, M., Dawson, V. L., and Dawson, T. M. (1998) Nitric oxide mediates *N*-methyl-D-aspartate receptor-induced activation of p21ras. *Proc. Natl. Acad. Sci. USA* **95**, 5773–5778.
80. Stamler, J. S., Toone, E. J., Lipton, S. A., and Sucher, N. J. (1997) (*S*)NO signals: translocation, regulation, and a consensus motif. *Neuron* **18**, 691–696.
81. Mott, H. R., Carpenter, J. W., and Campbell, S. L. (1997) Structural and functional analysis of a mutant Ras protein that is insensitive to nitric oxide activation. *Biochemistry* **36**, 3640–3644.
82. Finkbeiner, S. and Greenberg M. E. (1996) Ca(2+)-dependent routes to Ras: mechanisms for neuronal survival, differentiation, and plasticity? *Neuron* **16**, 233–236.
83. Dawson, T. M., Dawson, V. L., and Snyder, S. H. (1992) A novel neuronal messenger molecule in brain: the free radical, nitric oxide. *Ann. Neurol.* **32**, 297–311.
84. Choi, D. W. (1988) Glutamate neurotoxicity and diseases of the nervous system. *Neuron* **1**, 623–634.

85. Dawson, V. L., Dawson, T. M., London, E. D., Bredt, D. S., and Snyder, S. H. (1991) Nitric oxide mediates glutamate neurotoxicity in primary cortical cultures. *Proc. Natl. Acad. Sci. USA* **88,** 6368–6371.

86. Choi, D. W. and Rothman, S. M. (1990) The role of glutamate neurotoxicity in hypoxic-ischemic neuronal death. *Annu. Rev. Neurosci.* **13,** 171–182.

87. Dawson, V. L., Kizushi, V. M., Huang, P. L., Snyder, S. H., and Dawson, T. M. (1996) Resistance to neurotoxicity in cortical cultures from neuronal nitric oxide synthase-deficient mice. *J. Neurosci.* **16,** 2479–2487.

88. Samdani, A. F., Newcamp, C., Resink, A., Facchinetti, F., Hoffman, B. E., Dawson, V. L., and Dawson, T. M. (1997) Differential susceptibility to neurotoxicity mediated by neurotrophins and neuronal nitric oxide synthase. *J. Neurosci.* **17,** 4633–4641.

89. Hewett, S. J., Csernansky, C. A., and Choi, D. W. (1994) Selective potentiation of NMDA-induced neuronal injury following induction of astrocytic iNOS. *Neuron* **13,** 487–494.

90. Koh, J. Y., Gwag, B. J., Lobner, D., and Choi, D. W. (1995) Potentiated necrosis of cultured cortical neurons by neurotrophins. *Science* **268,** 573–575.

91. Choi, D. W. (1994) Glutamate receptors and the induction of excitotoxic neuronal death. *Prog. Brain Res.* **100,** 47–51.

92. Ferrante, R. J., Kowall, N. W., Beal, M. F., Richardson, E. P., Jr., Bird, E. D., and Martin, J. B. (1985) Selective sparing of a class of striatal neurons in Huntington's disease. *Science* **230,** 561–563.

93. Hyman, B. T., Marzloff, K., Wenniger, J. J., Dawson, T. M., Bredt, D. S., and Snyder, S. H. (1992) Relative sparing of nitric oxide synthase-containing neurons in the hippocampal formation in Alzheimer's disease. *Ann. Neurol.* **32,** 818–820.

94. Uemura, Y., Kowall, N. W., and Beal, M. F. (1990) Selective sparing of NADPH-diaphorase-somatostatin-neuropeptide Y neurons in ischemic gerbil striatum. *Ann. Neurol.* **27,** 620–625.

95. Gonzalez-Zulueta, M., Ensz, L. M., Mukhina, G., Lebovitz, R. M., Zwacka, R. M., Engelhardt, J. F., Oberley, L. W., Dawson, V. L., and Dawson, T. M. (1998) Manganese superoxide dismutase protects nNOS neurons from NMDA and nitric oxide-mediated neurotoxicity. *J. Neurosci.* **18,** 2040–2055.

96. Radi, R., Beckman, J. S., Bush, K. M., and Freeman, B. A. (1991) Peroxynitrite-induced membrane lipid peroxidation: the cytotoxic potential of superoxide and nitric oxide. *Arch. Biochem. Biophys.* **288,** 481–487.

97. Bonfoco, E., Krainc, D., Ankarcrona, M., Nicotera, P., and Lipton, S. A. (1995) Apoptosis and necrosis: two distinct events induced, respectively, by mild and intense insults with *N*-methyl-D-aspartate or nitric oxide/superoxide in cortical cell cultures. *Proc. Natl. Acad. Sci. USA* **92,** 7162–7166.

98. Keller, J. N., Kindy, M. S., Holtsberg, F. W., St Clair, D. K., Yen, H. C., Germeyer, A., Steiner, S. M., Bruce-Keller, A. J., Hutchins, J. B., and Mattson, M. P. (1998) Mitochondrial manganese superoxide dismutase prevents neural apoptosis and reduces ischemic brain injury: suppression of peroxynitrite production, lipid peroxidation, and mitochondrial dysfunction. *J. Neurosci.* **18,** 687–697.

99. Dawson, D. A. (1994) Nitric oxide and focal cerebral ischemia: multiplicity of actions and diverse outcome. *Cerebrovasc. Brain Metab. Rev.* **6,** 299–324.

100. Huang, Z., Huang, P. L., Panahian, N., Dalkara, T., Fishman, M. C., and Moskowitz, M. A. (1994) Effects of cerebral ischemia in mice deficient in neuronal nitric oxide synthase. *Science* **265,** 1883–1885.

101. Dalkara, T., Yoshida, T., Irikura, K., and Moskowitz, M. A. (1994) Dual role of nitric oxide in focal cerebral ischemia. *Neuropharmacology* **33,** 1447–1452.
102. Yoshida, T., Limmroth, V., Irikura, K., and Moskowitz, M. A. (1994) The NOS inhibitor, 7-nitroindazole, decreases focal infarct volume but not the response to topical acetylcholine in pial vessels. *J. Cereb. Blood Flow Metab.* **14,** 924–929.
103. Huang, Z., Huang, P. L., Ma, J., Meng, W., Ayata, C., Fishman, M. C., and Moskowitz, M. A. (1996) Enlarged infarcts in endothelial nitric oxide synthase knockout mice are attenuated by nitro-L-arginine. *J. Cereb. Blood Flow Metab.* **16,** 981–987.
104. Iadecola, C., Zhang, F., Xu, S., Casey, R., and Ross, M. E. (1995) Inducible nitric oxide synthase gene expression in brain following cerebral ischemia. *J. Cereb. Blood Flow Metab.* **15,** 378–384.
105. Iadecola, C. (1997) Bright and dark sides of nitric oxide in ischemic brain injury. *Trends Neurosci.* **20,** 132–139.
106. Oury, T. D., Ho, Y. S., Piantadosi, C. A., and Crapo, J. D. (1992) Extracellular superoxide dismutase, nitric oxide, and central nervous system O_2 toxicity. *Proc. Natl. Acad. Sci. USA* **89,** 9715–9719.
107. Schulz, J. B., Matthews, R. T., Jenkins, B. G., Ferrante, R. J., Siwek, D., Henshaw, D. R., Cipolloni, P. B., Mecocci, P., Kowall, N. W., Rosen, B. R., et al. (1995) Blockade of neuronal nitric oxide synthase protects against excitotoxicity in vivo. *J. Neurosci.* **15,** 8419–8429.
108. Schulz, J. B., Matthews, R. T., Muqit, M. M., Browne, S. E., and Beal, M. F. (1995) Inhibition of neuronal nitric oxide synthase by 7-nitroindazole protects against MPTP-induced neurotoxicity in mice. *J. Neurochem.* **64,** 936–939.
109. Przedborski, S., Jackson-Lewis, V., Yokoyama, R., Shibata, T., Dawson, V. L., and Dawson, T. M. (1996) Role of neuronal nitric oxide in 1-methyl-4-phenyl-1,2,3,6-tetrahydropyridine (MPTP)-induced dopaminergic neurotoxicity. *Proc. Natl. Acad. Sci. USA* **93,** 4565–4571.
110. Hantraye, P., Brouillet, E., Ferrante, R., Palfi, S., Dolan, R., Matthews, R. T., and Beal, M. F. (1996) Inhibition of neuronal nitric oxide synthase prevents MPTP-induced parkinsonism in baboons. *Nat. Med.* **2,** 1017–1021.
111. Dawson, V. L., Dawson, T. M., Uhl, G. R., and Snyder, S. H. (1993) Human immunodeficiency virus type 1 coat protein neurotoxicity mediated by nitric oxide in primary cortical cultures. *Proc. Natl. Acad. Sci. USA* **90,** 3256–3259.
112. Koka, P., He, K., Zack, J. A., Kitchen, S., Peacock, W., Fried, I., Tran, T., Yashar, S. S., and Merrill, J. E. (1995) Human immunodeficiency virus 1 envelope proteins induce interleukin 1, tumor necrosis factor alpha, and nitric oxide in glial cultures derived from fetal, neonatal, and adult human brain. *J. Exp. Med.* **182,** 941–951.
113. Adamson, D. C., Wildemann, B., Sasaki, M., Glass, J. D., McArthur, J. C., Christov, V. I., Dawson, T. M., and Dawson, V. L. (1996) Immunologic NO synthase: elevation in severe AIDS dementia and induction by HIV-1 gp41. *Science* **274,** 1917–1921.
114. Zhang, J., Dawson, V. L., Dawson, T. M., and Snyder, S. H. (1994) Nitric oxide activation of poly(ADP-ribose) synthetase in neurotoxicity. *Science* **263,** 687–689.
115. Wallis, R. A., Panizzon, K. L., Henry, D., and Wasterlain, C. G. (1993) Neuroprotection against nitric oxide injury with inhibitors of ADP-ribosylation. *NeuroReport* **5,** 245–248.

116. Heller, B., Wang, Z. Q., Wagner, E. F., Radons, J., Burkle, A., Fehsel, K., Burkart, V., and Kolb, H. (1995) Inactivation of the poly(ADP-ribose) polymerase gene affects oxygen radical and nitric oxide toxicity in islet cells. *J. Biol. Chem.* **270,** 11,176–11,180.

117. Eliasson, M. J., Sampei, K., Mandir, A. S., Hurn, P. D., Traystman, R. J., Bao, J., Pieper, A., Wang, Z. Q., Dawson, T. M., Snyder, S. H., and Dawson, V. L. (1997) Poly(ADP-ribose) polymerase gene disruption renders mice resistant to cerebral ischemia. *Nat. Med.* **3,** 1089–1095.

118. Brown, G. C. (1995) Nitric oxide regulates mitochondrial respiration and cell functions by inhibiting cytochrome oxidase. *FEBS Lett.* **369,** 136–139.

119. MacMillan-Crow, L. A., Crow, J. P., Kerby, J. D., Beckman, J. S., and Thompson, J. A. (1996) Nitration and inactivation of manganese superoxide dismutase in chronic rejection of human renal allografts. *Proc. Natl. Acad. Sci. USA* **93,** 11,853–11,858.

120. Strijbos, P. J., Leach, M. J., and Garthwaite, J. (1996) Vicious cycle involving Na^+ channels, glutamate release, and NMDA receptors mediates delayed neurodegeneration through nitric oxide formation. *J. Neurosci.* **16,** 5004–5013.

7

Cellular Signaling Pathways in Neuronal Apoptosis

Role in Neurodegeneration and Alzheimer's Disease

Carl W. Cotman, Haoyu Qian, and Aileen J. Anderson

INTRODUCTION

Apoptosis is a fundamental biological process that occurs during development and continues throughout life, serving to eliminate excess or injured cells. Whereas controlled apoptosis is of absolute necessity and benefit to an organism, aberrant regulation of this process is involved in a variety of diseases. In particular, recent evidence suggests that apoptosis may be reinitiated pathologically during aging by a variety of stimuli, causing the loss of a significant number of neurons and leading to dementia.

Neurons are particularly vulnerable to a variety of apoptotic inducers that can operate through several distinct apoptotic pathways based on the nature of the particular inducer and specific neuron type. The goal of this chapter is to describe what we know about the molecular pathways regulating neuronal apoptosis with respect to the uniqueness of neurons as nondividing cells with long processes. We also evaluate the hypothesis that apoptosis is a mechanism that contributes to neuronal death in Alzheimer's disease (AD) and discuss various findings in AD brain in relationship to molecular apoptotic mechanisms. As nondividing cells, neurons appear to have a set of strategies that may initiate "checkpoints" to allow for possible damage repair and promote survival. This "apoptosis checkpoint cascade" is evident in the Alzheimer's disease (AD) brain and involves the parallel induction of a series of antiapoptotic factors and repair proteins. We will discuss the possibility that distinct apoptotic pathways and an apoptosis checkpoint cascade might help to explain the evolution of neuropathology in AD brain (i.e., neurofibrillary tangle formation).

From: *Cerebral Signal Transduction: From First to Fourth Messengers*
Edited by: M. E. A. Reith © Humana Press Inc., Totowa, NJ

Neurons are unusual as cells, in that they extend processes over long distances. As a result, neuronal processes may encounter local apoptotic inducers and microenvironments that may place the cell at risk. Neurons may be able to control damaged processes by activating local "neuritic apoptosis", a mechanism that appears to involve caspase activation and other features common to apoptosis. Recent studies in cell culture in parallel with those in postmortem AD brain are beginning to define the characteristics of neuronal apoptosis and the nature of the molecular effector pathways activated. This strategy is essential for developing new therapeutics and refining existing ones.

INDUCERS OF APOPTOSIS EXIST IN THE AD BRAIN

Apoptosis can be induced in most neurons by a variety of stimuli, a number of which appear to accumulate in the aging and AD brain. In the aging and AD brain, the β-amyloid protein (Aβ), a 40–42 amino acid peptide, accumulates in the extracellular space as small deposits and senile plaques. Based on the observation that neurites surrounding Aβ deposits show both sprouting and degenerative responses, we proposed that this peptide is not metabolically inert, but rather possesses biological activity. Our findings established two key principles: that Aβ induces neurotoxicity in a conformation-specific manner and that apoptotic mechanisms underlie this toxicity *(1,2)*; these observations have since been confirmed by many others. Interestingly, prior to causing cell death, Aβ also induces the formation of dystrophic-like neurite morphology in cultured neurons *(3,4)*. Oxidative insult readily initiates apoptosis *(5)*, which is known to occur in the aging and AD brain *(6)*. Similarly, reductions in glucose metabolism have been suggested to contribute to neurodegeneration in AD *(7)*. Furthermore, excitotoxic damage can, under some conditions, initiate apoptosis, and many investigators have suggested that excitotoxic damage contributes to neurodegenerative diseases, including AD *(9)*. Recent studies have also shown that glutamate transport proteins may be greatly reduced in the AD brain *(10,11)*, which could exacerbate excitotoxic mechanisms. The profile of initiating factors strongly suggests that in the course of aging and age-related neurodegenerative disease, neurons are increasingly subjected to apoptotic inducers. In some cases, these factors could act synergistically. For example, neuronal apoptosis can be significantly potentiated by the addition of subthreshold doses of Aβ and either excitotoxic or oxidative insult or hypoglycemia *(8, 12–14)*. Finally, mitochondrial damage may contribute to apoptosis as an intracellular effector. Mitochondria are a major source of free radicals and the release of cytochrome-*c* is a potent inducer of caspase activation. Indeed, this organelle may be a prime target of aging-related insults in AD brain and thus a contributor to the apoptosis ascade.

Clearly, ample potential for apoptotic injury can be anticipated in the AD brain. In addition to the risk factors discussed, genetic risk factors associated with AD may increase neuronal risk for apoptosis.

EVIDENCE FOR APOPTOSIS IN SPORADIC AD

It is now well established that there is dramatic neuronal loss during the course of AD. A key question that has emerged in the last several years is whether the mechanism for that loss is apoptotic, necrotic, or proceeds by an entirely alternative set of pathways, virtually unique to neurons.

Apoptosis is generally defined on the basis of strict morphological criteria; the basic features include cell shrinkage, membrane blebbing, chromatin condensation, and nuclear fragmentation. Additionally, the cleavage of DNA into oligonucleosome-length fragments detectable by gel electrophoresis occurs in many, but not all models of apoptosis *(15–18)*. The process of DNA degradation produces a series of oligonucleosome-length DNA fragments that have newly generated 3'-OH ends. These strand breaks can be labeled by methods such as terminal transferase-mediated dUTP nick-end labeling (TUNEL). We have recently reported that cells in the AD brain exhibit labeling for DNA-strand breaks using TUNEL, and that many TUNEL-labeled cells exhibit an apoptoticlike morphological distribution of DNA strand breaks, including granulated and marginated patterns of intense TUNEL labeling, shrunken, irregular cellular shape, and the presence of apoptoticlike bodies consistent with apoptosis *(19,20)*. We observe low levels of TUNEL labeling in control cases, but frequent TUNEL-positive nuclei in brain tissue from AD cases. Significantly, we also observe numerous examples of cells with similar morphologies in both rapid autopsy samples and in tissue from cases in which AD pathology has not yet progressed sufficiently to meet the Consortium to Establish a Registry for Alzheimer Disease (CERAD) criteria for AD *(19,21)*. Consistent with our results, several investigators have reported similar findings in AD *(22–24)*.

Altered expression levels of several proteins in AD brain further implicates apoptosis as a mechanism in neurodegeneration. For example, the immediate early gene c-*Jun* is selectively induced in cultured neurons undergoing Aβ-amyloid-mediated apoptosis; additionally, immunoreactivity for c-Jun is elevated in AD and is colocalized with TUNEL-positive neurons *(19)*. Other proteins that have been shown to be pro-apoptotic in cultured neurons such as p53, Fas, and Bax all have been found to have an increased level of expression in AD brain in association with pathology *(25,26)*.

The transduction of apoptotic signals and the orchestration of apoptosis requires the coordinated action of several caspases, aspartate-specific cysteine proteases of which 13+ homologs have been identified to date. In addition to the presence of elevated pro-apoptotic markers, AD brain also exhibits evidence for caspase cleavage products. It is known from cell-culture experiments that actin is one of the targets of caspases. If apoptosis occurs in the AD brain, then caspase-dependent actin breakdown products may be generated and be detectable. As predicted, neurons in the AD brain stain for caspase cleavage products of actin (27). As our molecular tool base expands, it will be feasible to determine in more detail the nature of the caspase activation patterns in the AD brain. These will provide great insight into the molecular pathways in the apoptotic cascade. It is predicted that this approach is a productive and major growth area for the future.

CASPASE ACTIVATION AND DISTINCT
PATHWAYS IN APOPTOTIC CELL DEATH

Based on structural similarities, caspases can be classified into three subfamilies: the caspase 1 (ICE) subfamily, the caspase 2 (Nedd-2/Ich-1) subfamily, and the caspase 3 (CPP32/Yama/apopain) subfamily All caspases are expressed in normal cells in an enzymatically inactive form (pro-caspase) and are activated when cleaved after aspartate residues in conserved processing sites C-terminal to the catalytic cysteine residue. A principle in the field is that these proteases operate in cascades with various specific activators and functions in the death program. Examination of the caspase family and interactions between these and other cell death effector or inhibitor proteins, such as the Bcl-2 family of cell-death-associated proteins, has led to a clearer understanding of the molecular pathways governing apoptosis.

Staurosporine (STS) and activation of the Fas/Apo-1 receptor are examples of inducers of apoptosis that depend on caspase proteases. Current research on these two prototypical apoptotic stimuli suggest that at least two divergent pathways of apoptosis can be identified in a variety of cells. For example, the cowpox virus protein CrmA is known to block caspase 8 activity. Caspases 8 and 9 are upstream proteases that first receive apoptotic signals, then activate multiple other downstream effector caspases, including caspase 3, 6, and 7, as well as interact with each other to generate protease amplification cycles (28). CrmA inhibits Fas/Apo-1, tumor necrosis factor (TNF), and growth factor withdrawal-induced apoptosis (29,30), as well as loss of substrate attachment-mediated apoptosis (anoikis) (31); however, CrmA does not inhibit apoptosis induced by staurosporine or DNA-damaging agents (32,33). These data suggest the caspase 8 activation is selectively involved in some apoptotic cell death pathways but not others. Thus, it appears that, as illustrated in Fig. 1, there are at least two caspase-driven pathways of apoptosis that are stimulus-specific and show differential sensitivities to inhibition by CrmA and the Bcl-2 family.

We have recently examined these selective pathways in neuronal cell lines. Using stable transfectants of PC12 cells, we have demonstrated the existence of apoptotic pathways activated by concanavalin A and staurosporine, which are selectively blocked by CrmA and Bcl-2, respectively *(34)*. Importantly, CrmA-expressing cells were resistant to apoptosis induced by Aβ. Furthermore, gene transfer of CrmA into primary cultures of hippocampal neurons also conferred resistance to Aβ toxicity, but not to toxicity induced by STS (Fig. 1). These results suggest that similar apoptotic pathways exist in neurons and indicate that apoptosis induced by Aβ is likely to involve cross-linking of cell-surface receptors and activation of caspase 8.

Alternative downstream pathways may also operate and drive apoptosis, depending on the nature of the stimulus and the cell type. It is increasingly clear that cells can use many pathways in downstream cascades to accomplish and regulate apoptosis. Regulation of this process in different cell types may be as specialized as the process of differentiation. Thus, for example, apoptosis of PC12 cells or sympathetic neurons induced by trophic factor withdrawal requires caspase 2 activation but not caspase 1 or 3 activity. Even though the induction of caspase-3-like activity was observed in this paradigm, it is neither sufficient nor necessary for cell death. On the other hand, apoptosis initiated by superoxide dismutase (SOD1) downregulation is mediated by caspase 1 activity *(35,36)*. In other studies, caspase 3 seems to be the preferential executioner of Fas-mediated apoptosis in vivo. Whereas wild-type and caspase 1 (−/−) hepatocytes show typical apoptotic features such as cytoplasmic blebbing and nuclear fragmentation within 6 h under the presence of FasL, caspase 3 (−/−) cells do not exhibit the morphological features of apoptosis, and DNA fragmentation in these cells was significantly delayed *(37)*. Caspase 3 has also been reported to be the central player in ceramide-mediated apoptosis of AK-5 tumor cells *(38)*. Taken together, these results suggest that distinct apoptotic pathways can be initiated in a stimulus-specific manner and that different cell types may regulate apoptosis differently. However, in many cases, crosstalk across multiple pathways in a single cell type is likely.

Despite differences in the pathways of apoptosis induction, the end products of apoptotic caspase activity appear to be invariant. In particular, the generation of active caspase fragments from their pro-caspase isoforms, cleavage of various targets of activated caspases (e.g., actin and fodrin, which

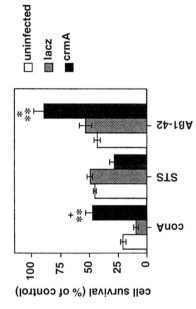

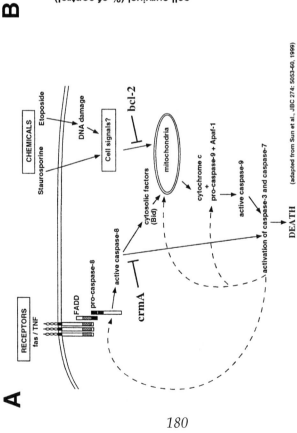

Fig. 1. (**A**) CrmA- and Bcl-2-dependent apoptotic pathways. (Adapted from ref. 28). (**B**) CrmA is protective against apoptotic cell death induced by Aβ1–42 in cultured PC12 cells. Data shown are mean ± S.E. ($n = 4$); + $p < 0.005$ vs ConA, * $p < 0.01$ vs Aβ1-4L, ** $p < 0.05$ versus uninfected.

generate specific breakdown products), and disruption of lipid membrane symmetry and phosphatidylserine exposure *(39,40)* appear to be important and reproducible events. These markers can, in turn, be used to make predictions and test hypotheses regarding the induction of specific and differentiable apoptotic protease cascades. Importantly, these markers represent a series of interconnected events in cellular apoptotic pathways.

BCL-2 FAMILY PROTEINS AND APOPTOSIS

As we have discussed, caspases are the effectors of cellular apoptosis, however, additional apoptotic regulatory families have been identified. Of these, the Bcl-2 family is the largest and perhaps best described. Overexpression of Bcl-2 protects a wide variety of cell types from apoptosis triggered by a variety of different stimuli, such as factor deprivation, irradiation, oxidative stress, or viral infection. Importantly, Bcl-2 is a member of a family of structurally related genes, including Bcl-2, Bcl-xL, Bcl-w, and Mcl-1, which inhibit apoptosis, and Bax, Bik, Bad, Bid, Bak, and Bcl-xs, which promote apoptosis.

Bcl-2 proteins can form homodimers or heterodimers with other family members, with one monomer antagonizing or enhancing the function of the other. Thus, the ratio of inhibitors versus activators may determine the likelihood of a cell to execute apoptosis *(41)*. For example, dimerization of the apoptosis-inducing members Bad or Bax with either Bcl-2 or Bcl-xL lowers the ability of these latter members to promote survival *(42,43)*. One mechanism by which the Bcl-2 family members can modulate dimerization is through phosphorylation. For example, one pro-apoptotic member of the family, Bad, can be phosphorylated by Akt, a kinase that can be activated by growth factors. Phosphorylated Bad loses its ability to heteromerize with Bcl-xL, thus permitting Bcl-xL to exert its survival-promoting effect even in the presence of Bax *(44)*. There are many hypothesis regarding exactly how Bcl-2 family homodimers and heterodimers act to regulate cell death. They include inhibition of antioxidant pathways, regulation of calcium homeostasis, participation in protein transport across membrane, interaction with signaling molecules such as Ras and Raf, and interaction with the mitochondria to regulate the release of cytochrome-*c (45)*.

The expression of Bcl-2 family members is correlated with neuronal viability. In sympathetic neurons, Bcl-2 expression decreases upon NGF deprivation; conversely, overexpression of Bcl-2 protects these neurons against NGF-deprivation-induced apoptosis *(46)*. Also, Bax mRNA and protein expressions increase in vulnerable regions of the brain after injury, implying a role for Bax in injury-induced neuronal generation in the brain *(47)*. Thus, proteins of the Bcl-2 family are critical regulators in the execution of cell death mechanisms.

In the AD brain, Bcl-2 might be expected to be downregulated and thus contribute to neurodegeneration. On the contrary, most neurons with DNA damage show an upregulation of Bcl-2. It is not uncommon to find cases where there is a strong colocalization of Bcl-2 expression with DNA damage in nearly all neurons in the field *(48)*. The situation with Bax is less clear-cut. In general there is an upregulation of Bax in the AD brain, but the pattern appears more variable. For example, some neurons with DNA damage show higher levels relative to surrounding nondamaged neurons, whereas others with DNA damage show lower levels *(26)*. The ratio of Bcl-2/Bax is probably a key determinate of cellular fate and may vary over time, depending on the success of the cell in repairing injuries.

CELL DIVISION CYCLE AND APOPTOSIS

Although the regulatory pathways of apoptosis discussed earlier are beginning to be elucidated, there are additional cellular pathways that influence cell survival and cell death via apoptosis. In particular, cell-cycle entry is known to be a critical point for the initiation of apoptosis. This mechanism exists as a part of a series of damage checkpoints in dividing cells. Interestingly, recent data have suggested that some cell-cycle proteins may be regulated by apoptosis regulatory proteins including Bax and Bcl-2 *(49)*.

The cell cycle (Fig. 2) is generally divided into four phases: the S phase associated with DNA replication, the M phase associated with mitosis, and two gap phase—G1 and G2—separating the S and M phases. Each phase of the cell cycle is associated with the expression of specific proteins. In particular, cyclins and cyclin-dependent kinases (CDK) are proteins that promote the synthesis of new proteins required for cell-cycle progression, whereas CDK inhibitor (CKI) proteins can inhibit cyclin/CDK complexes and prevent cell-cycle progression. Cell-cycle events are monitored during cell-cycle checkpoints that occur at the boundaries between phases. Cell-cycle progression can be blocked at these checkpoints in response to changes in both extracellular and intracellular signals. In addition, cycle arrest can also be induced upon detection of DNA damage. A normal dividing cell either re-enters and proceeds through the cell cycle after successfully repairing damage, or undergoes apoptosis if the damage is beyond repair. Thus, signals for cell division and apoptosis must be tightly coupled to each other to ensure a correct final outcome.

Neurons are postmitotic, terminally differentiated cells. Thus, in healthy adult neurons, cell-cycle markers are absent. A number of investigators have

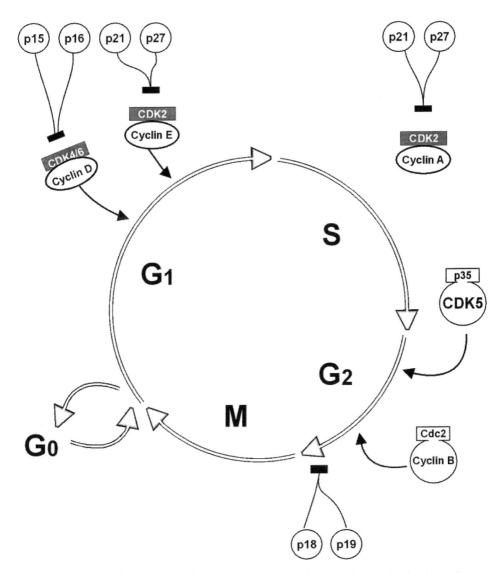

Fig. 2. The cell-division cycle is divided into four phases: DNA synthesis phase (S), mitotic phase (M), and two gap phases—G1 and G2. Transition from one phase of the cycle to the next is promoted by distinct cyclin and cyclin-dependent kinase (CDK) complexes and inhibited by cyclin kinase inhibitors such as p16, p21, and so forth.

proposed that neuronal apoptosis might result if neurons reenter the cell cycle and fail to complete it. This is supported by the observation that cells undergoing apoptosis display some morphological features of mitotic cells, such as chromatin condensation, breakdown of the nuclear envelope, and rounding of the cell body. Furthermore, pharmacological blockers of the G1/S phase, but not of the S, G2, and M phases of the cell-cycle protect neuronal cells against nerve growth factor (NGF)-deprivation-induced apoptosis *(50)* and inhibit caspase activation *(51)*. Conversely, overexpression of cyclin D1 induces apoptosis and transcript levels of cyclin D1 increase in NGF-deprived sympathetic neurons *(52)*. In addition, dominant negative mutants of CDK4 and CDK6, but not CDK2 and CDK3, can also inhibit neuronal apoptosis induced by NGF withdrawal *(53)*. Together, these data are consistent with the hypothesis that neurons could use cell-cycle regulatory components of the G1/S phase to initiate apoptosis. However, there are also data that seem to contradict this hypothesis. In particular, antisense cyclin D1 does not protect sympathetic neurons against NGF-deprivation-induced apoptosis *(54)*. Also, cyclin D1 expression is not increased in cerebellar granule neurons undergoing low-K^+ or serum-deprivation-induced apoptosis, implying that at least some populations of central nervous system (CNS) neurons do not require cyclin D1 for the induction of apoptosis *(55)*. An alternative hypothesis is that cyclins and CDKs are induced as a result of various insults to neurons, but whether or not they go on to induce apoptosis might depend on their interaction with CKIs and cell-cycle checkpoint proteins as well as the expression of other interacting proteins associated with apoptosis, such as Bcl-2. In dividing cells, CKIs and checkpoint proteins orchestrate cell-cycle arrest and repair, whereas in postmitotic cells such as neurons, they might also be able to delay apoptosis. Thus, cyclin D-induced apoptosis is prevented by coexpression of the CKI p16 *(56)*. Similarly, expression of the CKIs p16, p21, or p27 protects neurons from apoptotic death induced by NGF deprivation *(53)*. Conversely, the p21 antisense enhances death in neuroblastoma cells *(57)* and overexpression prevents death during myocyte differentiation *(58)*.

Cell-Cycle Markers Are Induced in AD Neurons and Appear to Anticipate Apoptosis

In the AD brain, there is accumulating evidence that some of the markers associated with the cell cycle are induced in those brain regions vulnerable to cell loss and neurofibrillary tangle (NFT) formation. Cyclin D in complex with CDK4 is an early event in the normal cell cycle. In the hippocampus, approximately 1% of neurons are cyclin D positive. The majority of these cells are found in regions CA1 and CA3/4. Proliferating cell nuclear antigen (PCNA), a auxiliary protein of DNA polymerase that is involved in DNA replication and repair, is normally expressed from the late G1 to the late G2/early M phase. This marker stains approximately 9% of hippocampal neurons and is particularly prominent in CA1 and CA3/4. Cyclin B1 is normally elevated in the cell cycle from G2 into mitosis. Interestingly, like PCNA, cyclin B1 is also increased in about 9% of hippocampal cells, and many neurons coexpressed these markers. The subcellular location of cyclin B1 and PCNA was most often cytoplasmic *(59)*. This largely cytoplasmic localization is not understood and could reflect a breakdown in intracellular trafficking, an as yet undefined mechanism or an artifact of postmortem conditions. Nonetheless, these findings suggest that cell-cycle-associated proteins are present in AD in vulnerable neuronal populations.

Cell-Cycle Inhibitors Are Induced in the AD Brain

There is good evidence that p16 is overexpressed in neurons in AD relative to control brain. NFTs and neurites show strong p16 immunoreactivity. In fact, on average, over 90% of NFT are p16 positive *(60)*. Neurons without overt fibrils are also labeled for p16. Imunoreactivity in these neurons reflects both diffuse cytoplasmic staining as well as granulovacular-type structures *(61)*. In addition to p16, other CKIs that bind directly to CDK4/6 also are elevated in the AD brain *(60)*. These data may indicate that neurons in the AD brain are being held in check *(61)*; that is, they enter the cell cycle, but rather than progressing through mitotic division, remain suspended in a state of damage alert and delayed, or checked, apoptosis. In correspondence with this checkpoint hypothesis of prolonged injury and delayed neuronal apoptosis, it is interesting that these observations parallel those of elevated Bcl-2 expression in AD neurons.

APOPTOSIS CHECKPOINT CASCADE IN THE AD BRAIN

Taken together, these data may suggest that neurons could activate an "apoptosis checkpoint cascade" in which injured neurons may regulate the activation of pro-apoptotic proteins such as Bax with antiapoptotic proteins such as Bcl-2. In addition, it can be hypothesized that damaged neurons could reenter the cell cycle and perhaps employ checkpoint molecules in a similar pro-apoptotic and antiapoptotic regulation point. This concept is illustrated in Fig. 3.

The presence of a mechanism to hold degeneration in check may provide an explanation for one of the seeming controversies in the AD literature. As we have described, TUNEL labeling provides evidence for active apoptosis in a large subset of neurons in the AD brain. However, many more AD neurons exhibit evidence for DNA damage in the absence of morphological changes indicative of terminal apoptosis (e.g., the formation of nuclear apoptotic bodies). In classical apoptosis, cells die within a few hours or

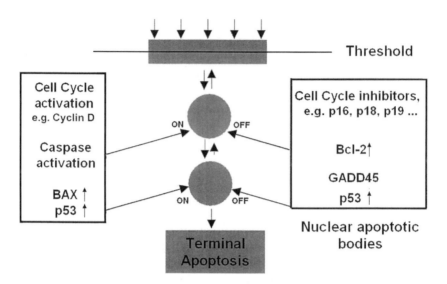

Fig. 3. The AD brain exhibits evidence for the presence of chronic apoptotic inducers and the accumulation of DNA damage. We propose that, in response to insult, both pro-apoptotic (cyclin D, p53, Bax, and caspases) and antiapoptotic (Bcl-2, p16, GADD45) pathways are set in motion. The ratio and activity of these molecules is what determines neuronal cell fate.

at best a few days of the initial insult. If TUNEL labeling in most cells reflected the true initiation of classical apoptosis, then it follows that most TUNEL-positive neurons would die in a few days. However, in many mild AD cases (mini mental state exam [MMSE] above 16) the majority of neurons exhibit TUNEL labeling for DNA damage. Thus, most neurons should have degenerated within a few days if apoptosis is actively in progress in these cells. However, this is inconsistent with the progression of neuronal loss in AD. Additionally, most TUNEL-positive neurons do not express terminal markers of apoptosis such as nuclear apoptotic bodies and other key molecular factors *(20,62)* For example, apoptosis-specific protein (ASP) is found in only a few of the neurons that show DNA fragmentation. This has lead some to the conclusion that neurons die primarily by necrosis in AD *(63)*.

On the other hand, it is possible that as nondividing cells neurons have developed a series of counteractive measures to repair damage and delay death; in other words, a kind of molecular counterattack and attempt to delay cell death in order to minimize unnecesary cell loss. This concept of an apoptosis checkpoint or decision cascade may help to understand an apparent puzzle in the neuronal apoptosis literature: The prolonged presence of indices of DNA damage and apoptotic regulatory protein expression may be a result of a counteractive strategy that neurons mobilize to hold apoptosis in check, delay death, and attempt repair. In this context, it is possible that cell-cycle proteins could contribute to cellular repair in neurons. Repair of DNA damage may be a particularly key example of such a mechanism. For example, many neurons in vulnerable regions of the AD brain show an upregulation of GADD45, a protein that is involved in DNA checkpoint repair at the G1 transition. AD neurons that express GADD45 often also show DNA damage and increased levels of Bcl-2 protein. Additionally, in support of a role in promoting cell survival, GADD45 transfected cells show improved survival after DNA damage *(64)*. Thus, Bcl-2, GADD45, and other protective molecules, such as PCNA, could serve to help repair DNA damage and assist in neuronal survival. Similarly, p16, p21, and other negative regulators of the cell cycle may serve to delay degeneration; that is, many checkpoints in the cell death pathway may exist and perhaps prevent the unnecessary loss of irreplaceable cells. This possibility may make the study of signal transduction pathways particularly critical because it may provide an opportunity for early interventions.

Cell Signaling Pathways for Apoptosis and Cell Cycle

From the above discussion, it is clear that neuronal apoptosis is an extremely complex process that can proceed via several distinct yet converging pathways. The execution of death is mediated by members of the Bcl-2 and caspase families. In these families of proteins, there is both redundancy and specialization in function across neuron types, and within single neurons, depending on the specific cell death stimulus. In addition, in many cell types, apoptotic cell signaling is probably tightly coupled to other essential cellular signaling pathways, such as proliferation and differentiation, to ensure that commitment to cell death is the correct response. Thus, a key issue is to identify central molecules in apoptotic signal transduction pathways and clarify their roles in apoptotic pathways. Toward this goal, much progress has been made in research using antisense and dominant negative mutants to show that apoptosis can be enhanced or arrested by alteration in the expression of multiple genes (Table 1). As more and more genes are identified, it is becoming clear that gene products that are important in regulating apoptosis often lie in the same signaling pathways that control other critical cellular functions such as proliferation, differentiation, and division. In this section and the next one, we will discuss several key Ras signaling pathways and their importance in mediating cell death versus survival. We also postulate a role for Ras in a neuronal "apoptotic checkpoint cascade," where Ras may be positioned to engage some cell-cycle checkpoint components in order to repair injured neurons and delay or prevent cell death.

Table 1
**Gene Manipulation in Multiple Signaling Pathways
Can Regulate Apoptosis**

Cell Line	Stimuli (ref.)	Expression	Protection
Sympathetic neuron	NGF deprivation *(53)*	CDK4 dom. neg. CDK6 dom. neg. p16, p21, p27 expression	Inhibit apoptosis
PC12	1 trophic factor w/d *(68)* 2 sidbis virus *(67)*	Ras dom. neg.	Inhibit apoptosis
Postmitotic striatal neurons	Dopamine induction *(70)*	SEK1 dom. neg.	Inhibit apoptosis
Sympathetic neuron	NGF deprivation *(69)*	SEK1 dom. neg.	Inhibit when coexpress w/ MEKK1
Sympathetic neuron	NGF deprivation *(69)*	c-Jun dom. neg.	Inhibit apoptosis
Hippocampal neuron	Kainic acid *(71)*	Jnk3 knock out	Inhibit apoptosis
Sympathetic neuron	NGF deprivation *(87)*	PI3K expression Akt expression	Inhibit apoptosis
PC12	NGF deprivation *(78)*	p38 dom. neg.	Inhibit apoptosis
DRG	Trophic factor w/d *(109)*	ICE 1β dom. neg.	Inhibit apoptosis
Sympathetic neuron PC12	Trophic factor w/d *(35)*	Nedd2/caspase2 antisense	Inhibit apoptosis
Cerebellar neurons	Ara-C *(110)*	GAPDH antisense	Inhibit apoptosis

Note. Manipulation of cell-cycle proteins (CDK4, 6, p16, p21, and p27), mitogenic signal transduction proteins (Ras and downstream effectors), and apoptotic program execution proteins (caspases) all regulate apoptosis in multiple cell types. Additional proteins in other cellular pathways (e.g., GAPDH) also appear to contribute to apoptosis in some cell types.

Ras belongs to a family of small GTPases that relay mitogenic signals to various intracelluar transduction cascades (Fig. 4). Although its involvement in cell differentiation and proliferation has been well defined, it is only recently been recognized for its role in apoptosis. The overexpression of Ras has been shown to increase cellular susceptibility to apoptosis *(65)* in response to apoptotic insults. A dominant negative mutant of Ras has been shown to inhibit Fas-induced apoptosis in Jurkat cells *(66)*, as well as apoptosis in neuronal cells induced by trophic factor withdrawal or sindbis virus *(67,68)*.

One signal transduction pathway downstream of Ras that seems to play a critical role in cellular apoptosis is the MEKK1/JNK/c-Jun cascade; that is, overexpression of MEKK1 induces apoptosis in sympathetic neurons.

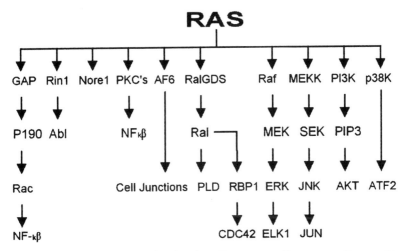

Fig. 4. A schematic diagram showing how Ras can mediate extracelluar and intracelluar signal transduction cascades via its many downstream effectors. Although the dominant downstream effector of Ras is Raf, which is believed to be involved in survival signaling of trophic factors, there are a plethora of other downstream effectors that could mediate a wide range of cellular functions such as apoptosis and cell junctions.

Coexpression of a dominant negative mutant of SEK1, an activator of JNK directly downstream of MEKK1, can inhibit MEKK1-induced apoptosis *(69)*. Expression of the mutant SEK1 alone can also prevent dopamine-induced apoptosis in postmitotic striatal neurons *(70)*. Also, mice lacking the JNK3 gene are resistant to excitotoxicity-induced apoptosis in the hippocampus *(71)* One of the major roles of JNK is phosphorylation and transcriptional regulation of c-Jun. c-Jun is an AP-1 transcription factor that is expressed prior to the commitment of NGF-deprived sympathetic neurons to apoptosis. A dominant negative c-Jun mutant has been shown to protect against neuronal apoptosis *(69,70,72)*. In addition to having a role in apoptosis, c-*Jun* and other immediate–early genes are also implicated in a wide range of cellular actions such as cell proliferation *(73)*, neuronal differentiation *(74)*, neuronal survival *(75)*, and learning and memory *(76)*. Thus, whether c-Jun induces apoptosis or some other cellular process might depend on upstream signaling pathways as well as interactions with other proteins.

Another stress-activated protein kinase downstream of Ras is the p38 kinase. It does not phosphorylate c-Jun but can increase c-*Jun* expression by phosphorylation of the ATF-2 trans-activation domain. Overexpression of upstream activators of p38—MKK3 and MKK6—was found to induce apoptosis in Jurkat cells *(77)*. p38 kinase was shown to be necessary in activation of apoptosis in PC12 cells after NGF withdrawal *(78)* and its activation by Erk inhibition triggers apoptosis in HeLa cells *(79)*

Interestingly, the Ras signaling pathway is known to directly regulate G1/S progression of the cell cycle (Fig. 5). Ras has been shown to induce the expression of cyclin D1, and cyclin D1 activation seems to be dependent on the Raf/MAPK pathway *(80)*. Thus, introduction of dominant negative Ras in cycling cells causes a decreased expression of D1 and G1 arrest *(81)*. This suggests that cell signaling in apoptosis and cell division are closely associated with each other in at least some cell types and that Ras might be a point of integration between these pathways.

Because of the potential catastrophe that inappropriate cell death could cause an organism, it comes as no surprise that apoptosis signaling is far from a single linear chain of events. Rather, it may involve the integration of many pathways, some of which can hold apoptosis in check. There are also several cellular signaling pathways inside the cell that generate antiapoptotic signals and promote cell survival. One such pathway downstream of Ras—the Raf1/MEK/Erk cascade—might be involved in generating survival signals. This pathway has been shown to have an opposing effect to the JNK proapoptotic pathway in PC12 cells *(78)*. Also, HeLa cells under hydrogen peroxide-induced oxidative stress are much more susceptible to apoptosis when Erk activation is inhibited *(79)*. Inhibition of Erk2 activation and synthesis also leads to increased apoptosis of RPE cells *(82)*. In addition to the antiapoptotic effect of activating Erk, Raf might also be able to suppress apoptosis in an Erk-independent fashion. In this context, Raf1 phosphorylates and inactivates BAD; further, there is evidence to suggest that Bcl-2 can target Raf1 to the mitochondria *(83)*, suggesting the existence of cellular trafficking mechanisms that would allow this interaction to occur in vivo.

Another pathway downstream of Ras—the PI3K/Akt pathway (Fig. 4)—has also been strongly implicated in the promotion of cell survival. PI3K can be activated via binding to tyrosine kinase receptors or via interaction with Ras *(84,85)*. Studies have shown that activation of PI3K and Akt pathways by Ras promotes cell survival, whereas inhibitors of PI3K induce apoptosis *(86)*. Expression of dominant negative forms of either PI3K or Akt is sufficient to induce apoptosis in sympathetic neurons in the presence of NGF. Moreover, the expression of permanently active mutants of PI3K and Akt can inhibit p38 activation and apoptosis *(87)*. Downstream targets of Akt that could play a role in apoptotic signaling include Bad and glycogen synthase kinase 3 (GSK-3) *(88)*.

In addition to inhibition of apoptosis through downstream effectors such as PI3K and Erk, Ras may also be able to hold apoptosis in check through a cell-cycle mechanism. As we mentioned earlier, Raf/Erk can signal the

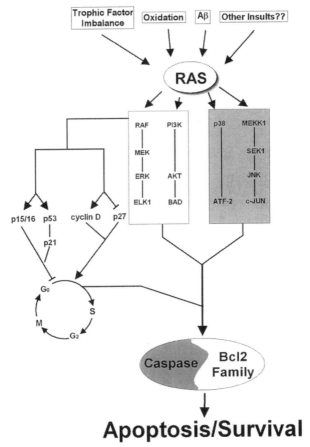

Fig. 5. Various insults to neurons are transduced through pro-apoptotic and anti-apoptotic pathways, which can cause the activation of cell-cycle markers that promote apoptosis, and cell-cycle inhibitors that inhibit apoptosis. Mitogenic signal transduction pathways, such as the ones mediated by Ras, also appear to regulate the actions of caspases and Bcl-2 family proteins, perhaps through modulation of gene transcription or protein phosphorylation. The exact mechanisms by which signaling transduction pathways may be able to regulate caspases and Bcl-2 proteins remains to be elucidated.

induction of cyclin D1. Additionally, stronger activation of Raf/Erk signaling can induce the activation of p21 *(89)*. Studies have also found that expression of Ras can upregulate other cell-cycle inhibitory molecules such as p53 and p16. p53 is a tumor-suppressor protein that has been implicated in the cell-cycle checkpoint at the G1/S phase. Its expression can be induced in response to DNA damage and various other sources of cellular stress inducers. p53 can function in cell fate decisions through its transcriptional activation capabilities. The cyclin kinase inhibitor p21 has been shown to be transcribed in a p53-dependent manner *(90)*. The Bax gene, which encodes a pro-apoptotic protein in the Bcl-2 family, is also p53 responsive *(91)*. Furthermore, GADD45, a protein associated with cell-cycle arrest and DNA repair, can also be induced by p53 *(92)*. Therefore, although weak Ras activation may promote cell-cycle progression and apoptosis, stronger Ras activation could cause cell-cycle arrest and initiate the checkpoint cascade.

Whether Ras activation actually promotes apoptosis or cell survival is also likely to be dependent on the physical state that the cell is in during Ras pathway activation *(93)*. Thus, for example, NGF posttreatment and fibroblast-growth factor (FGF) pretreatment can enhance peroxynitrite-induced apoptosis in PC12 cells. This enhancement can be completely blocked by dominant negative Ras. This finding may be particularly relevant to diseases such as AD, because AD neurons exhibit evidence for nitrated tyrosines and oxidative damage *(94)*.

Recently, aberrant expression of proteins in the Ras/MAPK pathway has also been identified in the AD brain; specifically, elevated levels of Ras, as well as two upstream activators of Ras, Son-of-Sevenless-1 (SOS1) and growth factor receptor-bound protein 2 (Grb2), have been identified *(95,96)*. These data underscore the importance of identifying key signal transduction proteins in apoptosis and AD pathogenesis.

APOPTOSIS CHECKPOINT CASCADE
AND NEUROFIBRILLARY TANGLE FORMATION

In the previous sections, we have discussed evidence that apoptotic pathways are active in AD. Tau hyperphosphorylation and neurofibrillary tangle (NFT) formation are hallmark pathologies of AD. Tau hyperphosphorylation, and possibly other mechanisms related to NFT formation, disrupt the cytoskeletal system, intracelluar transport, and the maintenance of cellular function. Since their original description, they have been assumed to be the primary cause of neuronal dysfunction and death.

On the other hand, it is our view that NFT formation is unlikely to be the sole mechanism of death in AD. Whereas some neurons with DNA damage and caspase activation contain NFTs, others do not. Further, if NFTs are the sole mechanism of neuronal demise, then it should follow that the number of intact cells plus extracellular NFTs should add up to the normal number of neurons in the healthy brain. In fact, it appears as if there is a net loss of neurons in the AD brain that is not accounted for by the number of extracellular tangles *(97)*. This would argue that an alternative mechanism of neuronal death exists. Of course, this conclusion depends on the premise that NFTs, once formed, are not degraded; in this regard, it is generally believed that NFTs once formed intracellularly do not degrade but persist as extracellular NFTs. Observations from brain-stem regions of AD patients also support the hypothesis that NFT formation cannot be the only cause of cell death. For example, in the AD brain, 50% of the neurons in the dorsal raphe nucleus and 60% of the neurons in the locus coeruleus are lost, but few NFTs are identified at any stage of disease *(98)*.

We suggest that NFT formation might be one of the possible outcomes of an apoptosis checkpoint cascade rather than the initial and sole cause of cell death. In vitro studies have found that tau can be phorsphorylated by cyclin-dependent kinases CDK2 and CDK5, both kinases that are usually associated with the G2/M phase of the normal cell cycle. Immunostaining studies have found both CDK2 and CDK5 to be closely associated with NFT-bearing neurons and neurons vulnerable to NFT formation in the AD brain. In many cases, these kinases are also colocalized with cyclin B1, the major cyclin present during the G2/M phase of the cell cycle *(99,100)*. Also consistent with this hypothesis is the finding that CKIs of the INK4 family, such as p16, are strongly expressed in NFT-bearing neurons. Thus, some vulnerable neurons could be caught between two opposing forces, that of degeneration and death versus that of partial repair and survival. One outcome of this battle could be NFT formation and survival suspended in a pathological state. Of course, this fate is not shared by all neurons. Some neurons may complete an apoptotic program long before NFTs have a chance to fully develop; conversely, cell death in some neurons may be independent of NFT formation altogether. Additional analyses on well-defined AD cases are clearly needed to resolve this hypothesis.

Can Apoptogenic Insults Act Locally on Neurites to Cause Neurodegeneration?

In the previous sections, we have discussed evidence for the existence of apoptogenic insults and apoptotic death in the AD brain. It is also clear, however, that there is compelling evidence for the early vulnerability of neuronal processes to synapse loss and to degeneration in the AD brain. Thus, it could be argued that the concept of apoptosis does not address the mechanism associated with the earliest and most significant early changes in AD. In addition to the presence of insults that affect the entire cell, it is possible that in the brain, insults accumulate in the local environment (Fig. 6). Neurons often project over long distances and, thus, the processes may encounter adverse microenvironments not experienced by the soma.

In the following section, we ask whether there is evidence for the existence of these apoptotic pathways in neurites whether does this perhaps

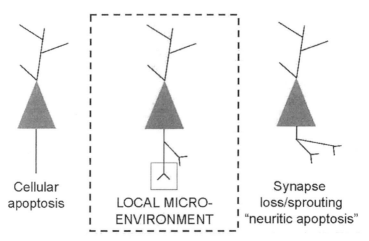

Cellular apoptosis | LOCAL MICRO-ENVIRONMENT | Synapse loss/sprouting "neuritic apoptosis"

Fig. 6. Neurons have processes that often project long distances and thus exist in a local microenvironment that is separate from the soma. We propose that distal neurites can undergo "neuritic apoptosis," independent of somal cell death. Such a mechanism could be involved in neuritic plasticity and remodeling during AD and also provide an explanation for early synaptic loss associated with this disease.

provide a new concept for understanding the basis of early synaptic loss. Several cell-free systems have been developed that display essentially all of the known events associated with apoptosis *(101,102)*. These cell-free systems are being used to elucidate the molecular events and establish their order during apop-tosis. Most recently, Ellerby and co-workers have established a cell-free system of neuronal apoptosis which can subdivide the cascade into premitochondrial, mitochondrial, and postmitochondrial stages *(103)*. Particularly critical to the activation of caspases is the release of cytochrome-*c* from mitochondria, which could easily occur locally in neuronal processes. Our approach complements these cell-free systems by looking at the events taking place specifically in the neurites and also represents a novel model that may be relevant to AD.

Axons Appear to Use Apoptosis-Related Mechanisms to Degenerate

Is there evidence for the existence of apoptotic pathways in neurites? To test the hypothesis that apoptotic insults can act directly on neurites to cause local neurodegeneration, it is essential to apply the apoptotic stimulus selectively to neuronal processes. We have recently developed a method to culture hippocampal neurons in modified Campenot chambers *(104)*. These chambers allow neurons cultured in one compartment (the somal compartment) to project neurites underneath a grease-sealed cover-slip barrier into a separate compartment (the neuritic compartment) where they can be selectively exposed to apoptotic stimuli. Neurites from hippocampal neurons cultured in these chambers on poly-D-lysine/laminin-2 penetrate the grease seal and enter the neuritic compartment within 2–4 d of plating. The neurites in the neuritic compartment of these chambers stained with a MAP5 antibody, suggesting that they represent axons. When apoptogenic stimuli including β-amyloid were applied only to the neuritic compartment of these cultures, neurites exposed to the inducers degenerated, whereas more proximal portions of the same neurites appeared to remain healthy.

To determine whether local biochemical changes indicative of an apoptotic program were present in these neurites, cultures were assayed for caspase activation and extracellular exposure of phosphatidylserine. Biotinylated VAD–CH$_2$F, an irreversible active-site directed caspase inhibitor, labels the cell bodies and neurites of dissociated hippocampal neurons exposed to apoptotic insults and can be used to assay caspase activation. Using this compound, activated caspases were detected in neurites locally exposed to the apoptogenic insults (Fig. 7).

Redistribution of phosphatidylserine from the inner leaflet to the outer leaflet of the plasma membrane, measured by the binding of annexinV to intact cells, has been shown to occur during apoptosis induced by multiple apoptogenic stimuli, including etoposide, Fas ligation, staurosporine, and trophic factor withdrawal, in many types of cells, including NGF-differentiated PC12 cells *(105)* and hippocampal neurons *(106)*. Extracellular exposure of phosphatidylserine during apoptosis requires caspase activation *(107)*. In compartmented cultures of hippocampal neurons, exposure of neurites to β-amyloid or other apoptotic stimuli for 3–4 h increased the binding of annexinV to these neurites. Furthermore, the induction of annexinV binding sites on neurites required caspase activation, because annexinV binding to neurites was blocked by prior exposure of the neurites to the nonselective caspase inhibitor z VAD-fmk.

Based on these results, we have suggested that neurite degeneration induced by Aβ and other apoptotic stimuli may reflect the local activation of an apoptotic program in neurites. The signal transduction pathways for this form of apoptosis are unknown but would be predicted to involve selective caspases as part of upstream non-nuclear events. Both the receptor and chemical pathways illustrated in Fig. 1 may be initiators of the program. This should prove to be an exciting new area of research for future study. The potential significance of this mechanism may be broad in scope.

In support of our data, others have described activation of caspases by β-amyloid in cortical synaptosomal preparations and in dendritic spines of cultured hippocampal neurons *(108)*. Caspase activation was accompanied

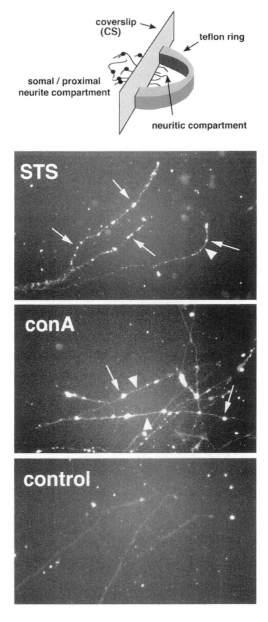

Fig. 7. Flourescent labeling for caspase activation in cultured hippocampal neurons. Exposure of the neuritic compartment of compartmented cultures to STS or Con A insult produces selective caspase activation in neuronal processes (*see* arrows), but not cell bodies in the somal compartment.

by mitochondrial membrane depolarization and by an apparent loss of plasma membrane symmetry in synaptosomes, and the term "synaptic apoptosis" was coined to describe the activation of apoptotic biochemical cascades in a "synaptic" cellular compartment.

Neuritic Apoptotic Mechanisms May Exist in Processes, Protect the Soma, and Be a Mechanism for Synaptic Loss

These data demonstrate that exposure of distal neurites to apoptotic insults, including β-amyloid, causes local neurite degeneration with morphological and biochemical characteristics of apoptosis. We suggest that these neurites undergo "neuritic apoptosis" (i.e., neurite degeneration using mechanisms common to classical apoptosis). Such mechanisms could conceivably be involved in early disease onset, as processes may be particularly vulnerable to apoptogenic stimuli. The loss of connectivity to areas where inducers are elevated may represent a mechanism to isolate the neuronal from the insult. Such mechanism, however, may not be restricted to disease but play a role in development and the turnover of synapses throughout life. The mechanism ironically becomes the target of synaptic loss and the decline of brain function in late-life degenerative conditions.

CONCLUSION

Over the life-span, it appears that the brain uses apoptosis to control cell fate. This has long been accepted as a principle in the development of the nervous system. Recently, as discussed in this chapter, we and others have suggested that this mechanism is applicable throughout life and may play a major role in aging and AD pathogenesis. As the brain ages, multiple risk factors that can contribute to apoptotic cell death accumulate. This is a particular challenge to neurons, which are differentiated postmitotic cells with extremely limited capacity for renewal. We suggest that neurons have special mechanisms embedded in a complex array of signal transduction pathways to counteract apoptotic processes, overcome cellular injury, and delay or prevent apoptosis. In fact, cell-cycle checkpoint proteins and anti-apoptotic markers are evident in the aging and AD brain. In this light, it can be hypothesized that neurons may re-enter the cell cycle to use checkpoint molecules for damage repair. These signal transduction processes, referred to as the "apoptosis checkpoint cascade," are a new area of research in the evolving field of age-related neurodegenerative diseases such as Alzheimer's disease. Ironically, however, some pathological mechanisms could result from the activation of such checkpoint mechanisms. For example, there may be evidence to suggest that NFT formation could be a result of the prolonged activation of cell cycle or other upregulated kinases. Thus, the strategy to delay cell death may come at a cost.

Neurons are unusual because they can send processes over extremely long distances. In the aged brain and that of individuals with AD, local conditions develop in the neuronal microenvironment that could selectively affect neuronal processes while leaving the soma relatively spared. Local degeneration of neuronal processes is evident in AD as one of the earliest signatures of dysfunction (e.g., synapse loss is an early sign of AD). We suggest that local apoptotic mechanisms may be key in early synaptic loss in AD. These involve local caspase activation, inversion of phosphatidylserine in the plasma membrane and the blebbing and disconnection of the process from the soma.

Apoptosis is a complex process involving multiple signal transduction pathways, multiple cellular compartments, and, possibly, the evolution of other secondary outcomes that ultimately compromise cellular function in some neurons. The challenge is to identify the key steps at the signal transduction level and on this basis develop early counteractive therapeutic strategies.

REFERENCES

1. Loo, D. T., Copani, A., Pike, C. J., Whittemore, E. R., Walencewicz, A. J., and Cotman, C.W. (1993) Apoptosis is induced by beta-amyloid in cultured central nervous system neurons. *PNAS* **90**, 7951–7955.
2. Pike, C. J., Burdick, D., Walencewicz, A., Glabe, C. G., and Cotman, C. W. (1993) Neurodegeneration induced by β-amyloid peptides in vitro: the role of peptide assembly state. *J. Neurosci.* **13**, 1676–1687.
3. Pike, C. J., Cummings, B. J., and Cotman, C. W. (1992) β-amyloid induces neuritic dystrophy in vitro: similarities with Alzheimer pathology. *NeuroReport* **3**, 769–772.
4. Watt, J. A., Pike, C. J., Walencewicz-Wasserman, A. J., and Cotman, C. W. (1994) Ultrastructural analysis of beta-amyloid-induced apoptosis in cultured hippocampal neurons. *Brain Res.* **661**, 147–156.
5. Whittemore, E. R., Loo, D. T., and Cotman, C. W. (1994) Exposure to hydrogen peroxide induces cell death via apoptosis in cultured rat cortical neurons. *NeuroReport* **5**, 1485–1488.
6. Benzi, G. and Moretti, A. (1995) Are reactive oxygen species involved in Alzheimer's disease? [see comments]. *Neurobiol. Aging* **16**, 661–674.
7. Hoyer, S., Oesterreich, K., and Wagner, O. (1988) Glucose metabolism as the site of the primary abnormality in early-onset dementia of Alzheimer type? *J. Neurol.* **235**, 143–148.
8. Copani, A., Koh, J., and Cotman, C. W. (1991) β-amyloid increases neuronal susceptibility to injury by glucose deprivation. *NeuroReport* **2**, 763–765.
9. Dodd, P. R., Scott, H. L., and Westphalen, R. I. (1994) Excitotoxic mechanisms in the pahtogenesis of dementia. *Neurochem. Int.* **25**, 203–219.

10. Masliah, E., Alford, M., DeTeresa, R., Mallory, M., and Hansen, L. (1996) Deficient glutamate transport is associated with neurodegeneration in Alzheimer's disease. *Ann. Neurol.* **40**, 759–766.
11. Simpson, I. A., Chundu, K. R., Davies-Hill, T., Honer, W. G., and Davies, P. (1994) Decreased concentrations of GLUT1 and GLUT3 glucose transporters in the brains of patients with Alzheimer's disease. *Ann. Neurol.* **35**, 546–551.
12. Koh, J. Y., Yang, L. L., and Cotman, C. W. (1990) β-amyloid protein increases the vulnerability of cultured cortical neurons to excitotoxic damage. *Brain Res.* **533**, 315–320.
13. Mattson, M. P., Cheng, B., Davis, D., Bryant, K., Lieberburg, I., and Rydel, R. E. (1992) Beta-Amyloid peptides destabilize calcium homeostasis and render human cortical neurons vulnerable to excitotoxicity. *J. Neurosci.* **12**, 376–389.
14. Pike, C. J., Ramezan-Arab, N., and Cotman, C. W. (1997) Beta-amyloid neurotoxicity in vitro: evidence of oxidative stress but not protection by antioxidants. *J. Neurochem.* **69**, 1601–1611.
15. Duke, R. C., Chervenak, R., and Cohen, J. J. (1983) Endogenous endonulclease-induced DNA fragmentation: an early event in cell-mediated cytolysis. *Proc. Natl. Acad. Sci. USA* **80**, 6361–6365.
16. Wyllie, A. H., Morris, R. G., Smith, A. L., and Dunlop, D. (1984) Chromatin cleavage in apoptosis: association with condensed chromatin morphology and dependence on macromolecular synthesis. *J. Pathol.* **142**, 67–77.
17. Tepper, C. G. and Studzinski, G. P. (1992) Teniposide induces nuclear but not mitochondrial DNA degradation. *Cancer Res.* **52**, 3384–3390.
18. Zakeri, Z. F., Quaglino, D., Latham, T., and Lockshin, R. A. (1993) Delayed internucleosomal DNA fragmentation in programmed cell death. *FASEB J.* **7**, 470–478.
19. Anderson, A. J., Su, J. H., and Cotman, C. W. (1996) DNA damage and apoptosis in Alzheimer's disease: colocalization with c-Jun immunoreactivity, relationship to brain area, and effect of postmortem delay. *J. Neurosci.* **16**, 1710–1719.
20. Su, J. H., Anderson, A. J., Cummings, B. J., and Cotman, C. W. (1994) Immuno-histochemical evidence for apoptosis in Alzheimer's disease. *NeuroReport* **5**, 2529–2533.
21. Cotman, C. W. and Su, J. H. (1996) mechanisms of neuronal death in Alzheimer's disease. *Brain Pathol.* **6**, 493–506.
22. Dragunow, M., Faull, R. L., Lawlor, P., Beilharz, E. J., Singleton, K., Walker, E. B., and Mee, E. (1995) In situ evidence for DNA fragmentation in Huntington's disease striatum and Alzheimer's disease temporal lobes. *NeuroReport* **6**, 1053–1057.
23. Lassmann, H., Bancher, C., Breitschopf, H., Wegiel, J., Bobinski, M., Jellinger, K., and Wisniewski, H. M. (1995) Cell death in Alzheimer's disease evaluated by DNA fragmentation in situ. *Acta Neuropathol. (Berlin)* **89**, 35–41.
24. Thomas, L. B., Gates, D. J., Richfield, E. K., O'Brien, T. F., Schweitzer, J. B., and Steindler, D. A. (1995) DNA end labeling (TUNEL) in Huntington's disease and other neuropathological conditions. *Exp. Neurol.* **133**, 265–272.
25. de la Monte, S. M., Sohn, Y. K., and Wands, J. R. (1997) Correlates of p53 and Fas (CD95)-mediated apoptosis in Alzheimer's disease. *J. Neurol. Sci.* **152**, 73–83.

26. Su, J. H., Deng, G., and Cotman, C. W. (1997) Bax protein expression is increased in Alzheimer's brain: correlations with DNA damage, Bcl-2 expression, and brain pathology. *J. Neuropathol. Exp. Neurol.* **56**, 86–93.

27. Yang, F., Sun, X., Beech, W., Teter, B., Wu, S., Sigel, J., Vinters, H. V., Frautschy, S. A., and Cole, G. M. (1998) Antibody to caspase-cleaved actin detects apoptosis in differentiated neuroblastoma and plaque-associated neurons and microglia in Alzheimer's disease. *Am. J. Pathol.* **152**, 379–389.

28. Sun, X.-M., McFarlane, M., Zhuang, J., Wolf, B. B., Green, D. R., and Cohen, G. M. (1999) Distinct caspase cascades are initiated in receptor-mediated and chemical-induced apoptosis. *J. Biol. Chem.* **274**, 5053–5060.

29. Tewari, M. and Dixit, V. (1995) Fas- and tumor necrosis factor-induced apoptosis is inhibited by the poxvirus crmA gene product. *J. Biol. Chem.* **270**, 3255–3260.

30. Gagliardini, V., Fernandez, P.-A., Lee, R. K. K., Drexler, H. C. A., Rotello, R. J., Fishman, M. C., and Yuan, J. (1994) Prevention of vertebrate neuronal death by the crmA gene [see comments] erratum **264** (1994) 1388. *Science* **263**, 826–828.

31. Frisch, S., Vuori, K., Kelaita, D., and Sicks, S. (1996) A role for Jun-N-terminal kinase in anoikis; suppression by bcl-2 and crmA. *J. Cell Biol.* **135**, 1377–1382.

32. Chinnaiyan, A. M., Orth, K., O'Rourke, K., Duan, H., Poirier, G. G., and Dixit, V. M. (1996) Molecular ordering of the cell death pathway. Bcl-2 and Bcl-xL function upstream of the CED-3-like apoptotic proteases. *J. Biol. Chem.* **271**, 4573–4576.

33. Orth, K., Chinnaiyan, A. M., Garg, M., Froelich, C. J., and Dixit, V. M. (1996) The CED-3/ICE-like protease Mch2 is activated during apoptosis and cleaves the death substrate lamin A. *J. Biol. Chem.* **271**, 16,443–16,446.

34. Ivins, K. J., Ivins, J. K., Sharp, J. P., and Cotman, C. W. (1999) Multiple pathways of apoptosis in PC12 cells. CrmA inhibits apoptosis induced by beta-amyloid. *J. Biol. Chem.* **274**, 2107–2112.

35. Troy, C. M., Stefanis, L., Greene, L. A., and Shelanski, M. L. (1997) Nedd2 is required for apoptosis after trophic factor withdrawal, but not superoxide dismutase (SOD1) downregulation, in sympathetic neurons and PC12 cells. *J. Neurosci.* **17**, 1911–1918.

36. Stefanis, L., Troy, C. M., Qi, H., Shelanski, M. L., and Greene, L. A. (1998) Caspase-2 (Nedd-2) processing and death of trophic factor-deprived PC12 cells and sympathetic neurons occur independently of caspase-3 (CPP32)-like activity. *J. Neurosci.* **18**, 9204–9215.

37. Zheng, T. S., Schlosser, S. F., Dao, T., Hingorani, R., Crispe, I. N., Boyer, J. L., and Flavell, R. A. (1998) Caspase-3 controls both cytoplasmic and nuclear events associated with Fas-mediated apoptosis in vivo. *Proc. Natl. Acad. Sci. USA* **95**, 13,618–13,623.

38. Anjum, R., Ali, A. M., Begum, Z., Vanaja, J., and Khar, A. (1998) Selective involvement of caspase-3 in ceramide induced apoptosis in AK-5 tumor cells. *FEBS Lett.* **439**, 81–84.

39. Martin, S. J., Reutelingsperger, C. P. M., McGahon, A. J., Rader, J. A., van Schie, R. C. A. A., LaFace, D. M., and Green, D. R. (1995) Early redistribution of plasma membrane phosphatidylserine is a general feature of apoptosis regardless of the initiating stimulus: inhibition by overexpression of bcl-2 and abl. *J. Exp. Med.* **182**, 1545–1556.

40. Zamzami, N., Marchetti, P., Castedo, M., Decaudin, D., Macho, A., Hirsch, T., Susin, S. A., Petit, P. X., Mignotte, B., and Kroemer, G. (1995) Sequential reduction of mitochondrial transmembrane potential and generation of reactive oxygen species in early programmed cell death. *J. Exp. Med.* **182**, 367–377.
41. Yang, E. and Korsmeyer, S. J. (1996) Molecular thanatopsis: a discourse on the BCL2 family and cell death. *Blood* **88**, 386–401.
42. Oltvai, Z. N. and Korsmeyer, S. J. (1994) Checkpoints of dueling dimers foil death wishes. *Cell* **79**, 189–192.
43. Yang, E., Zha, J., Jockel, J., Boise, L. H., Thompson, C. B., and Korsmeyer, S. J. (1995) Bad, a heterodimeric partner for Bcl-XL and Bcl-2, displaces Bax and promotes cell death. *Cell* **80**, 285–291.
44. Zha, J., Harada, Z., Yang, E., Jockel, J., and Korsmeyer, S. J. (1996) Serine phosphorylation of death agonist BAD in response to survival factor results in binding to 14-3-3 not BCL-X. *Cell* **87**, 619–628.
45. Reed, J. C. (1998) Bcl-2 family proteins. *Oncogene* **17**, 3225–3236.
46. Allsopp, T. E., Wyatt, S., Paterson, H. F., and Davies, A. M. (1993) The proto-oncogene bcl-2 can selectively rescue neurotrophic factor-dependent neurons from apoptosis. *Cell* **73**, 295–307.
47. Krajewski, S., Mai, J. K., Krajewska, M., Sikorska, M., Mossakowski, M. J., and Reed, J. C. (1995) Upregulation of bax protein levels in neurons following cerebral ischemia. *J. Neurosci.* **15**, 6364–6376.
48. Su, J. H., Satou, T., Anderson, A. J., and Cotman, C. W. (1996) Up-regulation of Bcl-2 is associated with neuronal DNA damage in Alzheimer's disease. *NeuroReport* **7**, 437–440.
49. Gil-Gomez, G., Berns, A., and Brady, H. J. (1998) A link between cell cycle and cell death: Bax and Bcl-2 modulate Cdk2 activation during thymocyte apoptosis. *EMBO J.* **17**, 7209–7218.
50. Farinelli, S. E. and Greene, L. A. (1996) Cell cycle blockers mimosine, ciclopirox, and deferoxamine prevent the death of PC12 cells and postmitotic sympathetic neurons after removal of trophic support. *J. Neurosci.* **16**, 1150–1162.
51. Stefanis, L., Park, D. S., Yan, C. Y., Farinelli, S. E., Troy, C. M., Shelanski, M. L., and Greene, L. A. (1996) Induction of CPP32-like activity in PC12 cells by withdrawal of trophic support. Dissociation from apoptosis. *J. Biol. Chem.* **271**, 30,663–30,671.
52. Freeman, R. S., Estus, S., and Johnson, E. M. (1994) Analysis of cell cycle-related gene expression in postmitotic neurons: selective induction of Cyclin D1 during programmed cell death. *Neuron* **12**, 343–355.
53. Park, D. S., Levine, B., Ferrari, G., and Greene, L. A. (1997) Cyclin dependent kinase inhibitors and dominant negative cyclin dependent kinase 4 and 6 promote survival of NGF-deprived sympathetic neurons. *J. Neurosci.* **17**, 8975–8983.
54. Greenlund, L. J., Deckwerth, T. L., and Johnson, E. M. (1995) Superoxide dismutase delays neuronal apoptosis: a role for reactive oxygen species in programmed neuronal death. *Neuron* **14**, 303–315.
55. Miller, T. M. and Johnson, E. M. (1996) Metabolic and genetic analyses of apoptosis in potassium/serum-deprived rat cerebellar granule cells. *J. Neurosci.* **16**, 7487–7495.
56. Kranenburg, O., van der Eb, A. J., and Zantema, A. (1996) Cyclin D1 is an essential mediator of apoptotic neuronal cell death. *EMBO J.* **15**, 46–54.

57. Poluha, W., Poluha, D. K., Chang, B., Crosbie, N. E., Schonhoff, C. M., Kilpatrick, D. L., and Ross, A. H. (1996) The cyclin-dependent kinase inhibitor p21 (WAF1) is required for survival of differentiating neuroblastoma cells. *Mol. Cell Biol.* **16**, 1335–1341.

58. Wang, J. and Walsh, K. (1996) Resistance to apoptosis conferred by Cdk inhibitors during myocyte differentiation. *Science* **273**, 359–361.

59. Busser, J., Geldmacher, D. S., and Herrup, K. (1998) Ectopic cell cycle proteins predict the sites of neuronal cell death in Alzheimer's disease brain. *J. Neurosci.* **18**, 2801–2807.

60. Arendt, T., Holzer, M., and Gartner, U. (1998) Neuronal expression of cycline dependent kinase inhibitors of the INK4 family in Alzheimer's disease. *J. Neural. Transm.* **105**, 949–960.

61. McShea, A., Harris, P. L., Webster, K. R., Wahl, A. F., and Smith, M. A. (1997) Abnormal expression of the cell cycle regulators P16 and CDK4 in Alzheimer's disease. *Am. J. Pathol.* **150**, 1933–1939.

62. Lucassen, P. J., Chung, W. C., Kamphorst, W., and Swaab, D. F. (1997) DNA damage distribution in the human brain as shown by in situ end labeling; area-specific differences in aging and Alzheimer disease in the absence of apoptotic morphology. *J. Neuropathol. Exp. Neurol.* **56**, 887–900.

63. Stadelmann, C., Bruck, W., Bancher, C., Jellinger, K., and Lassmann, H. (1998) Alzheimer disease: DNA fragmentation indicates increased neuronal vulnerability, but not apoptosis. *J. Neuropathol. Exp. Neurol.* **57**, 456–464.

64. Torp, R., Su, J. H., Deng, G., and Cotman, C. W. (1998) GADD45 is induced in Alzheimer's disease, and protects against apoptosis in vitro. *Neurobiol. Dis.* **5**, 245–252.

65. Kauffmann-Zeh, A., Rodriguez-Viciana, P., Ulrich, E., Gilbert, C., Coffer, P., Downward, J., and Evan, G. (1997) Suppression of c-Myc-induced apoptosis by Ras signalling through PI(3)K and PKB. *Nature* **385**, 544–548.

66. Gulbins, E., Coggeshall, K. M., Brenner, B., Schlottmann, K., Linderkamp, O., and Lang, F. (1996) Fas-induced Apoptosis is mediated by activation of a Ras and Rac protein-regulated signaling pathway. *J. Biol. Chem.* **271**, 26,389–26,394.

67. Joe, A. K., Ferrari, G., Jiang, H. H., Liang, X. H., and Levine, B. (1996) Dominant inhibitory Ras delays sindbis virus-induced apoptosis in neuronal cells. *J. Virol.* **70**, 7744–7751.

68. Ferrari, G. and Greene, L. A. (1994) Proliferative inhibition by dominant-negative Ras rescues naive and neuronally differentiated PC12 cells from apoptotic death. *EMBO J.* **13**, 5922–5928.

69. Eilers, A., Whitfield, J., Babij, C., Rubin, L. L., and Ham, J. (1998) Role of the Jun kinase pathway in the regulation of c-Jun expression and apoptosis in sympathetic neurons. *J. Neurosci.* **18**, 1713–1724.

70. Luo, Y., Umegaki, H., Wang, X., Abe, R., and Roth, G. S. (1998) Dopamine induces apoptosis through an oxidation-involved SAPK/JNK activation pathway. *J. Biol. Chem.* **273**, 3756–3764.

71. Yang, D. D., Kuan, C. Y., Whitmarsh, A. J., Rincon, M., Zhen, T. S., Davis, R. J., Rakic, P., and Flavell, R. A. (1997) Absence of excitotoxicity induced apoptosis in the hippocampus of mice lacking the Jnk3 gene. *Nature* **389**, 865–870.

72. Ham, J., Babij, C., Whitfield, J., Pfarr, C. M., Lallemand, D., Yaniv, M., and Rubin, L. L. (1995) A c-Jun dominant negative mutant protects sympathetic neurons against programmed cell death. *Neuron* **14**, 927–939.

73. Kovary, K. and Bravo, R. (1991) The jun and fos protein families are both required for cell cycle progression in fibroblasts. *Mol. Cell Biol.* **11**, 4466–4472.
74. Bartel, D. P., Sheng, M., Lau, L. F., and Greenberg, M. E. (1989) Growth factors and membrane depolarization activate distinct programs of early response gene expression: dissociation of fos and jun induction. *Gene Dev.* **3**, 304–313.
75. Haas, C. A., Deller, T., Naumann, T., and Frotscher, M. (1996) Selective expression of the immediate early gene c-jun in axotomized rat medial septal neurons is not related to neuronal degeneration. *J. Neurosci.* **16**, 1894–1903.
76. Dragunow, M. (1996) A role for immediate-early transcription factors in learning and memory. *Behav. Genet.* **26**, 293–299.
77. Raingeaud, J., Whitmarsh, A. J., Barrett, T., Derijard, B., and Davis, R. J. (1996) MKK3 and MKK6 regulated gene expression is mediated by the p38 mitogen-activated protein kinase signal transduction pathway. *Mol. Cell Biol.* **16**, 1247–1255.
78. Xia, Z., Dickens, M., Raingeaud, J., Davis, R. J., and Greenberg, M. E. (1995) Opposing effects of ERK and JNK-p38 MAP kinases on apoptosis. *Science* **270**, 1326–1331.
79. Berra, E., Diaz-Meco, M. T., and Moscat, J. (1998) The activation of p38 and apoptosis by the inhibition of Erk is antagonized by the phosphoinositide 3-kinase/Akt pathway. *J. Biol. Chem.* **273**, 10,792–10,797.
80. Cheng, M., Sexl, V., Sherr, C. J., and Roussel, M. F. (1998) Assembly of cyclin D-dependent kinase and titration of p27Kip1 regulated by mitogen-activated protein kinase kinase (MEK1). *Proc. Natl. Acad. Sci. USA* **95**, 1091–1096.
81. Peeper, D. S., Upton, T. M., Ladha, M. H., Neuman, E., Zalvide, J., Bernards, R., DeCaprio, J. A., and Ewen, M. E. (1997) Ras signalling linked to the cell-cycle machinery by the retinoblastoma protein. *Nature* **386**, 177–181.
82. Guillonneau, X., Brychaer, M., Launay-Longo, C., Courtois, Y., and Mascarelli, F. (1998) Endogenous FGF1-induced activation and synthesis of extracellular signal-regulated kinase 2 reduce cell apoptosis in retinal-pigmented epithelial cells. *J. Biol. Chem.* **273**, 22,367–22,373.
83. Wang, H. G., Rapp, U. R., and Reed, J. C. (1996) Bcl-2 targets the protein kinase Raf-1 to mitochondria. *Cell* **87**, 629–638.
84. Rodriguez-Viciana, P., Warne, P. H., Dhand, R., Vanhaesebroeck, B., Gout, I., Fry, M. J., Waterfield, M. D., and Downward, J. (1994) Phosphatidylinositol-3-OH kinase as a direct target of Ras. *Nature* **370**, 527–532.
85. Kaplan, D. R. and Stephens, R. M. (1994) Neurotrophin signal transduction by the Trk receptor. *J. Neurobiol.* **25**, 1404–1417.
86. Yao, R. and Cooper, G. M. (1995) Requirement for phosphatidylinositol-3 kinase in the prevention of apoptosis by nerve growth factor. *Science* **267**, 2003–2006.
87. Crowder, R. J. and Freeman, R. S. (1998) Phosphatidylinositol 3-kinase and Akt protein kinase are necessary and sufficient for the survival of nerve growth factor dependent sympathetic neurons. *J. Neurosci.* **18**, 2933–2943.
88. Datta, S. R., Dudek, H., Tao, X., Masters, S., Fu, H., Gotoh, Y., and Greenberg, M. E. (1997) Akt phosphorylation of BAD couples survival signals to the cell-intrinsic death machinery. *Cell* **91**, 231–241.
89. Woods, D., Parry, D., Cherwinski, H., Bosch, E., Lees, E., and McMahon, M. (1997) Raf-induced proliferation or cell cycle arrest is determined by the level of Raf activity with arrest mediated by p21Cip1. *Mol. Cell Biol.* **17**, 5348–5358.
90. El-Deiry, W. S., Tokino, T., Velculescu, V. E., Levy, D. B., Parsons, R., Trent, J. M., Lin, D., Mercer, W. E., Kinzler, K. W., and Vogelstein, B. (1993) WAF1, a potential mediator of p53 tumor suppression. *Cell* **75**, 817–825.

91. Xiang, H., Kinoshita, Y., Knudson, C. M., Korsmeyer, S.J., Schwartzkroin, P. A., and Morrison, R. S. (1998) Bax involvement in p53-mediated neuronal cell death. *J. Neurosci.* **18**, 1363–1373.
92. Fan, J. and Bertino, J. R. (1997) K-ras modulates the cell cycle via both positive and negative regulatory pathways. *Oncogene* **14**, 2595–2607.
93. Spear, N., Estevez, A. G., Johnson, G. V., Bredesen, D. E., Thompson, J. A., and Beckman, J. S. (1998) Enhancement of peroxynitrite-induced apoptosis in PC12 cells by fibroblast growth factor-1 and nerve growth factor requires p21Ras activation and is suppressed by Bcl-2. *Arch. Biochem. Biophys.* **356**, 41–45.
94. Su, J. H., Deng, G., and Cotman, C. W. (1997) Neuronal DNA damage precedes tangle formation and is associated with up-regulation of nitrotyrosine in Alzheimer's disease brain. *Brain Res.* **774**, 193–199.
95. Gartner, U., Holzer, M., Heumann, R., and Arendt, T. (1995) Induction of p21ras in Alzheimer pathology. *NeuroReport* **6**, 1441–1444.
96. McShea, A., Zelasko, D. A., Gerst, J. L., and Smith, M. A. (1999) Signal transduction abnormalities in Alzheimer's disease: evidence of a pathogenic stimuli. *Brain Res.* **815**, 237–242.
97. Gomez-Isla, T., Hollister, R., West, H., Mui, S., Growdon, J. H., Petersen, R. C., Parisi, J. E., and Hyman, B. T. (1997) Neuronal loss correlates with but exceeds neurofibrillary tangles in Alzheimer's disease. *Ann. Neurol.* **41**, 17–24.
98. Zweig, R. M., Ross, C. A., Hedreen, J. C., Steele, C., Cardillo, J. E., Whitehouse, P. J., Folstein, M. F., and Price, D. L. (1988) The neuropathology of aminergic nuclei in Alzheimer's disease. *Ann. Neurol.* **24**, 233–242.
99. Vincent, I., Jicha, G., Rosado, M., and Dickson, D. W. (1997) Aberrant expression of mitotic cdc2/cyclin B1 kinase in degenerating neurons of Alzheimer's disease brain. *J. Neurosci.* **17**, 3588–3598.
100. Pei, J. J., Grundke-Iqbal, I., Iqbal, K., Bogdanovic, N., Winblad, B., and Cowburn, R. F. (1998) Accumulation of cyclin-dependent kinase 5 (cdk5) in neurons with early stages of Alzheimer's disease neurofibrillary degeneration. *Brain Res.* **797**, 267–277.
101. Lazebnik, Y. A., Takahashi, A., Poirie, R. G. G., Kaufmann, S. H., and Earnshaw, W. C. (1995) Characterization of the execution phase of apoptosis in vitro using extracts from condemned-phase cells. *J. Cell Sci.* **19**(Suppl.), 41–49.
102. Newmeyer, D. D., Farschon, D. M., and Reed, J. C. (1994) Cell-free apoptosis in Xenopus egg extracts: inhibition by Bcl-2 and requirement for an organelle fraction enriched in mitochondria. *Cell* **79,** 353–364.
103. Ellerby, H. M., Martin, S. J., Ellerby, L. M., Naiem, S. S., Rabizadeh, S., Salvesen, G. S., Casiano, C. A., Cashman, N. R., Green, D. R., and Bredesen, D. E. (1997) Establishment of a cell-free system of neuronal apoptosis: comparison of pre-mitochondrial, mitochondrial, and postmitochondrial phases. *J. Neurosci.* **17**, 6165–6178.
104. Ivins, K. J., Bui, E. T., and Cotman, C. W. (1998) Beta-amyloid induces local neurite degeneration in cultured hippocampal neurons: evidence for neuritic apoptosis. *Neurobiol. Dis.* **5**, 365–378.
105. Rimon, G., Bazenet, C. E., Philpott, K. L., and Rubin, L. L. (1997) Increased surface phosphatidylserine is an early marker of neuronal apoptosis. *J. Neurosci. Res.* **48**, 563–570.

106. Holtsberg, F. W., Steiner, M. R., Keller, J. N., Mark, R. J., Mattson, M. P., and Steiner, S. M. (1998) Lysophosphatidic acid induces necrosis and apotosis in hippocampal neurons. *J. Neurochem.* **70**, 66–76.
107. Martin, S. J., Finucane, D. M., Amarante-Mendes, G. P., O'Brien, G. A., and Green, D. R. (1996) Phosphatidylserine externalization during CD95-induced apoptosis of cells and cytoplasts requires ICE/CED-3 protease activity. *J. Biol. Chem.* **271**, 28,753–28,756.
108. Mattson, M., Partin, J., and Begley, J. (1998) Amyloid beta peptide induces apoptosis-related events in synapses and dendrites. *Brain Res.* **807**, 167–176.
109. Friedlander, R. M., Gagliardini, V., Hara, H., Fink, K. B., Li, W., MacDonald, G., Fishman, M. C., Greeberg, A. H., Moskowitz, M. A., and Yuan, J. (1997) Expression of a dominant negative mutant of interleukin-1beta converting enzyme in transgenic mice prevents neuronal cell death induced by trophic factor withdrawal and ischemic brain injury. *J. Exp. Med.* **185**, 933–940.
110. Ishitani, R. and Chuang, D.-M. (1996) Glyceraldehyde-3-phosphate dehydrogenase antisense oligodeoxynucleotides protect against cytosine arabinonucleoside-induced apoptosis in cultured cerebellar neurons. *Proc. Natl. Acad. Sci. USA* **93**, 9937–9941.

8

The Role of Second Messengers in Neurodegeneration

Aase Frandsen, Jette Bisgaard Jensen, and Arne Schousboe

INTRODUCTION

Integration of environmental cues and signals in the central nervous system is dependent on fine-tuning of excitatory and inhibitory nervous activity. The vast majority of excitatory neurotransmission processes are mediated by the excitatory amino acid glutamate and aspartate acting on a variety of excitatory amino acid receptors *(1–4)*. These receptors are either forming ion channels or are coupled to G-proteins and, hence, they are named ionotropic and metabotropic receptors, respectively (*see* ref. *3*).

Classically, the ionotropic receptors have been classified on the basis of pharmacological properties. Based on *N*-methyl-D-aspartate (NMDA) and a number of naturally occurring glutamate analogs of restricted conformation, these receptors were originally termed NMDA and non-NMDA receptors, the latter being activated by, for example, quisqualate, ibotenate, and kainate *(5,6)*. The latter receptors were later divided into AMPA [2-amino-3-(3-hydroxy-5-methylisoxazol-4-yl)propionate] and kainate receptors based on binding studies using [^{3}H]-AMPA and [^{3}H]-kainate, respectively *(7,8)*. These classifications have subsequently been sustained and elaborated by the cloning of a variety of excitatory amino acid receptor subunits that form receptors and ionic channels corresponding to the pharmacologically defined NMDA, AMPA, and kainate receptors (*see* refs. *3* and *4*). Metabotropic glutamate receptors have been functionally classified in three major groups exhibiting somewhat different pharmacological properties *(9)*. The characterization has, moreover, been borne out by the advent of molecular cloning of eight subtypes of these receptors *(10–15)*. Group I receptors consisting of three splice variants of mGluR1 and two splice variants of mGluR5 mediate an increase in phosphoinositide hydrolysis and these receptors are potently activated by quisqualate and *t*-ACPD. Groups II and III are coupled to cAMP

From: *Cerebral Signal Transduction: From First to Fourth Messengers*
Edited by: M. E. A. Reith © Humana Press Inc., Totowa, NJ

homeostasis and mediate a decrease in the intracellular cAMP content. Group II comprising mGluR2 and mGluR3 is characterized by activation by L-CCG1 [(2S,1'S,2'S)-2-(carboxycyclopropyl)glycine] and t-ACPD [1S, 3R-1-amino-cyclopentane-1,3-dicarboxylate] and group III receptors consisting of mGluR4, -6, -7, and -8 are activated by L-AP4 (L-2-amino-4-phosphonobutyrate) and L-SOP (L-serine-0-phosphate).

In addition to the physiological action of the excitatory amino acids as neurotransmitters, they have been identified as powerful neurotoxins causing neurodegeneration when their extracellular concentration in the brain exceeds a critical level *(16–18)*. Although it is well established that activation of the ionotropic glutamate receptors play a fundamental role in the neurotoxic action of glutamate and aspartate *(18,19)*, it is less clear what may be the exact involvement of the metabotropic receptors *(18)*. It should, however, be kept in mind that metabotropic glutamate receptors can enhance synaptic release of glutamate *(20,21)*, which is likely to facilitate and possibly overactivate ionotropic receptors, thereby leading to excitotoxic actions *(see* ref. *9)*.

The common denominator in these toxic actions appears to be Ca^{2+}. The ionic channel formed by NMDA receptors is Ca^{2+} permeable and the channels activated by AMPA and kainate are under certain conditions permeable for Ca^{2+} *(see* ref. *3)*. The metabotropic receptors belonging to group I indirectly control intracellular Ca^{2+} levels via stimulation of IP_3 production (see below) and thus mediate an increase in intracellular Ca^{2+} *(9)*. Because aberrations in intracellular Ca^{2+} are likely to lead to neuronal degeneration *(see below)*, this appears to link both receptor subtypes to excitotoxicity via a disturbance of Ca^{2+} homeostasis.

CALCIUM HOMEOSTASIS

Correlation Between Cell Death and $[Ca^{2+}]_i$

For several decades, it has been realized that a connection exists between disturbances in intracellular Ca^{2+} homeostasis and cell death *(22,23)*. This is based on the repeated observations that a pronounced increase in the intracellular Ca^{2+} concentration in most cases will produce cell death, and such increases in intracellular Ca^{2+} almost always precede cell death brought about by pathophysiological stimuli. Additionally, it has been shown that prevention of an increase in the intracellular Ca^{2+} concentration can inhibit cell death *(24)*.

Although it seems clear that increases in intracellular Ca^{2+} in neurons play a role in neuronal death induced by agents such as excitatory amino acids *(see* ref. *25)*, attempts to correlate cell death to a specific increase in

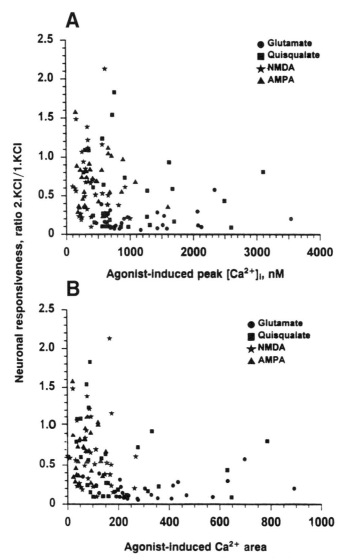

Fig 1. Scatterplot of ratio values obtained from single cortical neurons as a function of the peak nM $[Ca^{2+}]_i$ values (**A**) or as a function of the Ca^{2+} load (Ca^{2+} area; calculated as nM $[Ca^{2+}]_i \times$ time h) and (**B**) after exposure for 20 min to 100 μM glutamate, 300 μM quisqualate, 100 μM NMDA, or 100 μM AMPA. Each datum point represents the ratio value of a single cell obtained 90 min after removal of the agonist. (From ref. *26*, with permission, copyright 1994 National Academy of Sciences, USA.)

the intracellular Ca^{2+} concentration have largely failed. An example is provided in Fig. 1, which shows experiments performed in cerebral cortical neurons exposed to different excitatory amino acids acting on different receptor subtypes. It is seen that at moderately high increases in $[Ca^{2+}]_i$ expressed either as peak increases (Fig. 1A) or as total increases (Fig. 1B), it cannot be predicted whether a neuron will die. However, if a certain treshold for $[Ca^{2+}]_i$ is exceeded, neurons will enevitably be irreversibly damaged (26). These results are well in line with those of other investigators (27–29), but it should be mentioned that others have observed some correlation between the amplitude of the increase in $[Ca^{2+}]_i$ or the duration of the increase and subsequent cell death (30–33).

Ca²⁺ Entry and Ca²⁺ Pools

One reason why it has been difficult if not impossible to establish a meaningful relationship between $[Ca^{2+}]_i$ and neuronal cell death may be that an increase in $[Ca^{2+}]_i$ reflects events occurring at the site of Ca^{2+} entry (i.e., the plasma membrane), the mode of entry (i.e., Ca^{2+} channels activated by voltage or by receptors), and release of Ca^{2+} from a number of intracellular Ca^{2+} stores (i.e., endoplasmic reticulum and mitochondria). Additionally, cells are equipped with mechanisms such as the Ca^{2+}-ATPase and the Na^+/Ca^{2+} exchanger, which are able to extrude Ca^{2+} from the cell, thus ameliorating a moderate Ca^{2+} overload (34.35)

The Ca^{2+} entry in neurons occurs mainly via voltage-activated Ca^{2+} channels and through excitatory amino acid-activated receptors (36,37). Membrane depolarization either following action potentials or activation of excitatory amino acid receptors may open different subtypes of Ca^{2+} channels and both N- and L-type Ca^{2+} channels have been implicated in neuronal cell death (38–42). It appears, however, that influx of Ca^{2+} through excitatory amino acid receptors may be quantitatively more important (39,43,44). Because the Ca^{2+} permeability of NMDA receptors is much higher than that of AMPA and kainate receptors (see ref. 3), the NMDA receptor has been thought to be primarily responsible for the increase in $[Ca^{2+}]_i$ observed after exposure of neurons to glutamate, which activates all receptor subtypes (45). Although the Ca^{2+} permeability of AMPA/kainate receptors depends strongly on their subunit composition (see ref. 3), several studies have shown that these receptors can mediate Ca^{2+} influx (26,35,37,46–49). It should, however, be pointed out that in the case of AMPA- and kainate-mediated neurotoxicity, the Ca^{2+} entry is not the only determining factor, as it has recently been shown that the desensitization state of these receptors has great influence on the outcome of exposure of neurons to these agonists (50–53).

The intracellular Ca^{2+} concentration is determined not only by influx and efflux but also by the capacity for Ca^{2+} sequestering in different intracellular stores such as the endoplasmic reticulum, the mitochondria, and the cell nucleus *(54–56)*. The endoplasmic reticulum has two sets of receptors: the IP_3 and the ryanodine receptor, which trigger release of Ca^{2+} into the cytosol following stimulation with their respective agonists (*see* refs. *57* and *58*). These stores can also be affected by the drug dantrolene, which is known to block Ca^{2+} release from the sarcoplasmic reticulum *(59)*, and which has been proposed to interact with the ryanodine-sensitive Ca^{2+} channel in the reticulum *(37)*. Dantrolene has been shown to reduce the glutamate induced increase in $[Ca^{2+}]_i$ in cultured neurons *(60,61)* and, additionally, it inhibits a NMDA-mediated increase in $[Ca^{2+}]_i$ but not that mediated by AMPA/KA or K^+ depolarization *(39,61)*. As these latter effects were dependent on external Ca^{2+}, it was suggested that dantrolene may affect intracellular Ca^{2+} pools, of which at least one is stimulated to release Ca^{2+} by Ca^{2+} itself *(37)*. Moreover, it was suggested that in case of glutamate- and NMDA-induced cytotoxicity in neocortical neurons, these dantrolene-sensitive Ca^{2+} stores play a major role in Ca^{2+}-induced cell death. This probably explains at least some of the difficulties correlating cell death with an overall increase in $[Ca^{2+}]_i$.

Mitochondria

The ability of mitochondria to maintain a high Ca^{2+} level is dependent on an intact membrane potential and it is mediated mainly by the Ca^{2+} uniporter and the Na^+/Ca^{2+} exchanger *(62)*. During energy failure, Ca^{2+} may be released via the permeability transition pore *(63)* and this may have severe consequences with regard to subsequent triggering of processes leading to apoptosis or necrosis *(64)*.

EXCITOTOXICITY

Phenomenologically, activation of each of the individual EAA-receptor subtypes can cause neuronal cell death *(65,66)*. The finding that glutamate had a neurotoxic action at developmental stages at which NMDA was not toxic *(66,67)* indicates that toxicity exclusively mediated via activation of NMDA receptors cannot explain EAA-related toxicity in contrast to the original suggestion made by Choi et al. *(19)*. Likewise, the ability of EAA-receptor antagonists for either NMDA or non-NMDA receptors to reduce but not abolish glutamate toxicity supports this notion *(67)*. Despite more than 40 yr of intense research in the field, it is presently not clear to which extent the mechanisms mediating toxicity initiated by activation of the individual receptor subtypes have similarities or are identical. Judged from the conclusions of the involvement of calcium ions (cf. above text and ref. *37*), it seems reasonable to anticipate that at least part of the mechanisms involved are diverse for the different receptor subtypes involved. Further-

more, so far it has also not been possible to elucidate whether the pathological actions of EAAs are mediated via an exaggeration/dysregulation of the
mechanisms mediating the physiological actions.

Cell Death

Cell death can occur via two principally different mechanisms, namely
necrosis and *apoptosis* (or programmed cell death). Necrosis normally appears
traumatically, is of pathological nature, and may be described as a chaotic
(i.e., an unregulated process). The process involves gross areas of tissue and
is developing relatively slowly (days).

Apoptosis describes the fast (hours) but controlled destruction of single
cells after activation of cell death proteins. The process is in principle of
physiological nature.

Necrosis

Necrotic or accidential cell death occurs when the cell is exposed to extremely unphysiological processes. Necrosis appear to start with the loss of
control over the cellular homeostasis, leading to influx of water and extracellular ions. Intracellular organelles, primarily mitochondria, undergo swelling with subsequently cell swelling terminating in cell lysis. Because of the
ultimative destruction of the cell membrane, the cytoplasmic content, lyso-
somal enzymes included, is released to the extracellular space. Therefore,
necrotic cell death in vivo is often accompanied by extensive tissue damage,
resulting in an intense inflammatory response. Necrotic cell death may be in
vitro conveniently monitored by measuring the activity of the cytoplasmic
enzyme lactate dehydrogenase (LDH) or by means of MTT staining, which
is based on mitochondrial function *(19,37,52,65–69)*.

Apoptosis

In contrast to necrosis, apoptosis or programmed cell death occurs under
normal physiological curcumstances with the active participation of the cell
in its own destruction ("cellular suicide"). Morphologically, apoptotic cells
are characterized by pycnotic nuclei, formation of apoptotic bodies, "membrane blebbing," and cellular condensation *without* loss of cellular integrity
and with intact organelles. Biochemical characteristics of apoptosis are, among
other things, attenuation of ATP-dependent processes, which may lead
to impaired mitochondrial function and, in turn, decreased dehydrogenase
activity. Furthermore, apoptosis is often, but not always, accompanied by
formation of oligonucleosomal DNA fragmentation (DNA ladder after aga-
rose gel electrophoresis) (e.g., ref. *70*).

Assessment of Type of Cell Death

In order to establish the means by which the cell has died (i.e., by necrosis or apotosis), it is of importance to realize the existence of some pitfalls that might hamper the conclusion drawn. First, positive identification of apoptosis requires the use of more than one of the parameters characteristic for this form of cell death because the individual signs can occur as a response to other stimuli than apoptosis. Second, the very nature of apoptosis may blur the identification of the process. It is difficult to study the phenomenon as a result of neurodegenerative diseases in vivo partly because even though terminally the cell death is of tremendous dimensions, the process occurs over many years. Thus, at any given time in the process, only very few dying cells would be present. In connection with acute neuropathological conditions such as stroke, the neuronal damage occurs within days, whereas a possible apoptosis will occur within hours *(71–73)*. Furthermore, apoptotic cells are removed by phagocytosis *(74)*, thus leaving no trace for the investigator to discover. In vitro (with no phagocytes present) the apoptotic bodies and cellular debris ultimatively will swell and lyse. This terminal phase of apoptosis is called *secondary necrosis (75)*.

When the issue is to clarify whether cell death has occurred by necrosis or by apoptosis, it is also of importance to consider the capabilities and limitations of the cell viability test choosen. As examples can be mentioned that because regular apoptosis occurs with intact cell membranes, measurement of the release of the cytoplasmic enzyme LDH does not allow one to distinguish between true necrosis or secondary necrosis initiated by apoptosis. It would also be predicted that the LDH leakage method would not be positive for cell death in primary apoptosis. Futhermore, because of the possible mitochondrial dysfunction in early apoptosis, using the MTT test (which is based on intact mitochondrial function) might lead to underestimation of cell death induced by apoptosis. The recent evidence suggesting that loss of mitochondrial integrity and release of caspase-activating mitochondrial factors promote apoptosis *(76–78)* opens the possibility of a certain cross reactivity between necrosis and apoptosis.

The scenario just outlined makes it understandable why only lately it has been realized that apoptosis might play a role in the excitatory amino acid-induced neurodegeneration that traditionally has been considered to be of necrotic nature.

Early Response Genes

As discussed earlier, EAA-mediated neurotoxicity has been viewed as being of necrotic nature, but a wealth of evidence now supports the notion that transsynaptic regulation of gene expression is of crucial importance when considering molecular mechanisms of neurotoxicity in addition to the more physiological role in neural development and long-term adaptive changes in the mature nervous system *(79–83)*. An important issue of this hypothesis is transcriptional regulation by EAAs of early response genes, especially c-*jun* and c-*fos*. These genes encode transcription factors regulating the subsequent transcriptional activity of certain late response genes that might be of importance for cell survival or death *(84)*. Therefore, the recent findings of abnormal expression of c-*fos* *(63,83,85)* are of interest, although the functional bearings of such aberration is controversial. Thus, sustained and/or elevated c-*fos* expression is associated with neuronal apoptosis and developmental failures both in vivo and in vitro *(86–90)*. Also, in animal models of e.g. epilepsy and other neurodegenerative disorders, for example, disturbances of the normal expression pattern have been observed *(89,91)*. It has been suggested that the expression pattern of early response genes such as c-*fos* might be used as a predictive index for excitotoxic neuronal death *(83)*.

Clinical Perspectives

A greater mechanistic knowledge of the processes leading to cell damage related to exposure to excitatory amino acids is a prerequisite for the rational development of strategies for the treatment of the severe neurodegenerative diseases or conditions associated with with such damage. Basically, there are two possible strategies that may be followed. One is based on conventional receptor antagonists, and the other more novel approach is founded on mechanistic knowledge that, only with difficulty could be obtained without the prior development of specific receptor antagonists. Because excitatory amino acids are involved not only in undesirable processes but also in essential cognitive functions, a chronic treatment with exclusively based on administration of full antagonists is not attractive. It may, though, be acceptable to use such agents in short-lasting treatment in the acute phase of neurological disorders such as stroke. On the other hand, it may be of interest to focus on the development of partial antagonists because the administration of such compounds probably would tend to normalize the excessive stimulation with excitatory amino acid stimulation seen prior to or in combination with these conditions. Because partial antagonists (or agonists) would allow some excitatory activity, they may be acceptable as treatment on a more permanent basis. Another avenue of intervention for the treatment of chronic neurodegenerative diseases would be the attempt to identify "functional antagonists." Thus, research focused on the mechanisms has hitherto led to the identification of a range of pharmaca under evaluation for a possible therapeutic potential *(37,39,61,92)*. One of these putative drugs is dantrolene, preventing release of Ca^{2+} from intracellular stores *(37,39,61)*. Because the neuroprotective effect of these compounds is the result of interference with processes distally to the excitatory amino acid receptor activation, it is possible that some of the normal function mediated by excitatory amino acid receptors may be operative in the presence of such drugs. Accordingly, it is likely that therapy with drugs developed according to the mechanistic strategy in combination with partial blockage of excitatory amino acids receptors may have less severe side effects than therapeutic strategies exclusively based on excitatory amino acid receptor antagonists and, as such, might provide the basis for an acceptable and effective treatment of the relevant neuropathological diseases requiring permanent medication. It is likely that the use of cultured neurons of various types as a model system will facilitate such work. Brain damage caused by severe ischemia is presumably mediated via all types of excitatory amino acid receptors. Since dantrolene has no effect on toxicity induced by kainate or AMPA in vitro, the future of this compound presumably will lie in its use as an instrumental tool for experimental research rather than a therapeutic agent. The current knowledge of the deeper mechanisms in the neuronal stress response is too insufficient to provide a solid basis for the development of a rational program with a view of a pharmacological intervention. The most promising avenues to follow presumably will be connected with mitochondrial dysfunction (i.e., a systemic defect with initial symptoms being in the central nervous system and regulation of (early response) gene expression, including apoptosis. Because of their initial nature, targeting of these processes could be expected to be promising for the development of an effective treatment within the therapeutic window of acute neurological diseases, thus minimizing the severe sequela to occur.

REFERENCES

1. Watkins, J. C., Krogsgaard-Larsen, P., and Honoré, T. (1990) Structure–activity relationships in the development of excitatory amino acid receptor agonists and competitive antagonists. *Trends Pharmacol. Sci.* **11,** 25–33.
2. Schoepp, D. D. and Conn, P. J. (1993) Metabotropic glutamate receptors in brain function and pathology. *Trends Pharmacol. Sci.* **14,** 13–20.
3. Hollmann, M. and Heinemann, S. (1994) Cloned glutamate receptors. *Annu. Rev. Neurosci.* **17,** 31–108.
4. Lodge, D. (1997) Subtypes of glumate receptors, in *The Ionitropic Glutamate Receptors* (Monaghan, D. T. and Wenthold, R. J., eds.), Humana, Totowa, NJ, pp. 1–38.
5. Johnston, G. A. R., Curtis, D. R., Davies, J., and McCulloch, R. M. (1974) Spinal interneurone excitation by conformationally restricted analogues of L-glutamic acid. *Nature* **248,** 804–805.

6. Watkins, J. C. and Evans, R. H. (1981) Excitatory amino acid transmitters. *Annu. Rev. Pharmacol. Toxicol.* **21,** 165–204.
7. Honoré, T., Lauridsen, J., and Krogsgaard-Larsen, P. (1982) The binding of [^{3}H]-AMPA, a structural analogue of glutamic acid, to rat brain membranes. *J. Neurochem.* **38,** 173–178.
8. Foster, A. C. and Fagg, G. E. (1984) Acidic amino acid binding sites in mammalian membranes: their characteristics and relationship to synaptic receptors. *Brain Res. Rev.* **7,** 103–164.
9. Schoepp, D. D. (1994) Novel functions for subtypes of metabotropic glutamate receptors. *Neurochem. Int.* **24,** 439–449.
10. Masu, M., Tanabe, Y., Tsuchida, K., Shigemoto, R., and Nakanishi, S. (1991) Sequence and expression of a metabotropic glutamate receptor. *Nature* **349,** 760–765.
11. Huamed, K. M., Kuijper, J. L., Gilbert, T. L., Haldeman, B. A., O'Hara, P. J., Mulvihill, E. R., Almers, W., and Hagen, F. S. (1991) Cloning, expression, and gene structure of a G protein-coupled glutamate receptor from the rat brain. *Science* **252,** 1318–1321.
12. Nakanishi, S. (1992) Molecular diversity of glutamate receptors and implications for brain function. *Science* **258,** 597–603.
13. Pin, J. P. and Duvorisin, R. (1995) The metabotropic glutamate receptors: structure and functions. *Neuropharmacology* **34,** 1–26.
14. Saugstad, J. A., Kinzie, J. M., Mulvihill, E. R., Segerson, T. P., and Westbrook, G. (1994) Cloning and expression of a new member of the L-2-amino-4-phosphonobutyric acid-sensitive class of metabotropic glutamate receptors. *Mol. Pharm.* **45,** 367–372.
15. Saugstad, J. A., Kinzie, J. M., Shinohara, M. M., Segerson, T. P., and Westbrook, G. L. (1997) Cloning and expression of rat metabotropic glutamate-receptor 8 reveals a distinct pharmacological profile. *Mol. Pharmacol.* **51,** 119–125.
16. Lucas, D. R. and Newhouse, J. P. (1957) The toxic effect of sodium-L-glutamate on the inner layers of retina. *AMA Arch. Ophthalmol.* **58,** 193–201.
17. McGeer, E. G. and McGeer, P. L. (1976) Duplication of biochemical changes of Huntington's chorea by intrastriatal injections of glutamic and kainic acids. *Nature* **263,** 517–519.
18. Schousboe, A. and Frandsen, A. (1995) Glutamate receptors and neurotoxicity, in *CNS Neurotransmitters and Neuromodulators: Glutamate* (Stone, T. W., ed.), CRC, Boca Raton, FL, pp. 239–251.
19. Choi, D. W. (1988) Calcium-mediated neurotoxicity: relationship to specific channel types and role in ischemic damage. *TINS* **11,** 465–469.
20. Herrero, I., Miras-Portugal, M. T., and Sanchez-Prieto, J. (1992a) Activation of protein kinase C by phorbol esters and arachidonic acid required for optimal potentiation of glutamate exocytosis. *J. Neurochem.* **59,** 1574–1577.
21. Herrero, I., Miras-Portugal, M. T., and Sanchez-Prieto, J. (1992b) Positive feedback of glutamate exocytosis by metabotropic presynaptic receptor stimulation. *Nature* **360,** 163–166.
22. Schanne, F. A. X., Kane, A. B., Young, E. E., and Farber, J. L. (1979) Calcium dependence of toxic cell death. *Science* **206,** 700–702.
23. Siesjö, B. K. and Bengtsson, F. (1989) Calcium fluxes, calcium antagonists, and calcium-related pathology in brain ischemia, hypoglycemia, and spreading depression: a unifying hypothesis. *J. Cereb. Blood Flow Metab.* **9,** 127–140.
24. Dubinsky, J. M. (1993) Examination of the role of calcium in neuronal death. *Ann. NY Acad. Sci.* **679,** 34–42.

25. Choi, D. W. (1995) Calcium: still center-stage in hypoxic-ischemic neuronal death. *Trends Neurosci.* **18,** 58–60.
26. Witt, M.-R., Dekermendjian, K., Frandsen, A., Schousboe, A., and Nielsen, M. (1994) Complex correlation between excitatory amino acid induced increase in the intracellular Ca^{2+} concentration and subsequent loss of neuronal function in individual neocortical neurons in culture. *Proc. Natl. Acad. Sci. USA* **91,** 12,303–12,307.
27. Randall, R. D. and Thayer, S. A. (1992) Glutamate-induced calcium transient triggers delayed calcium overload and neurotoxicity in rat hippocampal neurons. *J. Neurosci.* **12,** 1882–1895.
28. Michaels, R. L. and Rothman, S. M. (1990) Glutamate neurotoxicity in vitro: antagonist pharmacology and intracellular calcium concentrations. *J. Neurosci.* **10,** 283–292.
29. Dubinsky, J. M. and Rothman, S. M. (1991) Intracellular calcium concentrations during "chemical hypoxia" and excitotoxic neuronal injury. *J. Neurosci.* **11,** 2545–2551.
30. Mattson, M. P., Guthrie, P. B., Hayes, B. C., and Kater, S. B. (1989) Roles for mitotic history in the generation and degeneration of hippocampal neuroarchitecture. *J. Neurosci.* **9,** 1223–1232.
31. Milani, D., Guidolin, D., Facci, L., Pozzan, T., Buso, M., Leon, A., and Skaper, S. D. (1991) Excitatory amino acid-induced alterations of cytoplasmic free Ca^{2+} in individual cerebellar granule neurons: role in neurotoxicity. *J. Neurosci. Res.* **28,** 434–441.
32. Tymianski, M., Charlton, M. P., Carlen, P. L., Tator, C. H. (1993) Source specificity of early calcium neurotoxicity in cultured embryonic spinal neurons. *J. Neurosci.* **13,** 2085–2104.
33. Eimerl, S. and Schramm, M. (1993) Potentiation of ^{45}Ca uptake and acute toxicity mediated by the *N*-methyl-D-asparatate receptor: The effect of metal binding agents and transition metal ions. *J. Neurochem.* **61,** 518–525.
34. Mattson, M. P., Guthrie, P. B., and Kater, S. B. (1989) A role for Na^+-dependent Ca^{2+} extrusion in protection against neuronal excitotoxicity. *FASEB J.* **3,** 2519–2526.
35. Brorson, J. R., Bleakman, D., Chard, P. S., and Miller, R. J. (1992) Calcium directly permeates kainate/(α-amino-3-hydroxy-5-methyl-4-isoxazole-propionic acid receptors in cultured cerebellar Purkinje neurons. *Mol. Pharmacol.* **41,** 603–608.
36. Miller, R. J. (1991) The control of neuronal Ca^{2+} homeostasis. *Prog. Neurobiol.* **37,** 255–285.
37. Frandsen, A. and Schousboe, A. (1993) Excitatory amino acid mediated cytotoxicity and calcium homeostasis in cultured neurons. *J. Neurochem.* **60,** 1202–1211.
38. Weiss, J. H., Hartley, D. M., Koh, J., and Choi, D. W. (1990) The calcium channel blocker nifedipine attenuates slow excitatory amino acid neurotoxicity. *Science* **247,** 1474–1477.
39. Frandsen, A. and Schousboe, A. (1992) Mobilization of dantrolene-sensitive intracellular calcium pools is involved in the cytotoxicity induced by quisqualate and N-methyl-D-aspartate but not by 2-amino-3-(3-hydroxy-5-methylisoxazol-4-yl) propionate and kainate in cultured cerebral cortical neurons. *Proc. Natl. Acad. Sci. USA* **89,** 2590–2594.
40. Valentino, K., Newcomb, R., Gadbois, T., Singh, T., Bowersox, S., Bitner, S., Justice, A., Yamashiro, D., Hoffman, B. B., Ciaranello, R., Miljanich, G., and Ramachandran, J. (1993) A selective N-type calcium channel antagonist protects against neuronal loss after global cerebral ischemia. *Proc. Natl. Acad. Sci. USA* **90,** 7894–7897.

41. Sucher, N. J., Lei, S. Z., and Lipton, S. A. (1991) Calcium channel antagonists attenuate NMDA receptor-mediated neurotoxicity of retinal ganglion cells in culture. *Brain Res.* **297,** 297–302.
42. Dreyer, E. B., Kaiser, P. K., Offermann, J. T., and Lipton, S. A. (1990) IIIV-1 coat protein neurotoxicity prevented by calcium channel antagonists. *Science* **248,** 364–367.
43. Holopainen, I., Enkvist, M. O. K., and Åkerman, K. E. O. (1989) Glutamate receptor agonists increase intracellular Ca^{2+} independently of voltage-gated Ca^{2+} channels in rat cerebellar granule cells. *Neurosci. Lett.* **98,** 57–62.
44. Lasarewicz, J. W., Lehmann, A., and Hamberger, A. (1987) Effects of Ca^{2+} entry blockers on kainate-induced changes in extracellular amino acids and Ca $^{2+}$ in vivo. *J. Neurosci. Res.* **18,** 341–344.
45. Choi, D. W. (1992) Excitotoxic cell death. *J. Neurobiol.* **23,** 1261–1276.
46. Brorson, J. R., Manzolillo, P. A., and Miller, R. J. (1994) Ca^{2+} entry via AMPA/ KA receptors and excitotoxicity in cultured cerebellar purkinje cells. *J. Neurosci.* **14,** 187-197.
47. Turetsky, D. M., Canzoniero, L. M. T., Sensi, S. L., Weiss, J. H., Goldberg, M. P., and Choi, D. W. (1994) Cortical neurones exhibiting kainate-activated cobalt uptake are selectively vulnerable to AMPA/kainate receptor-mediated toxicity. *Neurobiol. Dis.* **1,** 101–110.
48. Lu, Y. M., Yin, H. Z., Chiang, J., and Weiss, J. H. (1996) Ca^{2+}-permeable AMPA/ Kainate and NMDA channels: high rate of Ca^{2+} influx underlies potent induction of injury. *J. Neurosci.* **16,** 5457–5465.
49. Jensen, J. B., Schousboe, A., and Pickering, D. S. (1998) Development of Ca^{2+} permeable AMPA receptors in cultured neocortical neurons visualized by cobalt staining. *J. Neurosci. Res.* **54,** 273–281.
50. Savidge, J. R., Bleakman, D., and Bristow, D. R. (1997) Identification of kainate receptor-mediated intracellular calcium increases in cultured rat cerebellar granule cells. *J. Neurochem.* **69,** 1763–1766.
51. May, P. C. and Robison, P. M. (1993) Cyclothiazide treatment unmasks AMPA excitotoxicity in rat primary hippocampal cultures. *J. Neurochem.* **60,** 1171–1174.
52. Jensen, J. B., Schousboe, A., and Pickering, D. S. (1998) AMPA receptor mediated excitotoxicity in neocortical neurons is developmentally regulated and dependent upon receptor desensitization. *Neurochem. Int.* **32,** 505–513.
53. Jensen, J. B., Schousboe, A., and Pickering, D. S. (1999) Role of desensitization and subunit expression for kainate receptor-mediated neurotoxicity in murine neocortical cultures. *J. Neurosci. Res.* **55,** 208–217.
54. Carafoli, E. (1987) Intracellular calcium homeostasis. *Ann. Rev. Biochem.* **56,** 395–433.
55. Kiedrowski, L. and Costa, E. (1995) Glutamate-induced destabilization of intracellular calcium concentration homeostasis in cultured cerebellar granule cells: role of mitochondria in calcium buffering. *Mol. Pharmacol.* **47,** 140–147.
56. Gerasimenko, O. V., Gerasimenko, J. V., Tepikin, A. V., and Petersen, O. H. (1995) ATP-dependent accumulation and inositol triphosphate- or cyclic ADP-ribose-mediated release of Ca^{2+} from the nuclear envelope. *Cell* **80,** 439–444.
57. Wilcox, R. A., Primrose, W. U., Nahorski, S. R., and Challiss, R. A. J. (1998) New developments in the molecular pharmacology of the *myo*-inositol 1,4,5-trisphosphate receptor. *Trends Pharmacol. Sci.* **19,** 467–475.

58. Valdivia, H. H. (1998) Modulation of intracellular Ca^{2+} levels in the heart by sorcin and FKBP12, two accessory proteins of ryanodine receptors. *Trends Pharmacol. Sci.* **19**, 479–482.

59. Ward, A., Chaffman, M. O., and Sorkin, E. M. (1986) Dantrolene: a review of its pharmacodynamic and pharmacokinetic properties and therapeutic use in malignant hyperthermia, the neuroleptic malignant syndrome and an update of its use in muscle spasticity. *Drugs* **32**, 130–168.

60. Bouchelouche, P., Belhage, B., Frandsen, Aa., Drejer, J., and Schousboe, A. (1989) Glutamate receptor activation in cultured cerebellar granule cells increases cytosolic free Ca^{2+} by mobilization of cellular Ca^{2+} and activation of Ca^{2+} influx. *Exp. Brain Res.* **76**, 281–291.

61. Frandsen, Aa. and Schousboe, A. (1991) Dantrolene prevents glutamate cytotoxicity and Ca^{2+} release from intracellular stores in cultured cerebral cortical neurons. *J. Neurochem.* **56**, 1075–1078.

62. Gunter, T. E. and Pfeiffer, D. R. (1990) Mechanisms by which mitochondria transport calcium. *Am. J. Physiol.* **258**, C755–C786.

63. Gorman, A. M., Scott, M. P., Rumsby, P. C., Meredith, C., and Griffiths, R. (1995) Excitatory amino acid-induced cytotoxicity in primary cultures of mouse cerebellar cells correlates with with elevated, sustained *c-fos* proto-oncogene expression. *Neurosci. Lett.* **191**, 116–120.

64. Montal, M. (1998) Mitochondria, glutamate neurotoxicity and the death cascade. *Biochim. Biophys. Acta* **1366**, 113–126.

65. Frandsen, Aa. and Schousboe, A. (1987) Time and concentration dependency of the toxicity of excitatory amino acids on cerebral neurones in primary culture. *Neurochem. Int.* **4**, 77–81.

66. Frandsen, Aa. and Schousboe, A. (1990) Development of excitatory amino acid induced cytotoxicity in cultured neurons. *Int. J. Dev. Neurosci.* **2**, 209–216.

67. Frandsen, Aa., Drejer, J., and Schousboe, A. (1989) Glutamate-induced $^{45}Ca^{2+}$ uptake into immature cerebral cortex neurons show a distinct oharmacological profile. *J. Neurochem.* **53**, 1959–1962.

68. Frandsen, Aa., Quistorff, B., and Schousboe, A. (1990) Phenobarbital protects cerebral cortex neurones against toxicity induced by kainate but not by other excitatory amino acids. *Neurosci. Lett.* **111**, 233–238.

69. Frandsen, Aa., Krogsgaard-Larsen, P., and Schousboe, A. (1991) Novel glutamate receptor antagonists selectively protect against kainic acid neurotoxicity in cultured cerebral cortex neurons. *J. Neurochem.* **55**, 1821–1823.

70. Warner, H. R. (1997) Aging and regulation of apoptosis. *Curr. Topics Cell Regul.* **35**, 107–121.

71. Diemer, N. H., Valente, E., Bruhn, T., Berg, M., Jørgensen, M. B., and Johansen, F. F. (1993) Glutamate receptor transmission and ischemic nerve cell damage: evidence for the involvement of excitotoxic mechanisms. *Prog. Brain Res.* **96**, 105.

72. Waters, C. M. (1996) Mechanisms of neuronal cell death. An overview. *Mol. Chem. Neuropathol.* **28**, 145–151.

73. Rubin, L. L. (1997) Neuronal cell death: when, why and how. *Br. Med. Bull.* **53**, 617–631.

74. Savill, J. S., Wylie, A. H., Henson, J. E., Walport, M. J., Henson, M. P., and Haslett, C. I. (1989) Macrophage phagocytosis of aging neutrophils in inflammation: programmed cell death in the neutrophil leads to its recognition by macrophages. *J. Clin. Invest.* **83**, 865–875.

75. Lavin, M. and Watters, D. (1996) Programmed cell death. Harwood Academic, Switzerland.
76. Liu, X., Kim, C. N., Yang, J., Jemmerson, R., and Wang, X. (1996) Induction of apoptotic program in cell-free extract: requirement for ATP and cytochrome c. *Cell* **86**, 147–157.
77. Kluck, R. M., Bossy-Wetzel, E., Green, D. R., and Newmeyer, D. D. (1997) The release of cytochrome c from mitochondria: a primary site for Bcl-2 regulation of apoptosis. *Science* **275**, 1132–1136
78. Yang, J., Liu, X., Bhalla, K., Kim, C. N., Ibrado, A. M., Cai, J., Peng, T. I., Jones, D. P., and Wang, X. (1997) Prevention af apoptosis bu Bcl-2: release of cytochrome c from mitochondria blocked. *Science* **275**, 1129–1132.
79. Ginty, D. D., Bading, H., and Greenberg, M. M. (1992) Trans-synptic regulation of gene-expression. *Curr. Opin. Neurobiol.* **2**, 312–316.
80. Gass, P., Herdegen, T., Bravo R., and Kiessling, M. (1993) Induction and suppression of immediate early genes in specific rat brain regions by the non-competitive *N*-methyl-D-aspartate receptor antagonist MK-801. *Neuroscience* **53**, 749–758.
81. Morgan, J. I. and Curran, T. (1995) The immediate-early gene response and neuronal death and regeneration. *Neuroscientist* **1**, 68–75.
82. Herdegen, T. (1996) Jun, Fos and CREB/ATF transcription factors in the brain: control of gene expression under normal and patho-physiological conditions. *Neuroscientist* **2**, 153–161.
83. Griffiths, R., Malcolm, G., Ritchie, L., Frandsen, A., and Schousboe, A. (1997) Association of *c-fos* mRNA expression and excitotoxicity in primary cultures of mouse neocortical and cerebellar neurons. *J. Neurosci. Res,* **48**, 533–542.
84. Tatter, S. B., Galpern, W. R., and Isacson, O. (1995) Neurotrophic factor protection against excitotoxic neuronal death. *Neuroscientist* **1**, 286–297.
85. Meredith, C., Scott, M., Rumsby, P., Frandsen, A., Schousboe, A., Gorman, A., and Griffiths, R. (1995) Susatined c-fos expression is associated with excitotoxicity during the development of neuronal cells in vitro. *Biochem. Soc. Trans.* **24**, 6S.
86. Holt, J. T., Venkat Gopal, T., Moulton, A. D., and Nienhuis, A. W. (1986) Inducible production of c-fos anti-sense RNA inhibits 3T3 cell proliferation. *Proc. Natl. Acad. Sci. USA* **83**, 4794–4798.
87. Rüther, U., Garber, C., Komitowski, D., Müller, R., and Wagner, E. F. (1987) Deregulated c-fos expression interferes with normal bone development in transgenic mice. *Nature* **325**, 412–416.
88. Edwards, S. A., Rundell, A. Y. K., and Adamson, A. D. (1988) Expression of *c-fos* antisense RNA inhibits the differentiation of F9 cells to parietal endoderm. *Dev. Biol.* **129**, 91–102.
89. Smeyne, R. J., Vendrell, M., Hayward, M., Baker, S. J., Miao, G. G., Schilling, K., Robertson, L. M., Curran, T., and Morgan, J. I. (1993) Continous c-fos expression precedes programmed cell death in vivo. *Nature* **363**, 401–408.
90. Chen, S.-C. and Morgan, J. I. (1995) Apoptosis in the nervous system:new revelations from old transgenic mice. *J. Clin. Pathol.* **48**, 7–12.
91. Pennypacker, K. R., Thai, L., Hong, J.-S., and McMillian, M. K. (1994) Prolonged expression of AP-1 transcription factors in the rat hippocampus after systemic kainate treatment. *J. Neurosci.* **14**, 3998–4006.
92. Schehr, R. S. (1996) New treatments for stroke. *Nat. Biotechnol.* **14**, 1549–1554.

Part IV

Depression

9

Molecular and Cellular Determinants of Stress and Antidepressant Treatment

Ronald S. Duman

Studies to characterize the complex actions of stress and antidepressant treatment on brain function have made significant progress over the past 20 yr. This work demonstrates that in addition to regulation of neurotransmitter and neuroendocrine pathways, stress and antidepressant treatment also exert potent effects on neuronal morphology, cytoarchitecture, and intracellular signal transduction pathways. Moreover, progress is being made in understanding the molecular and cellular actions of stress in the context of the cellular and molecular systems that are thought to be involved in depression. The major first messengers involved in the action of stress and antidepressant treatment are the adrenal-glucocorticoids and monoamines. In addition, a role for excitatory amino acids and neurotrophic factors has been demonstrated. This chapter will briefly review the intracellular signaling pathways for each of these first messengers and then provide a more detailed analysis of these intracellular pathways in the context of stress and antidepressant treatment. Although a great deal of work still remains, this work has already begun to elucidate how alterations of these pathways may contribute to the pathophysiology and treatment of depression.

INFLUENCE OF STRESS ON NEURONAL FUNCTION

Acute stress results in activation of pathways that are designed to help an organism mount a response to stress, but repeated or severe stress results in adverse effects that contribute to many neurobiological disorders. In addition, other environmental factors or genetic determinants may make certain individuals more vulnerable to stress. Recent studies have begun to reveal the molecular and cellular actions of stress, particularly the role of excess glucocorticoids and neurotrophic factors, on cell survival and function.

From: *Cerebral Signal Transduction: From First to Fourth Messengers*
Edited by: M. E. A. Reith © Humana Press Inc., Totowa, NJ

Activation of the HPA Axis by Stress

Activation of the hypothalamic–pituitary–adrenal (HPA) axis and increased release of glucocorticoids is the hallmark of an animal's stress responses. This pathway includes release of corticotrophin-releasing factor (CRF) from the hypothalamus, which stimulates the release of adrenocorticotrophic hormone (ACTH) from the pituitary, which, in turn, stimulates the release of glucocorticoid from the adrenal gland. There are multiple feedback loops in this pathway that are designed to maintain homeostasis and return the HPA axis to normal, prestress levels. Glucocorticoids feed back to inhibit the HPA axis at the most proximal sites in the pathway, the pituitary and hypothalamus. In addition, the hippocampus represents a site of strong negative feedback for glucocorticoid regulation of the HPA axis *(1,2)*. The hippocampal feedback pathway is particularly relevant for understanding the deleterious effects of chronic stress and excess glucocorticoid levels. Certain populations of hippocampal neurons become damaged when exposed to repeated stress or high levels of glucocorticoids, and this damage could result in loss of the inhibitory control of the HPA axis. This could lead to a cycle where the damaging effects of stress are further amplified by the ever-increasing levels of glucocorticoid as hippocampal negative feedback is reduced.

Overactivation of the HPA axis, possibly as a result of reduced hippocampal inhibition, could contribute to the memory deficits, vegetative abnormalities, and, possibly, other aspects of affective illnesses. Hippocampal negative feedback is reported to be reduced in depressed patients *(3)*. In addition, loss of feedback inhibition and rising levels of glucocorticoids are associated with hippocampal atrophy and memory deficits that occur during natural aging (*see* refs. *4* and *5*). The cellular and molecular basis for the damaging effects of stress on hippocampal neurons are discussed in the following section.

Regulation of Hippocampal Neuronal Atrophy and Survival by Stress

The damaging influence of stress on hippocampal neurons occurs in different ways in the different populations of hippocampal neurons. The CA3 pyramidal neurons appear to be one of the populations of cells most sensitive to stress and glucocorticoids. These cells are influenced in three general ways by stress and glucocorticoid treatments (Fig. 1) (for a review, *see* ref. *6*). First, the CA3 pyramidal neurons are reported to undergo atrophy in response to these treatments *(7–9)*. Atrophy of CA3 neurons occurs in the form of decreased number and length of the apical dendrites and has been observed in rodents and nonhuman primates. Second, death or neurotoxicity of CA3 neurons is reported to occur in response to severe and prolonged

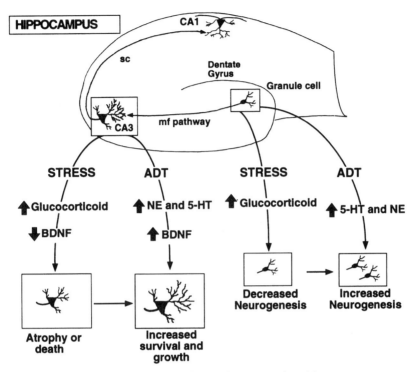

Fig. 1. A cellular model for the actions of stress and antidepressant treatment. Stress and excess levels of glucocorticoid cause atrophy or death of CA3 pyramidal neurons and decrease the neurogenesis of granule cells in the dentate gyrus in the hippocampus. Recent studies demonstrate that stress decreases the expression of brain derived neurotrophic factor (BDNF) in both of these cell layers of hippocampus, an effect that could contribute to the damaging effects of stress. Moreover, these effects of stress could contribute to the reduction in the volume of the hippocampus in patients with depression or posttraumatic stress disorder (*see text* for a discussion). In contrast, antidepressant treatment (ADT) increases the expression of BDNF and prevents the downregulation of BDNF elicited by stress. Increased levels of BDNF could have trophic actions that protect hippocampal neurons from further damage or, possibly, reverse the stress-induced damage. Induction of BDNF appears to result from sustained elevation of 5-HT and norepinephrine (NE) transmission and activation of the cAMP-CREB cascade (*see* Fig. 4).

stress or glucocorticoid treatments *(10–12)*. Third, stress and glucocorticoid treatments produce a state of neuroendangerment of CA3 pyramidal neurons. In this case, only the short-term stress or glucocorticoid exposure, which alone has no effect, is reported to exacerbte the damage induced by other types of neuronal insults, including excitotoxins, hypoglycemia, and hypoxia–ischemia *(6,10,13,14)*.

The dentate gyrus granule cells in the hippocampus are influenced in a different manner by stress and glucocorticoids. These cells exhibit a relatively unique characteristic in brain, in that they continue to undergo neurogenesis in adult animals. Neurogenesis, or continued cell birth of granule cells, together with cell death, is a normal process in adult rodents and has also been demonstrated in nonhuman primates (*see* refs. *15* and *16*), although another study failed to observe neurogenesis in adult nonhuman primates *(17)*. Several studies demonstrate that stress or excess levels of glucocorticoid reduce the rate of neurogenesis in rodents, and one of these also reports a similar effect in nonhuman primates (Fig. 1) *(15,16)*. Decreased neurogenesis is observed after a single, short-term exposure to stress. This overall decrease in the number of newly born cells could contribute to the functional and morphological effects of stress on hippocampus. In contrast, the rate of neurogenesis is reported to be increased in adult mice exposed to an enriched environment (i.e., increased exposure to interesting inanimate objects and social stimulation) compared to littermates housed in standard cages *(18)*. This suggests that positive stimuli can enhance the rate of neurogenesis. Although the functional consequences have not been determined, it is interesting to speculate that increased neurogenesis would increase the function of the mossy fiber pathway and the hippocampus (e.g., increased inhibitory control of the HPA axis and enhanced learning and memory).

Cells in other brain regions do not appear to be as sensitive to the damaging effects of stress as observed in the hippocampus. The reason for this regional difference is not completely understood but may be related to the higher levels of glucocorticoids and their receptors in hippocampus (*see* ref. 6). However, there have been fewer studies directed at brain regions other than the hippocampus, and it is possible that once a more thorough analysis is conducted, the damaging effects of stress and glucocorticoid treatments will be found to be more widespread.

Relevance to the Pathophysiology of Depression: Clinical Studies of Brain Volume

Although it is difficult to determine the relevance of these basic research studies in rodents and nonhuman primates, clinical reports have provided evidence that neuronal atrophy or cell death of hippocampal neurons may play a role in the pathophysiology of depression and other stress-related illnesses. First, as noted earlier, neuroendocrine studies demonstrate that hippocampal fast-feedback inhibition of the HPA axis is reduced in depressed patients, indicating that the function of the hippocampus is decreased *(3)*. Second, brain-imaging studies demonstrate a reduction in the volume of hippocampus, determined by magnetic resonance imaging (MRI), in patients with depression or posttraumatic stress disorder (*19,20*; *also see* ref. *21*). The volume of the hippocampus is also reported to be decreased in patients with Cushings disease *(22)*, lending further support to the hypothesis that increased glucocorticoid levels may be involved in the reduction of hippocampal volume. In addition to these studies in hippocampus, there are also reports that the function and size of the subgenual prefrontal cortex is reduced in patients with depression and bipolar disorder *(23)*. A decrease in the number of glia in this area of the prefrontal cortex has also been reported in patients with familial depression or bipolar disorder, indicating that death of non-neuronal cells also occurs in affective illness *(24)*.

MOLECULAR ACTIONS
OF STRESS AND GLUCOCORTICOIDS

The mechanisms that underlie the influence of stress and glucocorticoid exposure are likely to involve multiple cellular actions, as well as regulation of different neurotransmitter and neuroendocrine factors. Some of the best characterized effects that are relevant to neuronal damage are discussed in this section. The primary mechanism by which glucocorticoids influence cellular function is via binding to their cytoplasmic receptors, referred to as mineralocorticoid or glucocorticoid receptors, which are then translocated to the nucleus and act as gene transcription factors. Both receptor subtypes are expressed at very high levels in hippocampus, but it is the glucocorticoid receptor that is reported to mediate the damaging effects of stress and glucocorticoid treatments (*25*; *see* ref. *6*). In addition to the high level of receptors, the hippocampus contains higher levels of glucocorticoids, which may result from an enzymatic process that converts glucocorticoid metabolites back to the active steroid *(26)*.

Regulation of Energy Metabolism
by Stress and Glucocorticoids

One of the mechanisms by which stress and glucocorticoids are thought to influence neuronal survival and function is by reducing the energy capacity of neurons (*see* ref. *6*). Glucocorticoids are known to decrease the uptake of glucose in peripheral tissues, including fat cells and fibroblasts, and a similar effect is observed in the brain (Fig. 2). The mechanisms responsible for the decreased glucose uptake include both transcriptional and posttranscriptional modifications. Long-term exposure of adipocytes to glucocorticoid decreases the expression of the glucose transporter and thereby decreases the ability of cells to accumulate glucose *(27)*. This occurs via glucocorticoid

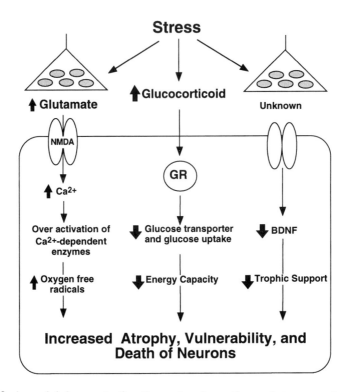

Fig. 2. A model demonstrating the molecular actions of stress on signal transduction. The damaging influence of stress is due to dysregulation of several factors, most notably excess levels of adrenal-glucocorticoids. Glucocorticoids bind to cytoplasmic receptors (GR) and are then translocated to the nucleus where they act as gene transcription factors. GR is the receptor subtype that is thought to mediate the damaging actions of stress. One effect of excess glucocorticoid levels is to reduce glucose uptake and thereby compromise the energetic capacity of cells. This is thought to occur via decreased expression of the glucose transporter, as well as translocation of the transporter away from the cell membrane to intracellular compartments. Stress is also reported to increase extracellular levels of glutamate, the major excitatory neurotransmitter in brain. This could lead to overactivation of glutamate receptors that gate Ca^{2+}, such as the NMDA receptor. This could result in excess levels of intracellular Ca^{2+}, which is known to regulate many cellular enzymes, and could lead to increased levels of oxygen-free radicals. Stress is also reported to decrease the expression BDNF in hippocampus. The mechanism underlying this effect is unknown, although there is preliminary evidence that the 5-HT and GABA neurotransmitter systems may contribute to the decreased expression of BDNF (*see* Fig. 3). A reduction of BDNF would be expected to decrease trophic support of hippocampal neurons. In addition to the damaging effects of excess glucocorticoids and glutamate and reduced BDNF on cell function, these pathways may exacerbate each other. For example, a reduction in cell energetics would make it more difficult to buffer excess intracellular Ca^{2+}. Taken together, these effects of stress result in atrophy, vulnerability, and death of neurons.

inhibition of transporter expression. In addition, short-term exposure to glucocorticoid induces a translocation of the glucose transporter from the cell membrane to intracellular sites in adipocytes via a posttranscriptional mechanism
(28). Although these effects of glucocorticoids have been demonstrated in
peripheral tissues, it is possible that they have similar actions in brain. Evidence for this hypothesis comes from reports demonstrating that glucocorticoids regulate glucose utilization in the brain *(29)* and decrease glucose
uptake in primary neuronal cultures *(30,31)*. This reduction in energy capacity is likely to increase the vulnerability of neurons to other types of insults,
such as excitotoxins, hypoxia–ischemia, and hypoglycemia. In addition,
over extended periods of time, a reduction in glucose could eventually lead
to neurotoxic effects. This possibility is supported by studies of Sapolsky
and colleagues on neuronal endangerment, as well as studies from others on
the influence of aging and glucocorticoids on neuronal loss (for a review,
see refs. *5* and *6*).

Regulation of Glutamate and Calcium by Stress and Glucocorticoids

Excessive activation of excitatory amino acid neurotransmitter systems
and influx of calcium are known to be critical determinants of neuronal
toxicity in response to many types of neuronal insult. Glutamate, the major
excitatory neurotransmitter system in brain, stimulates the influx of calcium
by activation of NMDA and non-NMDA ionotropic receptors (*see* ref. *32*).
Excessive activation of glutamatergic ionotropic receptors is known to contribute to the neurotoxicity that occurs in response to many types of neuronal insult, including repeated seizures and ischemia. Excess intracellular
calcium leads to sustained activation of multiple intracellular enzymes, which
have negative effects on cell survival and function (*see* ref. *33*). This may
include oxidative damage and generation of free radicals.

Stress or glucocorticoid treatment is reported to increase the extracellular
levels of glutamate in hippocampus as measured by dialysis, suggesting that
activation of glutamate receptors and influx of calcium may contribute to
the damaging effects of stress (Fig. 2) *(34,35)*. However, stress leads to similar increases in extracellular levels of glutamate in other brain regions that
are not damaged by stress, and there is evidence that the elevation of glutamate may result in part from the insertion of the dialysis probe (*see* ref. *35*).
It is also surprising that stress does not induce early response genes, such as
c-*Fos*, that are known to be induced by activation of glutamate systems. In
fact, expression of BDNF, which is known to be induced by glutamate, is
significantly decreased by stress *(36)*. These findings raise some questions
about the direct role of glutamate in the actions of stress and suggest that

there may be other factors that contribute to the atrophy and death of hippo-campal neurons (see the following). However, there is clear evidence that glucocorticoids enhance glutamate excitoxicity in response to other types of insult (*see* ref. *6*). Moreover, it is possible that the cumulative effects of an overactive glutamate system could contribute to the damaging effects of repeated stress.

Regulation of Neurotrophic Factors by Stress

A role for neurotrophic factors in the actions of stress, as well as antide-pressant treatments, is provided by several lines of evidence. BDNF belongs to the nerve growth factor family, which also includes nerve growth factor (NGF), neurotrophin-3 (NT-3), and neurotrophin-4 (NT-4). Although these neurotrophic factors were originally characterized for their role in the devel-opment and growth of neurons, it is now clear that they are vital to the main-tenance, survival, and function of neurons in the mature nervous system (*see* refs. *37–39*). The actions of neurotrophic factors are mediated by Trk receptors, transmembrane receptors that contain intrinsic tyrosine kinase activity *(40)*. The Trk receptors are relatively specific for each neurotrophic factor, although some receptors also bind to more than one neurotrophic factor. For example, TrkA is relatively selective for NGF, TrkB binds BDNF and NT-4 but has higher affinity for BDNF, and TrkC binds primarily NT-3. Activation of Trk receptors leads to autophosphorylation of the receptor and then coupling to intracellular signaling pathways, including the mitogen-activated protein (MAP) kinase pathway, phosphatidylinositol-3 kinase and phospholipase C-γ (*see* ref. *40*).

Smith and colleagues have demonstrated that stress induces a rapid and long-lasting decrease in the expression of BDNF in hippocampus (Fig. 2) *(36)*. Exposure to immobilization stress for 1–2 h decreases the expression of BDNF mRNA in the dentate gyrus granule cell layer and the CA3 pyra-midal cell layer. This effect is still observed after repeated (7 d) stress. Given the vital role that neurotrophic factors play in the survival and function of neurons, it is possible that decreased expression of BDNF could contribute to the damaging influence of stress on hippocampal neurons. If this were the case, then it should be possible to protect these neurons from damage by administration of BDNF. Conversely, reduction or depletion of BDNF should make these neurons more vulnerable to damage. These possibilities are currently being examined using local infusions of BDNF and BDNF mutant mice.

Repeated, but not acute, stress was also found to increase the expression of NT-3 in the dentate gyrus, but not in the CA3 pyramidal cell layer. In a subsequent study, expression of NT-3 in the locus coeruleus was also reported to be increased by chronic stress *(41)*. It is possible that upregulation of NT-3 contributes to the protection of dentate gyrus granule cells from damage. NT-3 is also reported to increase the survival and function of locus coeruleus neurons *(42)*, and upregulation of this neurotrophic factor may contribute to the increased function of these neurons in response to repeated stress (i.e., increased expression of tyrosine hydroxylase, the rate-limiting enzyme in the synthesis of norepinephrine).

Pathways Underlying the Regulation of BDNF

The pathways underlying the regulation of BDNF by stress could involve endocrine as well as neurotransmitter systems. The role of glucocorticoids in the downregulation of BDNF by stress was investigated in the study of Smith et al. *(36)*.They found that acute administration of high levels of glucocorticoids, similar to those induced by stress, did not lead to downregulation of BDNF expression. Chronic glucocorticoid treatment did lead to a reduction of BDNF expression, but this effect was not of the same magnitude as that observed with acute or chronic stress. In addition, adrenalectomy (supplemented with a low dose of glucocorticoid) did not block the downregulation of BDNF. These results indicate that stress-induced elevation of glucocorticoids cannot account for the downregulation of BDNF expression. Adrenalectomy, without glucocorticoid supplement, did reduce the basal expression of NT-3, although expression of this neurotrophic factor was not influenced by either acute or chronic glucocorticoid treatment.

The rapid nature of the stress-induced downregulation of BDNF expression suggests that a fast acting neurotransmitter system may be involved. The expression of BDNF is reported to be influenced by the major inhibitory and stimulatory neurotransmitters systems and appears to be linked, at least in part, to the level of neuronal activity (*see* ref. *39*). Activation of the major inhibitory system, GABA, decreases and activation of the major excitatory system, glutamate, increases the expression of BDNF. As discussed earlier, stress is reported to increase extracellular levels of glutamate but this would be expected to increase, not decrease, the expression of BDNF. It is possible that the downregulation of BDNF is mediated by increased GABAergic neurotransmission. One possibility is that stress activates GABAergic interneurons that are known to induce inhibitory postsynaptic potentials (IPSPs) in dentate gyrus granule cells (Fig. 3) *(43)*. Pharmacological experiments to test this hypothesis are complicated because antagonists of the $GABA_A$ receptor, which mediates the major inhibitory effects of the GABA system, result in generalized seizures.

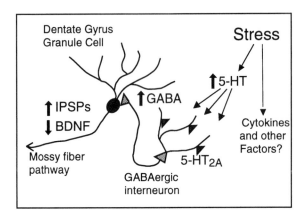

Fig. 3. A model demonstrating a potential mechanism for decreased expression of BDNF in response to stress. Stress could induce a rapid downregulation of BDNF expression via activation of 5-HT$_{2A}$ receptors located on GABAergic interneurons. Activation of these interneurons is known to increase IPSPs in the granule cells. Increased inhibition of dentate gyrus granule cells could lead to decreased activation of Ca^{2+}-dependent protein kinases and decreased expression of BDNF. Blockade of the 5-HT$_{2A}$ receptor reduces the stress-induced downregulation of BDNF. However, the blockade is partial, indicating that there are other factors that contribute to the influence of stress on BDNF expression. One possibility that is being investigated is that a cytokine, such as IL-1β, mediates this effect.

Monoamine neurotransmitters are also known to be regulated by stress and could be involved in the downregulation of BDNF expression. One possibility is the serotonin system, which is known to be activated by stress *(44,45)*. Support for this possibility is provided by a study demonstrating that activation of 5-HT$_{2A}$ receptors decreases the expression of BDNF in the dentate gyrus granule cell layer in a manner similar to that observed in response to stress *(46)*. In addition, pretreatment with a 5-HT$_{2A}$ antagonist partially blocks the downregulation of BDNF by stress (Fig. 3) *(46,47)*. Regulation of BDNF expression by 5-HT$_{2A}$ receptors could occur via activation of GABAergic interneurons, which inhibit the firing of granule neurons as discussed earlier in this chapter *(43)*.

The downregulation of BDNF is only partially blocked by 5-HT$_{2A}$ receptors, indicating that there are other systems that contribute to the stress response. We have also tested the influence of several other receptor antagonists, including those for 5-HT$_{1A}$, β-adrenergic and α_1-adrenergic, and CRF-R1 receptors, but none of these appear to alter the stress response *(47)*. It is also possible that cytokines, many of which are activated by stress, could mediate the stress response. One possibility is interleukin-1β (IL-1β), which is increased in the hippocampus and other brain regions in response to stress.

IL-1β also inhibits the release of glutamate in hippocampus and has been implicated in stress-induced impairment of long-term potentiation (LTP) and age-related damage of hippocampal neurons *(48,49)*. The role of IL-1β in the downregulation of BDNF by stress is currently under investigation.

INFLUENCE OF ANTIDEPRESSANT TREATMENTS ON NEURONAL MORPHOLOGY AND SURVIVAL

If neuronal atrophy and survival play a significant role in the effects of stress that eventually lead to depression, it is conceivable that treatments for depression may reverse these neuronal deficits or, at the very least, prevent further damage. Although this area of research is still being studied, early reports have provided evidence that antidepressant treatments are capable of influencing synaptic plasticity and neuronal morphology. The results of these studies are briefly discussed.

Influence of Antidepressant Treatments on CA3 Neuronal Morphology

There are surprisingly few studies on the influence of antidepressant treatments on the stress-sensitive CA3 pyramidal neurons. This is in part the result of the difficulty of performing these studies, which includes staining the neurons, obtaining camera lucida drawings of the processes of multiple cells, and measurement of the number and length of dendritic branch points. However, McEwen and colleagues have examined the influence of two antidepressant agents on the atrophy of CA3 pyramidal cells *(50)*. One of the drugs was an atypical antidepressant, tianeptine, which increases the reuptake of 5-HT, and the other was a 5-HT selective reuptake inhibitor, fluoxetine. In this study, they found that the atypical antidepressant blocked the dendritic atrophy of CA3 pyramidal cells that occurs in response to stress. They also found that when administered alone, tianeptine had no effect on the dendritic arborization of CA3 neurons. Chronic administration of fluoxetine did not influence the stress-induced atrophy of CA3 neurons, nor did it have any effect when administered alone. It is possible that different antidepressants or treatment regimens, including doses or times of treatment other, than those used for these studies, could result in a different outcome. The potential role of CA3 atrophy in the current models of stress-related illnesses warrants further testing of these possibilities.

Influence of Antidepressant Treatment on Dentate Gyrus Granule Cell Neurogenesis and Sprouting

Stress is reported to decrease the number of newly born dentate gyrus granule cells, an effect that could also contribute to the damaging effects on hippocampal function. Recent studies provide exciting evidence that antidepressant treatment increases neurogenesis of granule cells. In preliminary studies, we have found that chronic administration of electroconvulsive seizures (ECS) or a monoamine oxidase inhibitor, tranylcypromine, increases the number of newly born granule cells by nearly twofold over vehicle-treated controls (Fig. 1) *(51)*. In addition, studies conducted by Jacob and Gould (personal communication) demonstrate that chronic administration of fluoxetine increases neurogenesis of granule cells. Moreover, they have found that administration of a 5-HT_{1A} receptor agonist is sufficient to increase neurogenesis, suggesting that the influence of fluoxetine is mediated by activation of this receptor subtype *(52)*. Interestingly, administration of a 5-HT_{1A} receptor antagonist reduces neurogenesis, indicating that the basal rate of cell birth is influenced by the 5-HT system.

Upregulation of neurogenesis may represent a novel and significant mechanism of antidepressant action, particularly in stress-related illnesses. However, there are several points that must be addressed before the relevance of neurogenesis can be determined. First, do other types of antidepressants increase neurogenesis? Do antidepressants block or reverse the negative effects of stress? This information will indicate if the enhancement of neurogenesis is a common action of antidepressants or if this effect is specific to certain types of treatments. Second, the controversy about the occurrence of neurogenesis in adult primates must be addressed. Third, the consequences of altered neurogenesis, either increased or decreased, to the function of the hippocampus must be determined.

In addition to neurogenesis, the influence of antidepressant treatments on the sprouting of granule cells has been examined. Granule cell sprouting can be studied using a histochemical technique that selectively stains zinc-containing dentate gyrus granule cells in hippocampus. We have found that chronic administration ECS increases the sprouting of granule cells in rat hippocampus *(53)*. This effect is dependent on repeated ECS treatment and is long lasting in that it is still observed 180 d after the last seizure. In contrast to the increased sprouting reported in response to excitotoxin treatment or models of kindling, ECS does not result in cell damage or death of neurons. This implies that other factors may mediate the sprouting response. In this regard, we found that the ECS-induced sprouting is reduced in BDNF mutant mice that express approximately 50% less BDNF than their wildtype littermates. However, chronic administration of several types of antidepressant drugs did not influence granule sprouting. This indicates that if sprouting has any relevance to the action of antidepressants, it is specific to ECS. Alternatively, sprouting may be any epiphenomenon of ECS and is unrelated to antidepressant actions.

Alterations in neuronal atrophy or survival also suggest that antidepressants and stress may influence synaptic strength and/or function of hippocampal neurons. There are several reports demonstrating that stress has inhibitory effects in behavioral and cellular models of learning and memory *(54)*. However, there are fewer studies on the influence of antidepressant treatments. Studies of LTP, a cellular model of learning and memory, have reported both an increase and decrease in response to antidepressants *(55, 56)*. Behavioral studies have been more consistent and suggest that antidepressants enhance cognitive function in rats and depressed patients *(57–59)*, which is consistent with the possibility that hippocampal function is increased. However, further studies of more selective antidepressant drugs (e.g., 5-HT and norepinephrine-selective reuptake inhibitors) are required to fully characterize the influence of these treatments on learning and memory.

MOLECULAR ACTIONS
OF ANTIDEPRESSANT TREATMENTS

It is reasonable to assume that the actions of antidepressant treatment involve more than one, and possibly many, signal transduction pathways. Although the commonly used antidepressant agents act primarily via increasing synaptic levels of 5-HT and norepinephrine, each of these monoamines is capable of interacting with multiple receptor subtypes, which utilize different signal transduction pathways. In addition, these proximal effects may then lead to regulation of secondary or tertiary cascades of signaling molecules and target genes. Identification of these cascades of intracellular signaling systems will ultimately be necessary to fully understand the action of antidepressants. One system that has been extensively investigated is the cAMP intracellular cascade. The work regarding this system and its relevance to the actions of stress are discussed in the following section. However, it will be important to continue studies to define the additional regulatory pathways that are involved in the mechanism of action of antidepressant treatments.

Regulation of the cAMP System by Antidepressant Treatment

One of the first signal transduction systems reported to be influenced by antidepressant treatments is the cAMP signal transduction cascade. These early studies demonstrated that the ability of β-adrenergic receptor (βAR) agonists to stimulate the formation of cAMP in limbic rat brain regions was downregulated by repeated antidepressant treatment *(60,61)*. This work contributed to the monoamine subsensitivity hypothesis of depression, which states that depression is caused by overactivation of the βAR-cAMP system and that downregulation of this system underlies the action of antidepressant treatment (*see* ref. *62*). However, there are several problems with this hypothesis. First, not all antidepressants decrease levels of βAR-stimulated cAMP formation, suggesting either that this effect is not necessary for antidepressant efficacy or that different antidepressant agents have different mechanisms. Second, the time-course for downregulation of βAR-stimulated cAMP is more rapid than for the therapeutic action of antidepressants. Third, this hypothesis implies that antagonism of βAR-stimulated cAMP formation should have an antidepressant effect, but this is not the case *(63,64)*. In fact, treatments that enhance the βAR–cAMP system are reported to have therapeutic efficacy *(65)*.

In contrast to a subsensitivity hypothesis, further studies indicate that chronic antidepressant treatment actually upregulates the cAMP system at several postreceptor–intracellular sites (Fig. 4). First, studies by Rasenick and colleagues demonstrate that certain antidepressant treatments increase the coupling of Gs, the stimulatory G-protein, to adenylyl cyclase in rat cerebral cortex *(66,67)*. Second, chronic antidepressant administration is reported to increase levels of cAMP-dependent protein kinase (PKA) in particulate fractions of rat cerebral cortex *(68,69)*. One of these studies also provided preliminary evidence that there was an increase in levels of PKA in the nuclear fraction, suggesting that gene expression is regulated by PKA and antidepressant treatment *(68)*. Third, and relative to the latter point, chronic antidepressant treatment is reported to increase the expression of the cAMP response element binding protein (CREB), a transcription factor that mediates many of the actions of the cAMP system on gene expression *(70)*. Upregulation of CREB is observed with several different classes of antidepressants, including 5-HT and norepinephrine-selective reuptake inhibitors, and the time-course for this effect is consistent with that for the therapeutic action of antidepressant treatment. The possibility that antidepressant treatment increases cAMP–CREB regulated gene expression is supported by preliminary studies demonstrating that cAMP response element (CRE)-mediated reporter gene expression is increased by antidepressant treatment *(71)*.

Taken together, these findings suggest that the action of antidepressant treatment is mediated, at least in part, by upregulation of the cAMP–CREB cascade. These findings also raise the possibility that CREB is a common intracellular target of antidepressant treatment. CREB is an interesting common target because it can be regulated by Ca^{2+}-dependent pathways, as well as by the cAMP cascade (*see* Fig. 4). The transcriptional activity of CREB is regulated largely by its phosphorylation, which can occur via PKA, Ca^{2+}-calmodulin-dependent protein kinase (CAMK), and protein kinase C (PKC). Monoamine receptors coupled to the cAMP pathway include the βAR, as well as 5-$HT_{4,6,7}$ receptors, and receptors that influence Ca^{2+} pathways, and

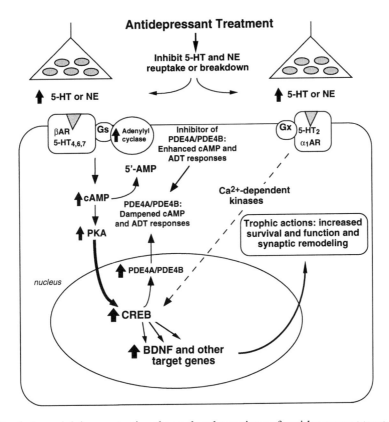

Fig. 4. A model demonstrating the molecular actions of antidepressant treatment on signal transduction. The therapeutic action of ADT is dependent on adaptations to increased synaptic levels of 5-HT and NE. One pathway that could contribute to the action of chronic ADT is the cAMP–CREB cascade. This cascade is reported to be activated by chronic ADT, including increased Gs–adenylyl cyclase coupling, increased particulate levels of PKA, and increased expression and function of CREB. This suggests that CREB is a common postreceptor target of ADT. CREB could underlie the actions of receptors that directly couple to the cAMP–PKA cascade (5-HT$_{4,6,7}$ and βAR) or receptors that couple to Ca^{2+}-dependent kinases (e.g., 5-HT$_2$ and α$_1$AR). Activation of the cAMP–CREB cascade indicates that target genes of this system are also regulated by ADT. This conclusion is supported by the finding that the expression of BDNF, a major neurotrophic factor in brain, is increased by chronic ADT. Increased levels of BDNF could have trophic actions on hippocampal neurons, including increased neuronal survival, function, and remodeling of synaptic or cellular architecture. The cAMP cascade is also regulated by cAMP phosphodiesterase-4 (PDE4), which catalyzes the breakdown of cAMP to 5'-AMP. Chronic ADT and activation of the cAMP–CREB cascade increases the expression of two PDE4 subtypes, PDE4A and PDE4B. Increased expression of PDE4A and PDE4B would dampen the activated cAMP response and the action of ADT (*see text* for discussion).

presumably CAMK and PKC, include the α_1-adrenergic and 5-HT$_{2A}$ receptors. The influence of antidepressant treatment on the phosphorylation state
of CREB has not yet been determined. The report that antidepressant treatment increases CRE-mediated gene expression is consistent with the hypothesis that CREB phosphorylation and function is also increased *(71)*. However,
analysis of the phosphorylation state of CREB will be critical to a more
complete understanding of the functional state of CREB and its role in the
action of antidepressant treatment.

Regulation of cAMP Phosphodiesterase by Antidepressant Treatment

Additional evidence that upregulation of the cAMP system is involved
in the action of antidepressant treatment is provided by studies of inhibitors
of phosphodiesterase, the enzyme responsible for the breakdown of cAMP.
Administration of a phosphodiesterase (PDE) inhibitor increases cellular
levels of cAMP and activates the cAMP–CREB cascade. There are at least
nine subfamilies of PDE, but one of these, the PDE4 subfamily, has relatively high affinity and selectivity for cAMP versus cGMP *(see* ref. *72)*. Basic
research studies in rodents demonstrate that administration of a selective
PDE4 inhibitor, rolipram, has antidepressant effects in behavioral models of
depression *(73,74)*. Moreover, clinical studies demonstrate that rolipram has
antidepressant efficacy when tested in depressed patients *(75–77)*. These
studies provide additional support for the hypothesis that activation of the
cAMP–CREB cascade is involved in the action of antidepressant treatment.

Although clinical studies demonstrate that PDE4 inhibitors have antidepressant efficacy, the clinical usefulness of these agents has been limited by
side effects, most notably nausea. This may be explained by the relatively
widespread distribution of PDE4 in brain and peripheral tissues. However,
the molecular identification of multiple PDE4 isozymes raises the possibility that a selective PDE4 inhibitor without side effects could be developed.
Within the PDE4 subfamily, there are four separate genes, referred to as
PDE4A–PDE4D, and each of these has multiple splice variants *(see* ref. *72)*.
Three of these subtypes, PDE4A, PDE4B, and PDE4D, are expressed at
relatively high levels in the brain and each has a unique pattern of distribution *(see* ref. *78)*. All three isozymes are expressed in limbic brain regions,
including cerebral cortex and hippocampus, although the pattern for each
varies within the different cell layers. In addition, PDE4B is expressed at higher
levels in the nucleus accumbens, a brain region involved in motivation and
reward and could be relevant to the anhedonia often seen in depressed
patients. Interestingly, preliminary data indicate that PDE4D is expressed at

high levels in the area postrema, a brain-stem region that mediates the central control of nausea and vomiting, but that expression of PDE4A and PDE4B is relatively low in this region *(78)*.

One approach to identify a PDE4 isozyme that could be a selective target for the development of antidepressants is to study the regulation of these isozymes by chronic treatments. This strategy is based on studies in cultured cells that demonstrate that activation of the cAMP cascade increases the expression of PDE4 isozymes *(see* refs. *72* and *78)*. This is thought to represent a compensatory adaptation to maintain homeostatic control of the cAMP system *(see* Fig. 4). There are now reports from our laboratory and others demonstrating that antidepressant treatment increases the expression of PDE4A and PDE4B, but not PDE4D in rat brain (Fig. 4) *(78–80)*. Upregulation of PDE4A and PDE4B was observed primarily in the cerebral cortex, although one study also reports regulation in hippocampus. In addition, we have observed an upregulation of PDE4B in the nucleus accumbens *(78)*. Although these findings implicate PDE4A and PDE4B, it is important to point out that the function of PDE4 isozymes is also regulated by phosphorylation and it is possible that PDE4D is regulated in this manner *(see* ref. *72)*. Further characterization of the role of these isozymes in the action of antidepressant treatment will be dependent on the development of selective inhibitors of the PDE4 subtypes. Another approach that is underway in several laboratories is to use homologous recombination to make knockout mice for each of the PDE4 genes and to study the phenotype of these mice in behavioral models of depression.

These findings suggest that specific inhibitors of PDE4A and PDE4B isozymes may have antidepressant effects, but without the side effects of rolipram (Fig. 4). In addition, the results also suggest that upregulation of PDE4A and PDE4B could reduce the response to traditional antidepressant treatments by reducing levels of cAMP. If this is the case, coadministration of a PDE4 inhibitor with another type of antidepressant, such as a 5-HT or norepinephrine-selective reuptake inhibitor, would be expected to enhance the response, including a more rapid response and greater overall effect. Preclinical studies support this possibility, demonstrating that coadministration of rolipram with a tricyclic antidepressant results in a more rapid induction of BDNF *(70)*.

Regulation of BDNF by Antidepressant Treatment

Regulation of the cAMP–CREB cascade suggests that gene expression is influenced by antidepressant treatment. One gene that may be relevant to the damaging influence of stress is BDNF. We have found that chronic antidepressant treatment increases the expression of BDNF in rat hippocampus (Fig. 4) *(70,81)*. Upregulation of BDNF is observed in response to several different classes of antidepressants, including 5-HT and norepinephrine-selective reuptake inhibitors. In addition, increased expression of BDNF is observed after chronic, but acute, antidepressant treatment. One exception is ECS, which induces BDNF mRNA after a single treatment, although chronic treatment leads to a more long-lasting and greater induction of BDNF *(81)*. Repeated ECS is also reported to increase levels of BDNF immunoreactivity in the dentate gyrus granule cell layer and mossy fiber pathway *(82)*. This suggests that there is anterograde transport of BDNF from the granule cell layer to provide trophic support for neurons in the CA3 pyramidal cell layer. BDNF expression is also increased by chronic administration of a PDE4 inhibitor, providing further support that activation of the cAMP–CREB cascade regulates the expression of BDNF *(70)*. The promoter region of the BDNF gene is also reported to have a CRE element that is regulated by the cAMP–CREB cascade *(83,84)*.

The role of BDNF in the action of antidepressant treatment is supported by several lines of evidence. First, local infusion of BDNF is reported to have antidepressant effects in two behavioral models of depression: the forced swim and learned helplessness paradigms *(85)*. Second, preliminary studies indicate that BDNF heterozygous knockout mice, which express lower levels of BDNF, display a depressivelike phenotype in the forced-swim test *(86)*. Third, there is an anecdotal report that intrathecal administration of BDNF to a patient with amyotrophic lateral sclerosis (ALS) has a behavioral activating effect. Finally, stress is reported to decrease the expression of BDNF and this effect could contribute to the atrophy and damage of hippocampal neurons *(36)*. The role of BDNF, as well as CREB, in the atrophy, survival, and neurogenesis of hippocampal neurons is currently being examined. These studies are being conducted using mutant mice that either overexpress or underexpress BDNF and CREB.

A Molecular and Cellular Hypothesis of Depression

Taken together, this work provides evidence for a molecular and cellular hypothesis of depression and the action of antidepressants (*see* Fig. 1). In some forms of depression, stress and elevated levels of glucocorticoids could lead to atrophy, and in severe cases, it could lead to death of CA3 sensitive hippocampal neurons. This could occur via decreased energy metabolism, increased glutamate and calcium excitotoxicity, and decreased expression of BDNF. In addition, exposure to stress during early stages of development could reduce neurogenesis and thereby compromise the function of the granule cell mossy fiber pathway. The atrophy and loss of CA3 neurons and decreased neurogenesis of granule cells could underlie the reduced hippocampal volume in patients with depression, as well as post-traumatic stress disorder (PTSD). This loss of hippocampal neuronal function could also underlie some of the vegetative and cognitive abnormalities observed in depressed patients. Antidepressant treatment, via activation of the cAMP–CREB cascade and increased expression of BDNF, may help to protect hippocampal neurons from further damage or possibly even reverse the atrophy induced by stress. These effects could then restore hippocampal function, including improved memory and inhibitory control of the HPA axis. Because the role of neurogenesis in adult humans has not been established, the role of increased cell birth in the action of antidepressant treatment in adults is uncertain.

This hypothesis could also explain the individual variability and susceptibility to stress-induced affective illness. Prior exposure of an individual to stress or some other type of neuronal insult, such as hypoxia–ischemia, hypoglycemia, or infection, could induce a relatively small degree of neuronal damage that is not sufficient to result in behavioral abnormalities. However, with subsequent exposures to stress or environmental insults over time, the damage to neurons may be cumulative and eventually lead to illness. This type of scenario is observed in Parkinson's disease, where up to 80% of the substantia nigra dopamine neurons are lost before the illness is expressed. It is also possible that there are genetic factors involved that increase the vulnerability hippocampal neurons to stress.

This model provides a framework for future preclinical and clinical studies to test this hypothesis. For example, it will be critical to determine if other types of stress, such as social stress, that are known to induce atrophy and death of neurons also decrease the expression of BDNF. In addition, the role of BDNF and CREB in the cellular and behavioral actions of stress and antidepressant treatment must be directly examined using mutant mouse models. Clinical imaging studies are also required to determine what percentage of patients exhibit reduced hippocampal volume and whether this effect is reversible with treatment. Analysis of BDNF in cerebral spinal fluid or postmortem brain tissue will also be necessary to determine the role of this neurotrophic factor in depression.

CONCLUSIONS AND FUTURE PERSPECTIVES

There are a few points that should be raised regarding the work outlined in this chapter. First, caution must be used when interpreting adaptive changes in response to antidepressant treatments. For example, although the level of βAR is decreased and PDE4 enzymes are increased, suggesting that the cAMP–CREB cascade is decreased, further analysis of the other cellular components, as well as the functional output, of this system indicate that antidepressant treatment activates this intracellular cascade. Second, a complete understanding of the actions of stress and psychotropic drugs will require a combination of molecular, cellular, and behavioral approaches. Although this has been started, a tremendous amount of work still remains in the area of stress and depression. Continued technical and conceptual advances in all areas of neurobiology will make this possible. Third, it is likely that the cAMP–CREB cascade and BDNF represent just a fraction of the intracellular pathways that are involved in the pathophysiology and treatment of depression. Work has already been conducted in certain areas, including the phosphatidylinositol and PKC pathway (*see* ref. *87*). The role of neurosteroids in stress and depression is also an area of interest *(88,89)*. In addition, the intracellular pathways that mediate signaling by Ca^{2+}, cytokines, and neurotrophic factors, as well as pathways controlling cell death and survival, are areas of interest. Continued development of conceptual and technical approaches in these areas of research, in conjunction with pharmacological and clinical studies, will ultimately lead to a better understanding of the molecular and cellular basis of depression. In addition, this work will result in the identification of novel targets for the development of faster acting and more effective therapeutic agents.

ACKNOWLEDGMENTS

This work was supported by US PHS grants MH45481, MH53199, and 2 PO1 MH25642, and by a Veterans Administration National Center Grant for PTSD, VA Medical Center.

REFERENCES

1. Herman, J. P. and Cullinan, W. E. (1997) Neurocircuitry of stress: central control of the hypothalamo-pituitary-adrenocortical axis. *Trends Neurosci.* **20**, 78–84.
2. Herman, J. P., Schafer, M. K. H., Young, E. A., Thompson, R. C., Douglass, J. O., Akil, H., and Watson, S. J. (1989) Hippocampal regulation of the hypothalamus–pituitary–adrenocortical axis: in situ hybridization analysis of CRF and vasopressin messenger RNA expression in the hypothalamic paraventricular nucleus following hippocampectomy. *J. Neurosci.* **9**, 3072–3082.
3. Young, E. A., Haskett, R. F., Murphy-Weinberg, V., Watson, S. J., and Akil, H. (1991) Loss of glucocorticoid fast feedback in depression. *Arch. Gen. Psychiatry* **48**, 693–699.
4. Lupien, S. J., De Leon, M., DeSanti, S., Convit, A., Tarnish, C., Nair, N. P. V., Thakur, M., McEwen, B. S., Hauger, R. L., and Meaney, M. J. (1998) Cortisol levels during human aging predict hippocampal atrophy and memory deficits. *Nature Neurosci.* **1**(1), 69–73.

5. Porter, N. M. and Landfield, P. W. (1998) Stress hormones and brain aging: adding injury to insult? *Nature Neurosci.* **1**(1), 3–4.

6. Sapolsky, R.M. (1996) Stress, glucocorticoids, and damage to the nervous system: the current state of confusion. *Stress* **1**, 1–19.

7. Magarinos, A. M., McEwen, B. S., Flugge, G., and Fuchs, E. (1996) Chronic psychosocial stress causes apical dendritic atrophy of hippocampal CA3 pyramidal neurons in subordinate tree shrews. *J. Neurosci.* **16**, 3534–3540.

8. Watanabe, Y., Gould, E., and McEwen, B. S. (1992) Stress induces atrophy of apical dendrites of hippocampal CA3 pyramidal neurons. *Brain Res.* **588**, 341–345.

9. Wooley, C. S., Gould, E., and McEwen, B. S. (1990) Exposure to excess glucocorticoids alters dendritic morphology of adult hippocampal pyramidal neurons. *Brain Res.* **531**, 225–231.

10. Sapolsky, R. (1985) A mechanisms for glucocorticoid toxicity in the hippocampus: increased neuronal vulnerabiligy to metabolic insults. *J. Neurosci.* **5**, 1227–1234.

11. Sapolsky, R.M., Uno, H., Robert, C. S., and Finsh, C. E. (1990) Hippocampal damage associated with prolonged glucocorticoid exposure in primates. *J. Neurosci.* **10**, 2897–2902.

12. Uno, H., Tarara, R., Else, J. G., Suleman, M. A., and Sapolsky, R. M. (1989) Hippocampal damage associated with prolonged and fatal stress in primates. *J. Neurosci.* **9**, 1705–1711.

13. Sapolsky, R. (1985) Glucocorticoid toxicity in the hippocampus: temporal aspects of neuronal vulnerability. *Brain Res.* **339**, 300–307.

14. Sapolsky, R. and Pulsinelli, W. (1986) Glucocorticoids potentiate ischemic injury to neurons: therapeutic implications. *Science* **229**, 1397–1400.

15. Gould, E., McEwen, B. S., Tanapat, P., Galea, L. A. M., and Fuchs, E. (1997) Neurogenesis in the dentate gyrus of the adult tree shrew is regulated by psychosocial stress and NMDA receptor activation. *J. Neurosci.* **17**(7), 2492–2498.

16. Gould, E., Tanapat, P., McEwen, B. S., Flugge, G., and Fuchs, E. (1998) Proliferation of granule cell precursors in the dentate gyrus of adult monkeys is diminished by stress. *Proc. Natl. Acad. Sci. USA* **95**, 3168–3171.

17. Eckenhoff, M. F. and Rakic, P. (1988) Nature and fate of proliferative cells in the hippocampal dentate gyrus during the life span of the rhesus monkey. *J. Neurosci.* **8**, 2729–2747.

18. Kempermann, G., Kuhn, H. G., and Gage, F. H. (1997) More hippocampal neurons in adult mice living in an enriched environment. *Nature* **386**, 493–495.

19. Bremner, J. D., Randall, P., Scott, T. M., Bronen, R. A., Seibyl, J. P., Southwick, S. M., Delaney, R. C., McCarthy, G., Charney, D. S., and Innis, R. B. (1995) MRI-based measurement of hippocampal volume in patients with combat-related post-traumatic stress disorder. *Am. J. Psychiatry* **152**, 973–981.

20. Sapolsky, R.M. (1996) Glucocorticoids and atrophy of the human hippocampus. *Science* **273,** 749–750.

21. Sheline, Y. I., Wany, P., Gado, M. H., Csernansky, J. G., and Vannier, M. W. (1996) Hippocampal atrophy in recurrent major depression. *Proc. Natl. Acad. Sci. USA* **93**, 3908–3913.

22. Starkman, M., Gebarski, S., Berent, S., and Schteingart, D. (1992) Hippocampal formation volume, memory dysfunction, and cortisol levels in patients with Cushing's syndrome. *Biol. Psychiatry* **32**, 756–762.

23. Drevets, W. C., Price, J. L., Simpson, J. R., et al. (1997) Subgenual prefrontal cortex abnormalities in mood disorders. *Nature* **386**, 824–827.

24. Ongur, D., Drevets, W. C., and Price, J. L. (1998) Glial loss in the subgenual prefrontal cortex in familial mood disorders. *Soc. Neurosci.* **24**, 990.

25. Packan, D. and Sapolsky, R. (1990) Glucocorticoid endangerment of the hippocampus: tissue, steroid and receptor specificity. *Neuroendocrinology* **51**, 613–620.

26. Rajan, V., Edwards, C., and Seckl, J. (1996) 11β-Hydroxysteroid dehydrogenase in cultured hippocampal cells reactivates inert 11-dehydrocorticosterone, potentiating neurotoxicity. *J. Neurosci.* **16**, 65–70.

27. Carter-Su, C. and Okamoto, K. (1985) Effects of glucocorticoids on hexose transport in rat adipocytes: evidence for decreased transporters in the plasma membrane. *J. Biol. Chem.* **260**, 11,091–11,098.

28. Garvey, W., Huecksteadt, T., Lima, F., and Birnbaum, M. (1989) Expression of a glucose transporter gene cloned from brain in cellular models of insulin resistance: dexamethasone decreases transporter mRNA in primary cultured adipocytes. *Mol. Endocr.* **3**, 1132–1137.

29. Kadekaro, M., Masanori, I., and Gross, P. (1988) Local cerebral glucose utilization is increased in acutely adrenalectomized rats. *Neuroendocrinology* **47**, 329.

30. Horner, H., Packan, D., and Sapolsky, R. (1990) Glucocorticoids inhibit glucose transport in cultured hippocampal neurons and glia. *Neuroendocrinology* **52**, 57–62.

31. Virgin, C., Ha, T., Packan, D., Tombaugh, G., Yang, S., Horner, H., and Sapolsky, R. (1991) Glucocorticoids inhibit glucose transport and glutamate uptake in hippocampal astrocytes: implications for glucocorticoid neurotoxicity. *J. Neurochem.* **57**, 1422–1428.

32. Cotman, C. W., Kahle, J. S., Miller, S. E., Ulas, J., and Bridges, R. J. (1995) Excitory amino acid neurotransmission, in *Psychopharmacology: The Fourth Generation of Progress* (Bloom, F. E. and Kupfer, D. J., eds.), Raven, New York, pp. 75–85.

33. Dugan, L. and Choi, D. (1994) Excitotoxicity, free radicals, and cell membrane changes. *Ann. Neurol.* **35**, S17–S21.

34. Bagley, J. and Moghaddam, B. (1995) Rapid sampling of extracellular glutamate in the prefrontal cortex and hippocampus in response to repeated stress: effect of diazepam. *Soc. Neurosci.* **21**, 189.12.

35. Lowy, M., Wittenberg, L., and Yamamoto, B. (1995) Effect of acute stress on hippocampal glutamate levels and spectrin proteolysis in young and aged rats. *J. Neurochem.* **65**, 268–274.

36. Smith, M. A., Makino, S., Altemus, M., Michelson, D., Hong, S.-K., Kvetnansky, R., and Post, R. M. (1995) Stress and antidepressants differentially regulate neurotrophin 3 mRNA expression in the locus coeruleus. *Proc. Natl. Acad. Sci. USA* **92**, 8788–8792.

37. Lindsay, R. M., Wiegand, S. J., Altar, A., and DiStefano, P. S. (1994) Neurotrophic factors: from molecule to man. *Trends Neurosci.* **17**, 182–190.

38. Lindvall, O., Kokaia, Z., Bengzon, J., Elmer, E., Kokaia, M. (1994) Neurotrophins and brain insults. *Trends Neurosci.* **17**, 490–496.

39. Thoenen, H. (1995) Neurotrophins and neuronal plasticity. *Science* **270**, 593–598.

40. Russell, D. S. (1995) Neurotrophins: mechanisms of action. *Neuroscientist* **1**, 3–6.

41. Smith, M. A., Makino, S., Kvetnansky, R., and Post, R. M. (1995) Stress alters the express of brain-derived neurotrophic factor and neurotrophin-3 mRNAs in the hippocampus. *J. Neurosci.* **15**, 1768–1777.

42. Sklair-Tavron, L. and Nestler, E. J. (1995) Opposing effects of morphine and the neurotrophins, NT-3, NT-4, and BDNF, on locus coeruleus neurons in vitro. *Brain Res.* **702**, 117–125.

43. Piguet, P. and Galvan, M. (1994) Transient and long-lasting actions of 5-HT on rat dentate gyrus neurones *in vitro*. *J. Physiol.* **481**, 629–639.

44. Chaouloff, F. (1993) Physiopharmacological interactions between stress hormones and central serotonergic systems. *Brain Res. Rev.* **18**, 1–32.

45. Vahabzadeh, A. and Fillenz, M., (1994) Comparison of stress-induced changes in noradrenergic and serotonergic neurons in the rat hippocampus using microdialysis. *Eur. J. Neurosci.* **6**, 1205–1212.

46. Vaidya, V. A., Marek, G. J., Aghajanian, G. A., and Duman, R. S. (1997) 5-HT$_{2A}$ receptor-mediated regulation of brain-derived neurotrophic factor mRNA in the hippocampus and the neocortex. *J. Neurosci.* **17**(8), 2785--2795.

47. Vaidya, V. A. and Duman, R. S. (1999) Role of 5-HT2A receptors in down-regulation of BDNF by stress. *Neurosci. Lett.* **287**, 1–4.

48. Murray, C. A. and Lynch, M. A. (1998) Evidence that increased hippocampal expression of the cytokine interleukin-1β is a common trigger for age- and stress-induced impairments in long-term potentiation. *J. Neurosci.* **18**(8), 2974–2981.

49. Murray, C. A., McGahon, B., McBennett, S., and Lynch, M. A. (1997) Interleukin-1b inhibits glutamate release in hippocampus of young, but not aged, rats. *Neurobiol. Aging* **18**(3), 343–348.

50. Watanabe, Y., Gould, E., Daniels, D. C., Cameron, H., and McEwen, B. S. (1992) Tianeptine attenuates stress-induced morphological changes in the hippocampus. *Eur. J. Pharmacol.* **222**, 157–162.

51. Duman, R. S. and Malberg, J. E. (1998) Neural plasticity in the pathophysiology and treatment of depression. *Am. Coll. Neuropsychopharmacol.* **37**, 89.

52. Jacobs, B. L., Tanapat, P., Reeves, A. J., and Gould, E. (1998) Serotonin stimulates the production of new hippocampal granule neurons via the 5HT1A receptor in the adult rat. *Soc. Neurosci.* **24**, 1992.

53. Vaidya, V. A., Siuciak, J., Du, F., and Duman, R. S. (1999) Mossy fiber sprouting and synaptic reorganization induced by chronic administration of electroconvulsive seizure: role of BDNF. *Neuroscience* **89**, 157–166.

54. McEwen, B. S. and Sapolsky, R. M. (1990) Stress and cognitive function. *Curr. Opin. Neurobiol.* **5**, 205–216.

55. Birnstiel, S. and Hass, H. L. (1991) Acute effects of antidepressant drugs on long-term potentiation (LTP) in rat hippocampal slices. *Naunyn Schmied. Arch. Pharmacol.* **344**, 79–83.

56. Massicotte, G., Bernard, J., and Ohayon, M. (1993) Chronic effects of trimipramine, an antidepressant, on hippocampal synaptic plasticity. *Behav. Neural Biol.* **59**, 100–106.

57. Allain, H., Lieury, A., Brunet-Bourgin, F., Mirabaud, C., Trebon, P., Le Coz, F., and Gandon, J. M. (1992) Antidepressants and cognition: comparative effects of moclobemide, viloxazine and maprotiline. *Psychopharmacology* **106**, S56–S61.

58. Spring, B., Gelenberg, A. J., Garvin, R., and Thompson, S. (1992) Amitriptyline, clovoxamine and cognitive function: a placebo-controlled comparison in depressed outpatients. *Psychopharmacology* **108**, 327–332.

59. Yau, J. L. W., Olsson, T., Morris, R. G. M., Meaney, M. J., and Seckl, J. R. (1995) Glucocorticoids, hippocampal corticosteroid receptor gene expression and anti-

depressant treatment: relationship with spatial learning in young and aged rats. *Neuroscience* **66**, 571–581.

60. Banerjee, S. P., Kung, L. S., Riggi, S. J., and Chanda, S. K. (1977) Development of β-adrenergic receptor subsensitivity by antidepressants. *Nature* **268**, 455–456.

61. Vetulani, J. and Sulser, F. (1975) Action of various antidepressant treatments reduces reactivity of noradrenergic cAMP generating system in limbic forebrain. *Nature* **257**, 495–496.

62. Charney, D. S., Menkes, D. B., and Heninger, G. R. (1981) Receptor sensitivity and the mechanism of action of antidepressant treatment. *Arch. Gen. Psychiatry* **38**, 1160–1173.

63. Avorn, J., Everitt, D. E., and Weiss, S. (1986) Increased antidepressant use in patients prescribed β-blockers. *JAMA* **255**, 357–360.

64. Paykel, E. S., Fleminger, R., and Watson, J. P. (1982) Psychiatric side effects of antihypertensive drugs other than reserpine. *J. Clin. Psychopharmacol.* **2**, 14–39.

65. Goodwin, F. K., Prange, A. J., Post, R. M., Muscettola, G., and Lipton, M. A. (1982) Potentiation of antidepressant effect by triiodothyronine in tricyclic nonresponders. *Am. J. Psychiatry* **139**, 34–38.

66. Menkes, D. B., Rasenick, M. M., Wheeler, M. A., and Bitensky, M. W. (1983) Guanosine triphosphate activation of brain adenylate cyclase: enhancement by longterm antidepressant treatment. *Science* **129**, 65–67.

67. Ozawa, H. and Rasenick, M. M. (1991) Chronic electroconvulsive treatment augments coupling of the GTP-binding protein Gs to the catalytic moiety of adenylyl cyclase in a manner similar to that seen with chronic antidepressant drugs. *J. Neurochem.* **56**, 330–338.

68. Nestler, E. J., Terwilliger, R. Z., and Duman, R. S. (1989) Chronic antidepressant administration alters the subcellular distribution of cyclic AMP-dependent protein kinase in rat frontal cortex. *J. Neurochem.* **53**, 1644–1647.

69. Perez, J., Tinelli, D., Brunello, N., and Racagni, G. (1989) cAMP-dependent phosphorylation of soluble and crude microtubule fractions of rat cerebral cortex after prolonged desmethylimipramine treatment. *Eur. J. Pharmacol.* **172**, 305–316.

70. Nibuya, M., Nestler, E. J., and Duman, R. S. (1996) Chronic antidepressant administration increases the expression of cAMP response element binding protein (CREB) in rat hippocampus. *J. Neurosci.* **16**, 2365–2372.

71. Thome, J., Sakai, N., Shin, K. H., Steffen, C., Impey, S., Storm, D. R., and Duman, R. S. (1998) Regulation of cAMP-response-element promoter activity by antidepressant treatment in CRE-lacZ transgenic mice. *Soc. Neurosci.* **24**, 1844.

72. Conti, M. and Jin, S.-L. C. (1999) The molecular biology of cyclic nucleotide phosphodiesterases. *Prog. Nucl. Acid Res. Mol. Biol.* **63**, 1–38.

73. O'Donnell, J. M. (1993) Antidepressant-like effects of rolipram and other inhibitors of cyclic AMP phosphodiesterase on behavior maintained by differential reinforcement of low response rate. *J. Pharmacol. Exp. Ther.* **264**, 1168–1178.

74. Wachtel, S. R. and Schneider, H. D. (1986) Rolipram, a novel antidepressant drug, reverses the hypothermia and hypokinesia of monoamine-depleted mice by an action beyond postsynaptic monoamine receptors. *Neuropharmacology* **25**, 1119–1126.

75. Bobon, D., Breulet, M., Gerard-Vandenhove, M. A., Guilot-Goffioul, F., Plonteux, G., Sastre-Y-Hernandez, M., Schratzer, M., Troisfontaines, B., Vov-Frenckell, R., and Wachel, H. (1988) Is phosphodiesterase inhibition a new mechanism of antidepressant action? A double blind double-dummy study between rolipram and desipr-

amine in hospitalized major and/or endogenous depressives. *Eur. Arch. Psychiatry Neurol. Sci.* **238**, 2–6.

76. Fleischhacker, W. W., Hinterhuber, H., Bauer, H., Pflug, B., Berner, P., Simhandl, C., Wilf, R., Gerlach, W., Jaklitsch, H., Sastre-y-Hernandez, M., Schmeding-Wiegel, H., Sperner-Unterweger, B., Voet, B., and Schubert, H. (1992) A multicenter double-blind study of three different doses of the new cAMP-phosphodiesterase inhibitor rolipram in patients with major depresive disorder. *Neuropsychobiology* **26**, 59–64.

77. Horowski, R. and Sastre-Y-Hernandez, M. (1985) Clinical effects of the neurotrophic selective cAMP phosphodiesterase inhibitor rolipram in depressed patients: global evaluation of the preliminary reports. *Curr. Ther. Res.* **38**, 23–29.

78. Takahashi, M., Terwilliger, R., Lane, C., Mezes, P. S., Conti, M., and Duman, R. S. (1999) Chronic antidepressant administration increases the expression of cAMP phosphodiesterase 4A and 4B isoforms. *J. Neurosci.* **9**, 610–618.

79. Suda, S., Nibuja, M., Ishiguro, T., and Suda, H. (1998) Transcriptional and translational regulation of phosphodiesterase type IV isozymes in rat brain by electroconvulsive seizure and antidepressant drug treatments. *J. Neurochem.* **71**, 1554–1563.

80. Ye, Y., Conti, M., Houslay, M. D., Faroqui, S. M., Chen, M., and O'Donnell, J. M. (1997) Noradrenergic activity differentially regulates the expression of rolipram-sensitive, high-affinity cyclic AMP phosphodiesterase (PDE4) in rat brain. *J. Neurochem.* **69**, 2397–2404.

81. Nibuya, M., Morinobu, S., and Duman, R. S. (1995) Regulation of BDNF and trkB mRNA in rat brain by chronic electroconvulsive seizure and antidepressant drug treatments. *J. Neurosci.* **15**, 7539–7547.

82. Smith, M. A., Zhang, L. X., Lyons, W. E., and Mamounas, L. A. (1997) Anterograde transport of endogenous brain-derived neurotrophic factor in hippocampal mossy fibers. *NeuroReport* **8**, 1829–1834.

83. Shieh, P. B., Hu, S.-C., Bobb, K., Timmusk, T., and Ghosh, A. (1998) Identification of a signaling pathway involved in calcium regulation of BDNF expression. *Neuron* **20**, 727–740.

84. Tao, X., Finkbeiner, S., Arnold, D. B., Shaywitz, A. J., and Greenberg, M. E. (1998) Ca^{2+} Influx regulates BDNF transcription by a CREB family transcription factor-dependent mechanism. *Neuron* **20**, 709–726.

85. Siuciak, J. A., Lewis, D. R., Wiegand, S. J., and Lindsay, R. M. (1996) Antidepressant-like effect of brain derived neurotrophic factor (BDNF). *Pharmacol. Biochem. Behav.* **56**, 131–137.

86. Malberg, J. E., Messer, C. J., Nestler, E. J., and Duman, R. S. (1998) Analysis of BDNF knockout mice in the forced swim test: a potential model for depression? *Soc. Neurosci.* **24**, 1200.

87. Manji, H. K., Potter, W. C., and Lenox, R. H. (1995) Signal transduction pathways. *Arch. Gen. Psychiatry* **52**, 531–543.

88. Uzunov, D. P., Cooper, T. B., Costa, E., and Guidotti, A. (1996) Fluoxetine-elicited changes in brain neurosteroid content measured by negative ion mass fragmentography. *Proc. Natl. Acad. Sci. USA* **93**, 12,599–12,604.

89. Uzunova, V. Sheline, Y., Davis, J. M., Rasmusson, A., Uzunov, D. P., Costa, E., and Guidotti, A. (1998) Increase in the cerebrospinal fluid content of neurosteroids in patients with unipolar major depression who are receiving fluoxetine or fluvoxamine. *Proc. Natl. Acad. Sci. USA* **95**, 3239–3244.

10

Long-Term Bidirectional Hormonal and Neuroplastic Responses to Stress

Implications for the Treatment of Depression

Robert M. Post, Susan R. B. Weiss, Li-Xin Zhang, and M. Guoqiang Xing

INTRODUCTION

As far back as the works of Emil Kraepelin—who separated the affective disorders from schizophrenia—a prominent role of stressors in the precipitation of early episodes of affective illness was noted *(1)*. Kraepelin spoke of the complex and variegated course of bipolar illness with its waxing and waning of patterns and episodes, and he was among the first to report on a general tendency for increased rapidity of episode cycling (i.e., the well interval between episodes shortened as a function of the number of successive episodes) (Fig. 1). Kraepelin wrote:

> ...we must regard all alleged injuries as possibly sparks for the discharge of individual attacks, but that the real cause of the malady must be sought in *permanent internal changes*, which at least very often, perhaps always, are innate.

> ...in spite of the removal of the discharging cause, the attack follows its independent development. But, finally, the appearance of wholly similar attacks on wholly dissimilar occasions or quite without external occasion shows that even there where there has been external influence, it must not be regarded as a necessary presupposition for the appearance of the attack.

> Unfortunately the powerlessness of our efforts to cure must only too often convince us that the attacks of manic–depressive insanity may be to an astonishing degree *independent of external influences. (1*, pp. 180–181)

In subsequent studies in untreated cohorts or in those patients receiving only acute treatment and not long-term prophylaxis, the observation of cycle acceleration has been almost universally replicated *(2,3)*. Moreover, even in the modern psychopharmacological era, there is evidence for this pattern of illness evolution in patients with treatment-refractory affective illness *(4–9)*. In

From: *Cerebral Signal Transduction: From First to Fourth Messengers*
Edited by: M. E. A. Reith © Humana Press Inc., Totowa, NJ

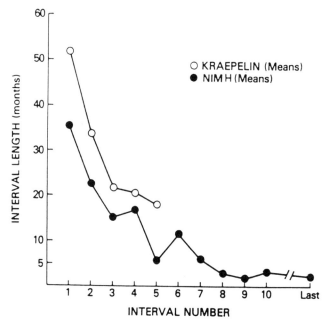

Fig. 1. Decreasing well intervals in recurrent affective illness are representative of a pattern of cycle acceleration that is typical for many patients with inadequately treated mood disorders. Remarkably, patients studied by Kraepelin in the prepsychopharmacologic era showed patterns similar to those refractory to medications studied at the National Institute of Mental Health (for a further review of relevant studies, *see* ref. 2)

the largest study of its kind, 20,350 Danish first-admission patients hospitalized for either unipolar or bipolar depression were followed by Kessing et al. *(10,11)*, who confirmed this tendency toward cycle acceleration even in patients receiving naturalistic treatment in the community; that is, the incidence of and latency to relapse with another depressive hospitalization varied in proportion to the number of prior hospitalizations for depression (Fig. 2).

Thus, there appear to be several types of sensitization processes at work in the recurrent affective disorders, two of which are stressor and episode sensitization *(12,13)*. Stressor sensitization refers to the fact that increasing affective responsivity occurs with repeated exposures to psychosocial stressors *(14,15)*, leading to more symbolic or conditioned stimuli capable of initiating affective episodes and, ultimately, the emergence of autonomous illness occurrences *(12)*. Empirically, this is observed as stressors being more evident in the precipitation of early compared to later episodes of affective illness. This does not imply that stressors are no longer capable of precipitating episodes of affective illness *(16,17)*, just that they are no longer critical (Table 1).

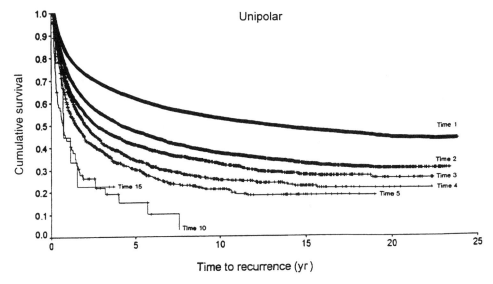

Recurrence at successive episodes in unipolar affective disorder

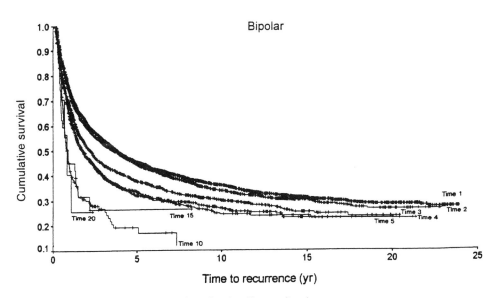

Recurrence at successive episodes in bipolar affective disorder

Fig. 2. Cumulative survival (probability of remaining well) was calculated using the Kaplan–Meier method for estimation with censored observations. Eight different index admissions (1, 2, 3, 4, 5, 10, 15, and 20) represent the number of prior hospitalizations for depression. For both unipolar (**top**) and bipolar (**bottom**) depressed patients, incidence of and latency to relapse varied as a function of the number of prior depressions. (Reproduced from ref. *11* with permission from Kessing et al.)

Table 1
Studies of Association Between Life Events and First versus Subsequent Episodes of Affective Disorder

Author	Disorder	Number of Episodes	N	Percent of Patients for Whom Major Life Events Preceded Episode		P	Assessment
				First Episode	Later Episode		
Matussek et al. (1965)	Depression	1	242	44		—	Stressors (138 psychologic; 58 somatic) had to clearly precede onset of episode
		2	135		34	—	
		3	82		24	—	
		4	119		19	—	
Angst (1966)	Depression	1	103	60		—	No Inventory
		≥4			38	—	
Okuma and Shimoyama (1972)	Manic Depression	1	134	45		—	Any event (3 mo prior)
		2	134		26	—	
		3	134		13	—	
Glassner et al. (1979)	Manic Depression	1	25	75		—	Event rated stressful by patient and on Holmes and Rahe Scale (1 yr prior; usually 2–24 d); role loss critical in patients and comparison subjects
		>1[a]			56	—	
Ambelas (1979)[b]	Mania	1	14	50		<0.01	Paykel Life Events Scale (4 wk prior); one-third of cases followed bereavement
		≥2		67	28		
Guttierrez et al. (1981)	Depression	1	43	55.8		<0.05	Social and somatic stressors; patients with late onset had more events than did those with early onset
		2	35		40.0		
		3	18		38.8		
		≥4	47		29.7		
Perris (1984)	Depression	1	37	62	50[c]	<0.02	Semistructured interview; 56 item inventory (3 mo prior)
		≥2	112	43	19[d]	<0.001	

252

Study	Diagnosis					p	Instrument
Dolan et al. (1985)	Depression	1 / ≥2	21 / 57	62	29	<0.05	Bedford College-Life Events and Difficulties Schedule (6 mo prior) (Brown, Harris, 1978)
Ezquiaga et al. (1987)	Depression	<3 / ≥3	52 / 45	50	16	<0.01	Semistructured interview (Brown, Harris); no effect of chronic stress
Ambelas (1987)	Mania	1 / ≥2	50 / 40	66	20	<0.001	Paykel Life Events Scale (4 wk prior)
Ghaziuddin et al. (1990)	Depression	1 / ≥2	33 / 40	91	50	<0.05	Paykel Life Events Scale (6 mo prior)
Cassano et al. (1989)	Depression	1 / ≥2	94 / 173	66.0	49.4	<0.05	Paykel Life Events Scale
Wistedt et al. (1997)	Unipolar	6.8	81	72		R = −.52	Paykel Life Events Scale
	Bipolar	8.2	191	63		R = −.75	
Hammend & Gitlin	Bipolar	Depressive-7.4 Manic - 7.6	52				
Castine et al. (1998)	Schizophrenia	≤3	32	more recent life events		p = 0.01	Paykel Life Events Scale
Nierenberg et al. (1998)	Depression	1st versus >3 episodes	176	1st episode had more stressful negative life events compared with recurrent		p = 0.037	Paykel Life Events Scale, Perceived Stress Scale (p = .037)

[a] For this group, the most recent hospitalization was preceded by a life event resulting in role loss.
[b] Of surgical comparison subjects, 6.6% had experienced recent major life events.
[c] Percentage for negative or undesirable events.
[d] Percentage for events involving psychological conflict.

In a parallel fashion, episode sensitization refers to the idea that with each episode occurrence, an increasing vulnerability to subsequent episodes occurs *(11)*. This is likely related to the biological residues left by each episode through its effects on gene expression and protein synthesis. The observation of episodes begetting other episodes appears true when attempts at prophylaxis are not successful, and it suggests that episode repetitions may have prognostic significance for the development of treatment resistance. It has become well established that patients with rapid cycling, for example (four or more episodes in a year), are less responsive to lithium *(18)*, carbamazepine *(19,20)*, and perhaps valproate *(21)* than those without such frequent recurrences. More recently, ultrafast cycling patterns have been classified *(22,23)* such as ultrarapid cycling (defined as four or more episodes occurring within a month's period of time), and ultra-ultrarapid or ultradian cycling (dramatic mood shifts within a single day on 4 or more days within the same week). These ultrafast patterns appear to be particularly refractory to pharmacological intervention as well, and they often require complex combination therapy *(22,24–26)*.

Given the relatively consistent observations of stressor and episode sensitization in the course of recurrent unipolar and bipolar affective illness, it becomes necessary to conceptualize the types of possible long-term neurobiological mechanisms underlying these processes. This chapter focuses on the emerging evidence of such processes and the potential candidate mechanisms involved, such as experiential changes in gene expression. When a family history of bipolar affective disorder is not present [as is the case in more than one-half of all bipolar patients *(27)*], and early environmental and psychosocial factors appear important to the onset of illness, some type of long-term neurobiological vulnerability factor must be involved. In addition, patients with recurrent affective disorders are at high risk for alcohol and substance abuse *(28)*, and these factors have negative prognostic implications for the treatment of affective illness *(29)*. This led us to the further consideration that alcohol and drugs themselves impart additional long-term effects on illness vulnerability, most likely also through their effects on gene expression *(30)*.

Some types of adult life experience can leave immutable memory traces and associated alterations in biology and neurochemistry *(31,32)*. This is evident in patients with posttraumatic stress disorder (PTSD). Early life experiences may be even more likely to lead to long-term psychobiological changes; for example, early emotional neglect can result in a range of effects from psychosocial dwarfism *(33–35)*, to the proneness to anaclitic depression *(36)*, to the blank stares and mal development of the children of the Romanian orphanages of the early 1990s *(37,38)*.

Thus, type of stressor, intensity, severity, frequency of occurrence, and, importantly, stage of development of the central nervous system (CNS) at the time of exposure are among the many variables that would appear to impart different outcomes on biochemistry and subsequent behavior. In general, the type of stressor appears crucial in determining the nature of later illness manifestations; for example, horrific stressors associated with threat of loss of life or limb to the individual or his/her immediate friends or family tend to be associated with PTSD, and stressors that involve losses of social or family relationships, or occupational context (e.g., loss of self-esteem) are more likely to result in affective disorders *(39–41)*. The temporal characteristics of these events in relation to the unfolding development of the CNS also appear to have major and differential consequences to the individual and their course of psychiatric dysfunction *(31)*.

LONG-TERM CONSEQUENCES OF SINGLE AND REPEATED MATERNAL SEPARATIONS

The elegant studies of Liu et al. *(42)*, Meany et al. *(43)*, and Plotsky *(44)* have demonstrated permanent effects of repeated maternal separation on neuroendocrine regulation and behavior. Repeated separations of neonatal rats from their mothers for 180 min compared with non-separated controls or those separated for 15 min have revealed that these animals are hypercortisolemic as adults, at least partly based on increased corticotropin releasing factor (CRF) excretion. Most interestingly, these animals show anxiety in open-field situations and are also more prone to the acquisition of alcohol and cocaine self-administration compared with their littermate controls. The hypercortisolemia, anxious behaviors, and vulnerability for substance abuse are all reversible with chronic treatment with serotonin-selective antidepressants. However, when these treatments are discontinued, the hypercortisolemia and proneness to substance abuse returns.

These data are of interest from a variety of perspectives. First, they provide strong evidence for experiential impact on gene expression and demonstrate that with the appropriate timing of the insult, these can have lasting and ethologically relevant impacts. In addition, many of the aspects of this type of neuroendocrine and behavioral dysregulation mirror those of the affective disorders and provide a number of interesting convergences with this syndrome *(45–47)*.

McKinney and Bunney *(48)* outlined four primary requirements for an ideal animal model of an illness, including (1) the induction of relevant homologous behaviors, (2) similar or parallel precipitating or inducing circumstances, (3) similar physiological and biochemical changes, and (4) reversibility with agents known to be effective in the treatment of the clinical syndrome being modeled. Many of these criteria are partially or completely satisfied by the maternal separation model *(44)*.

The implications of these long-term but treatable psychobiological alterations induced by maternal separation extend beyond the validation of any given animal model, however. They also demonstrate the modulatory (i.e., potentially bidirectional) impact of experience, depending on timing, quality, and magnitude of the inducing principle or stressor. This is most clearly demonstrated in the work of Meaney and associates in which the 15-min separated or "handled" control group is actually observed to experience protective effects against age-related cognitive decline and associated hippocampal cell loss *(43,49)*. Some of the mediating variables have been elucidated, including the fact that the briefly handled animals are tainted with a human scent, and this initiates increased licking and grooming of the rat pup by the mother *(42)*. Thus, this subsequent positive response of increased contact, attention, and maternal nurturance thus also appears to have life-long beneficial consequences to the developing mammal.

In a related model, Zhang et al. *(50)* and Levine et al. *(51)* have shown that a single 24-h period of maternal deprivation in 11- to 12-day-old rat pups can also leave indelible biochemical and behavioral imprints. These animals also exhibit hypercortisolemia and increased response to dopaminergic challenge (with apomorphine) as adults *(52)*. This 24-h period of maternal deprivation experience doubles the number of apoptotic cells in the CNS *(53)* when these animals are sacrificed on day 12. These apoptotic cells are widely dispersed throughout the brain, but are particularly concentrated in the external germinal layer of the cerebellum and in the supergranular layer of the hippocampus (which ultimately disappears in the adult). That these cells are undergoing apoptosis is evident not only from the fact that they are "apo-tagged" but also by the increased expression of the apoptotic gene *BAX* and *Bcl-X* (both short [*Bcl-Xs*, apoptotic] and long forms [*Bcl-Xl*, protective]). The increased ratio of apoptotic to protective and antiapoptotic genes is thought to lead the cells to programmed cell death *(54)*. Another transcription factor thought to be involved in the regulation of developmental apoptosis is c-*jun* and this, as well as nerve growth factor (NGF), also show increased expression in the maternally deprived compared with control animals *(55)*. In addition, however, nitric oxide synthase (NOS) and calcium/calmodulin kinase II (CaMKII) are dramatically downregulated *(56)* in the maternally deprived rat pups, potentially providing a route to impaired CNS development and learning and memory.

Whether these increases and decreases in gene expression represent an absolute alteration or merely an acceleration or retardation in normal devel-

NEUROANATOMY OF EMOTION AND AFFILIATION

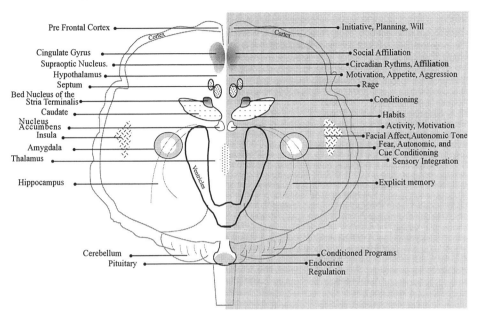

Fig. 3. Areas of the brain (**left labels**) are linked in a highly preliminary way with some of the emotional and affiliative functions (**right labels**) they modulate. Further and more precise definition of the neuroanatomy of the emotional homunculus and its neuroplasticity (similar to that revealed for the distorted representations of the body surface for sensory function in the parietal cortex) should help in the delineation of the physio-anatomy of the major psychiatric illnesses.

opmental sequences remains to be delineated. Nonetheless, these changes brought about by a 24-h period of separation from the mother are clearly demonstrable at the level of not only brain biochemistry but also its cellular elements and, thus, begin to outline the potential substrates for the sustained impact of deprivation on subsequent brain development and behavior.

With more detailed studies of the timing, incidence, duration, and quality of insult, a range of transient to permanent biochemical and physiological effects of early environmental and maternal deprivation could be evaluated in the medial and paralimbic parts of the brain involved in the modulation of emotion and affiliation (Fig. 3). This might be similar to what has already been elegantly described for the anatomy and function of the visual cortex based on deprivation or alteration of early visual input *(57)*. In the case of emotional function, the transduction mechanisms and modulatory influences of highly interrelated pathways involving connections with secondary, tertiary, and higher-order association areas are likely to be vastly more complicated than for a single sensory modality such as vision. The situation is also likely to be exquisitely more complex in the human because of the potential for vast numbers of compensatory adaptations, such as others in the environment substituting for the maternal loss. This can occur in an adequate or inadequate fashion leading to subsequent vulnerability to depression or other psychiatric complications *(58)*. However, by developing an understanding of the polymodal experiences involved in the maternal–neonatal rat pup interactions, we can begin to assess how some of these systems can be affected by early experience *(59)*.

Conversely, one can imagine appropriate or ideal environmental circumstances favoring superior brain development and protection against age-related decline. This has been shown to occur in the early interventions studies of Meaney and associates *(43,49)* described above and in the environmental enrichment studies of Greenough and colleagues *(60,61)*. Recently, Kempermann et al. *(62,63)* have demonstrated the potential for the adult nervous system to respond to environmental enrichment by increasing the number of progenitor neurons differentiating into neurons compared with glia, again providing a prototype for understanding how environmental impact can affect gene expression and neuronal development.

STRESSOR AND PSYCHOMOTOR STIMULANT-INDUCED BEHAVIORAL SENSITIZATION IN ADULT ANIMALS

Antelman et al. *(64,65)*, Robinson et al. *(66,67)*, and Kalivas and Stewart *(68)* have shown the potential bidirectional cross-sensitization between sensitization induced by psychomotor stimulants and some types of environmental stressors. This is of considerable interest in relation to the high comorbidity of substance abuse and the affective disorders *(28,69,70)*, both of which have been linked to stressful life experiences in their initiation, progression, and in the precipitation of relapse. Thus, cocaine sensitization can be used both as a model for psychomotor stimulant abuse with long-term effects on gene expression and as a potential model of the effects of recurrent stressors on these and related neural systems.

We have been particularly interested in the observation that cocaine-induced behavioral sensitization has an important conditioned component *(71–73)*. Repeated administration of the psychomotor stimulant in the same environment comes to evoke increasing behavioral responses over time, in contrast to animals pretreated with equal amounts of drug and rechallenged in a different environment. The demonstration of prominent conditioned effects in this model also puts the accompanying changes in biochemistry and gene expression into a psychologically or psychiatrically relevant context; that is, the resultant brain alterations are associated not simply with exposure to the drug. In some cases, they may occur only under the combined circumstances of drug exposure and the specific context previously paired with drug; in others, drug exposure may be sufficient; and in still other circumstances, exposure to the environment or conditioned drug-related stimuli may trigger many of the same biochemical and behavioral consequences as the drug itself. Thus, each of these components is capable of tapping into similar biological substrates and altering behavior based on prior experience. In addition, as stated earlier, nonpharmacological stressors appear to impinge on some of the same neural circuits as drugs of abuse and alcohol leading to multiple converging roads toward psychopathology and possibly therapeutic intervention.

Evolving Neural Substrates

Some of the preliminary biochemical, neuroanatomical, and pharmacological blueprints for this conditioned component of cocaine sensitization have begun to be elucidated. There is increased dopamine release in the nucleus accumbens, but not the striatum *(71)*. Neuroleptics block the development, but not the expression, of cocaine-conditioned sensitization *(73)*. Both the amygdala and nucleus accumbens are involved in this context dependency, as lesions of these two areas inhibit the development of behavioral sensitization *(71,74,75)*. However, if repeated exposures to drug and context occur, conditioned sensitization can develop, even with a damaged amygdala.

These data may relate to the differential neuroanatomy of representational or one-trial memory versus habit acquisition, which requires repeated pairings or exposures *(76)*. Representational memory is thought to be highly dependent on the structures in the medial part of the temporal lobe including the perirhinal and entorhinal cortices and the hippocampus, whereas, at least in some paradigms, the striatum is critical for habit memory and intact medial–temporal structures are not required *(76,77)*.

Similarly, Kalivas et al. *(78,79)*, Robinson et al. *(67)*, and others have emphasized that the cell body area of the ventral–tegmental region appears necessary for the induction of stimulant-induced behavioral sensitization, but its expression and long-term maintenance appear to require other neural substrates, including the accumbens and perhaps beyond *(80–82)*.

Taken together, these data, suggesting that the substrates mediating cocaine- and stressor-induced behavioral sensitization are not static, are very much in keeping with the similar observations noted later relating to another model of learning and memory, amygdala kindling (*see* Amygdala Kindling and the Search for Quenching Paradigms). In the kindling model, changes are observed not only in biochemistry but also in the synaptic and cellular micro-structure of the brain *(83)*. Moreover, as observed with sensitization, the progression from developing to completed seizures and eventually to spontaneity is also accompanied by an evolving neuroanatomical substrate. Using c-*fos*, Clark et al. *(84)* demonstrated that initial amygdala stimulations resulted in unilateral expression of c-*fos* mRNA in limbic cortical areas surrounding the amygdala and the hippocampus (if a long afterdischarge was evoked). With completed seizures, bilateral induction of c-*fos* occurred that included nonlimbic cortical regions as well, and, following a spontaneous seizure (in a single animal), c-*fos* was induced unilaterally in the contralateral cortex, suggesting the possibility of a mirror focus involvement in the initiation of the spontaneous convulsion. A similar neuroanatomical progression was observed using other markers of gene expression, such as peptide mRNA expression for thyrotropin-releasing hormone (TRH) and enkephalin *(85)*.

Role of Glutamate

In the kindling model, conditioned components do not appear to play as prominent a role as in the cocaine sensitization model, and in many other characteristics, the two models of memory show very different underlying biochemical and pharmacological components. In some cases, they are even opposing, wherein stresses can inhibit amygdala-kindled seizures and yet facilitate cocaine-induced sensitization and, similarly, catecholaminergic effects can enhance sensitization but inhibit kindling, and so forth (Table 2). Nonetheless, there are some convergences in relation to the effects of gluta-mate in which the glutamate N-methyl-D-aspartate (NMDA) receptor antagonist, MK-801, blocks the development but not the expression of cocaine *(86,87)* and kindling-induced *(88–90)* behavioral and convulsive endpoints *(91,92)*. These data suggest that NMDA receptors are involved in both of these models of learning and memory, as they are in many other paradigms as well. However, once the neural systems for such effects are in place, it appears that non-NMDA receptors, such as those involving AMPA-type glutamate receptors *(93;* Pert et al., unpublished data), become more prominent in the maintenance of the sensitized or kindled substrate, similar to what is observed in the maintenance as opposed to the development of certain forms of long-term potentiation *(94,95)*.

Conditioned Components of Sensitization

Cocaine *(96–98)*, as well as many stressors *(49,99–101)* has been demonstrated to increase CRF release, and the conditioned component of cocaine sensitization appears to also involve a conditioned corticosterone response *(102)*. Adrenalectomy can block sensitization in amphetamine and cocaine

Table 2
Phenomena in Course of Affective Disorders
Modeled by Kindling and Behavioral Sensitization

Descriptors	Kindling (K)	Sensitization (S)	Phenomenon in affective illness
Stressor vulnerability	—	++	Initial stressors early in development may be without effect but predispose to greater reactivity upon rechallenge
Stressor precipitation	++	++	Later stress may precipitate full-blown episode
Conditioning may be involved	—	++	Stressors may become more symbolic
Episode Autonomy	++	—	Initially precipitated episodes may occur spontaneously
Cross-sensitization with stimulants	++	++	Co-morbidity with drug abuse may work in both directions affective illlness ↔ drug abuse
Vulnerability to relapse	++	++	S and K demonstrate long term increases in responsivity
Episodes may:			
a. become more severe	++	++	S and K both show behavioral evolution in severity or stages
b. show more rapid onsets		++	Hyperactivity and stereotypy show more rapid onsets
c. become spontaneous	+		K seizures evolve to spontaneous seizures
Anatomical and biochemical substrates evolve	++	+	K memory-trace evolves from unilateral to bilateral; S memory trace evolves from midbrain to n. accumbens
IEGs involved	++	++	Immediate Early Genes (IEGs) such as c-fos induced
Alterations in gene expression occur	++	++	IEGs may change later gene expression, e.g., peptides over longer time domains
Change in synaptic microstructure occurs	++	—	Neuronal sprouting and cell loss indicate structural changes
Pharmacology differs as function or stage of evolution	++	++	K differs as a function of stage; S differs as a function of development versus expression

paradigms *(103,104)* and intracerebroventricular (icv) administration of the CRF antagonist alpha-helical CRH has also been reported to block amphetamine sensitization *(105)* but not cocaine sensitization conditioned (Weiss et al., *unpublished data*).

The studies of Brown et al. *(106)* on conditioned effects of stimulants indicate that it is not the striatal or accumbens systems that are direct targets of cocaine that show a conditioned increase in c-*fos* expression. Rather, it is the limbic and cortical structures that appear to be affected by exposure to an environment previously paired with drug. CRF itself, administered intracerebroventricularly, can increase c-*fos* expression unilaterally on the side of the injection and bilaterally if a seizure is induced *(107)*. Thus, the effects of stressors, their associated stimuli, CNS peptides, and their downstream hormonal targets all appear to be involved in a complex chain of events that mediate some components of the long-lasting effects in this paradigm.

Thus, this conceptual perspective has direct relevance to stressor sensitization in recurrent affective illness and for the related problem of substance abuse and its prominent conditioned components, which are capable of triggering relapse *(108,109)*. A series of recent studies in humans have, similarly, identified an important role of mediotemporal (amygdala) and ventral striatal (nucleus accumbens) structures in the conditioned components of cocaine craving *(110)*. Thus, just as the amygdala appears crucial for the formation of many aspects of conditioned emotional and autonomic responses in animals *(111–113)*, there are convergent human data now linking the amygdala and related neural substrates to these phenomena in man. To the extent that stimulant use and abuse is so highly comorbid with bipolar illness, it is hoped that a better understanding of the stress-related and stress-independent mechanisms and conditioned components of sensitization that are involved in both cocaine use and recurrent affective illness may lead to a better understanding and treatment approach to both syndromes.

The amygdala also appears to be a crucial substrate involved in the recognition of facial emotion in animals and man. Affectively ill patients appear to have deficiencies in appreciation of facial emotions, particularly those related to happiness, sadness, anger, and fear *(114,115)*. At rest, affectively ill patients demonstrate a hypermetabolic amygdala *(116)*, and studies involving pharmacological provocation of the amygdala with the local anesthetic procaine *(117)* further suggest abnormalities in the responsiveness of the amygdala in patients with affective disorders. This area of the brain also appears to be importantly involved in the physical and psychic traumas of PTSD and their conditioned components *(118–122)*. As such, understanding some of the details of amygdala physiology and plasticity may assist in uncovering therapeutic approaches to dysfunction in this area of the brain.

LONG-TERM POTENTIATION AND LONG-TERM DEPRESSION IN THE AMYGDALA SLICE

Whereas long-term potentiation (LTP) and long-term depression (LTD) have been most prominently studied in the hippocampus and, in particular in the CA1 NMDA-dependent synapse *(123,124)*, little work has been done in the amygdala because of the increased complexity of the neuronal architecture and the related methodological and technical difficulties that this evokes. However, recently, several groups have begun to study plasticity in amygdala neurons and have demonstrated both convergences and divergences from that revealed in the hippocampus.

As indicated in Fig. 4, whereas a single high-frequency burst in the hippocampus induces an NMDA-dependent LTP, such a burst produces only short-term potentiation (STP) in the amygdala and, instead, two temporally separated volleys are required to produce LTP. Bidirectional modulation of synaptic plasticity is evident in the hippocampus, with low-frequency stimulation inducing LTD, high-frequency stimulation inducing LTP, and each capable of reversing the other *(125)*.

The situation is more complicated in the amygdala because the same parameters of stimulation (1 Hz for 15 min) can induce either LTP or LTD, depending on the prior stimulation history of the neuron. These data are of particular interest because they reveal the phenomenon of metaplasticity, where the effects of a stimulus may not be apparent until subsequent stimulation occurs. In this case, it is the direction of the plasticity that is affected by the history of prior stimulation of the amygdala neuron. When axons in the external capsule are stimulated using a low-frequency train (1 Hz for 15 min), neurons in the basolateral amygdala demonstrate a gradual developing and prolonged LTP. However, when the same low-frequency stimulation is preceded by a high-frequency train (100 Hz for 1 s), then LTD results *(126)*. Li and associates have demonstrated that the type-2 metabotropic glutamate receptors are important for this priming effect of the high-frequency stimulation; that is, when an antagonist of the type-2 metabotropic glutamate receptor is used, the low-frequency stimulation after priming is no longer capable of evoking LTD.

To the extent that similar processes and mechanisms are at play in the amygdala of the primate brain, they raise the interesting possibility that the same low-frequency stimulation applied in vivo might also have differential effects in normal versus pathological substrates. As such, a highly primed or sensitized amygdala substrate (perhaps as occurs in the affective disorders or PTSD) could show differential long-term adaptation or depotentiation to the same low-frequency stimulation that might evoke an increased response in

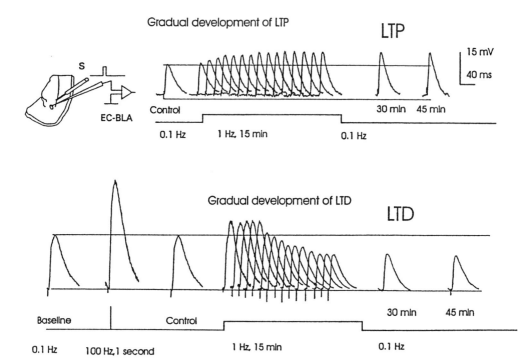

Fig. 4. History-dependent bidirectional modulation of synaptic transmission by low-frequency stimulation of external capsule. Traces were intracellularly recorded synaptic responses evoked by stimulation of external capsule. (**Top panel**) Low-frequency stimulation (LFS, 1 Hz for 15 min) induces a gradual development of synaptic LTP lasting for more than 30 min after the termination of LFS. (**Bottom panel**) In contrast, the same low-frequency stimulation given 10 min after a prior high-frequency stimulation (HFS) induces an initial transient synaptic facilitation followed by enduring synaptic LTD lasting for more than 30 min after the termination of LFS. Thus, LFS produces either LTP or LTD, depending on the prior history of neural stimulation. LTP: long-term potentiation; LTD: long-term depression.

the nonprimed state. To the extent that metabotropic type-2 glutamate receptors are involved in such a process, one could wonder whether a type-2 agonist might be helpful in changing an LTP type of response to low-frequency stimulation to one showing long-lasting decrements similar to LTD.

Whereas 15 min of 1-Hz stimulation is capable of reversing LTP in the hippocampus, the stimulation characteristics that reverse LTP in the amygdala (which is induced by two separate high-frequency bursts) have not yet been elucidated. The amygdala thus appears less amenable to depotentiation than the hippocampus. It is hoped that once the stimulation characteristics that can reverse LTP in the amygdala slice in vitro are uncovered, they might lead to the utilization of similar parameters in vivo and in the clinic that could parallel this ability to reverse the long-lasting increases in amygdala synaptic excitability.

These data are of interest not only from the pharmacological perspective but also in relation to physiological manipulations, which may be possible in humans using repeated transcranial magnetic stimulation (rTMS). Can amygdala stimulation characteristics be determined and utilized such that long-lasting alterations toward normal amygdala function can be achieved in neuropsychiatric patients who demonstrate pathological dysregulation in this and related structures?

AMYGDALA KINDLING AND THE SEARCH FOR QUENCHING PARADIGMS

Kindling Progression

From the outset, Goddard and associates *(127)* recognized the importance of the kindling model, not only for the study of epileptogenesis but also as a model of learning and memory *(128)*. In kindling, brief 1-s high-frequency stimulation not only evokes increasing duration, complexity, and spread of amygdala afterdischarges, but it also drives sequential stages of seizure evolution culminating in the appearance of generalized motor seizures to the same 1-s stimulation that was previously subthreshold. Moreover, the studies of Pinel et al. *(129,130)* and Wada and associates *(131,132)* have further documented that extensive numbers of amygdala stimulations eventually leads to the phenomenon of spontaneity in which generalized seizures occur in the absence of amygdala stimulation.

Changes in Gene Expression

The complex cascade of effects on gene expression and on the biochemistry and microstructure of the brain that accompany kindling are just beginning to be clarified. It is clear, however, that there is an extraordinary spatiotemporal evolution of changes in gene expression in the kindling model as a function of number of stimulations. Initial stimulations evoke largely localized and unilateral changes, whereas increased numbers of stimulations evoke bilateral changes in more extensive systems with immediate–early genes triggering a host of changes in late effector gene (LEG) expression *(85,133–136)*. These, in turn, program changes in synaptic biochemistry and microstructure, some of which are transient and others that may be long-lasting. For example, seizures evoke transient upregulation in the benzodiazepine–GABA receptor–chloride ionophore complex in peptides such as CRH and TRH, and in neurotrophic factors such as brain-derived neurotrophic factor (BDNF). Seizures also produce long-lasting changes in brain micro-structure, such as sprouting of dentate granule cells and neuronal loss, prob-ably via apoptosis in the dentate hilus area *(137)*.

Pathological Versus Adaptive Changes in Gene Expression

Recent work in our laboratory has suggested that many of these changes in gene expression are adaptive and compensatory, whereas others appear to be related to the primary pathophysiology of kindling and the maintenance of the increased excitability or neuronal memory trace *(136,138,139)*. Thus, we postulate that many types of stimulation, including those evoking amyg-dala-kindled seizures, elicit not only primary pathological effects on gene expression but also those related to adaptive and compensatory mechanisms programmed in an attempt to restore or maintain homeostasis.

As such, the identification and differentiation of these types of changes in clinical epilepsy and other neuropsychiatric conditions, including depres-sion, may lead to dual targets of therapeutics. For example, in terms of anti-epileptogenic and anticonvulsant mechanisms, not only could one attempt to inhibit the primary pathological changes but also to enhance adaptive and compensatory mechanisms. We, and others, have postulated that a variety of neuropeptide and neurotransmitter substances are part of this endogenous anticonvulsant mechanism. Not only are adenosine and other modulatory sub-stances released by kindled seizures that can limit seizure and afterdischarge duration *(140)*, but also the mRNA expression for TRH and other putative anticonvulsant peptides such as neuropeptide Y (NPY) are increased *(138)*.

When TRH is administered intracerebrally into the hippocampus or amyg-dala, anticonvulsant effects are achieved *(141,142)*. These and a variety of other data provide initial but compelling evidence that some peptides and related substances may be part of an endogenous anticonvulsant response system. In fact, one genetic strain of seizure-prone rats has been found to be deficient in cholecystokinin (CCK) in the hippocampus. In one of the first demonstra-tions of the physio-logical potential of gene therapy for CNS diseases, increas-ing CCK gene expression in the hippocampus was shown to reverse this seizure proneness *(143)*.

Quenching with Low-Frequency Stimulation vs Low-Level Direct Current

Given the relative permanence of kindled-induced changes in CNS excita-bility in the amygdala kindling model of neuronal learning and memory, Weiss and Post and associates began to explore potential paradigms for reversing these changes. Although it became increasingly apparent that not only syn-aptic *(144)* but also cellular events at the level of axonal and dendritic sprouting were involved *(145,146)*, we were nonetheless encouraged by the work of McEwen *(147)* and others, indicating that many structural modifications of the CNS were themselves highly variable and neuroplastic and, thus, potentially reversible. Such a perspective was obviously also forthcoming from the model of the reversible structural basis of the neural substrates involved in certain birds' song production *(148–150)*.

Our initial attempts for several years at kindling reversal only appeared to potentiate kindled excitability or induce status epilepticus until we adopted some of the stimulation characteristics utilized in hippocampal LTD (i.e., low frequency and more chronic stimulation) *(151,152)*. Although initial studies appeared highly provocative in suggesting that these parameters would also be capable of quenching amygdala-kindled seizures in vivo, subsequent studies revealed that it was a low-level direct current (DC) that was crucial to the quenching effect *(153)*.

Implication for Frequency-Dependent Effects of rTMS in Mood and Neural Activity

Nonetheless, the notion of bidirectional changes in synaptic excitability achieved in the LTP/LTD paradigms of amygdala and hippocampal slices and, preliminarily, in the kindling/quenching phenomena, led us to explore the potential frequency-dependence of rTMS in attempts to find optimal therapeutic parameters of this type of brain stimulation in man. Following initial observations of our group, in collaboration with Wassermann and Hallett of the National Institute of Neurological Disorders and Stroke (NINDS), that 20-Hz stimulation of the left frontal cortex of depressed patients appeared able to induce antidepressant effects in some subjects *(154)*, we, and others, conducted more controlled studies demonstrating weak to substantial antidepressant effects of higher-frequency (10- to 20-Hz) stimulation (over the left but not right frontal cortex or occiput) in severely depressed patients *(155,156)*. Whereas Pascual-Leone et al. *(156)* reported rather dramatic effects on mood with this type of rTMS, less consistent and dramatic effects were evident in our studies and the search was begun for more optimal parameters *(152,157)*.

Thus, studies were initiated of the potential differential effect of 1-Hz versus 20-Hz rTMS versus sham stimulation. Initial data of Kimbrell and associates *(158)* using these parameters at 80% of motor threshold demonstrated differential effects on mood of 1- and 20-Hz within individual subjects. Those responding positively to one modality worsened when receiving the other, as revealed in Hamilton depression ratings ($r = -0.80$, $n = 10$, $p < 0.004$) *(158)*. This inverse relationship has been replicated in a second study comparing clinical changes following 1-Hz versus 20-Hz rTMS at 100% of motor threshold ($r = -0.59$, $n = 15$, $p < 0.02$) *(159)*.

Thus, the initial prediction of differential effects as a function of frequency was preliminarily supported and further data have now emerged that are convergent with this perspective. In the study of Kimbrell et al. *(158)*, those with the lowest cerebral glucose utilization at baseline appeared to improve with the 20-Hz stimulation, whereas those who had high cerebral glucose utilization compared with normal volunteer controls appeared to improve with the 1-Hz stimulation.

Moreover, when we examined changes in blood flow with 2 wk of the 1-Hz versus 20-Hz paradigm (this time at 100% of motor threshold), Speer and associates observed opposite effects of these two frequencies on regional cerebral blood flow measured by positron emission tomography (PET) using $^{15}0$ water. These assessments revealed that 20-Hz rTMS for 2 wk was associated with widespread increases in cerebral blood flow and 1 Hz with more restricted decreases, and each of these was capable of reversing the direction of changes in blood flow induced by the other *(159)*.

The Search for Robust and Sustained Effects of rTMS

The biochemical and molecular events underlying these apparent opposing effects on physiology remain to be explained. However, when we attempted to use 1-Hz stimulation to reverse the hyperactivity of frontal temporal structures revealed with FDG PET in patients with PTSD, the associated clinical effects did not appear to be long-lasting, and patients relapsed shortly after the discontinuation of the rTMS *(119)*. Similarly, a number of depressed patients appeared to lose their improvement shortly after rTMS was discontinued *(154,155,158)*. Whatever neurochemical events were underlying the changes in metabolism and blood flow observed with rTMS and the improvements in mood or PTSD symptoms, they did not appear to be of a sustained variety.

Furthermore, the incidence and magnitude of treatment effects in depression appeared to vary widely from study to study *(160)*, with some reports suggesting approximately equal efficacy of rTMS and electroconvulsive therapy (ECT) in nondelusional depressed patients *(161)* and other open and controlled studies reporting minimal to no therapeutic effects of closely related parameters *(162)*. Therefore, it is apparent that much work remains before rTMS can become a clinically relevant therapeutic tool in the treatment of affective disorders and PTSD. Although one hopes that this can ultimately be achieved, the statistically significant (and sometimes clinically relevant) therapeutic effects of active rTMS compared with sham raises questions about potential mechanisms of this nonconvulsant stimulation compared with the well-recognized therapeutic effects of ECT *(160)*.

Potential Mechanism of Nonconvulsive Stimulation with rTMS

It had been paradigmatic that, at least in ECT, a seizure was the necessary prerequisite for therapeutic effects *(163,164)*. Although this may be the case for ECT, the emerging data suggest that this is not the case for rTMS. We have postulated that whereas seizures are necessary for the induction of the adaptive changes in ECT that are the therapeutic principles in depression, rTMS may be able to evoke some of these adaptive mechanisms more directly without the requirement of a seizure *(157)*. Thus, it is well known that ECT in man *(165,166)* and electroconvulsive seizures (ECS) in animals *(167,168)* increases TRH mRNA and TRH protein, which has been postulated to at least partially relate to ECT's therapeutic effects *(164,169)*. Perhaps alterations in TRH mRNA or the other critical neuroadaptive changes pertinent to the effects of ECT could eventually be induced by appropriate nonconvulsive brain stimulation parameters with rTMS. To the extent that this becomes possible, it would obviously have great importance for therapeutics. ECT is cumbersome, costly, and stigmatized *(170)* because it requires anesthesia and the induction of a seizure, which can be accompanied by moderate to severe memory loss *(171)*.

Repeated transcranial magnetic stimulation (rTMS), in contrast, can be administered with relatively little discomfort to the awake individual without the induction of a seizure and, given the typical stimulation parameters so far utilized, evokes few side effects on learning and memory *(172)*. The ability to stimulate the awake brain also raises the ultimate potential that one could make use of experience-dependent neuroplasticity in applying relatively more selective therapeutic treatments pertinent to a given pathological overactivated or underactivated neural substrate *(173)*.

IMPLICATIONS FOR THE FUTURE

Investigators have long speculated and surmised that the antidepressant drugs required a cascade of neurobiological events, perhaps involving changes in gene transcription, based on the long lag in achieving maximal therapeutic response *(174,175)*. Changes were postulated in the downregulation of beta receptors *(177,178)*, the regulation of adenylate cyclase *(176)*, and, more recently, the induction of binding to the cyclic AMP response element (CRE) *(179,180)* and changes in neurotrophic factor gene expression *(181)*.

Smith and associates *(182)* demonstrated that whereas some types of stressors appeared to decrease BDNF in the hippocampus, treatment with the antidepressants did the opposite, and cotreatment with antidepressants during stressor induction could block at least some of the stressor-induced changes in neurotrophic factor gene expression. These data were subsequently replicated and extended by Duman *(179)* and now it is apparent that there are a host of changes in gene expression induced by the antidepressants including alterations in glucocorticoid receptors *(183,184)* and neurotrophic factors *(179,181)* that could be implicated in the psychotropic effects of these agents. It is noteworthy that ECS is also capable of transiently increasing BDNF expression in the hippocampus *(185)* in a manner similar to that of many antidepressants and opposite to the effects of some stressors.

Given the discussion of compensatory mechanisms in the previous section, one could wonder about the differential effects of different frequencies of brain stimulation achieved with rTMS over different locations on neurotrophic factor gene expression as well. The presence of BDNF is required for the induction of some types of learning and memory, as revealed through knockout strategies in mice *(186)*; that is, in hippocampal slices prepared from BDNF-deficient mice, LTP cannot be induced. Moreover, these mice show impaired spatial learning in a Morris water maze *(186)*. Could appropriate rTMS stimulation increase low levels of BDNF or other neurotrophic factors that have been decreased by stressors or preprogrammed at a low or high level based on genetic vulnerability? We predict that these questions will be answered in the not too distant future and that rTMS and other nonconvulsive brain stimulation paradigms will begin to be utilized for therapeutic purposes. These may be particularly relevant in the area of psychiatric illnesses, where, by definition, the lesions tend to be more "functional" and, therefore, potentially more neuroplastic and amenable to modulation than many types of neurological illnesses in which cellular elements are lost altogether by either excitotoxic or apoptotic mechanisms.

Thus, as we move rapidly toward the 21st century, we hope that the long-term changes in biochemistry, physiology, and neuroanatomy evoked by either genetic vulnerability or experiential modulation of gene expression will be able to be partially countered by a variety of current and rapidly emerging pharmacological and physiological techniques. To the extent that this is possible, the further elucidation of the duration and long-term consequences of stress and depression may come to have increasing therapeutic as well as pathophysiological implications.

REFERENCES

1. Kraepelin, E. (1921) *Manic-Depressive Insanity and Paranoia.* E. S. Livingstone, Edinburgh.
2. Cutler, N. R. and Post, R. M. (1982) Life course of illness in untreated manic-depressive patients. *Compr. Psychiatry* **23**, 101–115.
3. Grof, P., Alda, M., and Ahrens, B. (1995) Clinical course of affective disorders: were Emil Kraepelin and Jules Angst wrong? *Psychopathology* **28**(Suppl 1), 73–80.

4. Goldberg, J. F., Harrow, M., and Grossman, L. S. (1995) Course and outcome in bipolar affective disorder: a longitudinal follow-up study. *Am. J. Psychiatry* **152,** 379–384.

5. Maj, M., Pirozzi, R., and Kemali, D. (1989) Long-term outcome of lithium prophylaxis in patients initially classified as complete responders. *Psychopharmacology (Berlin)* **98,** 535–538.

6. Maj, M., Veltro, F., Pirozzi, R., Lobrace, S., and Magliano, L. (1992) Pattern of recurrence of illness after recovery from an episode of major depression: a prospective study. *Am. J. Psychiatry* **149,** 795–800.

7. Paykel, E. S. (1994) Historical overview of outcome of depression. *Br. J. Psychiatry* **26,** 6–8.

8. Roy-Byrne, P., Post, R. M., Uhde, T. W., Porcu, T., and Davis, D. (1985) The longitudinal course of recurrent affective illness: life chart data from research patients at the NIMH. *Acta Psychiatr. Scand.* **317**(Suppl), 1–34.

9. Squillace, K., Post, R. M., Savard, R., and Erwin, M. (1984) Life charting of the longitudinal course of recurrent affective illness, in *Neurobiology of Mood Disorders* (Post, R. M. and Ballenger, J. C., eds.), Williams and Wilkins, Baltimore, MD, pp. 38–59.

10. Kessing, L. V. (1998) Recurrence in affective disorder. II. Effect of age and gender. *Br. J. Psychiatry* **172,** 29–34.

11. Kessing, L. V., Andersen, P. K., Mortensen, P. B., and Bolwig, T. G. (1998) Recurrence in affective disorder. I. Case register study. *Br. J. Psychiatry* **172,** 23–28.

12. Post, R. M. (1992) Transduction of psychosocial stress into the neurobiology of recurrent affective disorder. *Am. J. Psychiatry* **149,** 999–1010.

13. Post, R. M. and Ballenger, J. C., eds. (1984) *Neurobiology of Mood Disorders.* Williams & Wilkins, Baltimore, MD.

14. Castine, M. R., Meador-Woodruff, J. H., and Dalack, G. W. (1998) The role of life events in onset and recurrent episodes of schizophrenia and schizoaffective disorder. *J. Psychiatr. Res.* **32,** 283–288.

15. Nierenberg, A. A., Pingol, M. G., Baer, H. J., Alpert, J. E., Pava, J., Tedlow, J. R., and Fava, M. (1998) Negative life events initiate first but not recurrent depressive episodes. *APA New Res. Program Abstr.* 132 (abstract).

16. Ambelas, A. (1987) Life events and mania. A special relationship. *Br. J. Psychiatry* **150,** 235–240.

17. Swendsen, J., Hammen, C., Heller, T., and Gitlin, M. (1995) Correlates of stress reactivity in patients with bipolar disorder. *Am. J. Psychiatry* **152,** 795–797.

18. Post, R. M., Kramlinger, K. G., Altshuler, L. L., Ketter, T. A., and Denicoff, K. (1990) Treatment of rapid cycling bipolar illness. *Psychopharmacol. Bull.* **26,** 37–47.

19. Denicoff, K. D., Smith-Jackson, E. E., Disney, E. R., Ali, S. O., Leverich, G. S., and Post, R. M. (1997) Comparative prophylactic efficacy of lithium, carbamazepine, and the combination in bipolar disorder. *J. Clin. Psychiatry* **58,** 470–478.

20. Okuma, T. (1993) Effects of carbamazepine and lithium on affective disorders. *Neuropsychobiology* **27,** 138–145.

21. Calabrese, J. R., Rapport, D. J., Kimmel, S. E., Reece, B., and Woyshville, M. J. (1993) Rapid cycling bipolar disorder and its treatment with valproate. *Can. J. Psychiatry* **38,** S57–S61.

22. Kramlinger, K. G. and Post, R. M. (1996) Ultra-rapid and ultradian cycling in bipolar affective illness. *Br. J. Psychiatry* **168,** 314–323.

23. Leverich, G. S. and Post, R. M. (1996) Life charting the course of bipolar disorder. *Curr. Rev. Mood Anxiety Disord.* **1,** 48–61.

24. Frye, M. A. (1999) The increasing use of polypharmacotherapy for refractory mood disorders: twenty-two   years of study. *J. Clin. Psychiatry,* in press.

25. Post, R. M., Frye, M. A., Leverich, G. S., Denicoff, K. D. (1998) The role of complex combination therapy in the treatment of refractory bipolar illness. *CNS Spectrums* **3,** 66–86.

26. Post, R. M., Ketter, T. A., Pazzaglia, P. J., Denicoff, K., George, M. S., Callahan, A., Leverich, G., and Frye, M. (1996) Rational polypharmacy in the bipolar affective disorders. *Epilepsy Res.* **11**(Suppl), 153–180.

27. Suppes, T., Leverich, G. S., Keck, P. E., Jr., Nolen, W., Denicoff, K. D., Altshuler, L. L., McElroy, S. L., Rush, A. J., Kupka, R., Bickel, M., and Post, R. M. (1999) The Stanley Foundation Bipolar Treatment Outcome Network: II. Demographics and illness characteristics of the first 261 patients. *J. Affect. Disord.,* in press.

28. Kessler, R. C., Crum, R. M., Warner, L. A., Nelson, C. B., Schulenberg, J., and Anthony, J. C. (1997) Lifetime co-occurrence of DSM-III-R alcohol abuse and dependence with other psychiatric disorders in the National Comorbidity Survey. *Arch. Gen. Psychiatry* **54,** 313–321.

29. Brady, K. T. and Lydiard, R. B. (1992) Bipolar affective disorder and substance abuse. *J. Clin. Psychopharmacol.* **12,** 17S–22S.

30. Post, R. M., Weiss, S. R. B., and Leverich, G. S. (1994) Recurrent affective disorder: roots in developmental neurobiology and illness progression based on changes in gene expression. *Dev. Psychopathol.* **6,** 781–813.

31. Post, R. M. (1996) Impact of psychosocial stress on gene expression: implications for PTSD and recurrent affective disorder, in *Theory and Assessment of Stressful Life Events* (Miller, T. W., ed.), International University Press, New York, pp. 37–91.

32. Post, R. M., Weiss, S. R. B., and Smith M. (1995) Sensitization and kindling: implications for the evolving neural substrate of PTSD, in *Neurobiology and Clinical Consequences of Stress: From Normal Adaptation to PTSD* (Friedman, M. J., Charney, D. S., and Deutch, A. Y., eds.), Lippincott-Raven, Philadelphia, pp. 203–224.

33. Green, W. H., Campbell, M., and David, R. (1984) Psychosocial dwarfism: a critical review of the evidence. *J. Am. Acad. Child Psychiatry* **23,** 39–48.

34. Powell, G. F., Brasel, J. A., and Blizzard, R. M. (1967) Emotional deprivation and growth retardation simulating idiopathic hypopituitarism. I. Clinical evaluation of the syndrome. *N. Engl. J. Med.* **276,** 1271–1278.

35. Powell, G. F., Brasel, J. A., Raiti, S., and Blizzard, R. M. (1967) Emotional deprivation and growth retardation simulating idiopathic hypopituitarism. II. Endocrinologic evaluation of the syndrome. *N. Engl. J. Med.* **276,** 1279–1283.

36. Emde, R. N., Polak, P. R., and Spitz, R. A. (1965) Anaclitic depression in an infant raised in an institution. *J. Am. Acad. Child Psychiatry* **4,** 545–553.

37. Carlson, M. and Earls, F. (1997) Psychological and neuroendocrinological sequelae of early social deprivation in institutionalized children in Romania. *Ann. NY Acad. Sci.* **807,** 419–428.

38. Kaler, S. R. and Freeman, B. J. (1994) Analysis of environmental deprivation: cognitive and social development in Romanian orphans. *J. Child Psychol. Psychiatry* **35,** 769–781.

39. Kendler, K. S., Kessler, R. C., Neale, M. C., Heath, A. C., and Eaves, L. J. (1993) The prediction of major depression in women: toward an integrated etiologic model. *Am. J. Psychiatry* **150,** 1139–1148.

40. Paykel, E. S., Myers, J. K., Dienelt, M. N., Klerman, G. L., Lindenthal, J. J., and Pepper, M. P. (1969) Life events and depression. A controlled study. *Arch. Gen. Psychiatry* **21,** 753–760.

41. Roy, A., Breier, A., Doran, A. R., and Pickar, D. (1985) Life events in depression. Relationship to subtypes. *J. Affect. Disord.* **9,** 143–148.

42. Liu, D., Diorio, J., Tannenbaum, B., Caldji, C., Francis, D., Freedman, A., Sharma, S., Pearson, D., Plotsky, P. M., and Meaney, M. J. (1997) Maternal care, hippocampal glucocorticoid receptors, and hypothalamic–pituitary–adrenal responses to stress. *Science* **277,** 1659–1662.

43. Meaney, M. J., Aitken, D. H., Van Berkel, C., Bhatnagar, S., and Sapolsky, R. M. (1988) Effect of neonatal handling on age-related impairments associated with the hippocampus. *Science* **239,** 766–768.

44. Plotsky, P. M. (1997) Long-term consequences of adverse early experience: a rodent model. *Biol. Psychiatry* **41,** 77S (abstract).

45. Plotsky, P. M., Owens, M. J., and Nemeroff, C. B. (1995) Neuropeptide alterations in mood disorders, in *Psychopharmacology: The Fourth Generation of Progress* (Bloom, F. E. and Kupfer, D. J., eds.), Raven, New York, pp. 971–981.

46. Post, R. M. and Weiss, S. R. B. (1997) Emergent properties of neural systems: how focal molecular neurobiological alterations can affect behavior. *Dev. Psychopathol.* **9,** 907–929.

47. Post, R. M., Weiss, S. R. B., Li, H., Smith, M., Zhang, L. X., Xing, G., Osuch, E., and McCann, U. (1998) Neural plasticity and emotional memory. *Dev. Psychopathol.* **10,** 829–855.

48. McKinney, W. T., Jr. and Bunney, W. E., Jr. (1969) Animal model of depression. I. Review of evidence: implications for research. *Arch. Gen. Psychiatry* **21,** 240–248.

49. Plotsky, P. M. and Meaney, M. J. (1993) Early, postnatal experience alters hypothalamic corticotropin-releasing factor (CRF) mRNA, median eminence CRF content and stress-induced release in adult rats. *Mol. Brain Res.* **18,** 195–200.

50. Zhang L. X., Xing G . Q., Levine S., Post R. M., Smith M. A. (1997) Maternal deprivation induces neuronal death. *Soc. Neurosci. Abstr.* **23,** 1113 (abstract).

51. Levine, S., Huchton, D. M., Wiener, S. G., and Rosenfeld, P. (1991) Time course of the effect of maternal deprivation on the hypothalamic- pituitary-adrenal axis in the infant rat. *Dev. Psychobiol.* **24,** 547–558.

52. Rots, N. Y., De Jong, J., Workel, J. O., Levine, S., Cools, A. R., and De Kloet, E. R. (1996) Neonatal maternally deprived rats have as adults elevated basal pituitary-adrenal activity and enhanced susceptibility to apomorphine. *J. Neuroendocrinol.* **8,** 501–506.

53. Smith, M. A., Zhang, L. X., Dent, G., Zhan, Y., Post, R. M., and Levine, S. (1998) Maternal deprivation increases apoptosis in the infant rat brain. *ACNP Abstr.*

54. Korsmeyer, S. J. (1995) Regulators of cell death. *Trends Genet.* **11,** 101–105.

55. Zhang, L. X., Xing, G. Q., Levine, S., Post, R. M., and Smith, M. A. (1998) Effects of maternal deprivation on neurotrophic factors and apoptosis-related genes in rat pups. *Soc. Neurosci. Abstr.* **24,** 451 (abstract).

56. Xing, G. Q., Smith, M. A., Levine, S., Yang, S. T., Post, R. M., and Zhang, L. X. (1998) Suppression of CaMKII and nitric oxide synthase by maternal deprivation in the brain of rat pups. *Soc. Neurosci. Abstr.* **24,** 452 (abstract).

57. Hubel, D. H. and Wiesel, T. N. (1979) Brain mechanisms of vision. *Sci. Am.* **241,** 150–162.
58. Breier, A., Kelsoe, J. R., Kirwin, P. D., Beller, S. A., Wolkowitz, O. M., and Pickar, D. (1988) Early parental loss and the development of adult psychopathology. *Arch. Gen. Psychiatry* **45,** 987–993.
59. Suchecki, D., Mozaffarian, D., Gross, G., Rosenfeld, P., and Levine, S. (1993) Effects of maternal deprivation on the ACTH stress response in the infant rat. *Neuroendocrinology* **57,** 204–212.
60. Camel, J. E., Withers, G. S., and Greenough, W. T. (1986) Persistence of visual cortex dendritic alterations induced by postweaning exposure to a "superenriched" environment in rats. *Behav. Neurosci.* **100,** 810–813.
61. Wallace, C. S., Kilman, V. L., Withers, G. S., and Greenough, W. T. (1992) Increases in dendritic length in occipital cortex after 4 days of differential housing in weanling rats. *Behav. Neural Biol.* **58,** 64–68.
62. Kempermann, G., Kuhn, H. G., and Gage, F. H. (1997) More hippocampal neurons in adult mice living in an enriched environment. *Nature* **386,** 493–495.
63. Kempermann, G., Kuhn, H. G., and Gage, F. H. (1998) Experience-induced neurogenesis in the senescent dentate gyrus. *J. Neurosci.* **18,** 3206–3212.
64. Antelman, S. M. (1988) Stressor-induced sensitization to subsequent stress: implications for the development and treatment of clinical disorders, in *Sensitization in the Nervous System* (Kalivas, P. W. and Barnes, C. D., eds.), Telford, Caldwell, NJ, pp. 227–254.
65. Antelman, S. M., Kocan, D., Knopf, S., Edwards, D. J., and Caggiula, A. R. (1992) One brief exposure to a psychological stressor induces long-lasting, time-dependent sensitization of both the cataleptic and neurochemical responses to haloperidol. *Life Sci.* **51,** 261–266.
66. Robinson, T. E., Angus, A. L., and Becker J. B. (1985) Sensitization to stress: the enduring effects of prior stress on amphetamine-induced rotational behavior. *Life Sci.* **37,** 1039–1042.
67. Robinson, T. E., Becker, J. B., Young, E. A., Akil, H., and Castaneda, E. (1987) The effects of footshock stress on regional brain dopamine metabolism and pituitary beta-endorphin release in rats previously sensitized to amphetamine. *Neuropharmacology* **26,** 679–691.
68. Kalivas, P. W. and Stewart, J. (1991) Dopamine transmission in the initiation and expression of drug- and stress-induced sensitization of motor activity. *Brain Res. Rev.* **16,** 223–244.
69. Brady, K. T. and Sonne, S. C. (1995) The relationship between substance abuse and bipolar disorder. *J. Clin. Psychiatry* **56**(Suppl 3), 19–24.
70. Regier, D. A., Farmer, M. E., Rae, D. S., Locke, B. Z., Keith, S. J., Judd, L. L., and Goodwin, F. K. (1990) Comorbidity of mental disorders with alcohol and other drug abuse. Results from the Epidemiologic Catchment Area (ECA) Study. *JAMA* **264,** 2511–2518.
71. Post, R. M., Weiss, S. R., Fontana, D., and Pert, A. (1992) Conditioned sensitization to the psychomotor stimulant cocaine. *Ann. NY Acad. Sci.* **654,** 386–399.
72. Post, R. M., Weiss, S. R. B., and Pert, A. (1987) The role of context in conditioning and behavioral sensitization to cocaine. *Psychopharmacol. Bull.* **23,** 425–429.
73. Weiss, S. R. B., Post, R. M., Pert, A., Woodward, R., and Murman, D. (1989) Context-dependent cocaine sensitization: differential effect of haloperidol on development versus expression. *Pharmacol. Biochem. Behav.* **34,** 655–661.

74. Pierce, R. C., Reeder, D. C., Hicks, J., Morgan, Z. R., and Kalivas, P. W. (1998) Ibotenic acid lesions of the dorsal prefrontal cortex disrupt the expression of behavioral sensitization to cocaine. *Neuroscience* **82,** 1103–1114.

75. Squillace, K. M., Post, R. M., and Pert A. (1982) Effect of lidocaine pretreatment on cocaine-induced behavior in normal and amygdala-lesioned rats. *Neuropsychobiology* **8,** 113–122.

76. Mishkin, M. and Appenzeller, T. (1987) The anatomy of memory. *Sci. Am.* **256,** 80–89.

77. Squire, L. R. and Zola-Morgan, S. (1991) The medial temporal lobe memory system. *Science* **253,** 1380–1386.

78. Kalivas, P. W. and Duffy, P. (1990) Effect of acute and daily cocaine treatment on extracellular dopamine in the nucleus accumbens. *Synapse* **5,** 48–58.

79. Kalivas, P. W., Duffy, P., DuMars, L. A., and Skinner, C. (1988) Behavioral and neurochemical effects of acute and daily cocaine administration in rats. *J. Pharmacol. Exp. Ther.* **245,** 485–492.

80. Henry, D. J. and White, F. J. (1991) Repeated cocaine administration causes persistent enhancement of D1 dopamine receptor sensitivity within the rat nucleus accumbens. *J. Pharmacol. Exp. Ther.* **258,** 882–890.

81. Henry, D. J. and White, F. J. (1992) Electrophysiological correlates of psychomotor stimulant-induced sensitization. *Ann. NY Acad. Sci.* **654,** 88–100.

82. Henry, D. J. and White, F. J. (1995) The persistence of behavioral sensitization to cocaine parallels enhanced inhibition of nucleus accumbens neurons. *J. Neurosci.* **15,** 6287–6299.

83. Post, R. M., Weiss, S. R. B., Ketter, T. A., Denicoff, K. D., George, M. S., Frye, M. A., Smith, M. A., and Leverich, G. S. (1997) The kindling model: implications for the etiology and treatment of mood disorders. *Curr. Rev. Mood Anxiety Disord.* **1,** 113–126.

84. Clark, M., Post, R. M., Weiss, S. R., and Nakajima, T. (1992) Expression of c-fos mRNA in acute and kindled cocaine seizures in rats. *Brain Res.* **582,** 101–106.

85. Rosen, J. B., Kim, S. Y., and Post, R. M. (1994) Differential regional and time course increases in thyrotropin-releasing hormone, neuropeptide Y and enkephalin mRNAs following an amygdala kindled seizure. *Brain Res. Mol. Brain Res.* **27,**71–80.

86. Carey, R. J., Dai, H., Krost, M., and Huston, J. P. (1995) The NMDA receptor and cocaine: evidence that MK-801 can induce behavioral sensitization effects. *Pharmacol. Biochem. Behav.* **51,** 901–908.

87. Schenk, S., Valadez, A., McNamara, C., House, D. T., Higley, D., Bankson, M. G., Gibbs, S., and Horger, B. A. (1993) Development and expression of sensitization to cocaine's reinforcing properties: role of NMDA receptors. *Psychopharmacology (Berlin)* **111,** 332–338.

88. Gilbert, M. E. (1988) The NMDA-receptor antagonist, MK-801, suppresses limbic kindling and kindled seizures. *Brain Res.* **463,** 90–99.

89. Hughes, P., Singleton, K., and Dragunow, M. (1994) MK-801 does not attenuate immediate-early gene expression following an amygdala afterdischarge. *Exp. Neurol.* **128,** 276–283.

90. Sato, K., Morimoto, K., and Okamoto, M. (1988) Anticonvulsant action of a noncompetitive antagonist of NMDA receptors (MK-801) in the kindling model of epilepsy. *Brain Res.* **463,** 12–20.

89. Post, R. M., Weiss, S. R. B., and Pert A. (1992) Sensitization and kindling effects of chronic cocaine administration, in *Cocaine: Pharmacology, Physiology, and Clini-*

cal Strategies (Lakoski, J. M., Galloway, M. P., and White, F. J., eds.), Telford, Caldwell, NJ, pp 115–161.

92. Racine, R. J. (1972) Modification of seizure activity by electrical stimulation. I. After-discharge threshold. *Electroencephalogr. Clin. Neurophysiol.* **32,** 269–279.

93. Wong, M.-L., Smith, M. A., Licinio, J., Doi, S. Q., Weiss, S. R. B., Post, R. M., and Gold, P. W. (1993) Differential effects of kindled and electrically induced seizures on a glutamate receptor (GluR1) gene expression. *Epilepsy Res.* **14,** 221–227.

94. Maren, S., Tocco, G., Standley, S., Baudry, M., and Thompson, R. F. (1993) Postsynaptic factors in the expression of long-term potentiation (LTP): increased glutamate receptor binding following LTP induction in vivo. *Proc. Natl. Acad. Sci. USA* **90,** 9654–9658.

95. Muller, D., Joly, M., and Lynch, G. (1988) Contributions of quisqualate and NMDA receptors to the induction and expression of LTP. *Science* **242,** 1694–1697.

96. Moldow, R. L. and Fischman, A. J. (1987) Cocaine induced secretion of ACTH, beta-endorphin, and corticosterone. *Peptides* **8,** 819–822.

97. Rivier, C. and Vale, W. (1987) Cocaine stimulates adrenocorticotropin (ACTH) secretion through a corticotropin-releasing factor (CRF)-mediated mechanism. *Brain Res.* **422,** 403–406.

98. Zhou, Y., Spangler, R., LaForge, K. S., Maggos, C. E., Ho, A., and Kreek, M. J. (1996) Corticotropin-releasing factor and type 1 corticotropin-releasing factor receptor messenger RNAs in rat brain and pituitary during "binge"-pattern cocaine administration and chronic withdrawal. *J. Pharmacol. Exp. Ther.* **279,** 351–358.

99. Coplan, J. D., Andrews, M. W., Rosenblum, L. A., Owens, M. J., Friedman, S., Gorman, J. M., and Nemeroff, C. B. (1996) Persistent elevations of cerebrospinal fluid concentrations of corticotropin-releasing factor in adult nonhuman primates exposed to early-life stressors: implications for the pathophysiology of mood and anxiety disorders. *Proc. Natl. Acad. Sci. USA* **93,** 1619–1623.

100. Cratty, M. S., Ward, H. E., Johnson, E. A., Azzaro, A. J., and Birkle, D. L. (1995) Prenatal stress increases corticotropin-releasing factor (CRF) content and release in rat amygdala minces. *Brain Res.* **675,** 297–302.

101. Erb, S., Shaham, Y., and Stewart, J. (1998) The role of corticotropin-releasing factor and corticosterone in stress- and cocaine-induced relapse to cocaine seeking in rats. *J. Neurosci.* **18,** 5529–5536.

102. DeVries, A. C., Taymans, S. E., Sundstrom, J. M., and Pert, A. (1998) Conditioned release of corticosterone by contextual stimuli associated with cocaine is mediated by corticotropin-releasing factor. *Brain Res.* **786,** 39–46.

103. Prasad, B. M., Ulibarri, C., Kalivas, P. W., and Sorg, B. A. (1996) Effect of adrenalectomy on the initiation and expression of cocaine-induced sensitization. *Psychopharmacology (Berlin)* **125,** 265–273.

104. Rivet, J. M., Stinus, L., LeMoal, M., and Mormede, P. (1989) Behavioral sensitization to amphetamine is dependent on corticosteroid receptor activation. *Brain Res.* **498,** 149–153.

105. Cole, B. J., Cador, M., Stinus, L., Rivier, J., Vale, W., Koob, G. F., and Le Moal, M. (1990) Central administration of a CRF antagonist blocks the development of stress-induced behavioral sensitization. *Brain Res.* **512,** 343–346.

106. Brown, E. E., Robertson, G. S., and Fibiger, H. C. (1992) Evidence for conditional neuronal activation following exposure to a cocaine-paired environment: role of forebrain limbic structures. *J. Neurosci.* **12,** 4112–4121.

107. Clark, M., Weiss, S. R., and Post, R. M. (1991) Expression of c-fos mRNA in rat brain after intracerebroventricular administration of corticotropin-releasing hormone. *Neurosci. Lett.* **132,** 235–238.
108. Childress, A. R., McLellan, A. T., Ehrman, R., and O'Brien, C. P. (1988) Classically conditioned responses in opioid and cocaine dependence: a role in relapse? *NIDA Res. Monogr.* **84,** 25–43.
109. Childress, A. R., McLellan, A. T., and O'Brien, C. P. (1986) The role of conditioning factors in the development of drug dependence. *Psychiatr. Clin. North. Am.* **9,** 413–426.
110. London, E. D., Stapleton, J. M., Phillips, R. L., Grant, S. J., Villemagne, V. L., Liu, X., and Soria, R. (1996) PET studies of cerebral glucose metabolism: acute effects of cocaine and long-term deficits in brains of drug abusers. *NIDA Res. Monogr.* **163,** 146–158.
111. Hitchcock, J. and Davis, M. (1986) Lesions of the amygdala, but not of the cerebellum or red nucleus, block conditioned fear as measured with the potentiated startle paradigm. *Behav. Neurosci.* **100,** 11–22.
112. Hitchcock, J. and Davis, M. (1987) Fear-potentiated startle using an auditory conditioned stimulus: effect of lesions of the amygdala. *Physiol. Behav.* **39,** 403–408.
113. Rosen, J. B., Hitchcock, J. M., Sananes, C. B., Miserendino, M. J., and Davis, M. (1991) A direct projection from the central nucleus of the amygdala to the acoustic startle pathway: anterograde and retrograde tracing studies. *Behav. Neurosci.* **105,** 817–825.
114. George, M. S., Huggins, T., McDermut, W., Parekh, P. I., Rubinow, D., and Post, R. M. (1998) Abnormal facial emotion recognition in depression: serial testing in an ultra-rapid-cycling patient. *Behav. Modif.* **22,** 192–204.
115. Rubinow, D. R. and Post, R. M. (1992) Impaired recognition of affect in facial expression in depressed patients. *Biol. Psychiatry* **31,** 947–953.
116. Drevets, W. C., Price, J. L., Simpson, J .R. J., Todd, R. D., Reich, T., Vannier, M., Raichle, M. E. (1997) Subgenual prefrontal cortex abnormalities in mood disorders. *Nature* **386,** 824–827.
117. Ketter, T. A., Andreason, P. J., George, M. S., Pazzaglia, P. J., Marangell, L. B., and Post, R. M. (1993) Blunted CBF response to procaine in mood disorders. *Abstracts of the 146th Annual Meeting of the American Psychiatric Association,* Abstract No. NR297.
118. LeDoux, J. E. (1994) Emotion, memory and the brain. *Sci. Am.* **270,** 50–57.
119. McCann, U. D., Kimbrell, T. A., Morgan, C. M., Anderson, T., Geraci, M., Benson, B. E., Wassermann, E. M., Willis, M. W., and Post, R. M. (1998) Repetitive transcranial magnetic stimulation for posttraumatic stress disorder. *Arch. Gen. Psychiatry* **55,** 276–279.
120. Rauch, S. L., Van der Kolk, B. A., Fisler, R. E., Alpert, N. M., Orr, S. P., Savage, C. R., Fischman, A. J., Jenike, M. A., and Pitman, R. K. (1996) A symptom provocation study of posttraumatic stress disorder using positron emission tomography and script-driven imagery. *Arch. Gen. Psychiatry* **53,** 380–387.
121. Semple, W. E., Goyer, P., McCormick, R., Morris, E., Compton, B., Muswick, G., Nelson, D., Donovan, B., Leisure, G., and Berridge, M. (1993) Preliminary report: brain blood flow using PET in patients with posttraumatic stress disorder and substance-abuse histories. *Biol. Psychiatry* **34,** 115–118.

122. Shin, L. M., Kosslyn, S. M., McNally, R. J., Alpert, N. M., Thompson, W. L., Rauch, S. L., Macklin, M. L., and Pitman, R. K. (1997) Visual imagery and perception in posttraumatic stress disorder. A positron emission tomographic investigation. *Arch. Gen. Psychiatry* **54**, 233–241.

123. Malenka, R. C. (1994) Synaptic plasticity in the hippocampus: LTP and LTD. *Cell* **78**, 535–538.

124. Oliet, S. H., Malenka, R. C., and Nicoll, R. A. (1997) Two distinct forms of long-term depression coexist in CA1 hippocampal pyramidal cells. *Neuron* **18**, 969–982.

125. Muller, D., Hefft, S., and Figurov, A. (1995) Heterosynaptic interactions between LTP and LTD in CA1 hippocampal slices. *Neuron* **14**, 599–605.

126. Li, H., Weiss, S. R., Chuang, D. M., Post, R. M., and Rogawski, M. A. (1998) Bidirectional synaptic plasticity in the rat basolateral amygdala: characterization of an activity-dependent switch sensitive to the presynaptic metabotropic glutamate receptor antagonist 2S-alpha- ethylglutamic acid. *J. Neurosci.* **18**, 1662–1670.

127. Goddard, G. V., McIntyre, D. C., and Leech, C. K. (1969) A permanent change in brain function resulting from daily electrical stimulation. *Exp. Neurol.* **25**, 295–330.

128. Goddard, G. V. and Douglas, R. M. (1975) Does the engram of kindling model the engram of normal long term memory? *Can. J. Neurol. Sci.* **2**, 385–394.

129. Pinel, J. P. and Rovner, L. I. (1978) Experimental epileptogenesis: kindling-induced epilepsy in rats. *Exp. Neurol.* **58**, 190–202.

130. Pinel, J. P., Skelton, R., and Mucha, R. F. (1976) Kindling-related changes in afterdischarge "thresholds." *Epilepsia* **17**, 197–206.

131. Wada, J. A. and Sata, M. (1974) Generalized convulsive seizures induced by daily electrical stimulation of the amygdala in cats. Correlative electrographic and behavioral features. *Neurology* **24**, 565–574.

132. Wada, J. A., Sato, M., and Corcoran, M. E. (1974) Persistent seizure susceptibility and recurrent spontaneous seizures in kindled cats. *Epilepsia* **15**, 465–478.

133. Clark, M., Smith, M., Weiss, S. R. B., and Post, R. M. (1994) Modulation of hippocampal glucocorticoid and mineralocorticoid receptor mRNA expression by amygdaloid kindling. *Neuroendocrinology* **59**(5), 451–456.

134. Smith, M. A., Weiss, S. R., Berry, R. L., Zhang, L. X., Clark, M., Massenburg, G., and Post, R. M. (1997) Amygdala-kindled seizures increase the expression of corticotropin-releasing factor (CRF) and CRF-binding protein in GABAergic interneurons of the dentate hilus. *Brain Res.* **745**, 248–256.

135. Smith, M. A., Weiss, S. R. B., Abedin, T., Kim, H., Post, R. M., and Gold, P. W. (1991) Effects of amygdala kindling and electroconvulsive seizures on the expression of corticotropin-releasing hormone in the rat brain. *Mol. Cell Neurosci.* **2**, 103–116.

136. Weiss, S. R., Clark, M., Rosen, J. B., Smith, M. A., and Post, R. M. (1995) Contingent tolerance to the anticonvulsant effects of carbamazepine: relationship to loss of endogenous adaptive mechanisms. *Brain Res. Brain Res. Rev.* **20**, 305–325.

137. Zhang, L. X., Smith, M. A., Li, X. L., Weiss, S. R. B., and Post, R. M. (1998) Apoptosis of hippocampal neurons after amygdala kindled seizures. *Brain Res. Mol. Brain Res.* **55**, 198–208.

138. Post, R. M. and Weiss, S. R. (1992) Ziskind-Somerfeld Research Award 1992. Endogenous biochemical abnormalities in affective illness: therapeutic versus pathogenic. *Biol. Psychiatry* **32**, 469–484.

139. Post, R. M. and Weiss, S. R. B. (1996) A speculative model of affective illness cyclicity based on patterns of drug tolerance observed in amygdala-kindled seizures. *Mol. Neurobiol.* **13,** 33–60.

140. Dragunow, M., Goddard, G. V., and Laverty, R. (1985) Is adenosine an endogenous anticonvulsant? *Epilepsia* **26,** 480–487.

141. Kubek, M. J., Liang, D., Byrd, K. E., and Domb, A. J. (1996) Prolonged seizure suppression by a single implantable polymeric TRH microdisk preparation. *Brain Res.* **809,** 189–197.

142. Wan, R.-Q., Noguera, E. C., and Weiss, S. R. B. (1998) Anticonvulsant effects of intrahippocampal injection of TRH in amygdala kindled rats. *Neuroreport* **9,** 677–682.

143. Zhang, L. X., Li, X. L., Smith, M. A., Post, R. M., and Han, J. S. (1997) Lipofectin-facilitated transfer of cholecystokinin gene corrects behavioral abnormalities of rats with audiogenic seizures. *Neuroscience* **77,** 15–22.

144. Geinisman, Y., Morrell, F., and de Toledo-Morrell, L. (1990) Increase in the relative proportion of perforated axospinous synapses following hippocampal kindling is specific for the synaptic field of stimulated axons. *Brain Res.* **507,** 325–331.

145. Cavazos, J. E. and Sutula, T. P. (1990) Progressive neuronal loss induced by kindling: a possible mechanism for mossy fiber synaptic reorganization and hippocampal sclerosis. *Brain Res.* **527,** 1–6.

146. Sutula, T. P., Cavazos, J. E., and Woodard, A. R. (1994) Long-term structural and functional alterations induced in the hippocampus by kindling: implications for memory dysfunction and the development of epilepsy. *Hippocampus* **4,** 254–258.

147. McEwen, B. S. (1994) Corticosteroids and hippocampal plasticity. *Ann. NY Acad. Sci.* **746,** 134–142.

148. Bottjer, S. W. and Arnold, A. P. (1997) Developmental plasticity in neural circuits for a learned behavior. *Annu. Rev. Neurosci.* **20,** 459–481.

149. Bottjer, S. W. and Johnson, F. (1992) Matters of life and death in the songbird forebrain. *J. Neurobiol.* **23,** 1172–1191.

150. Nottebohm, F. (1989) From bird song to neurogenesis. *Sci. Am.* **260,** 74–79.

151. Weiss, S. R., Li, X. L., Rosen, J. B., Li, H., Heynen, T., and Post, R. M. (1995) Quenching: inhibition of development and expression of amygdala kindled seizures with low frequency stimulation. *NeuroReport* **6,** 2171–2176.

152. Weiss, S. R. B., Li, X.-L., Heynen, T., and Post, R. M. (1997) Kindling and quenching: conceptual links to rTMS. *CNS Spectrums* **2,** 32–68.

153. Weiss, S. R. B., Eidsath, A., Li, X. L., Heynen, T., and Post, R. M. (1998) Quenching revisited: low level direct current inhibits amygdala-kindled seizures. *Exp. Neurol.* **154,** 185–192.

154. George, M. S., Wassermann, E. M., Williams, W. A., Callahan, A., Ketter, T. A., Basser, P., Hallett, M., and Post, R. M. (1995) Daily repetitive transcranial magnetic stimulation (rTMS) improves mood in depression. *NeuroReport* **6,** 1853–1856.

155. George, M. S., Wassermann, E. M., Kimbrell, T. A., Little, J. T., Williams, W. E., Danielson, A. L., Greenberg, B. D., Hallett, M., and Post, R. M. (1997) Mood improvement following daily left prefrontal repetitive transcranial magnetic stimulation in patients with depression: a placebo-controlled crossover trial. *Am. J. Psychiatry* **154,** 1752–1756.

156. Pascual-Leone, A., Rubio, B., Pallardo, F., and Catala, M. D. (1996) Rapid-rate transcranial magnetic stimulation of left dorsolateral prefrontal cortex in drug-resistant depression. *Lancet* **348,** 233–237.

157. Post, R. M., Kimbrell, T., Frye, M., George, M., McCann, U., Little, J., Dunn, R., Li, H., and Weiss, S. R. B. (1997) Implications of kindling and quenching for the possible frequency dependence of rTMS. *CNS Spectrums* **2,** 54–60.

158. Kimbrell, T. A., Little, J. T., Dunn, R. T., Frye, M. A., Greenberg, B. D., Wassermann, E. M., Repella, J. D., Danielson, A. L., Willis, M. W., Benson, B. E., Speer, A., Osuch, E., and Post, R. M. (1998) Frequency dependence of antidepressant response to left prefrontal repetitive transcranial magnetic stimulation (rTMS) as a function of baseline cerebral glucose metabolism. *Biol. Psychiatry,* in press.

159. Speer, A. M., Kimbrell, T. A., Dunn, R. T., Osuch, E. A., Frye, M. A., Willis, M. W., and Wassermann E. M. (1998) Differential changes in rCBF with one versus 20 Hz rTMS in depressed patients. *APA New Res. Program Abstr.* 82 (abstract).

160. Post, R. M., Kimbrell, T., McCann, U., Dunn, R., Osuch, E., Speer, A., and Weiss, S. R. B. (1998) Repetitve transcranial magnetic stimulation as a neuropsychiatric tool: present status and future potential. *J. ECT* **15,** 39–59.

161. Grunhaus, L., Dannon, P., and Schreiber, S. (1998) Effects of transcranial magnetic stimulation on severe depression. Similarities with ECT. *Biol. Psychiatry* **43,** 76S (abstract).

162. Markwort, S., Cordes, P., and Aldenhoff, J. (1997) Die transkranielle Magnetstimulation als Behandlungs-alternative zur Elektrokrampf-therapie bei therapieresistenten Depressionen—Eine Literaturubersicht. (Transcranial magnetic stimulation as an alternative to electroshock therapy in treatment resistant depressions. A literature review.) *Fortschr. Neurol. Psychiatr.* **65,** 540–549.

163. Fink, M. (1979) Convulsive Therapy: Theory and Practice. Raven, New York.

164. Fink, M. (1990) How does convulsive therapy work? *Neuropsychopharmacology* **3,** 73–82.

165. Lykouras, L., Markianos, M., Hatzimanolis, J., and Stefanis, C. (1991) Effects of ECT course on TSH and prolactin responses to TRH in depressed patients. *J. Affect. Disord.* **23,** 191–197.

166. Nerozzi, D., Graziosi, S., Melia, E., Aceti, F., Magnani, A., Fiume, S., Fraioli, F., and Frajese, G. (1987) Mechanism of action of ECT in major depressive disorders: a neuroendocrine interpretation. *Psychiatry Res.* **20,** 207–213.

167. Kubek, M. J., Meyerhoff, J. L., Hill, T. G., Norton, J. A., and Sattin, A. (1985) Effects of subconvulsive and repeated electroconvulsive shock on thyrotropin-releasing hormone in rat brain. *Life Sci.* **36,** 315–320.

168. Sattin, A., Hill, T. G., Meyerhoff, J. L., Norton, J. A., and Kubek, M. J. (1987) The prolonged increase in thyrotropin-releasing hormone in rat limbic forebrain regions following electroconvulsive shock. *Regul. Peptides* **19,** 13–22.

169. Post, R. M. (1990) ECT: the anticonvulsant connection. Comments on "How does convulsive therapy work?" *Neuropsychopharmacology* **3,** 89–92.

170. Hay, D. P., Hay, L., and Spiro, H. (1989) The enigma of the stigma of ECT: 50 years of myth and misrepresentation. *Wis. Med. J.* **88,** 4–10.

171. Pollina, D. A. and Calev, A. (1997) Amnesia associated with electroconvulsive therapy: progress in pharmacological prevention and treatment. *CNS Drugs* **7,** 381–387.

172. Post, R. M., Kimbrell, T. A., McCann, U., Dunn, R. T., George, M. S., and Weiss, S. R. (1997) Are convulsions necessary for the antidepressive effect of electroconvulsive therapy: outcome of repeated transcranial magnetic stimulation? *Encephale* **23**(Spec No. 3), 27–35.

173. Post, R. M. and Speer, A. M. (1999) Speculations on the future of rTMS and related therapeutic modalities, in *Transcranial Magnetic Stimulation in Neuropsychiatry* (George, M. S. and Belmaker, R. H., eds.), American Psychiatric Press, Washington, DC, in press.

174. Ackenheil, M. (1990) The mechanism of action of antidepressants revised. *J. Neural. Transm.* **32**(Suppl), 29–37.

175. Post, R. M., Uhde, T. W., Rubinow, D. R., and Huggins, T. (1987) Differential time course of antidepressant effects following sleep deprivation, ECT, and carbamazepine: clinical and theoretical implications. *Psychiatry Res.* **22**, 11–19.

176. Sulser, F. (1984) Antidepressant treatments and regulation of norepinephrine-receptor-coupled adenylate cyclase systems in brain. *Adv. Biochem. Psychopharmacol.* **39**, 249–261.

177. Sanders-Bush, E., Conn, J., and Sulser, F. (1985) The serotonin/norepinephrine-linked beta-adrenoceptor system and the mode of action of antidepressants. *Psychopharmacol. Bull.* **21**, 373–378.

178. Sulser F. (1987) Serotonin–norepinephrine receptor interactions in the brain: implications for the pharmacology and pathophysiology of affective disorders. *J. Clin. Psychiatry* **48**(Suppl), 12–18.

179. Duman, R. S. (1998) Novel therapeutic approaches beyond the serotonin receptor. *Biol. Psychiatry* **44**, 324–335.

180. Duman R. S., Heninger, G. R., and Nestler, E. J. (1997) A molecular and cellular theory of depression. *Arch. Gen. Psychiatry* **54**, 597–606.

181. Smith, M. A., Makino, S., Altemus, M., Michelson, D., Hong, S. K., Kvetnansky, R., and Post, R. M. (1995) Stress and antidepressants differentially regulate neurotrophin 3 mRNA expression in the locus coeruleus. *Proc. Natl. Acad. Sci. USA* **92**, 8788–8792.

182. Smith, M. A., Makino, S., Kvetnansky, R., and Post, R. M. (1995) Stress and glucocorticoids affect the expression of brain-derived neurotropic factor and neurotrophin–3 mRNAs in the hippocampus. *J. Neurosci.* **15**, 1768–1777.

183. Gold, P. W., Licinio, J., Wong, M. L., and Chrousos, G. P. (1995) Corticotropin releasing hormone in the pathophysiology of melancholic and atypical depression and in the mechanism of action of antidepressant drugs. *Ann. NY Acad. Sci.* **771**, 716–729.

184. Holsboer, F. and Barden, N. (1996) Antidepressants and hypothalamic–pituitary–adrenocortical regulation. *Endocr. Rev.* **17**, 187–205.

185. Zetterstrom, T. S., Pei, Q., and Grahame-Smith, D. G. (1998) Repeated electroconvulsive shock extends the duration of enhanced gene expression for BDNF in rat brain compared with a single administration. *Brain Res. Mol. Brain Res.* **57**, 106–110.

186. Korte, M., Staiger, V., Griesbeck, O., Thoenen, H., and Bonhoeffer, T. (1996) The involvement of brain-derived neurotrophic factor in hippocampal long-term potentiation revealed by gene targeting experiments. *J. Physiol. (Paris)* **90**, 157–164.

11
Signal Transduction Abnormalities in Bipolar Affective Disorder

Peter P. Li, Stavroula Andreopoulos, and Jerry J. Warsh

INTRODUCTION

Bipolar affective disorder (BD), or manic–depressive illness, is a chronic, severe, and debilitating mental disorder characterized by episodes of mania or hypomania and depression, which afflicts an estimated 1% of the general population (1). Evidence from family, genetic, and psychobiological studies indicates a major role for biological factors in the genesis, expression, and perpetuation of BD (1). Recent advances arising from research on transmembrane signal transduction processes and intracellular second messengers have reshaped current thinking about the pathophysiological basis of BD. A synthesis is emerging from postmortem brain, clinical and preclinical studies that recognizes disturbances in brain intracellular signaling mechanisms may play an important role in the development of this disorder (2). In this chapter, we critically review and appraise the body of current observations implicating disturbances in an array of intracellular signaling disturbances, including adenosine 3', 5'-cyclic monophosphate (cAMP) and calcium and phosphoinositide (PPI) signaling cascades in BD as depicted in Fig. 1. We also describe results from complementary investigations on the pharmacological mechanisms of mood-stabilizing agents, which, in most cases, appear to target specific "nodes" in postreceptor signal transduction cascades normalizing the postulated intracellular signaling abnormalities that disrupt neuronal function in BD.

cAMP SIGNALING DISTURBANCES IN BIPOLAR DISORDER

Early interest in cAMP signaling in BD stemmed from findings indicating that this second messenger mediated the activity of several monoamine neurotransmitters (e.g., norepinephrine, serotonin and dopamine, etc.) that

From: *Cerebral Signal Transduction: From First to Fourth Messengers*
Edited by: M. E. A. Reith © Humana Press Inc., Totowa, NJ

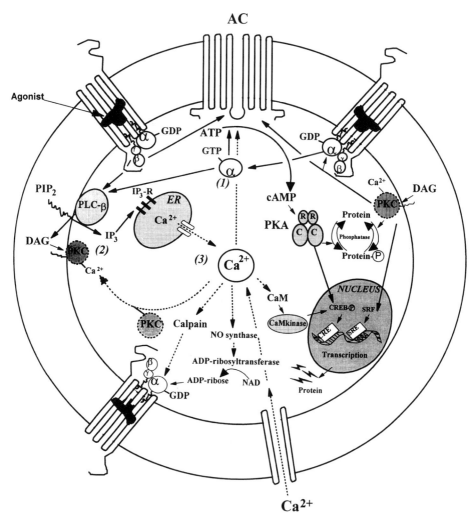

Fig. 1. Schematic representation of the major signaling pathways implicated in the pathophysiology of bipolar affective disorder: (1) cAMP signaling. (2) PPI signaling. (3) Ca^{2+} signaling. Abbreviations: AC, adenylyl cyclase; ATP, adenosine triphosphate; CaM, calmodulin; cAMP, cyclic adenosine monophosphate; CRE, cAMP response element; CREB, cAMP response element-binding protein; DAG, diacylglycerol; GDP, guanosine diphosphate: GTP, guanosine triphosphate; IP_3, inositol-1,4,5-triphosphate; IP_3-R, IP_3 receptor; NO, nitric oxide; PIP_2, phosphatidylinositol-4,5-bisphosphate; PKA, protein kinase A (with C, catalytic subunit and R, regulatory subunit) ; PKC, protein kinase C; PLC, phospholipase C; SRE, serum response element; SRF, serum response factor.

were already implicated in mood disorders *(3,4)*. Studies of plasma cAMP levels *(5)* and agonist-stimulated cAMP formation in mononuclear leukocytes (MNL) from unipolar and BD patients *(6,7)* were among the first lines of evidence suggesting disturbances in the cAMP signaling cascade in the pathophysiology of BD. Findings of blunted isoproterenol-stimulated cAMP response in MNL from BD patients *(6,7)* in the absence of consistent changes of the receptor densities or affinities *(8)* suggested that disturbances might occur at postreceptor sites in the cAMP signaling cascade in BD, including such protein components as guanine nucleotide binding proteins (G-proteins), adenylyl cyclase (AC), and cAMP-dependent protein kinase (PKA). Observations of increased agonist-stimulated binding of $[^{3}H]Gpp(NH)p$ (a stable nonhydrolyzable analog of guanosine triphosphate [GTP]) in MNL membranes from manic but not euthymic (normal mood) BD patients compared with controls *(9)* pointed to disturbances at the level of G-proteins as a potential site of dysregulated signaling in BD. These latter findings provided the first clinical observations directly implicating G-protein-mediated signal transduction abnormalities in BD and set an important focus for subsequent research activities that followed in this field.

Two complementary lines of human studies have figured prominently in the study of G-protein-mediated signaling systems to date. These include postmortem studies, in which the concentration and functionality of various G-protein subunits have been measured, and clinical studies, in which G-proteins and their function have been measured in peripheral cells such as platelets, leukocytes, and transformed B lymphoblasts from BD patients. Postmortem brain investigations provide the ability to measure the concentration and function of the signaling proteins directly in brain tissue. However, there are a number of limitations and pitfalls to this approach. Prominent among these are the effects of factors such as the following: the instability of proteins and mRNA during postmortem delay in cooling, removal, and freezing of tissue; antemortem agonal status, cause of death, deficiencies in documentation of clinical diagnostic features, outcome, and drug histories; and comorbid age-related medical conditions. Although some of these potential confounding factors can be minimized in case-controlled studies matched on these variables, it is difficult to address such issues as the relationship between any observed biochemical changes and the severity of illness using this research strategy. In this regard, parallel clinical studies using peripheral blood cells provide an alternative complementary approach to further address the clinical relevance of any postmortem findings. Epstein–Barr-transformed B-lymphoblasts derived from BD patients can also serve as a model for investigation of cellular and molecular basis of signal transduction abnormalities in this disorder. B-Lymphoblasts can be grown for a number of generations under controlled cell-culture conditions isolated from the potential effects of hormones and metabolic-related factors associated with the ill state, thus circumventing the confounding effects of medications and state-dependent factors. Thus, they provide another tool to address the question of trait-related versus state-related issues, which cannot be resolved in postmortem studies.

Postmortem Brain Studies

G-Protein Studies in BD

Despite the problems alluded to in the above, the ability to measure the level and function of proteins, such as G-proteins, in postmortem brain provides an important strategy for investigating directly intracellular signal transduction processes in neuropsychiatric disorders. Young et al. *(10,11)* were the first to use this approach to study possible signal transduction disturbances through the stimulatory G-protein (Gs)-coupled cAMP signaling pathway in the brain of patients with an antemortem diagnosis of BD. Compared with nonpsychiatric controls matched on age, postmortem delay, and brain pH, significantly higher levels of the long splice variant (52 kDa) of the Gs α-subunit, $G\alpha_{s-L}$, were found in frontal, temporal, and occipital cortex, but not in parietal cortex, hippocampus, caudate, and cerebellum. Moderately increased $G\alpha_{s-L}$ levels were observed in the thalamus. In comparison, the levels of the short spliced variant (45 kDa) of $G\alpha_s$ ($G\alpha_{s-S}$) were markedly increased in hippocampus and caudate, whereas its concentrations were significantly reduced in occipital cortex. On the one hand, these findings suggest that changes in $G\alpha_s$ levels may occur more diffusely throughout the brain in BD. However, the lack of significant differences in $G\alpha_s$ levels in parietal cortex *(12)* and cerebellum compared with controls appear at odds with this notion. Interestingly, in those brains regions showing increased $G\alpha_s$ levels, only one, not both, splice variants was affected. These observations suggested that differential regulation of the alternative splicing mechanism(s) for the $G\alpha_s$ gene in affected regions of the BD brain might account for the relative increases in a specific $G\alpha_s$ spliced variant. On the other hand, differences in the turnover of the $G\alpha_s$ isoforms, mRNA stability, or translation could also explain the differential changes in $G\alpha_s$ isoforms. Because subtle differences have been reported between the two splice variants of $G\alpha_s$ in the interaction with receptor *(13)* and adenylyl cyclase *(14)*, regulation of the differential expression of the two $G\alpha_s$ spliced variants may have important consequences for the intracellular cAMP signaling.

In contrast to the changes observed for $G\alpha_s$, no statistically significant differences were found in $G\alpha_{i-1}$, $G\alpha_{i-2}$, $G\alpha_o$, $G\alpha_z$, $G\beta_{36}$, or $G\beta_{35}$ levels in any of the brain regions studied between BD and control groups *(10,11,15,16)*. As the mammalian adenylyl cyclase activities are also modulated by α_i-, α_o-, α_z-, and $\beta\gamma$-subunits of G-proteins *(17)*, these findings implicate the selective dysregulation of the stimulatory branch of the G-protein-mediated cAMP signaling mechanisms in the pathophysiology of BD.

Given the evidence for a BD vulnerability gene near the pericentromeric region of chromosome 18, on which the gene encoding $G\alpha_{olf}$ is located *(18, 19)*, the possibility was raised that expression of $G\alpha_{olf}$, which is highly homologous to $G\alpha_s$ isoforms *(20)*, may be altered in BD. In our previous study *(10,11)*, $G\alpha_{olf}$ (46 kDa) was not resolved from $G\alpha_{s-S}$. Using sodium dodecyl sulfate–polyacrylamide gel electrophoresis (SDS-PAGE) conditions, which resolve $G\alpha_{olf}$ from $G\alpha_{s-S}$, and an antiserum directed against the C-terminal decapeptide common to these two α subunits, we re-examined their levels in several brain regions in which $G\alpha_s$ protein levels were previously shown to be elevated *(10–12)*. No differences were observed, however, between control and BD groups in the levels of $G\alpha_{olf}$ in frontal cortex, temporal cortex, caudate, and hippocampus. Only a moderate increase in $G\alpha_{olf}$ levels was found in BD occipital cortex *(12)*. These results, together with the lack of association between BD and any of the observed $G\alpha_{olf}$ alleles in a linkage disequilibrium analysis study *(21)*, argue against a role for $G\alpha_{olf}$ in the pathogenesis of BD.

The mechanism(s) responsible for the higher $G\alpha_s$ levels found in the BD brain is still unknown. It is now clear that expression of G-protein α-subunits is regulated by a number of different processes, such as changes in gene transcription, mRNA translation, and/or posttranslational events (e.g., changes in $G\alpha_s$ protein turnover) *(22,23)*. Genetic studies examining a dinucleotide repeat polymorphism in intron 3 or a biallelic polymorphism in exon 5 of the $G\alpha_s$ gene have not revealed linkage between the $G\alpha_s$ gene and BD *(24,25)*. Similarly, no mutations have been identified in either the promoter or coding sequences of the $G\alpha_s$ gene in BD patients *(25)*. These negative results argue against a mutation of the $G\alpha_s$ gene itself contributing to the genetic susceptibility to BD.

The lack of concomitant changes in $G\alpha_s$ mRNA levels in the same cerebral cortical regions manifesting elevated $G\alpha_s$ levels in BD compared with controls *(26)*, however, suggests that the changes in $G\alpha_s$ levels are more likely related to processes governing $G\alpha_s$ turnover. In this regard, several important posttranslational processes can potentially regulate the levels of $G\alpha_s$ including receptor activation and ADP-ribosylation (both of which appear to increase $G\alpha_s$ susceptibility to degradation) *(23)*, lipid acylation *(27)*, Ca^{2+}-dependent (e.g., calpain) *(28)* and ATP-dependent (e.g., ubiquitin) proteolysis *(29)* and, of course, crosstalk regulatory mechanisms *(30)*. Several recent studies have examined the effect of ADP-ribosylation on $G\alpha_s$ levels, independent of any second-messenger mechanisms. Exposure of GH_3 cells to cholera toxin induced a significant reduction (74–95%) in immunoreactive $G\alpha_s$ levels, suggesting that this covalent modification marks the protein for accelerated degradation *(31)*. Similarly, cholera toxin reduced membrane $G\alpha_{s-L}$ levels in L6 skeletal myoblasts *(32)*, once again suggesting that rapid degradation/turnover of $G\alpha_s$ occurs following ADP-ribosylation of this subunit. This suggests that posttranslational modification of $G\alpha_s$ by ADP-ribosylation might contribute to the cellular pathophysiology of BD. That chronic treatment of rats with lithium increases ADP-ribosylation of substrate proteins, including $G\alpha_s$ *(33)*, adds still further support to this notion.

Recently, we have confirmed that $G\alpha_s$ in autopsied human brain is also a substrate for both endogenous and cholera toxin-catalyzed ADP-ribosylation *(34)*, as in other tissues. Furthermore, we have reported preliminary evidence demonstrating reduced endogenous ADP-ribosylation of $G\alpha_s$ isoforms in BD temporal cortex compared with nonpsychiatric controls *(35)*. These observations suggest that a disturbance(s) in ADP-ribosyltransferase and/or cofactors that regulate its activity could potentially reduce the turnover of $G\alpha_s$, promoting elevations in $G\alpha_s$ levels in key cerebral cortical regions in BD. This interesting possibility, as well as the other posttranslational processes regulating $G\alpha_s$ turnover noted, clearly merit closer scrutiny in understanding the basis for the $G\alpha_s$ alterations reported in BD.

Functional Significance of Altered $G\alpha_s$ Protein in BD

An important question arising from the $G\alpha_s$ immunolabeling findings is that of the functional relevance of the higher levels with respect to cAMP signaling processes in BD brain. On the basis of a series of biochemical studies, we and others have been able to piece together several lines of evidence indicating that such elevations are likely to be attended by increased signaling through the $G\alpha_s$-mediated cAMP signaling cascade. A significant increase in forskolin-stimulated adenylyl cyclase activity was first observed in the temporal and occipital, but not in the frontal cortex of postmortem BD brain. The enhanced forskolin-stimulated cAMP response could reflect increased concentrations of specific adenylyl cyclase subtypes (types II and IV) measured under our assay conditions, in which Ca^{2+}/calmodulin was absent. However, direct measurement of the immunoreactive levels of various adenylyl cyclase subtypes (types I, IV, and V/VI) revealed no differences in BD cortical regions compared with controls *(36)*. On the other hand, a significant correlation found between forskolin-stimulated cAMP response and the levels of $G\alpha_{s-L}$ across these cortical regions *(11)* suggested a direct relationship between higher $G\alpha_s$ levels and forskolin response in BD cerebral cortex. Other evidence for functional disturbance associated with higher $G\alpha_s$ levels comes from the study of Friedman and Wang *(15)*. They showed that basal and isoproterenol-stimulated [^{35}S]GTPγS binding to $G\alpha_s$ were enhanced in BD frontal cortex mediated, in part, via elevated $G\alpha_s$ levels in this region. Their findings in immunoprecipitation studies also suggested that a significantly greater proportion of $G\alpha_s$ proteins may exist in the heterotrimeric (αβγ) state. These observations, taken together with the lack of significant changes in β-adrenoceptor densities in BD cortical regions compared with controls *(37)*, clearly support the notion that a functional alteration in $G\alpha_s$ proteins contributes to increased cAMP responses in BD brain.

The instability of cAMP postmortem precludes meaningful interpretation of direct measurements of this second messenger in the autopsied brain. Thus, to further assess the state of cAMP signaling in the BD brain, we reasoned that measurement of the levels and/or function of downstream targets, which are known to be regulated by cAMP, might reflect the state of cAMP signaling that existed in brain at the time of or just prior to death. We first measured the amount of regulatory subunits and catalytic activity of PKA, the primary target of intracellular cAMP signaling, in the postmortem BD brain. The expression of PKA regulatory subunits, estimated by the total number of [^{3}H]cAMP binding sites, was significantly reduced across all brain regions in cytosolic fractions of BD frontal, temporal, occipital and parietal cortex, cerebellum, and thalamus compared with matched controls *(38)*. In contrast, no differences were observed in [^{3}H]cAMP binding in the membrane fractions from these same regions. Given that sustained elevations in intracellular cAMP levels cause adaptive changes in the levels of PKA regulatory subunits *(39)*, the decrease in [^{3}H]cAMP binding in the cytosolic fractions from the BD brain likely reflects changes in response to upstream alterations in processes modulating cAMP levels. Of special interest in this regard, recent findings show that the reduction of regulatory subunits of PKA in the cytosolic fractions of BD temporal cortex is accompanied by a higher basal kinase activity and significantly lower apparent activation constant for cAMP in the cytosolic fractions of BD temporal cortex *(39a)*. Whether such changes disturb endogenous cAMP-stimulated phosphorylation, similar to that reported in platelets from euthymic BD patients *(40)* remains to be demonstrated. Nevertheless, these observed changes in PKA provide additional important evidence for dysregulation in the $G\alpha_s$-mediated cAMP cascade in BD.

Diagnostic Specificity of the Hyperfunctional $G\alpha_s$–cAMP Signaling in BD

The increases in the levels and function of $G\alpha_s$-protein identified in the postmortem BD brain appear to be specific to this disorder and not a general correlate of psychiatric or neurodegenerative disease. Our group first reported normal levels of $G\alpha_s$-protein in autopsied temporal cortex obtained from patients with schizophrenia, Alzheimer's disease *(11)*, or olivoponto-cerebellar atrophy *(41)*. Recent studies of G-protein levels and function in the postmortem brain obtained from subjects with diagnoses of depression *(42,43)*, schizophrenia *(44–49)*, substance abuse/dependence *(16,44,50,51)*, and Alzheimer's disease *(52–55)* also support the notion that increased $G\alpha_{s-L}$ levels and functionality are characteristic of BD.

In those studies which determined [^{3}H]cAMP binding or PKA activity in autopsied brain samples from patients with depression *(56)*, schizophrenia *(45)*, and Parkinson's *(46)* and Alzheimer's disease *(57)*, the profile of changes found in these PKA measures contrasts with those observed in the BD brain (i.e., reduced [^{3}H]cAMP binding in the cytosolic fraction accompanied by higher basal kinase activity and affinity of cAMP for PKA). These differences between BD and comparison neuropsychiatric disorders also support the diagnostic specificity of the hyperfunctional $G\alpha_s$–cAMP signaling pathway in BD.

Clinical Studies

G-Protein Studies in Peripheral Cell Models

The demonstration of increased agonist-stimulated [^{3}H]Gpp(NH)p in MNL membranes from manic but not euthymic BD patients *(9)*, together with a growing body of evidence showing coexpression of abnormal proteins in peripheral blood cells and brain tissues in several neurodegenerative disorders *(58–60)*, underscores the potential utility of peripheral blood cells to explore the clinical relevance of findings in the postmortem BD brain and the extent to which they reflect state-dependent versus trait-dependent changes. In line with postmortem brain findings, significantly higher levels of $G\alpha_{s-S}$, the major species in peripheral blood cells, were found in MNL from depressed *(61)* or manic *(62,63)* BD patients and in platelets of euthymic BD patients *(64)* compared with healthy subjects. Interestingly, these G-protein changes were observed irrespective of treatment *(62)*. In addition, the $G\alpha_{s-S}$ abnormalities appeared to be specific to BD, as no such differences were found in MNLs or platelets from patients with major depression *(61,65)*, panic disorder *(66)*, social phobia *(66)*, or abstinent alcoholics *(67)* compared with controls. Collectively, these observations extend on those obtained in the postmortem BD brain, confirming that the G-protein changes occur in the living affected individual and are expressed extracerebrally, at least in peripheral leukocytes and platelets. The results also suggest that the elevation in $G\alpha_{s-s}$ levels is, to some extent, trait dependent *(61,62,64)*. Indeed, the observation of higher $G\alpha_{s-s}$ immunolabeling in transformed B-lymphoblasts from bipolar II patients compared with healthy subjects *(68)* adds still stronger support to this notion.

Despite finding increased levels of $G\alpha_s$ in MNLs, in one study the functional response to Gs stimulation was blunted in bipolar depressed subjects. Significantly lower levels of GTPγS- and fluoride-stimulated adenylyl cyclase activities were found in depressed BD patients compared with matched controls *(12)*. In contrast, no significant differences were seen in either basal or forskolin-stimulated adenylyl cyclase activities between major depressive disorder patients and healthy subjects *(12)*. Because the levels of $G\alpha_i$ were also higher in MNL membranes from these subjects *(61)*, the blunted GTPγS- and fluoride-stimulated responses may reflect compensatory adaptations that occur in the inhibitory control of adenylyl cyclase activity in MNLs from BD patients. This may offset the cellular effects of elevated $G\alpha_s$ levels. In this regard, an imbalance in the coordinated regulation of stimulatory and inhibitory cAMP signaling systems may, therefore, be important to the pathophysiology of BD.

Recently, Avissar and co-workers *(63,69)* reported findings consistent with the notion that the G-protein changes observed in MNLs from bipolar and major depressive disorder patients are primarily state dependent. Agonist-stimulated [³H]GppNHp binding and the levels of $G\alpha_{s-s}$ and $G\alpha_i$ were found to be lower in MNLs from patients with either major depression or BD, with a major depressive episode, compared with healthy subjects. In addition, significant inverse correlations were found between Beck Depression Rating scores, $G\alpha$-protein levels, and the degree of agonist-stimulated [³H]GppNHp binding *(69)*. Whereas manic patients showed significantly higher levels of $G\alpha_{s-s}$ and $G\alpha_i$, bipolar depressed patients showed significant reduction in $G\alpha_{s-s}$ and $G\alpha_i$ levels compared with control subjects *(69)*.

It should be noted as well that some medications may also affect $G\alpha_s$-protein levels in blood cells, thus confounding the interpretation of any observed differences. For example, Mitchell et al. *(64)* found that platelet $G\alpha_{s-s}$ levels were significantly higher in BD patients receiving lithium only, compared with those taking carbamazepine only. Indeed, platelet $G\alpha_{s-s}$ levels in the latter group were indistinguishable from healthy comparison subjects, suggesting a confounding effect of treatment with some (i.e., carbamazepine) but not other (i.e., lithium) mood-stabilizing agents. We have also found that state of illness and medications affect MNL G-protein immunolabeling in

BD *(68)*. MNL Gα$_{s\text{-}S}$ immunolabeling levels were 27% lower in drug-free bipolar I patients compared with age- and sex-matched healthy subjects, whereas Gα$_{s\text{-}L}$ immunolabeling was lower in medicated compared with drug-free bipolar II patients, and higher in ill compared with remitted bipolar II patients. Gα$_i$ immunolabeling was 33% lower in drug-free remitted bipolar I patients compared with healthy subjects and showed a trend toward higher levels (35%) in drug-free ill BD patients compared with healthy subjects. Collectively, the findings of the foregoing studies clearly suggest that state dependence, medication, and other extraneous factors also affect MNL Gα$_s$ and Gα$_i$ levels in BD patients. On the one hand, it seems that the expression of underlying trait-dependent molecular abnormalities in MNLs from BD may be modified by state and medications. However, another possibility, yet to be completely excluded, is that the confounding effects of state and medications might result from changes in the subpopulations of leukocytes with quantitatively different levels of Gα$_s$ expression. This potential confound, although acknowledged by investigators *(63,69)*, has yet to be definitively addressed.

Very recently, significantly higher levels of Gα$_s$ mRNA were reported in neutrophils from BD patients compared with control subjects *(70)*, a difference that was evident irrespective of lithium treatment. Of interest, neutrophil Gα$_{i2}$ mRNA levels in unmedicated BD patients were indistinguishable from healthy controls, but they were significantly increased in BD patients treated with lithium *(70)*. Upregulation of Gα$_s$ mRNA in human neutrophils appears to be specific for BD, as no apparent differences in Gα$_s$ and Gα$_{i2}$ mRNA were found in neutrophils of unipolar depressed patients. Although these findings suggest that Gα$_s$ expression is increased in these cells in BD, the absence of confirmatory immunolabeling data from the same samples precludes such a conclusion because α-subunit mRNA levels can change without corresponding alterations in respective protein levels *(71)*.

Additional evidence indicating that abnormalities in the cAMP signaling cascade are expressed in peripheral bloods cells was obtained in recent studies showing higher cAMP-stimulated phosphorylation of the small G-protein Rap1, a Ras-like protein in the platelets from untreated euthymic BD patients *(40,72)*. Whether and how these disturbances are tied to alterations at upstream sites in this signal transduction chain needs to be systematically investigated. However, as Rap1-mediated signaling appears to modulate the mitogen-activated protein (MAP) kinase pathway *(73,74)*, the abnormality in cAMP-stimulated endogenous phosphorylation found in BD subjects may impact on other signaling cascades and processes through "crosstalk" mechanisms *(30,75,76)*. Indeed, evidence of alterations in PPI and Ca^{2+} signaling in BD, as described in the following, may be explained best on such an interactive basis. Regardless, the above findings highlight the multiplicity of postreceptor signaling changes that occur in BD involving processes affecting the levels and function of a number of components in the cAMP signaling pathway.

PPI Signaling Disturbances in Bipolar Disorder

It is well established that substantial crossregulation occurs between G-protein-coupled cAMP and PPI signaling pathways *(30,75)*. This, together with a large body of data showing that lithium regulates the PPI second-messenger system *(77)*, suggests alterations in the activity of this signal transduction system also may play a role in the pathophysiology of BD. Furthermore, the capacity for bidirectional crossregulation between cAMP and PPI signaling systems *(75)* provides a potential means for disturbances in either of these signaling systems to affect each other.

Evidence supporting disturbances in PPI signaling in BD has also been obtained using the postmortem brain and peripheral blood cells. Furthermore, important findings have been obtained using both directed immunolabeling and functional assays, similar to the approaches used in studies of cAMP signaling in BD. These are considered in the following subheadings.

Postmortem Brain Studies

Significantly higher levels of $G\alpha_{q/11}$ and a moderate elevation in the $G\alpha_{q/11}$–regulated phospholipase C-β_1 isozyme (PLC-β_1) have been reported in the occipital cortex of BD subjects compared with matched controls *(78)*. The increases in $G\alpha_{q/11}$ and PLC-β_1 appeared to be regionally specific, as no significant differences were found between BD and control subjects in the cerebral frontal and temporal cortex *(15,78)*. In addition, $G\alpha_{q/11}$ levels, expressed as a percent of respective postmortem delay and age-matched controls, correlated significantly with PLC-β_1 values in occipital cortex but not in the other regions examined *(78)*.

Parallel studies of $G\alpha_{q/11}$ functional activity, as revealed by GTPγS-stimulated [^{3}H]phosphatidylinositol (PI) hydrolysis, indicated a selective decrease in this response in the BD occipital cortex but not other cortical regions *(79)*. Because NaF-stimulated [^{3}H]PI hydrolysis was not different in the BD occipital cortex, it has been suggested that the function of $G\alpha_{q/11}$ may be impaired at the level of GDP–GTP exchange *(79)*. Furthermore, the functional impairment of $G\alpha_{q/11}$ signaling activity may lead to compensatory changes (i.e., enhanced synthesis and/or stability of the mRNA encoding these α-subunits), resulting in the higher $G\alpha_{q/11}$ levels observed in BD occipital cortex *(78)*.

The reciprocal changes in the levels and activity of $G\alpha_{q/11}$ in BD occipital cortex also appear to uniquely distinguish this disorder from Alzheimer's disease *(79a)*, schizophrenia *(44)*, alcoholism *(44)*, and major depression *(80)*, in which the activity of $G\alpha_{q/11}$ is either increased or decreased in the absence of changes in $G\alpha_{q/11}$ and PLC-β_1 levels. It is also noteworthy that the largest differences in $G\alpha_s$ levels and forskolin-stimulated adenylyl cyclase activity were observed in the occipital cortex of BD subjects *(10,11)*. Given the substantial evidence of crossregulation between cAMP and PPI signaling pathways, the possibility that the observed differences in $G\alpha_{q/11}$ and PLC-β_1 levels may reflect the consequences of relatively greater disturbances in cAMP signaling in this brain region in BD cannot be excluded.

Dysregulation of $G\alpha_{q/11}$ levels and activity in BD patients may have important downstream effects in the PPI signaling cascade. This notion is supported by preliminary findings of higher basal membrane protein kinase C (PKC) activity and greater stimulus-induced redistribution of the enzymes in autopsied BD frontal cortex compared with matched controls *(81)*. Furthermore, the levels of cytosolic PKC-α and membrane-associated PKC-γ and PKC-ζ isozymes were elevated, whereas cytosolic PKC-ε was reduced in the BD frontal cortex. The above results, taken together, strongly implicate disturbances in $G\alpha_{q/11}$-mediated PPI signaling as playing a role in the pathophysiology of BD.

Postmortem investigations of *myo*-inositol levels and inositol monophosphate phosphomonoesterase (IMPase) activity in BD *(82)* have yielded little promising data in support of altered PPI signaling in BD, unfortunately. Furthermore, reduced *myo*-inositol levels were also found in depressed suicide subjects, suggesting these changes are not specific to BD. Although *myo*-inositol levels may regulate phospholipase C activity *(83)*, the pathophysiological relevance of lower *myo*-inositol levels in the postmortem BD frontal cortex *(82)* remains to be determined.

Clinical Studies

Although no significant differences in the $G\alpha_{q/11}$ levels have been found in platelets and leukocytes from BD subjects compared with matched controls *(62)*, other findings from studies of platelets do implicate abnormalities in the PPI signaling pathway in these cells in BD. Brown and co-workers *(84)* reported significantly higher platelet membrane phosphatidyl inositol 4,5-bisphosphate (PIP$_2$) concentrations in drug-free bipolar manic patients compared with healthy controls. In addition, platelet membrane PKC activity was significantly increased in unmedicated bipolar manic patients compared with healthy individuals *(85)*. Interestingly, the elevated PIP$_2$ levels and PKC activity in the manic state were both normalized following lithium therapy *(85,86)*, suggesting that the hyperactive PPI signaling pathway in BD represents another important target of lithium action in treating this disorder. It is not clear, however, whether these PPI signaling abnormalities arise *de novo* or are a consequence of disturbances in the Ca^{2+} and/or cAMP signal transduction systems through crossregulation.

Of interest from a therapeutic perspective, significantly lower IMPase activity has been found in immortalized lymphoblastoid cell lines derived from BD patients responsive to lithium therapy compared with either lithium nonresponsive BD patients or healthy controls *(87)*. In lymphoblastoid cell lines treated chronically with lithium, IMPase mRNA levels were significantly increased. These results suggest that in lithium-responsive BD patients who exhibit trait-dependent low IMPase activity, the transcriptional upregulation of IMPase mRNA levels by lithium may normalize IMPase enzyme levels and activity *(87)*. Such evidence further supports the idea that the effect of lithium on PPI signaling is at least one important component of its molecular actions, which accounts for its therapeutic efficacy in BD.

CALCIUM SIGNALING DISTURBANCES IN BIPOLAR DISORDER

Among the signal transduction and second-messenger disturbances implicated in the pathophysiology of BD, abnormalities in intracellular calcium (Ca^{2+}) signaling are of particular importance, given the role of this ion in regulating many neuronal processes *(88)*. Investigations of intracellular Ca^{2+} signaling disturbances in BD have been conducted using peripheral blood cells because viable neurons from affected and comparison subjects, necessary for such research, are inaccessible for direct analysis and manipulation.

Abnormalities of intracellular Ca^{2+} dynamics in BD were first suggested by the observation of transient increases in serum Ca^{2+} level during manic episodes *(89)*. This observation was further strengthened by findings of higher calmodulin-dependent Ca^{2+}-ATPase levels and Ca^{2+}-stimulated activity in erythrocyte membranes from bipolar depressed subjects compared with normal controls *(90,90a)*. Using the Ca^{2+}-sensitive fluorescent dyes Fura 2 or Quin 2 to measure intracellular Ca^{2+} concentration, it has been shown that the basal and agonist-stimulated intracellular Ca^{2+} levels are significantly higher in platelets and/or lymphocytes from untreated manic and depressed BD patients compared with control subjects *(91–96)*. Incubation with an ultrafiltrate of plasma from BD patients had no effect on platelet intracellular Ca^{2+} levels of healthy individuals, suggesting that an "endogenous" factor(s) in these cells is responsible for the high basal platelet Ca^{2+} concentration in BD *(97)*. Intracellular Ca^{2+} homeostasis is controlled by a complex array of processes that include inositol trisphosphate and ryanodine receptor-stimulated release from endoplasmic reticulum storage pools *(98)*, store-operated Ca^{2+} influx *(88,99)*, and Ca^{2+}–ATPase pumps *(100)*, to highlight a few. As noted earlier, some observations point to disturbances in Ca^{2+}–ATPase *(90)* and Na^{+}/K^{+} ATPase *(101)* and PPI signaling, which could contribute to the Ca^{2+} homeostasis abnormalities observed. Moreover, evidence of cAMP signaling in BD suggest the possibility that altered Ca^{2+} homeostasis could occur as a consequence of crosstalk regulation. Regardless of the specific mechanisms that account for the observed changes in basal and agonist-stimulated intracellular Ca^{2+} levels, these results provided compelling evidence implicating abnormal intracellular Ca^{2+} signaling in BD.

The pattern of Ca^{2+} changes in BD contrasts with that observed in patients with a major depressive disorder, suggesting the type of Ca^{2+} disturbance may be diagnostically specific. First, basal intracellular Ca^{2+} concentrations in platelets/lymphocytes from patients with a major depressive disorder do not differ from those in matched controls, whereas they are higher in BD patients *(96,102,103)*. Second, whereas serotonin-induced Ca^{2+} responses were enhanced in both BD and major depressive disorder patients *(104–106)*, higher thrombin-stimulated Ca^{2+} responses were evident only in the BD patients *(92,102)*. These observations, together with the finding of lower erythrocyte membrane Ca^{2+}–ATPase activity in patients with major depression *(90)*, suggest altered Ca^{2+} homeostasis also occurs in major depression, but likely in a different form and reflecting distinct pathophysiological mechanisms.

An important question arising out of these studies relates to the nature of the intracellular Ca^{2+} changes that occur in BD. Some studies suggest that the high basal Ca^{2+} concentration in platelets and/or lymphocytes from BD patients is state dependent and normalizes with improvement in mood *(92,95,107)*. However, there is also evidence indicating that the abnormalities in Ca^{2+} signaling in BD are trait dependent and persist with remission *(94,105)*. In this regard, we have recently shown that higher basal intracellular Ca^{2+} concentrations occur in transformed lymphoblasts from BD patients compared with healthy subjects. Interestingly, the basal intracellular Ca^{2+} concentration was significantly higher in the B-lymphoblasts from patients with bipolar I disorder compared with healthy subjects, but not in psychiatric patients with bipolar II disorder, major depressive disorder, or without primary mood disorders *(96)*. These findings suggest that the abnormality(ies) underlying altered Ca^{2+} homeostasis in BD are, at least in part, trait dependent. Equally important, such findings suggest the expression of this trait-related abnormality in Ca^{2+} regulation in transformed B-lymphoblasts from bipolar I subjects may represent a phenotypic marker characteristic of this patient group, or a subtype of bipolar I disorder. As such, it could provide an important aid to elucidate the pathogenetic mechanisms of BD *(108)*.

NEUROPHARMACOLOGICAL EVIDENCE FOR SIGNALING DISTURBANCES IN BD

Studies of the action of the mood-stabilizing agents lithium, carbamazepine, and valproate provide additional support that disturbances in CNS intracellular signal transduction pathways may be central to the pathophysiology of BD. There is now substantial evidence that mood-stabilizing drugs, at therapeutically effective concentrations, affect cAMP, PPI, and Ca^{2+} signaling systems *(77,109,110)*. As will be discussed, the modulatory effects of these agents on the intracellular signaling systems appear to act as an "homeostatic lever" to reset the postulated signaling abnormalities in BD back toward their normal functional range.

Mood-Stabilizing Agents and cAMP Signaling System

Lithium and the anticonvulsant mood stabilizers carbamazepine and valproate inhibit both receptor- and postreceptor- (e.g., GTP and its analog, fluoride, forskolin, and Ca^{2+}/calmodulin) stimulation of adenylyl cyclase activity in vitro and ex vivo *(111–115)*. The in vitro effect of lithium on adenylyl cyclase appears to be mediated by competitive displacement of Mg^{2+} from a regulatory site on the enzyme, whereas the ex vivo inhibitory effects of lithium and carbamazepine involve altered efficacy of receptor/G-protein and/or G-protein/adenylyl cyclase coupling *(113,116)*. In addition, chronic lithium administration to animals modifies transcriptional regulation of genes encoding $G\alpha_s$, $G\alpha_i$, and adenylyl cyclase subtypes *(71,117,118)*. It also affects posttranslational modification of G-proteins through processes such as ADP-ribosylation *(33)* and stabilizes G-proteins in the heterotrimeric form *(62,110)* thus influencing their functional capacity in receptor–G-protein–effector interactions.

There is also considerable evidence that mood-stabilizing agents modulate downstream signaling regulated by activation of adenylyl cyclase. Chronic lithium treatment altered in vitro cAMP-dependent protein phosphorylation in rat hippocampal particulate fractions *(119)* and increased the cortical levels of DARPP-32 [dopamine- and cAMP-regulated phosphoprotein *(120)*. Because the protein phosphatase 1 activity is regulated by DARPP-32, the phosphorylation status of other cellular proteins may conceivably be modified by lithium. Furthermore, chronic administration of lithium to rats has been shown to increase the immunoreactive levels of the regulatory and the catalytic subunits of PKA *(121)* and induce translocation of PKA from cytosol to nuclear fractions in fractionated preparations from frontal cortex *(120)*. These PKA changes may promote phosphorylation of specific nuclear proteins involved in the regulation of gene expression that are essential for the induction and maintenance of therapeutic response.

Mood-Stabilizing Agents and PPI Signaling System

In keeping with the hypothesis that altered PPI second-messenger systems may be important in the pathophysiology of BD, a large body of evidence has accumulated demonstrating that lithium exerts significant modulatory effects on the PPI signaling system in a wide variety of cell systems, including brain. At therapeutically relevant concentrations, lithium inhibits PPI resynthesis by blocking *myo*-inositol recycling, possibly leading to an attenuation of an overactive PPI-linked neuronal response. A number of studies also indicate that chronic lithium treatment decreases receptor- and post-receptor- (e.g., fluoride, GTPγS, etc.) activated PPI turnover, possibly by impairing receptor–$G\alpha_{q/11}$ coupling *(77,109,110)*.

Chronic lithium treatment also increases the levels of several PPI-derived lipid metabolites including diacylglycerol and the liponucleotides, in various cell types including brain *(122,123)*, suggesting that the diacylglycerol-regulated effector enzyme, PKC, could be a target for regulation by lithium. It has been shown that chronic lithium administration modulates PKC-mediated phosphorylation of several substrate proteins in rat hippocampus membrane and cytosolic fractions, in the absence of any effects on the activity or cellular redistribution of PKC *(119,124)*. Notably, the phosphorylation and levels of a major specific PKC substrate protein, MARCKS (myristoylated alanine-rich C kinase substrate), were significantly reduced *(124)*. As down-regulation of MARCKS is preceded by its phosphorylation by PKC *(125)*, the lithium-induced changes in hippocampal MARCKS may reflect an initial activation of PKC followed by degradation of the protein. Such a mechanism is consistent with data indicating that acutely lithium stimulates PKC, whereas, chronically, it attenuates PKC-mediated responses (reviewed in ref. *110*). In two recent studies, Manji et al. *(126,127)* showed that prolonged lithium administration reduces phorbol ester binding (a global index of PKC level) and selectively decreases the membrane levels of PKC-α and PKC-ε isozymes in rat hippocampus. Similar changes on the PKC isozymes *(128)* and MARCKS *(129)* were also evident with chronic valproate treatment. Taken together, these data strongly suggest that such changes in the level and/or activity of PKC isozymes may have significant effects on downstream targets regulated by PKC activation, which may be of particular importance in the therapeutic actions of these mood stabilizers.

Mood-Stabilizing Agents and Ca^{2+} Signaling

In addition to the panoply of effects on cAMP and PPI signaling mechanisms, there is also evidence suggesting that regulation of Ca^{2+} homeostasis may represent a molecular target for the therapeutic effects of mood-stabilizing agents. For example, chronic exposure of mouse astrocytes in primary cultures to therapeutically relevant concentrations of lithium decreases basal intracellular Ca^{2+} concentration and attenuates noradrenaline-induced increases in intracellular Ca^{2+} concentration *(130)*. Chronic treatment with lithium also reduces 5-HT-induced increases in intracellular Ca^{2+} concentration in C6 rat glioma cells *(131)*, whereas the agonist-evoked Ca^{2+} response is also modulated by therapeutic concentrations of carbamazepine and valproate *(132,133)*. Similarly, chronic lithium treatment reduces the N-methyl-D-aspartate (NMDA) receptor-evoked Ca^{2+} response in primary culture of rat cerebellar granule cells *(134)*. Significant inhibition of NMDA-induced Ca^{2+} response by clinically relevant concentrations of carbamazepine and valproate has also been reported *(135)*. Of particular note are the preliminary findings that calcium channel blockers such as nimodipine and verapamil were clinically effective in the treatment of acute mania *(136,137)*, further supporting the notion that regulation of Ca^{2+} dynamics might be important for the therapeutic effects of mood-stabilizing agents.

In summary, the mood-stabilizing agents currently in clinical use all appear to have neurobiological actions that attenuate or normalize disturbances in the same signaling systems implicated in BD by postmortem and peripheral cell studies. This clearly provides another important line of pharmacological evidence, albeit indirect, for the relevance of intracellular signal transduction abnormalities in the pathophysiology of BD.

CONCLUSION

Over the past decade, postreceptor second-messenger generating systems have become of great interest with regard to the psychobiology of BD and the neurobiological action of mood stabilizing agents, particularly lithium. Compelling evidence from postmortem, clinical, and pharmacological studies is solidifying the notion that disturbances in multiple signaling systems (e.g., cAMP, PPI and Ca^{2+}) as depicted in Fig. 1 are central to the pathophysiology of BD. It is presently unknown whether the signaling disturbances in BD are related to or independent of each other. As illustrated in Fig. 1, there is extensive evidence of crossregulation among these signaling pathways *(30,75,138)*. Thus, it is conceivable that the disturbances identified in BD may reflect dysregulation of the interaction and crosstalk among these signaling mechanisms, which are necessary for normal homeostatic function under physiological conditions. Among the more pressing questions regarding the signal transduction abnormalities in BD are that of the specific molecular defect(s) promoting the development of these disturbances and which signal transduction system is first affected. This knowledge will be critical in designing much more effective medications and novel biological interventions for the management of BD. The possibility that specific signal transduction disturbances are unique to subgroups of BD or major depression holds up the potential for their use as phenotypic and, perhaps, diagnostic markers. Indeed, the recent demonstration that some of these signal transduction abnormalities are stably expressed in B-lymphoblasts in subgroups of BD suggest the potential for such endophenotypic markers to aid in elucidating the molecular and genetic basis of these mood disorders. Hopefully, the recent advances in cellular and molecular biology will lead us farther along the path toward a better understanding of the role of these signal transduction disturbances in the pathogenesis of BD.

REFERENCES

1. Goodwin, F. K. and Jamison, K. R. (1990) *Manic-Depressive Illness*, Oxford University Press, New York, pp. 503–524.
2. Warsh, J. J. and Li, P. P. (1996) Second messenger systems and mood disorders. *Curr. Opin. Psychiatry* **9**, 23–29.
3. Bunney, W. E. and Davis, J. M. (1965) Norepinephrine in depressive reactions: a review. *Arch. Gen. Psychiatry* **13**, 483–494.
4. Schildkraut, J. J. (1965) The catecholamine hypothesis of affective illness: a review of supporting evidence. *Am. J. Psychiatry* **122**, 509–522.
5. Lykouras, E., Varsou, E., Garelis, E., Stefanis, C. N., and Malliaras, D. (1978) Plasma cyclic AMP in manic-depressive illness. *Acta Psychiat. Scand.* **57**, 447–453.
6. Extein, I., Tallman, J., Smith, C. C., and Goodwin, F. K. (1979) Changes in lymphocyte beta-adrenegic receptors in depression and mania. *Psychiatry Res.* **1**, 191–197.
7. Pandey, G. N., Dysken, M. W., Garver, D. L., and Davis, J. M. (1979) Beta-adrenergic receptor function in affective illness. *Am. J. Psychiatry* **136**, 675–678.
8. Jeanningros, R., Mazzola, P., Azorin, J. M., Samuelian-Massa, C., and Tissot, R. (1991) β-Adrenoceptor density of intact mononuclear leukocytes in subgroups of depressive disorders. *Biol. Psychiatry* **29**, 789–798.
9. Schreiber, G., Avissar, S., Danon, A., and Belmaker, R. (1991) Hyperfunctional G proteins in mononuclear leukocytes of patients with mania. *Biol. Psychiatry* **29**, 273–280.
10. Young, L. T., Li, P. P., Kish, S. J., Siu, K. P., and Warsh, J. J. (1991) Postmortem cerebral cortex Gs alphasubunit levels are elevated in bipolar affective disorder. *Brain Res.* **553**, 323–326.
11. Young, L. T., Li, P. P., Kish, S. J., Siu, K. P., Kamble, A., Hornykiewicz, O., and Warsh, J. J. (1993) Cerebral cortex $G_s\alpha$ protein levels and forskolin-stimulated cyclic AMP formation are increased in bipolar affective disorder. *J. Neurochem.* **61**, 890–898.

12. Warsh, J. J., Young, L. T., and Li, P. P. (1999) Guanine nucleotide binding (G) protein disturbances in bipolar affective disorder, in *Mechanisms of Antibipolar Treatments: Focus on Lithium, Carbamazepine and Valproic Acid* (Manji, H. K., Bowden, C. L., and Belmaker, R. H., eds.), American Psychiatric Press, Washington, DC, in press.

13. Seifert, R., Wenzel-Seifert, K., Lee, T. W., Gether, U., Sanders-Bush, E., and Kobilka, B.K. (1998) Differential effects of $G_s\alpha$ splice variants on β_2-adrenoceptor-mediated signaling: the β_2-adrenoceptor coupled to the long splice variant of $G_s\alpha$ has properties of a constitutive active receptor. *J. Biol. Chem.* **273**, 5109–5116.

14. Walseth, T. F., Zhang, H.-J., Olson, L. K., Schroeder, W. A., and Robertson, R. P. (1989) Increase in Gs and cyclic AMP generation in HIT cells: evidence that the 45-kDa α-subunit of G_s has greater functional activity than the 52-kDa α-subunit. *J. Biol. Chem.* **264**, 21,106–21,111.

15. Friedman, E. and Wang, H. Y. (1996) Receptor-mediated activation of G protein is increased in postmortem brains of bipolar affective disorder subjects. *J. Neurochem.* **67**, 1145–1152.

16. Warsh, J. J., McLeman, E., Li, P. P., and Kish, S. J. (1998) Reduced nucleus accumbens $G\alpha i$ immunolabeling in chronic drug users. *Biol. Psychiatry* **43**, 15S.

17. Sunahara, R. K., Dessauer, C. W., and Gilman, A. G. (1996) Complexity and diversity of mammalian adenylyl cyclases. *Annu. Rev. Pharmacol. Toxicol.* **36**, 461–480.

18. Berrettini, W. H., Ferraro, T. N., Goldin, L. R., Weeks, D. E., Detera-Wadleigh, S., Nurnberger, J. I., and Gershon, E. S. (1994) Chromsome 18 DNA markers and manic-depressive illness: evidence for a susceptibility gene. *Proc. Natl. Acad. Sci. USA* **91**, 5918–5921.

19. Berrettini, W. H. , Ferraro, T. N., Goldin, L. R., Detera-Wadleigh, S. D., Choi, H., Muniec, D., Guroff, J. J., Kazuba, D. M., Nurnberger, J. I., Hsieh, W. T., Hoehe, M. R., and Gershon, E. S. (1997) A linkage study of bipolar illness. *Arch. Gen. Psychiatry* **54**, 27–35.

20. Jones, D. T. and Reed, R. R. (1989) Golf: an olfactory neuron-specific G protein involved in odorant signal transduction. *Science* **244**, 790–793.

21. Tsiouris, S. J., Breschel, T. S., Xu, J., McInnis, M. G., and McMahon, F. J. (1996) Linkage disequilibrium analysis of G-olf$_\alpha$ (GNAL) in bipolar affective disorder. *Am. J. Med. Genet.* **67**, 491–494.

22. Spiegel, A. M., Shenker, A., and Weinstein, L. S. (1992) Receptor–effector coupling by G proteins: implications for normal and abnormal signal transduction. *Endocr. Rev.* **13**, 536–565.

23. Milligan, G. (1993) Agonist regulation of cellular G protein levels and distribution: mechanisms and functional implications. *Trends Pharmacol. Sci.* **14**, 413–418.

24. Le, F., Mitchell, P., Vivero, C., Waters, B., Donald, J., Selbie, L. A., Shine J., and Schofield P. (1994) Exclusion of close linkage of bipolar to the Gs-α subunit gene in nine Australian pedigrees. *J. Affect. Disord.* **32**, 187–195.

25. Ram, A., Guedj, F., Cravchik, A., Weinstein, L., Cao, Q., Badner, J. A., Goldin, L. R., Grisaru, N., Manji, H. K., Belmaker, R. H., Gershon, E. S., and Gejman, P. V. (1997) No abnormality in the gene for the G protein stimulatory α subunit in patients with bipolar disorder. *Arch. Gen. Psychiatry* **54**, 44–48.

26. Young, L. T., Asghari, V., Li, P. P., Kish, S., Fahnestock, M., and Warsh, J. J. (1996) Stimulatory G-protein α-subunit mRNA levels are not increased in autopsied cerebral cortex from patients with bipolar disorder. *Mol. Brain Res.* **42**, 45–50.

27. Wedegaetner, P. B., Wilson, P. T., and Bourne, H. R. (1995) Lipid modifications of trimeric G proteins. *J. Biol. Chemistry* **270**, 503–506.

28. Greenwood, A. and Jope, R. (1994) Brain proteolysis by calpain: enhancement by lithium. *Brain Res.* **636**, 320–326.

29. Madura, K. and Varshavsky, A. (1994) Degradation of Gα by the N-end rule pathway. *Science* **265**, 1454–1458.

30. Port, J. D. and Malbon, C. C. (1993) Integration of transmembrane signaling: cross-talk among G-protein-linked receptors and other signal transduction pathways. *Trends Cardiovasc. Med.* **3**, 85–92.

31. Chang, F. H. and Bourne, H. R. (1989) Cholera toxin induces cAMP-independent degradation of Gs. *J. Biol. Chem.* **264**, 5352–5357.

32. Milligan, G., Unson, C. G., and Wakelam, J. O. (1989) Cholera toxin treatment produces down-regulation of the α-subunit of the stimulatory guanine-nucleotide-binding protein (G_s). *Biochem. J.* **262**, 643–649.

33. Nestler, E. J., Terwilliger, R. Z., and Duman, R. S. (1995) Regulation of endogenous ADP-ribosylation by acute and chronic lithium in rat brain. *J. Neurochem.* **64**, 2319–2324.

34. Andreopoulos, S., Li, P. P., Siu, K. P., and Warsh, J. J. (1999) Characterization of α_s-immunoreactive ADP-ribosylated proteins in postmortem human brain. *J. Neurosci. Res.* **56**, 632–643.

35. Andreopoulos, S., Siu, K. P., Li, P. P., and Warsh, J. J. (1997) Reduced ADP-ribosylation of $G\alpha_s$ in postmortem bipolar disorder temporal cortex. *Biol. Psychiat.* **41**(Suppl), 61S.

36. Reiach, J. S., Li, P. P., Warsh, J. J., Kish, S. J., and Young, L. T. (1999) Reduced adenylyl cyclase immunolabeling and activity in postmortem temporal cortex of depressed suicide subjects. *J. Affect. Disord.*, in press.

37. Young, L. T., Li, P. P., Kish, S. J., and Warsh, J. J. (1994) Cerebral cortex β-adrenoceptor binding in bipolar affective disorder. *J. Affect. Disord.* **30**, 89–92.

38. Rahman, S., Li, P. P., Young, L. T., Kofman, O., Kish, S. J., and Warsh, J. J. (1997) Reduced [^{3}H]cyclic AMP binding in postmortem brain from subjects with bipolar affective disorder. *J. Neurochem.* **68**, 297–304.

39. Francis, S. H. and Corbin, J. D. (1994) Structure and function of cyclic nucleotide-dependent protein kinases. *Annu. Rev. Physiol.* **56**, 237–272.

39a. Fields, A, Li, P. P., Kish, S. J., and Warsh, J. J. (1999) Increased cyclic AMP-dependent protein kinase activity in postmortem brain from patients with bipolar affective disorder. *J. Neurochem.* **73**, 1704–1710.

40. Perez, J., Zanardi, R., Mori, S., Gasperini, M., Smeraldi, E., and Racagni, G. (1995) Abnormalities of cAMP-dependent endogenous phosphorylation in platelets from patients with bipolar disorder. *Am. J. Psychiatry* **152**, 1204–1206.

41. Kish, S. J., Young, L. T., Li, P. P., Siu, K. P., Robitaille, Y., Ball, M. J., Schut, L., and Warsh, J. J. (1992) Elevated stimulatory and reduced inhibitory G protein α-subunits in cerebellar cortex of patients with dominantly inherited olivopontocerebellar atrophy. *J. Neurochem.* **60**, 1816–1820.

42. Ozawa, H., Gsell, W., Frolich, L., Zochling, F., Pantucek, H., Beckmann, P., and Riderer, F. (1993) Imbalance of the G_s and $G_{i/o}$ function in post-mortem human brain of depressed patients. *J. Neural. Transm.* **94**, 63–69.

43. Cowburn, R. F., Marcusson, J. O., Eriksson, A., Wiehager, C., and O'Neill, C. (1994) Adenylyl cyclase activity and G-protein subunit levels in postmortem frontal cortex of suicide victims. *Brain Res.* **633**, 297–304.

44. Jope, R. S., Song, L., Grimes, C. A., Pacheco, M. A., Dilley, G. E., Li, X., Meltzer, H. Y., Overholser, J. C., and Stockmeier, C. A. (1998) Selective increases in phosphoinositide signaling activity and G protein levels in postmortem brain from subjects with schizophrenia or alcohol dependence. *J. Neurochem.* **70**, 763–771.

45. Nishino, N., Kitamura, N., Hashimoto, T., Kajimoto, Y., Shirai, Y., Murakami, N., Nakai, T., Komure, O., Shirakawa, O., Mita, T., and Nakai, H. (1993) Increase in [^{3}H]cAMP binding sites and decrease in $G_i\alpha$ and $G_o\alpha$ immunoreactivities in left temporal cortices from patients with schizophrenia. *Brain Res.* **615**, 41–49.

46. Nishino, N., Kitamura, N., Hashimoto, T., and Tanaka, C. (1993) Transmembrane signalling systems in the brain of patients with Parkinson's disease. *Rev. Neurosci.* **4**, 213–222.

47. Okada, F., Crow, T. J., and Roberts, G. W. (1991) G proteins (Gi, Go) in the medial temporal lobe in schizophrenia: preliminary report of a neurochemical correlate of structural change. *J. Neural Transm.* **84**, 147–153.

48. Okada, F., Tokumitsu, Y., Takahashi, N., Crow, T. J., and Roberts, G. W. (1994) Reduced concentrations of the α-subunit of GTP-binding protein Go in schizophrenic brain. *J. Neural Transm.* **95**, 95–104.

49. Yang, C. Q., Kitamura, N., Nishino, N., Shirakawa, O., and Nakai, H. (1998) Isotype-specific G protein abnormalities in the left superior temporal cortex and limbic structures of patients with chronic schizophrenia. *Biol. Psychiatry* **43**, 12–19.

50. Ozawa, H., Katamura, Y., Hatta, S., Satio, T., Katada, T., Gsell, W., Froelich, L., Takahata, N., and Riederer, P. (1993) Alterations of guanine nucleotide-binding proteins in post-mortem human brain in alcoholics. *Brain Res.* **62**, 174–179.

51. Escriba, P. V., Sastre, M., and Garcia-Sevilla, J. A. (1994) Increased density of guanine nucleotide-binding proteins in the postmortem brains of heroin addicts. *Arch. Gen. Psychiatry* **51**, 494–501.

52. Cowburn, R. F., O'Neill, C., Ravid, R., Alafuzoff, I., Winblad, B., and Fowler, C. J. (1992) Adenylyl cyclase activity in postmortem human brain: evidence of altered G protein mediation in Alzheimer's disease. *J. Neurochem.* **58**, 1409–1419.

53. Cowburn, R. F., O'Neill, C., Ravid, R., Winblad, B., and Fowler, C. J. (1992) Preservation of G_i-protein inhibited adenylyl cyclase activity in the brains of patients with Alzheimer's disease. *Neurosci. Lett.* **141**, 16–20.

54. McLaughlin, M., Ross, B. M., Milligan, G., McCulloch, J., and Knowler, J. T. (1991) Robustness of G proteins in Alzheimer's disease: an immunoblot study. *J. Neurochem.* **57**, 9–14.

55. Ohm, T. G., Bohl, J., and Lemmer, B. (1991) Reduced basal and stimulated (isoprenaline, GppNHp, forskolin) adenylate cyclase activity in Alzheimer's disease correlated with histopathological changes. *Brain Res.* **540**, 229–236.

56. Lowther, S., Katona, C. L. E., Crompton, M. R., and Horton, R. W. (1997) Brain [^{3}H]cAMP binding sites are unaltered in depressed suicides, but decreased by antidepressants. *Brain Res.* **758**, 223–228.

57. Bonkale, W. L., Fastbom, J., Wiehager, B., Ravid, R., Winblad, B., and Cowburn, R. F. (1996) Impaired G-protein-stimulated adenylyl cyclase activity in Alzheimer's disease brain is not accompanied by reduced cyclic-AMP-dependent protein kinase A activity. *Brain Res.* **737**, 155–161.

58. Trottier, Y., Lutz, Y., Stevanin, G., Imbert, G., Devys, D., Cancel, G., Saudou, F., Weber, C., David, G., Tora, L., Agid, Y., Brice, A., and Mandel, J. L. (1995) Polyglutamine expansion as a pathological epitope in Huntington's disease and four dominant cerebellar ataxias. *Nature* **378**, 403–406.

59. Polymeropoulos, M. H., Lavedan, C., Leroy, E., Ide, S. E., Dehejia, A., Dutra, A., Pike, B., Root, H., Rubenstein, J., Boyer, R., Stenroos, E. S., Chandrasekharappa, S., Athanassiadou, A., Papapetropoulos, T., Johnson, W. G., Lazzarini, A. M., Duvoisin, R. C., Di Iorio, G., Golbe, L. I., and Nussbaum, R. L. (1997) Mutation in the α-synuclein gene identified in families with Parkinson's disease. *Science* **276**, 2045–2047.

60. Wang, G., Ide, K., Nukina, N., Goto, J., Ichikawa, Y., Uchida, K., Sakamoto, T., and Kanazawa, I. (1997) Machado–Joseph disease gene product identified in lymphocytes and brain. *Biochem. Biophys. Res. Commun.* **233**, 476–479.

61. Young, L. T., Li, P. P., Kamble, A., Siu, K. P., and Warsh, J. J. (1994b) Mononuclear leukocyte G protein levels in depressed patients with bipolar or major depressive disorder. *Am. J. Psychiatry* **151**, 594–596.

62. Manji, H. K., Chen, G., Shimon, H., Hsiao, J. K., Potter, W. Z., and Belmaker, R. (1995) Guanine nucleotide-binding proteins in bipolar affective disorder. Effects of long-term lithium treatment. *Arch. Gen. Psychiatry* **52**, 135–144.

63. Avissar, S., Nechamkin, Y., Barki-Harrington, L., Roitman, G., and Schreiber, G. (1997) Differential G protein measures in mononuclear leukocytes of patients with bipolar mood disorder are state dependent. *J. Affect. Disorders* **43**, 85–93.

64. Mitchell, P. B., Manji, H. K., Chen, G., Jolkovsky, L., Smith-Jackson, E., Denicoff, K., Schmidt, M., and Potter, W. Z. (1997) High levels of $G_s\alpha$ in platelets of euthymic patients with bipolar affective disorder. *Am. J. Psychiatry* **154**, 218–223.

65. Garcia-Sevilla, J. A., Walzer, C., Busquets, X., Escriba, P. V., Balant, L., and Guimon, J. (1997) Density of guanine nucleotide-binding proteins in platelets with major depression: increased abundance of the $G\alpha_{i2}$ subunit and down-regulation by antidepressant treatment. *Biol. Psychiatry* **42**, 704–712.

66. Stein, M. B., Chen, G., Potter, W. Z., and Manji, H. K. (1996) G-protein level quantification in platelets and leukocytes from patients with panic disorder. *Neuropsychopharmacology* **15**, 180–186.

67. Waltman, C., Levine, M. A., McCaul, M. E., Svikis, D. S., and Wand, G. S. (1993) Enhanced expression of the inhibitory protein $G_{i2}\alpha$ and decreased activity of adenylyl cyclase in lymphocytes of abstinent alcoholics. *Alcohol Clin. Exp. Res.* **17**, 315–320.

68. Emamghoreishi, M., Li, P., Schlichter, L., Parikh, S., Cooke, R., Sibony, D., and Warsh, J. J. (1998) Elevated Gαs immunolabeling in lymphoblasts from bipolar II subjects. *Biol. Psychiatry* **43**, 36S.

69. Avissar, S., Nechamkin, Y., Roitman, G., and Schreiber, G. (1997) Reduced G protein functions and immunoreactive levels in mononuclear leukocytes of patients with depression. *Am. J. Psychiatry* **154**, 211–217.

70. Spleiss, O., van Calker, D., Scharer, L., Adamovic, K., Berger, M., and Gebicke-Haerter, P. J. (1998) Abnormal G protein α_s- and α_{i2}-subunit mRNA expression in bipolar affective disorder. *Mol. Psychiatry* **3**, 512–520.

71. Li, P. P., Young, L. T., Tam, Y. K., and Warsh, J. J. (1993) Effects of chronic lithium and carbamazepine treatment on G-protein subunit expression in rat cerebral cortex. *Biol. Psychiatry* **34**, 162–170.

72. Zanardi, R., Racagni, G., Smeraldi, E., and Perez, J. (1997) Differential effects of lithium on platelet protein phosphorylation in bipolar patients and healthy subjects. *Psychopharmacology* **129**, 44–47.

73. Cook, S. J., Rubinfeld, B., Albert, I., and McCormick, F. (1993) Rap V12 antagonizes Ras-dependent activation of ERK1 and ERK2 by LPA and EGF in Rat-1 fibroblasts. *EMBO J.* **12**, 3475–3485.
74. York, R. D., Yao, H., Dillon, T., Ellig, C. L., Eckert, S. P., McCleskey, E. W., and Stork, P. J. S. (1998) Rap1 mediates sustained MAP kinase activation induced by nerve growth factor. *Nature* **392**, 622–626.
75. Hill, S. J. and Kendall, D. A. (1989) Cross-talk between different receptor-effector systems in the mammalian CNS. *Cell Signalling* **1**, 135–141.
76. Gutkind, J. S. (1998) The pathways connecting G protein-coupled receptors to the nucleus through divergent mitogen-activated protein kinase cascades. *J. Biol. Chem.* **273**, 1839–1842.
77. Jope, R. S. and Williams M. B. (1994) Lithium and brain signal transduction systems. *Biochem. Pharmacol.* **47,** 429–441.
78. Mathews, R., Li, P. P., Young, L. T., Kish, S. J., and Warsh, J. J. (1997) Increased $G\alpha_{q/11}$ immunoreactivity in postmortem occipital cortex from patients with bipolar affective disorder. *Biol. Psychiatry* **411**, 649–656.
79. Jope, R. S., Song, L., Li, P. P., Young, L. T., Kish, S. J., Pacheco, M. A., and Warsh, J. J. (1996) The phosphoinositide signal transduction system is impaired in bipolar affective disorder brain. *J. Neurochem.* **66**, 2402–2409.
79a. Jope, R. S., Song, L., and Li, X. (1994) Impaired phosphoinositide hydrolysis in Alzheimer's brain. *Neurobiol. Aging* **15,** 221–226.
80. Pacheco, M. A., Stockmeier, C., Meltzer, H. Y., Overholser, J. C., Dilley, G. E., and Jope, R. S. (1996) Alterations in phosphoinositide signaling and G-protein levels in depressed suicide brain. *Brain Res.* **723**, 37–45.
81. Wang, W. Y. and Friedman, E. (1996) Enhanced protein kinase C activity and translocation in bipolar affective disorder brain. *Biol. Psychiatry* **40**, 568–575.
82. Shimon, H., Agam, G., Belmaker, R. H., Hyde, T. M., and Kleinman, J. E. (1997) Reduced frontal cortex inositol levels in postmortem brain of suicide victims and patients with bipolar disorder. *Am. J. Psychiatry* **154**, 1148–1150.
83. Batty, I. H. and Downes, C. P. (1995) The mechanism of muscarinic receptor-stimulated phosphatidylinositol re-synthesis in 1321N1 astrocytoma cells and its inhibition by Li^+ ions. *J. Neurochem.* **65**, 2279–2289.
84. Brown, A. S., Mallinger, A. G., and Renbaum, L. C. (1993) Elevated platelet membrane phosphatidylinositol-4,5-bisphosphate in bipolar mania. *Am. J. Psychiatry* **150**, 1252–1254.
85. Friedman, E., Wang, H. Y., Levinson, D., Connell, T. A., and Singh, H. (1993) Altered platelet protein kinase C activity in bipolar affective disorder, manic episode. *Biol. Psychiatry* **33**, 520–525.
86. Soares, J. C., Dippold, C. S., Mallinger, A. G., Frank, E., and Kupfer, D. J. (1998) Effects of lithium on platelet membrane phosphoinositides in bipolar disorder patients. *Biol. Psychiatry* **43**,70S.
87. Shamir, A., Ebstein, R. P., Nemanov, L., Zohar, A., Belmaker, R. H., and Agam, G. (1998) Inositol monophosphatase in immortalized lymphoblastoid cell lines indicates susceptibility to bipolar disorder and response to lithium therapy. *Mol. Psychiatry* **3**, 481–482.
88. Clapham, D. E. (1995) Calcium signaling. *Cell* **80**, 259–268.
89. Carmen, J. S. and Wyatt, R. J. (1979) Calcium: bivalent action in the bivalent psychoses. *Biol. Psychiatry* **14**, 295–336.

90. Meltzer, H. L., Kassir, S., Goodnick, P. J., Fieve, R. R., Chrisomalis, L., Feliciano, M., and Szypula, D. (1988) Calmodulin-activated calcium ATPase in bipolar illness. *Neuropsychobiology* **20**, 169–173.

90a. Bowden, C. L., Huang, L. G., Javors, M. A., Johnson, J. M., Seleshi, E., McIntyre, K., Contreras, S., and Maas, J. W. (1988) Calcium function in affective disorders and healthy controls. *Biol. Psychiatry* **23**, 367–376.

91. Dubovsky, S. L., Christiano, J., Daniell, L. C., Franks, R. D., Murphy, J., Adler, L., Baker, N., and Harris, R. A. (1989) Increased platelet intracellular calcium concentration in patients with bipolar affective disorders. *Arch. Gen. Psychiatry* **46**, 632–638.

92. Dubovsky, S. L., Lee, C., Christiano, J., and Murphy, J. (1991) Elevated platelet intracellular calcium concentration in bipolar depression. *Biol. Psychiatry* **29**, 441–450.

93. Dubovsky, S. L., Murphy, J., Thomas, M., and Rademacher, J. (1992) Abnormal intracellular calcium ion concentration in platelets and lymphocytes of bipolar patients. *Am. J. Psychiatry* **149**, 118–120.

94. Tan, C. H., Javors, M. A., Seleshi, E., Lowrimore, P. A., and Bowden, C. L. (1990) Effects of lithium on platelet ionic intracellular calcium concentration in patients with bipolar (manic-depressive) disorder and healthy controls. *Life Sci.* **46**, 1175–1180.

95. Berk, M., Bodemer, W., Oudenhove, T. V., and Butkow, N. (1995) The platelet intracellular calcium response to serotonin is augmented in bipolar manic and depressed patients. *Human Psychopharmacol.* **10**, 189–193.

96. Emamghoreishi, M., Schlichter, L., Li, P. P., Parikh, S., Sen, J., Kamble, A., and Warsh, J. J. (1997) High intracellular calcium concentrations in transformed lymphoblasts from subjects with bipolar I disorder. *Am. J. Psychiatry* **154**, 976–982.

97. Dubovsky, S. L., Thomas, M., Hijazi, A., and Murphy, J. (1994) Intracellular calcium signalling in peripheral cells of patients with bipolar affective disorder. *Eur. Arch. Psychiatry Clin. Neurosci.* **243**, 229–234.

98. Ehrlich, B. E. (1995) Functional properties of intracellular calcium-release channels. *Curr. Opin. Neurobiol.* **5**, 304–309.

99. Putney, J. W. (1990) Capacitative calcium entry revisited. *Cell Calcium* **11**, 611–624.

100. Rasmussen, H. (1989) The cycling of calcium as an intracellular messenger. *Sci. Am.* **261**, 66–73.

101. Cherry, L. and Swann, A. C. (1994) Cation transport mediated by Na+, K+-adenosine triphosphatase in lymphobalst cells from patients with bipolar I disorder, their relatives, and unrelated control subjects. *Psychiatry Res.* **53**, 111–118.

102. Eckert, A., Gann, H., Riemann, D., Aldenhoff, J., and Muller, W. E. (1994) Platelet and lymphocyte free intracellular calcium in affective disorders. *Eur. Arch. Psychiatry Clin. Neurosci.* **243**, 235–239.

103. Vollmayr, B., Sulger, J., Gabriel, P., and Aldenhoff, J. B. (1995) Mitogen stimulated rise of intracellular calcium concentration in single T lymphocytes from patients with major depression is reduced. *Prog. Neuro-Psychopharmacol. Biol. Psychiatry* **19**, 1263–1273.

104. Mikuni, M., Kagava, A., Takahashi, K., and Meltzer, H. Y. (1992) Serotonin but not norepinephrine-induced calcium mobilization of platelets is enhanced in affective disorders. *Psychopharmacology (Berlin)* **106**, 311–314.

105. Kusumi, I., Koyama, T., and Yamashita, I. (1994) Serotonin-induced platelet intra-cellular calcium mobilization in depressed patients. *Psychopharmacology (Berlin)* **113**, 322–327.
106. Okamoto, Y., Kagaya, A., Shinno, H., Motohashi, N., and Yamawaki, S. (1995) Serotonin-induced platelet calcium mobilization is enhanced in mania. *Life Sci.* **56**, 327–332.
107. Dubovsky, S. L., Lee, C., Christiano, J., and Murphy, J. (1991b) Lithium decreases platelet intracellular calcium concentration in bipolar patients. *Lithium* **2**, 167–174.
108. Leboyer, M., Bellivier, F., Nosten-Bertrand, M., Jouvent, R., Pauls, D., and Mallet, J. (1998) Psychiatric genetics: search for phenotypes. *Trends Neurosci.* **21**, 102–105.
109. Hudson, C., Young, L. T., Li, P. P., and Warsh, J. J. (1993) CNS signal transduc-tion in the pathophysiology and pharmacotherapy of affective disorders and schi-zophrenia. *Synapse* **13**, 278–293.
110. Manji, H. K., Potter, W. Z., and Lenox, R. H. (1995b) Signal transduction path-ways: molecular targets for lithium's actions. *Arch. Gen. Psychiatry* **52**, 531–543.
111. Ebstein, R. P., Hermoni, M., and Belmaker, R. H. (1980) The effect of lithium on noradrenaline-induced cyclic AMP accumulation in rat brain: inhibition after chronic treatment and absence of supersensitivity. *J. Pharmacol. Exp. Ther.* **213**, 161–167.
112. Ebstein, R. P., Lerer, B., Bennett, E. R., Shapira, B., Kindler, S., Shemesh, Z., and Gerstenhaber, N. (1988) Lithium modulation of second messenger signal amplifi-cation in man: inhibition of phosphatidylinositol-specific phospholipase C and adenylate cyclase activity. *Psychiatry Res.* **24**, 45–52.
113. Mork, A., Geisler, A., and Hollund, P. (1992) Effects of lithium on 2nd messenger systems in the brain. *Pharmacol. Toxicol.* **71**, 4–17.
114. Chen, G., Pan, B., Hawver, D. B., Wright, C. B., Potter, W. Z., and Manji, H. K. (1996) Attenuation of cyclic AMP production by carbamazepine. *J. Neurochem.* **67**, 2079–2086.
115. Chen, G., Manji, H. K., Wright, C. B., Hawver, D. B., and Potter, W. Z. (1996) Effects of valproic acid on β-adrenergic receptors, G-proteins, and adenylyl cycl-ase in rat C6 glioma cells. *Neuropsychopharmacology* **15**, 271–280.
116. Avissar, S. and Schreiber, G. (1992) The involvement of guanine nucleotide bind-ing proteins in the pathogenesis and treatment of affective disorders. *Biol. Psychia-try* **31**, 415–459.
117. Colin, S. F., Chang, H. C., Mollner, S., Pfeuffer, T., Reed, R.R., Duman, R. S., and Nestler, E. J. (1991) Chronic lithium regulates the expression of adenylate cyclase and $G_i\alpha$-subunit in rat cerebral cortex. *Proc. Natl. Acad. Sci. USA* **88**, 10,634–10,637.
118. Li, P. P., Tam, Y. K., Young, L. T., and Warsh, J.J. (1991) Lithium decreases Gs, Gi1 and Gi2 α-subunit mRNA levels in rat cortex. *Eur. J. Pharmacol. Mol. Phar-macol.* **206**, 165–166.
119. Casebolt, T. L. and Jope, R. S. (1991) Effects of chronic lithium treatment on protein kinase C and cyclic AMP dependent protein phosphorylation. *Biol. Psy-chiatry* **29**, 233–243.
120. Guitart, X. and Nestler, E. J. (1992) Chronic administration of lithium or other antidepressants increases levels of DARPP-32 in rat frontal cortex. *J. Neurochem.* **59**, 1164–1167.

121. Mori, A., Tardito, D., Dorigo, A., Zanardi, R., Smeraldi, E., Racagni, G., and Perez, J. (1998) Effects of lithium on cAMP-dependent protein kinase in rat brain. *Neuropsychopharmacology* **19**, 233–240.

122. Godfrey, P. P. (1989) Potentiation by lithium of CMP-phosphatidate formation in carbachol-stimulated rat cerebral-cortical slices and its reversal by myo-inositol. *Biochem. J.* **258**, 621–624.

123. Brami, B. A., Leli, U., and Hauser, G. (1993) Elevated phosphatidyl-CMP is not the source of diacylglycerol accumulation induced by lithium in NG108–15 cells. *J. Neurochem.* **60**, 1137–1142.

124. Lenox, R. H., Watson, D. G., Patel, J., and Ellis, J. (1992) Chronic lithium administration alters a prominent PKC substrate in rat hippocampus. *Brain Res.* **570**, 333–340.

125. Lindner, D., Gschwendt, M., and Marks, F. (1992) Phorbol ester-induced down-regulation of the 80-kDa myristoylated alanine-rich C kinase substrate-related protein in Swiss 3T3 fibroblasts: inhibition by staurosporine. *J. Biol. Chem.* **267**, 24–26.

126. Manji, H. K., Etcheberrigaray, R., Chen, G., and Olds, J. L. (1993) Lithium decreases membrane-associated protein kinase C in hippocampus: selectivity for the α enzyme. *J. Neurochem.* **61**, 2303–2310.

127. Manji, H. K., Bersudsky, Y., Chen, G., Belmaker, R. H., and Potter, W. Z. (1996) Modulation of protein kinase C isozymes and substrates by lithium: the role of myo-inositol. *Neuropsychopharmacology* **15**, 370–381.

128. Chen, G., Manji, H. K., Hawver, D. B., Wright, C. B., and Potter, W. Z. (1994) Chronic sodium valproate selectively decreases protein kinase C α and ε in vitro. *J. Neurochem.* **63**, 2361–2364.

129. Lenox, R. H., McNamara, R. K., Watterson, J. M., and Watson, D. G. (1996) Myristoylated alanine-rich C kinase substrate (MARCKS): a molecular target for the therapeutic action of mood stabilizers in the brain? *J. Clin. Psychiatry* **57**(Suppl 13), 23–31.

130. Chen, Y. and Hertz, L. (1996) Inhibition of noradrenaline stimulated increase in $[Ca^{2+}]_i$ in cultured astrocytes by chronic treatment with a therapeutically relevant lithium concentration. *Brain Res.* **711**, 245–248.

131. Yamaji, T., Kagaya, A., Uchitomi, Y., Yokota, N., and Yamawaki, S. (1997) Chronic treatment with antidepressants, verapamil, or lithium inhibits the serotonin-induced intracellular calcium response in individual C6 rat glioma cells. *Life Sci.* **60**, 817–823.

132. Nilsson, M., Hansson, E., and Rönnback, L. (1992) Agonist-evoked Ca^{2+} transients in primary astroglial cultures—modulatory effects of valproic acid. *Glia* **5**, 201–209.

133. Yamaji, T., Kagaya, A., Uchitomi, Y., Yokata, N., and Yamawaki, S. (1996) Effects of carbamazepine and sodium valproate on 5-HT-induced calcium increase in individual C6 rat glioma cells. *Neuropsychobiology* **34**, 22–25.

134. Nonaka, S., Hough, C. J., and Chuang, D. M. (1998) Chronic lithium treatment robustly protects neurons in the central nervous system against excitotoxicity by inhibiting N-methyl-D-aspartate receptor-mediated calcium flux. *Proc. Natl. Acad. Sci. USA* **95**, 2642–2647.

135. Hough, C. J., Irwin, R. P., Gao, X. M., Rogawski, M. A., and Chuang, D. M. (1996) Carbamazepine inhibition of N-methyl-D-aspartate-evoked calcium flux in rat cerebellar granule cells. *J. Pharmacol. Exp. Ther.* **276**, 143–149.

136. Chou, J. C.-Y. (1991) Recent advances in treatment of acute mania. *J. Clin. Psychopharmacol.* **11**, 3–21.
137. Pazzaglia, P. J., Post, R. M., Ketter, T. A., George, M. S., and Maranggell, L. B. (1993) Preliminary controlled trial of nimodipine in ultra-rapid cycling affective dysregulation. *Psychiatry Res.* **49**, 257–272.
138. Dubovsky, S. L., Murphy, J., Christiano, J., and Lee, C. (1992) The calcium second messenger system in bipolar disorders: data supporting new research directions. *J. Neuropsychiatry* **4**, 3–14.

Part V

Drug Dependence

12

Messengers in Opioid Dependence
Gene Disruption Studies

Rafael Maldonado and Olga Valverde

INTRODUCTION

Opioid drugs are powerful therapeutic agents that have been used for a thousand years to relieve various pathological situations. At the present time, morphine and its derivatives still remain the most potent class of analgesics, and they have clear therapeutic indications despite a considerable number of adverse side effects. However, some opioid compounds also represent an important public health problem because heroin, a diacetylated morphine derivative, is one of the most prominent illicit drugs of abuse. Most of the mechanisms involved in acute opioid responses have been well clarified now. Nevertheless, the neurobiological processes underlying opioid addiction are still poorly understood and remain a field of intense research activity.

MOLECULAR AND BIOCHEMICAL
CHANGES INDUCED DURING OPIOID DEPENDENCE

Since the identification of opioid receptors *(1–3)* and their endogenous ligands *(4)* in the central nervous system, a rapid development of the neurobiology of opioid research has been observed. Opioid peptides derive from precursor proteins known as proopiomelanocortin, proenkephalin, and prodynorphin *(4,5)*. In addition, a novel peptide has recently been isolated from the rat brain, which may represent an endogenous opioid highly selective for mu opioid receptors *(6)*. Endogenous and exogenous opioids exert their pharmacological actions throught three main types of receptors: μ-, δ-, and κ-opioid receptors, which have been recently cloned (for a review, *see* ref. *7*). Opioid receptors are all proteins of approximately 60 kDa and belong to the family of G-protein-coupled receptors that have seven transmembrane

From: *Cerebral Signal Transduction: From First to Fourth Messengers*
Edited by: M. E. A. Reith © Humana Press Inc., Totowa, NJ

domains *(8)*. Three genes encoding respectively the μ, δ, and κ opioid receptors have been identified *(7,9)*. Each of these receptors has been shown to have a high degree of amino acid sequence similarity *(7,9,10)*. Indeed, the μ, δ, and κ receptors are homologous to one another at both the nucleic acid and amino acid levels, being highly conserved in regions spanning the transmembrane domains and intracellular loops *(9)*.

Upon activation, opioid receptors become associated with the G-protein, which interacts with many other cellular proteins to produce various biological effects. One target of these G-proteins is an action on ion channels. Thus, opioid agonists increase the activity of inward-rectifying K^+ channels and decrease the activity of voltage-dependent Ca^{2+} channels *(11)*. Both of these actions are considered inhibitory effects. Opioid-receptor activation also modifies the activity of several intracellular second-messenger systems, including cyclic AMP, Ca^{2+}-dependent protein kinases, phospholipase C, and mitogen-activated protein (MAP) kinases pathways *(12–14)*. The best known pathway for opioid signal transduction is the cyclic AMP cascade. Thus, acute opioid administration decreases intracellular cyclic AMP levels by acting through an inhibitory G-protein, which decreases adenylyl cyclase activity *(11)*. Compensatory changes on the cyclic AMP system have been hypothesized to be an important part of the adaptive changes occurring during chronic opioid administration. Thus, opioid inhibition of cyclic AMP pathways could lead to a compensatory upregulation in this system, which has been suggested to be directly linked to the development and expression of opioid dependence *(12)*. Several pharmacological findings support this hypothesis: (1) The injection of cyclic AMP intracerebroventricularly increases naloxone-precipitated morphine abstinence *(15)*; (2) a quasiwithdrawal syndrome can be induced in naive animals by treatment with phosphodiesterase inhibitors *(16)* or the local administration of a protein kinase activator *(17)*; (3) central administration of protein kinase inhibitors has been reported to decrease the severity of naloxone-precipitated morphine withdrawal syndrome *(17–19)*. However, adaptive changes to chronic opioid administration have also been reported in second messengers other than cyclic AMP, such as the phosphatidylinositol (for a review, *see* ref. *20*) and the MAP kinases pathways *(14)*.

The changes induced on the cyclic AMP pathway during opioid dependence have been mainly identified in vivo in the locus coeruleus, the major noradrenergic nucleus in the brain that mediates many of the somatic signs of opioid abstinence *(21,22)*. Thus, chronic morphine treatment upregulates the cyclic AMP system in this brain structure at every major step between receptor and response, including G-proteins *(24)*, adenylyl cyclase, protein kinase *(25)*, and the phosphorylation of proteins *(26)*. The activation of these second-messenger and protein phosphorylation pathways during opioid dependence lead to the stimulation of certain transcription factors, which then activates transcriptional activity and, ultimately, changes early gene expression. Thus, changes in the levels and phosphorylation state of the cAMP response element-binding (CREB) protein have been also reported during opioid dependence *(27–29)*.

Opioid exposure has also been found to produce modifications at the level of the opioid receptors. Thus, opioid receptors can be phosphorylated by some G-protein receptor kinases, and such phosphorylation produces a desensitization in receptor function *(30)*. Additionally, opioid exposure has been reported to induce sequestration and recycling of μ opioid receptors in neurons in vivo *(31,32)*. However, the functional relevance of these receptor adaptations on opioid tolerance and dependence in vivo remains to be elucidated.

Together with all these processes of homologous regulation produced on opioid receptors and their intracellular messengers, the development of opioid dependence includes processes of heterologous regulation affecting neurotransmitters other than those within the opioid system *(33)*. Thus, changes in several neurochemical systems, such as the noradrenergic and dopaminergic systems, and neuropeptides have been observed during chronic opioid administration and withdrawal. An important noradrenergic hyperactivity is produced during opioid withdrawal in the locus coeruleus and its projection areas, which seems to be related to the physical manifestations of abstinence *(34,35)*. In contrast, a profound and long-term depression of mesolimbic dopamine activity is observed during opioid abstinence, as well as during the withdrawal of other drugs of abuse *(36,37)*. This dopamine decrease is interpreted to be correlated to the disphoric states associated with opioid abstinence (for a review, *see* ref. *38*).

Recent advances in molecular biology techniques have provided useful tools for allowing a better understanding of the functional relevance of the different molecular and biochemical changes observed during opioid dependence. Indeed, the activity of known genes may be modified in vivo using gene-targeting technology, which is becoming a widely used approach to investigate gene function in the whole animal. Additionally, the recent molecular cloning of the opioid receptors *(10,39)* now allows the use of these techniques to address the functional study of these receptors by a genetic approach (for a review, *see* ref. *7*).

Mice lacking opioid receptors as well as other receptors and intracellular messengers related to opioid responses have been recently generated by homologous recombination. The observation of these mutant mice has shed a new light on the mode of action of opioids and on the neurobiological mechanisms underlying opioid dependence at the molecular level. The next sections of this chapter will summarize the recent advances on the neurobiological substrate of opioid dependence obtained from the behavioral analysis of mice deficient in opioid receptors, D2 dopaminergic receptors, and the transcription factor CREB.

PARTICIPATION OF THE OPIOID
RECEPTORS IN OPIOID DEPENDENCE

Several pharmacological approaches have been used to evaluate the particular contribution of each type of opioid receptor to the development and expression of opioid dependence. One of the most prominent approches has been the administration of selective opioid agonists. For example, repeated administration of the μ-selective agonist DAMGO induces a marked degree of dependence in rats *(40,41)*. In contrast, a moderate abstinence effect is observed after the chronic administration of the δ-selective agonists DPDPE and DSTBULET *(40,41)*, whereas chronic administration of a κ-selective agonist results in an even milder abstinence syndrome *(40,42)*. In all of these studies, the μ, δ, and κ receptor agonists were repeatedly administered at large doses, which might result in a significant loss of selectivity. Therefore, the different dependence categories observed in these studies may be the result of the stimulation of more than one type of opioid receptor. Another pharmacological procedure used with the same purpose consisted of inducing withdrawal by administering selective antagonists of the different opioid receptors to animals that have received chronic treatment with a large dose of morphine *(43)* (Fig. 1). In this case, CTAP, a selective μ opioid-receptor antagonist was the most effective and potent compound at inducing the morphine withdrawal syndrome, indicating a preferential participation of μ opioid receptors in the expression of morphine abstinence. The administration of high doses of the selective κ opioid-receptor antagonist norbinaltorphimine produced a moderate degree of withdrawal in morphine-dependent animals. The chronic administration of norbinaltorphimine during the development of morphine dependence increased the severity of naloxone-precipitated opioid withdrawal *(45)*. Additionally, the acute administration of the kappa receptor agonist U-50,488H failed to modify the severity of morphine withdrawal *(46)*. Taken together, these findings seem to indicate that κ receptors do not play a crucial role in the development of dependence associated with chronic morphine treatment. The selective δ opioid-receptor antagonist naltrindole did not produce any relevant manifestation of abstinence in morphine-dependent rats *(43)*. In contrast, the blockade of delta opioid

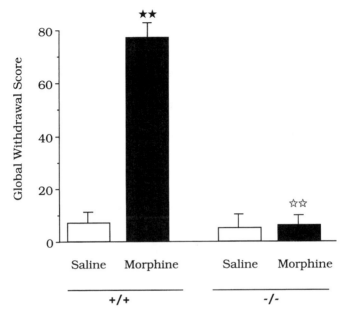

Fig. 1. Global withdrawal score in mutant mice lacking opioid μ opioid receptors and their wild-type littermates. Withdrawal syndrome was precipitated by naloxone administration in mice chronically treated with morphine. ★★ $p < 0.01$ compared with saline-treated mice; ☆☆ $p < 0.01$ comparison between wild-type and mutant groups receiving the same treatment (one-way ANOVA). (Reproduced from ref. *44*, with permission.)

receptors both prior to and during the duration of morphine treatment has been reported to attenuate the naloxone-precipitated morphine withdrawal syndrome *(47)*. Moreover, the administration of the δ receptor antagonist naltriben precipitated several somatic signs of abstinence in morphine-dependent mice *(47)*. However, certain doses of the delta opioid-receptor antagonists used in these previous studies have been reported to block some of the pharmacological responses induced by the stimulation of μ opioid receptors *(48)*. Therefore, the results obtained from these pharmacological studies using μ-, δ-, and κ-selective compounds have not permitted the clear separation of the specific roles mediated by each opioid receptor.

μ, κ, and, recently δ opioid receptors have each been inactivated by applying a similar strategy; that is, by disrupting the receptor gene in mice by homologous recombination *(7)*. These mice provide the opportunity to unambiguously characterize the physiological role of each type of receptor. Thus, to inactive the μ opioid-receptor gene, a neomycin-resistance gene (*Neo*) has been inserted into the gene encoding such a receptor *(49)*. Saturation binding with

the μ-receptor-selective ligand ³[H]DAMGO demonstrated a complete abolition of μ-binding sites in homozygous animals. The absence of μ opioid
receptors did not produce any compensatory mechanisms among other components of the endogenous opioid system. Thus, the total number of δ- and
κ-binding sites, as well as the expression of proopiomelanocortin, proenkephalin, and prodynorphin, were not modified in the brain of μ-deficient
animals. Autoradiographic mapping showed no major modification in the
distribution of δ and κ sites in these mutant mice *(49)*, although a recent
study has shown some regional specific changes in discrete brain areas *(49a)*.
Spontaneous behavior was not altered in mutant mice, except for a slight
decrease in locomotor activity after habituation *(49)*, and a higher sensitivity
to some specific nociceptive stimuli *(50)*. Acute antinociceptive responses
induced by morphine in the hot-plate and tail-immersion tests were completely abolished in these mutant mice. The development of physical morphine
dependence was also deeply investigated in μ-opioid-receptor-deficient
mice. Opioid dependence was induced by administering increasing doses of
morphine for 5 d *(49)*. The doses of morphine used (from 20 to 100 mg/kg,
ip, twice a day) were presumably high enough to produce a nonselective
activation of the different opioid receptors. A withdrawal syndrome was precipitated by the acute injection of naloxone at a dose (1 mg/kg, sc) reported
to block the pharmacological responses induced by μ, δ, and κ opioid-
receptor-selective agonists *(40,51)*. This dose of naloxone precipitated an
abstinence syndrome in wild-type mice that consisted of the classical signs
of withdrawal, including behavioral, vegetative, and biochemical manifestations. In contrast, none of the manifestations of opioid withdrawal were
observed in mutant mice, whose behavior after receiving naloxone was
exactly the same as that exhibited by saline-treated animals (Fig. 1). These
results clearly define the importance of μ opioid-receptors in the development and expression of morphine dependence *(49)*. The rewarding properties of morphine were also investigated in mutant mice and their wild-type
littermates using the place-conditioning paradigm. Morphine induced a clear
place preference in wild-type mice, whereas this conditioned behavior was
not observed in mutant mice. Mutant mice spent the same time in the mor-
phine-paired compartment during the preconditioning and the testing phase,
presumably because of the loss of the rewarding properties of morphine
(Fig. 2). The inability of morphine to induce place preference and opioid
dependence in μ-opioid-receptor-deficient mice indicates that the addictive
properties of this compound are mediated by the μ opioid receptor *(49)*.

Taken together, all these pharmacological and molecular results suggest
that κ and δ opioid receptors play a minor role, if any, in morphine depen-

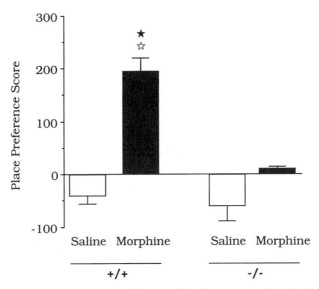

Fig. 2. Place preference induced by morphine in mutant mice lacking μ opioid receptors and their wild-type littermates. Place preference scores were calculated as the difference between postconditioning and preconditioning time spent in the compartment associated with the drug. ★ $p < 0.01$ compared with saline-treated animals: ☆ $p < 0.01$ comparisons between wild-type and mutant groups receiving the same treatment (one-way ANOVA). (Reproduced from ref. *44*, with permission.)

dence. Accordingly, we have also analyzed the development of opioid dependence in mutant mice deficient in κ opioid receptors. First, it is interesting to note that mice lacking κ opioid receptors appear normal under basal conditions *(52)*. Thus, the expression of endogenous opioid peptide genes (proopiomelanocortin, proenkephalin, and prodynorphin) and the total number and distribution of μ and δ opioid receptors were similar in mutant and wild-type mice. The presence of the κ opioid receptor was not essential for maintaining normal spontaneous responses under nonstressful or stressful conditions or for maintaining normal circadian rhythms and locomotor activity. The only modification found on spontaneous behavior of κ-opioid-receptor-deficient mice was a higher sensitivity to visceral pain induced by a chemical stimulus *(52)*. Moreover, antinociceptive responses elicited by morphine did not significantly changed in these mutant mice, at least in response to the presentation of thermal stimuli *(52)*. In contrast, morphine physical dependence was modified in κ-opioid-receptor-deficient mice, as revealed by a moderate reduction in several behavioral signs of naloxone-precipitated withdrawal (Fig. 3). Therefore, the κ-opioid-receptor gene, although not essential, also participates in the expression of morphine abstinence. The

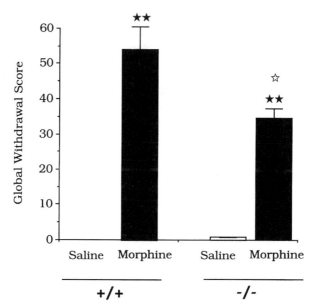

Fig. 3. Global withdrawal score in mutant mice lacking κ opioid receptors and their wild-type littermates. A withdrawal syndrome was precipitated by naloxone administration in mice chronically treated with morphine. ★★ $p < 0.01$ compared with saline-treated mice; ☆ $p < 0.05$ comparison between wild-type and mutant groups receiving the same treatment (one-way ANOVA). (Reproduced from ref. 52, with permission.)

receptor arising from the κ-opioid-receptor gene may functionally modulate this μ-opioid-receptor-mediated response in a manner similar to that of other heterologous neurotransmitter systems (for a review, *see* ref. 53). We have also observed that morphine-conditioned place preference is not altered in κ-opioid-receptor-deficient mice (Fig. 4). We anticipated that the absence of the κ opioid receptor might allow a facilitation of morphine reward in κ mutant mice, because of the reported opposing actions of μ and κ agonists in modulating the endogenous tone of mesolimbic dopaminergic neurons *(54)* and the ability of the κ agonist to block morphine reward *(55)*. However, our data indicate that the absence of κ sites has little influence on the neuronal mechanisms involved in the rewarding properties of morphine.

At present, only a few studies have been until now performed with knock-out mice deficient in δ opioid receptors *(56,57)*. These studies have examined antinociceptive responses and the development of opioid tolerance in δ-deficient mice but have not yet addressed the different aspects of morphine dependence in these animals. Thus, it has been reported that morphine antinociception is essentially unaffected in δ knockout mice, whereas the

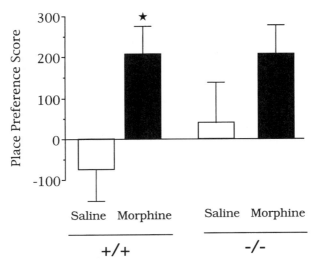

Fig. 4. Place preference induced by morphine in mutant mice lacking κ opioid receptors and their wild-type littermates. Place preference scores were calculated as the difference between postconditioning and preconditioning time spent in the compartment associated with the drug. ★ $p < 0.05$ compared with saline-treated animals (one-way ANOVA). (Reproduced from ref. *52*, with permission.)

development of tolerance to the morphine antinociceptive effects is completely abolished *(57)*. Further studies using δ-opioid-receptor-deficient mice will be performed in the near future in order to determine the exact contribution of this opioid receptor to the mechanisms involved in opioid addiction.

Knockout mice deficient in a specific opioid receptor represent an excellent model for clarifying the proposed functional interactions between the different opioid receptors *(58)*. These interactions have been suggested based on the results of biochemical and pharmacological approaches and could have important clinical implications. Thus, a recent report shows that several pharmacological responses, including antinociception and respiratory effects induced by the κ-selective agonist U-50,488H are not modified in mutant mice deficient in μ opioid receptors *(50)*. These results strongly suggest that functional properties of the κ receptor are maintained in mutant mice and that the κ opioid receptor mainly acts independent of the μ receptor. However, some pharmacological responses of selective δ opioid agonists are reduced in mice lacking the μ receptor *(50,59)*. For instance, the antinociceptive responses of deltorphin II and DPDPE are not modified when using the hot-plate test, but a reduction in the effectiveness of DPDPE is observed in the tail-flick test *(50)*. Therefore δ-receptor-mediated antinociception seems to be influenced by the presence of μ receptors, probably at the level of the spinal cord. A reduction of DPDPE antinociception in both tail-flick and hot-plate tests has been also reported *(59)*. Both studies support the notion of a μ/δ cooperativity in opioid analgesia, as previously suggested by pharmacological experiments *(60)*. Quantitative autoradiography studies revealed a regionally specific small level of downregulation in δ receptors in μ-deficient mice *(8)*. These changes in receptor expression could also have some influence on the reported functional impairments in δ-receptor-mediated behavior.

At present, knockouts for several genes in the opioid system have been constructed, including opioid receptor, β-endorphin, and proenkephalin knockouts. Homozygous mutant mice, which lack β-endorphin *(61)*, preproenkephalin *(62)*, the μ opioid receptor *(49,63,64)*, the δ opioid receptor *(56)*, or the κ opioid receptor *(52)* suggest that the absence of a single component of the opioid system does not markedly alter its development *(7)*. Recently, it became possible to generate double-mutant mice lacking two opioid receptors. Deficient mice for δ/μ, μ/κ, or δ/κ opioid receptors will provide a new tool for evaluating the role of each opioid receptor in the phenomenon of drug addiction by using animals containing only a single type of opioid receptor.

THE INVOLVEMENT OF THE TRANSCRIPTION FACTOR CREB IN OPIOID DEPENDENCE

As stated in the first subheading of this chapter, several intracellular messengers, including transcription factors, have been reported to be modified during opioid dependence. CREB is a member of the CRE/activating transcription factor family that has been reported to be modified during chronic opioid administration and withdrawal *(27)*. CREB was initially identified as the major nuclear factor mediating a transcriptional response to elevated levels of cAMP *(65)* and is one of the best characterized nuclear target proteins for phosphorylation by protein kinase A *(66–68)*. In addition, CREB has been identified as a substrate for other kinases, such as Ca^{2+}/calmodulin kinases *(69)*, a glycogen synthase kinase-3 *(70)*, and RSK2, a member of the ribosomalprotein S6 kinase family *(71)*.

Acute morphine administration decreases the level of CREB phosphorylation in the locus coeruleus, a state associated with decreased CREB activity. Following chronic morphine treatment, the levels of CREB phosphorylation are not different from control animals, but during opioid withdrawal, a substantial increase in CREB phosphorylation is observed *(27)*. The total levels of CREB immunoreactivity are not modified after acute morphine, whereas chronic morphine increases CREB protein in the locus coeruleus *(28)* but decreases it in the nucleus accumbens *(29)*. These adaptive changes of CREB in the locus coeruleus could hypothetically contribute to the upregulation of the other cAMP pathway proteins and thereby to the behavioral effects of chronic opioid administration.

The behavioral relevance of the biochemical changes occurring on CREB during opioid dependence has been now clarified by using genetic approaches. Indeed, a mutate allele of the CREB gene has been produced in mice by targeting the first translated exon of this gene, which includes the protein initiation codon *(72)*. However, the synthesis of the CREB protein was not completely prevented in these mice because this mutation leads to an alternative splicing event that gives rise to an increase in a relatively minor isoform of CREB, termed CREBβ. Therefore, whereas the two major isoforms of CREB, α and Δ, are missing the minor isoform, β is upregulated in mice carrying this hypomorphic allele of CREB *(73)*. When deleting the entire DNA binding domain of CREB, a functional null allele is obtained. Mice homozygous for the CREB null mutation lack all CREB protein but die at birth and therefore cannot be used for further studies (for a review, *see* ref. *74*).

Several responses to acute and chronic opioid administration have been evaluated in CREBαΔ mutant mice *(75)*. Acute morphine produced the same locomotor and antinociceptive responses in mutant mice and their wild-type littermates. No obvious differences were observed between these two groups during the course of chronic morphine treatment. However, the somatic expression of the withdrawal syndrome precipitated by naloxone was less severe in mutant than in wild-type morphine-dependent mice (Fig. 5). Thus, all the behavioral and vegetative signs of withdrawal were significantly lower in CREBαΔ mutant mice. Some specific signs of abstinence were completely abolished in these mutants, such as sniffing and ptosis. This decreased withdrawal was not due to an alteration in the total amount of opioid receptors because binding studies showed no modification in the number or affinity of opioid receptors in these mutant mice. During chronic opioid administration, mice develop tolerance to most of its acute pharmacological actions. The development of tolerance to the antinociceptive effects of morphine was significant in CREBαΔ mutant mice, but to a lesser extent than in wild-type mice *(74,75)*.

A dramatic increase in the expression of immediate early genes has been reported during opioid withdrawal in several brain structures, including the locus coeruleus, amygdala, nucleus accumbens, cortex, hypothalamus, and autonomic areas of the brain *(76,77)*. Opioid regulation of c-*fos* expression may be mediated by CREB because CRE is essential for the induction of this gene *(78)*. However, the immediate early genes c-*fos* and c-*jun* were

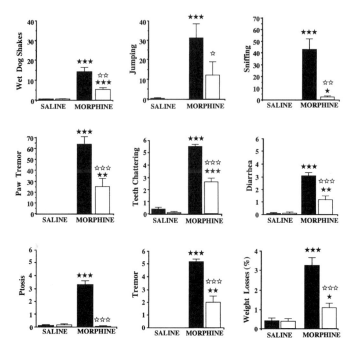

Fig. 5. Analysis of morphine-induced physical dependence. Incidence of behavioral signs of abstinence and weight loss measured during naloxone precipitated morphine-withdrawal syndrome in mutant mice with a mutation of the gene encoding CREB and their wild-type littermates. Withdrawal syndrome was precipitated by naloxone administration in mice chronically treated with morphine. ★★★ $p < 0.001$; ★★ $p < 0.01$; ★ $p < 0.05$ compared with saline-treated mice; ☆☆☆ $p < 0.001$; ☆☆ $p < 0.01$; ☆ $p < 0.05$ comparison between wild-type and mutant groups receiving the same treatment (one-way ANOVA). (Reproduced from ref. *75*, with permission.)

still induced following the precipitation of the morphine withdrawal syndrome in CREBαΔ mutant mice, despite the reduction in CREB protein and the dramatic attenuation of the behavioral expression of morphine abstinence *(75)*. One possible explanation for this maintained induction of c-*fos* could be that this response is not mediated solely by CREB binding to CRE. Indeed, other elements in the promotor region of this gene can also be important for c-*fos* induction *(79)*, and several members of the CRE family of transcription factors, such as cAMP response element modulator protein (CREM), activating transcription factor 1 (ATF-1), and the minor isoform of CREB, CREBβ, are still present and upregulated in CREBαΔ mutant mice *(72,73)*.

A recent study has shown that the local administration of antisense oligonucleotides directed against CREB into the locus coeruleus reduces local levels of CREB immunoreactivity, attenuates the somatic expression of morphine abstinence, prevents the morphine-induced upregulation of type VIII adenylyl cyclase and tyrosine hydroxylase in this brain structure, and completely suppresses a morphine-withdrawal-induced increase in spontaneous firing rates of locus coeruleus neurons in brain slices *(80)*. Interestingly, this blockade is reversed by addition of the protein kinase A activator 8-bromo-cAMP, but not by the adenylyl cyclase activator forskolin. These findings confirm the results obtained with CREBαΔ knockout mice indicating the contribution of CREB to the expression of physical opioid dependence and indicate a specific involvement of the locus coeruleus in this neurobiological mechanism. Thus, the increase in CREB transcription may be responsible for the upregulation of type VIII adenylyl cyclase and tyrosine hydroxylase in this brain structure during opioid withdrawal *(80)*.

The transcription factor CREB has also been recently shown to be related to the rewarding effects of another drug of abuse, cocaine. Indeed, overexpression of CREB in the nucleus accumbens decreases the rewarding effects of cocaine, whereas overexpression of a dominant-negative mutant CREB increases the rewarding effects of cocaine *(81)*. Therefore, the upregulation of CREB induced by cocaine administration may counteract positive feedback-type adaptations that tend to intensify drug reward *(81,82)*. Whether such molecular adaptations of the transcription factor CREB are involved in the rewarding effects of opioids remain to be elucidated.

DOPAMINERGIC SYSTEM AS A HETEROLOGOUS NEUROTRANSMITTER INVOLVED IN OPIOID DEPENDENCE

Changes in various neurotransmitters, different from the opioids, have been reported during chronic opioid administration and at the time of spontaneous or naloxone-precipitated morphine abstinence, which underlie a heterologous regulation of the opioid-dependence processes (for a review, *see* refs. *38* and *53*). One of the heterologous neurotransmitters closely related to the neurochemistry substrate of opioid addiction is the dopamine. Indeed, converging evidence suggests that many drugs of abuse act through mechanisms involving the brain neurotransmitter dopamine and the neural systems that it regulates *(83)*. The binding of opioids to their specific receptors increases the activity of mesolimbic dopaminergic neurons in the midbrain. The cell bodies of the dopamine neurons are located in the ventral tegmental area and project to the forebrain, nucleus accumbens, olfactory tubercle, frontal cortex, amygdala, and septal area. Rats will self-administer morphine directly into the dopaminergic midbrain structures *(84)*, and opioids administration has been reported to increase both the firing rate of ventral–

tegmental-area dopaminergic neurons and the release of dopamine in the nucleus accumbens *(85)*.

Morphine abstinence has been shown to be associated with a reduction in dopamine levels in the nucleus accumbens *(37)*, which seems to be the key zone that mediates the rewarding effects of opioids *(86)*. However, studies in which the dopamine function has been modified with drugs or lesions have proven contradictory results. The blockade of the D1 and D2 dopamine receptors with antagonists decreases the rewarding properties of opioids as measured using the place-preference paradigm and intracranial electrical self-stimulation techniques *(54,87,88,117)*. Morphine-induced place preference was preferentially blocked by D1 dopamine-receptor antagonists, whereas D2 dopamine-receptor antagonists were less effective *(89)*. However, the currently used D1 receptor antagonists also exhibit significant effects at nondopaminergic receptors *(90,91)*. Dopamine antagonists also reduced the response to opioids in the self-administration paradigm, although it has been suggested that this effect is related to a motor impairment or to the decrease in the incentive aspect of the opioid reward *(85)*. Along these lines, a dopamine-independent mechanism for the rewarding properties of opioids has also been proposed to exist at the level of the nucleus accumbens *(92)*. Thus, rats trained to self-administer cocaine and heroin and receiving 6-hydroxydopamine lesions of the nucleus accumbens showed a time-dependent decrease or extintion of cocaine self-administration, whereas heroin self-administration returned to near-normal levels *(93)*. A similar dopamine-independent effect for both heroin self-administration and heroin-induced place preference has been reported in a series of experiments using dopamine-receptor antagonists *(94)*.

Several studies have also evaluated the participation of the dopaminergic mesolimbic system in physical opioid dependence. The reported effects of dopaminergic compounds on the expression of physical opioid withdrawal are contradictory and depend on the route of administration, the compound used, and the animal species. When dopaminergic agents were given systemically, most studies found an increase in the expression of opioid withdrawal in response to several dopaminergic agonists, although some D1 and D2 agonists were able to decrease some specific manifestation of the abstinence (for a review, *see* ref. *38*). In contrast to most of the results obtained after peripheral administration of dopaminergic agents, it has been reported that the activation of D2 receptors within the accumbens prevents the somatic signs of naloxone-induced morphine withdrawal and, conversely, that blockade of nucleus accumbens D2 receptors in opioid-dependent animals elicits somatic withdrawal symptoms, suggesting an important role of dopamine via D2 receptor in the expression of morphine abstinence *(95)*. However, the administration of opioid antagonists into the nucleus accumbens failed to precipitate the withdrawal syndrome in morphine-dependent rats *(22)*. Consequently, although the nucleus accumbens is probably not a major site for the initiation of physical opioid withdrawal, it could participate in regulating circuits that drive somatic and aversive responses to opioid withdrawal.

All of these controversial points regarding the role of dopamine in drugs of abuse-induced rewarding effects and opioid dependence have been recently re-evaluated using genetic approaches. One focus of interest has been the study of mutant mice lacking one type of dopaminergic receptor, particularly D1, D2, and D4 dopamine receptors, which have been classically involved in the rewarding effects of opioids and psychostimulants. In this line, studies performed with D1 receptor knockout mice showed that these receptors are essential for the psychomotor stimulant effects and for the inhibitory responses on the generation of action potentials in the nucleus accumbens induced by cocaine *(96)*. Furthermore, mice lacking dopamine D4 receptors have been shown to be supersensitive to ethanol and psychostimulants *(97)*. Thus, locomotor activity in mice lacking D4 receptors was stimulated by ethanol, cocaine, and methamphetamine to a greater extent than in wild-type mice, revealing an involvement of the D4 receptor in drug-induced locomotor activity. In addition, mutant mice consistently showed elevated dopamine synthesis and turnover in dorsal striatum and outperformed on the rotarod *(97)*. To our knowledge, no data exist regarding the possible modification of opioid-induced dependence and rewarding effects within these D1 and D4 deficient mice.

By using knockout mice lacking dopamine D2 receptors *(98)*, we have recently established the role of dopamine D2 receptors in the mechanisms of opioid dependence and reward *(99)*. These mutant mice present a decrease in the spontaneous locomotor activity *(98)*, without major impairment in the exploratory behavior *(99)*. Furthermore, the acute effects induced by morphine in antinociception and locomotor activity were preserved *(99)* or even increased *(100)* in D2 receptor knockout mice. The rewarding properties of morphine were tested in these animals by using the conditioning place-preference paradigm. Two different doses of morphine (3 and 9 mg/kg, sc) failed to induce a place-preference, indicating that this conditioning behavior was completely absent in D2 mutant mice (Fig. 6). The absence of morphine place preference was not due to the locomotor impairment of mutant mice, because these mice preserve a motor response to endogenous and exogenous opioids and exhibited a behavior similar to that of wild-type mice during the preconditioning phase, including the number of visits to each compartment

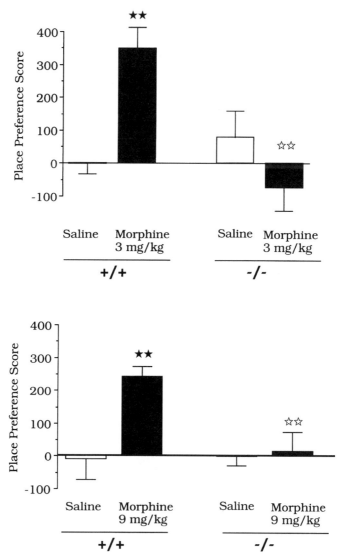

Fig. 6. Place preference induced by morphine in mutant mice lacking D2 dopamine receptors and their wild-type littermates. Place preference scores were calculated as the difference between postconditioning and preconditioning time spent in the compartment associated with the drug. ★★ $p < 0.01$ compared with saline- treated animals. ☆☆ $p < 0.01$ comparisons between wild-type and mutant groups receiving the same treatment (one-way ANOVA). (Reproduced from ref. *99*, with permission.)

(99). Thus, there seems to be a dissociation between opioid-mediated motor and motivational responses. The effects of another incentive stimulus, the response to palatable food, was also studied in the conditioning place pref-

erence in these D2-receptor-deficient mice. Food reward produced the same conditioned place preference in both wild-type and mutant mice, as indicated by a significant increase in the time spent in the food-associated compartment during the testing phase *(99)*. Therefore, the motivational deficit produced by the absence of dopamine D2 receptors selectively affects the rewarding properties of morphine, but it does not modify food-induced place preference. The role played by the D2 dopamine receptor in opioid physical dependence has also been investigated in D2-receptor-deficient mice. Opioid dependence was induced in wild-type and D2 mutant mice by chronic administration of increasing doses of morphine (Fig. 7), and the manifestation of the somatic signs of withdrawal was evaluated after naloxone administration. No differences between D2 mutant mice and their wild-type littermates were observed in the expression of the somatic signs, indicating that D2 receptors are not required to obtain a complete manifestation of the somatic signs of withdrawal *(99)*. These results clearly demonstrated the existence of a dissociation between the mechanisms involved in the rewarding properties of opioids and those involved on the expression of the somatic signs of abstinence. Therefore, dopamine D2 receptors are crucial for the expression of the rewarding properties of opioids but not for the development and expression of opioid physical dependence. A recent finding has also shown that mice lacking the dopamine D2 receptor exhibited a marked aversion to ethanol, relative to the higher preference and consumption exhibited by wild-type littermates evaluated in the two-bottle choice paradigm. Sensitivity to ethanol-induced locomotor impairment was also reduced in these mutant mice, although they showed a normal locomotor depressant response to a dopamine D1 antagonist, demonstrating that dopamine signaling via D2 receptors is an essential component of the molecular pathway determining ethanol self-administration and sensitivity *(101)*.

In spite of these results on morphine and ethanol motivational effects, a growing line of evidence suggests that dopamine is not the key to the rewarding actions of cocaine. Cocaine blocks the uptake of dopamine via neuronal plasma membrane transporters and it also blocks voltage-gated sodium channels. However, knockout mice lacking the dopamine transporter are able to establish cocaine-conditioned place preference *(102)* and self-administer this drug of abuse *(103)*, indicating the participation of other mechanisms in the rewarding effects of cocaine.

Other heterologous neurotransmitter systems have also been involved in mediating the addictive properties of opioids. Thus, the noradrenergic system in the locus coeruleus seems to play a crucial role in the expression of the different somatic signs of opioid withdrawal *(21,104)*. The possible participation of the serotonergic system in response to drugs of abuse and in

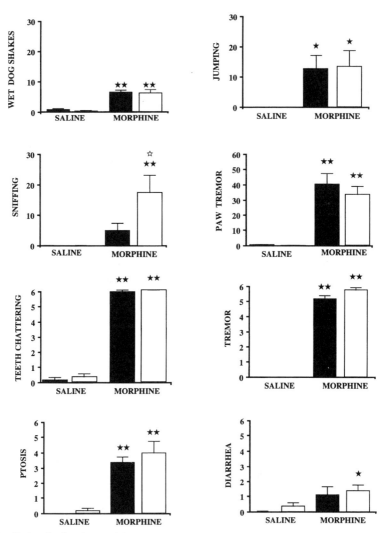

Fig. 7. Analysis of morphine-induced physical dependence. Incidence of behavioral signs of abstinence measured during naloxone-precipitated morphine withdrawal syndrome in mutant mice lacking dopamine D2 receptors (white columns) and their littermates (black columns). Withdrawal syndrome was precipitated by naloxone administration in mice chronically treated with morphine. ★★ $p < 0.01$; ★ $p < 0.05$ compared with saline-treated mice; ☆ $p < 0.05$ comparison between wild-type and mutant groups receiving the same treatment (one-way ANOVA). (Reproduced from ref. 99, with permission.)

morphine dependence is actually another important focus of research. This is a particularly controversial and relevant topic, given the often mutually inhibitory interaction between serotonin and dopamine. Along this line, a recent work has reported the existence of an increased vulnerability to cocaine in mice lacking the 5HT-1B serotonin receptor, as measured by the self-administration model *(105)*. GABA, the most widely distributed inhibitory neurotransmitter, is also involved in the effects of several drugs of abuse *(106)*, and a GABA-mediated pathway has been proposed as a common output for drug reward, including opioids *(92)*. Furthermore, the GABA system also seems to modulate the severity of the expression of the morphine withdrawal syndrome *(107–109)*. Finally, a within-adaptative model, in which chronic morphine would induce the generation of peptides with opioid antagonistic properties, has also been considered to explain the opioid dependence phenomenon *(110)*. This model allows consideration of the role of neuropeptides termed anti-opioid, including substances such as cholecystokinin, FMRF amide, MSH, and MIF-1/tyr-MIF-1-like peptides *(111–116)*.

CONCLUSION

Recent findings obtained using knockout mice deficient in the different opioid receptors have revealed that μ opioid receptors play a key role in the development and expression of opioid dependence, as well as in the rewarding properties of morphine. κ opioid receptors play only a minor role in morphine dependence. Thus, the κ-opioid-receptor gene, although not essential, may functionally modulate this μ-opioid-receptor-mediated response. Previous pharmacological studies suggest that the involvement of δ opioid receptors in morphine dependence is negligible. However, further studies using δ opioid-receptor-deficient mice need to be performed in order to determine the exact contribution of this opioid receptor to opioid dependence. The alterations in the cyclic AMP pathway and particularly in the activity of the transcription factor CREB seem to be responsible for the somatic expression of opiate abstinence. The involvement of these molecular adaptations of the transcription factor CREB in the rewarding effects of opioids remain to be elucidated. The use of genetic approaches has also permitted the further clarification of the involvement of dopamine receptors in the mechanisms of opioid addiction. Thus, dopamine D2 receptors are crucial for the expression of the rewarding properties of opioids but not for the development and expression of opioid physical dependence, showing a clear dissociation between the mechanisms involved in these two opioid responses.

The use of the mouse models developed with the recent advances in molecular biology techniques, as well as those that will be available in the near future, could provide a definitive step toward clarifying the molecular basis of the different biochemical and behavioral manifestations of opioid dependence. The complete elucidation of these mechanims would assist in developing a more rational therapy for opioid addiction.

REFERENCES

1. Pert, C. B. and Synder, S. H. (1973) Properties of opiate-receptor binding in rat brain. *Proc. Natl. Acad. Sci. USA* **70,** 2243–2247.
2. Simon, E. J., Hiller, J. M., and Edelman, I. (1973) Stereospecific binding of the potent narcotic analgesic [³H] etorphine to rat-brain homogenate. *Proc. Natl. Acad. Sci. USA* **70,** 1947–1949.
3. Terenius, L. (1973) Stereospecific intereaction between narcotic analgesics and a synaptic plasma menbrane fraction of rat cerebral cortex. *Acta Pharmacol. Toxicol.* **32,** 317–320.
4. Hughes, J., Smith, T. W., and Kosterlitz, H. W. (1975) Identification of two related pentapeptides from the brain with potent opiate agonist activity. *Nature* **258,** 577–579.
5. Rossier, J. (1982) Opioid peptides have found their roots. *Nature* **298,** 221.
6. Zadina, J. E., Hackler, L., Ge, L., and Kastin, A. J. (1997) A potent and selective endogenous agonist for the μ-opiate receptor. *Nature* **386,** 499–502.
7. Kieffer, B. (1999) Opioids: first lessons from knockout mice. *Trend Pharmacol. Sci.* **20,** 19–26.
8. Kieffer, B. L. (1997) Molecular aspects of opioid receptor, in *Pain: Handbook of Experimental Pharmacology* (Dickenson, A. and Besson, J. M., eds.), Springer-Verlag, Berlin, pp. 281–303.
9. Kieffer, B. L. (1995) Recent advances in molecular recognition and signal transduction of active peptides: receptors for opioid peptides. *Cell Mol. Neurobiol.* **15,** 615–635.
10. Evans, C. J., Keith, D. E., Morrison, H., Magendzo, K., and Edwards, R. H. (1992) Cloning of a delta receptor by functional expression. *Science* **258,** 1952–1955.
11. Nestler, E. J., Hope, B. T., and Widnell, K. L. (1993) Drug addiction: a model for the molecular basis of neural plasticity. *Neuron* **11,** 995–1006.
12. Nestler, E. J. (1997) Molecular mechanisms underlying opiate addiction: implications for medication development. *Neuroscience* **9,** 84–93.
13. García-Sevilla, J. A., Ventayol, P., Busquets, X., La Harpe, R., Walzer, C., and Guimón, J. (1997) Regulation of immunolabelled μ-opioid receptors and protein kinase C-α and ζ isoforms in the frontal cortex of human opiate addicts. *Neurosci. Lett.* **226,** 29–32.
14. Schulz, S. and Höllt, V. (1998) Opioid withdrawal activates MAP kinase in locus coeruleus neurons in morphine-dependent rats in vivo. *Eur. J. Neurosci.* **10,** 1196–1201.
15. Collier, H. O. J., Francis, D. L., and Mc Donald-Gibson, W. J. (1975) Prostaglandins, cyclic AMP and the mechanism of opiate dependence. *Life Sci.* **17,** 85–90.
16. Francis, D. L., Roy, A. C., and Collier, H. O. J. (1975) Morphine abstinence and quasi-abstinence effects after phosphodiesterase inhibitors and naloxone. *Life Sci.* **17,** 85–90.
17. Punch, L. J., Self, D. W., Nestler, E. J., and Taylor, J. R. (1997) Opposite modulation of opiate withdrawal behaviors on microinfusion of a protein kinase A inhibitor versus activator into the locus coeruleus of periaqueductal gray. *J. Neurosci.* **17,** 8520–8527.
18. Maldonado, R., Valverde, O., Garbay, C., and Roques, B. P. (1995) Protein kinases in the locus coeruleus and periaqueductal gray matter are involved in the expression of opiate withdrawal. *Naunyn Schmiedebergs Arch. Pharmacol.* **352,** 565–575.

19. Tokuyama, S., Feng, Y., Wakabayashi, H., and Ho, I. (1995) Possible involvement of protein kinases in physical dependence on opioids: studies using protein kinase inhibitors H-7 and H-8. *Eur. J. Pharmacol.* **284,** 101–107.

20. Harris, H. W. and Nestler, E. J. (1993) Opiate regulation of signal transduction pathways, in *The Neurobiology of Opiates* (Hammer, R., ed.), CRC, New York, pp. 301–331.

21. Aghajanian, G. K. (1978) Tolerance of locus coeruleus neurons to morphine and suppression of withdrawal response by clonidine. *Nature* **276,** 186–188.

22. Maldonado, R.,Stinus, L., Gold, L. H., and Koob, G. F. (1992) Role of different brain structures in the expression of the physical morphine withdrawal syndrome. *J. Pharmacol. Exp. Ther.* **261,** 669–677.

24. Nestler, E. J., Erdos, J. J., Terwilliger, R., Duman, R. S., and Tallman, J. F. (1989) Regulation of G-proteins by chronic morphine in the rat locus coeruleus. *Brain Res.* **476,** 230–239.

25. Nestler, E. J. and Tallman, J. F. (1988) Chronic morphine treatment increases cyclic AMP-dependent protein kinase activity in the rat locus coeruleus. *Mol. Pharmacol.* **33,** 127–132.

26. Guitart, X. and Nestler, E. J. (1989) Identification of morphine and cyclic AMP-regulated phosphoproteins (MARPPs) in the locus coeruleus and other regions of rat brain: regulation by acute and chronic morphine. *J. Neurosci.* **9,** 4371–4387.

27. Guitart, X., Thompson, M. A., Mirante, C. K., Greenberg, M. E., and Nestler, E. J. (1992) Regulation of cyclic AMP response element binding protein (CREB). Phosphorylation by acute and chronic morphine in the rat locus coeruleus. *J. Neurochem.* **58,** 1168–1171.

28. Widnell, K. L., Russell, D. S., and Nestler, E. J.(1994) Regulation of expression of cAMP response element-binding protein in the lcus coeruleus in vivo and in a locus coeruleus-like cell line in vitro. *Proc. Natl. Acad. Sci. USA* **91,** 10,947–10,951.

29. Widnell, K. L., Self, D. W., Lane, S. B., Russell, D. S., Vaidya, V. A., Miserendino, M. J. D., Rubin, C. S., Duman, R. S., and Nestler, E. J. (1996) Regulation of CREB expression: in vivo evidence for a functional role in morphine action in the nucleus accumbens. *J. Pharmacol. Exp. Ther.* **276,** 306–315.

30. Pei, G., Kieffer, B. L., Lefkowitz, R. J., and Freedman, N. J. (1995) Agonist-dependent phosphorylation of the mouse delta-opioid receptor: involvement of G-protein-coupled receptor kinases but not protein kinase C. *Mol. Pharmacol.* **48,** 173–177.

31. Sternini, C., Spann, M., Anton, B., Deith, Jr., D. E., Bunnett, N. W., Von Zastrow, M., Evans, C., and Brecha, N. C. (1996) Agonist selective endocytosis of μ opioid receptor by neurons in vivo. *Proc. Natl. Acad. Sci. USA* **93,** 9241–9246.

32. Bernstein, M. A. and Welch, S. P. (1998) Mu-opioid receptor down-regulation and c-AMP-dependent protein kinase phosphorylation in a mouse model of chronic morphine tolerance. *Mol. Brain Res.* **55,** 237–242.

33. Koob, G. F. and Bloom, F. E. (1988) Cellular and molecular mechanisms of drug dependence. *Science* **242,** 715–723.

34. Done, C., Silverstone, P., and Sharp, T. (1992) Effect of naloxone-precipitated morphine withdrawal on noradrenaline release in rat hippocampus in vivo. *Eur. J. Pharmacol.* **215,** 333–336.

35. Rossetti, Z. L., Longu, G., Mercuro, G., and Gessa, G. L. (1993) Extraneuronal noradrenaline in the prefrontal cortex of morphine-dependent rats: tolerance and withdrawal mechanisms. *Brain Res.* **609,** 316–320.

36. Acquas, E., Carboni, E., and Di Chiara, G. (1991) Profound depression of meso-limbic dopamine release after morphine withdrawal in dependent rats. *Eur. J. Pharmacol.* **193,** 133–134.
37. Rossetti, Z. L., Hmaidan, Y., and Gessa, G. L. (1992) Marked inhibition of meso-limbic dopamine release. A common feature of ethanol, morphine, cocaine and amphetamine abstinence in rats. *Eur. J. Pharmacol.* **221,** 227–234.
38. Maldonado, R., Stinus, L., and Koob, G. F. (1996b) *Neurobiological Mechanisms of Opiate Withdrawal.* Springer-Verlag, Heilderberg.
39. Kieffer, B. L., Befort, K., Gaveriaux-Ruff, C., and Hirth, C. G. (1992) The delta opioid receptor: isolation of a cDNA by expression cloning and pharmacological characterization. *Proc. Natl. Acad. Sci. USA* **89,** 12,048–12,052.
40. Cowan, A., Zhu, X. Z., Mosberg, H. I., Omnaas, J. R., and Porreca, F. (1988) Direct dependence studies in rats with agents selective for different types of opioid receptors. *J. Pharmacol. Exp. Ther.* **246,** 950–955.
41. Maldonado, R., Feger, J., Fournié-Zaluski, M. C., and Roques, B. P. (1990) Differences in physical dependence induced by selective mu or delta opioid agonists and by endogenous enkephalins protected by peptidase inhibitors. *Brain Res.* **520,** 247–254.
42. Aceto, M. D., Bowman, E. R., and Harris, L. S. (1990) Dependence studies of new compounds in the rhesus monkey, rat and mouse. *NIDA Res. Monogr.* **95,** 578–631.
43. Maldonado, R., Negus, S., and Koob, G. F. (1992a) Precipitation of morphine withdrawal syndrome in rats by administration of mu-, delta- and kappa-selective opioid antagonists. *Neuropharmacology* **31,** 1231–1241.
44. Valverde, O., Maldonado, R., and Kieffer, B. (1998) Recent findings on the mechanim of action of morphine. *CNS Drugs* **10,** 1–10.
45. Suzuki, T., Narita, M., Takahashi, Y., Misawa, M., and Nagase, H. (1992) Effects of nor-binaltorphimine on the development of analgesic tolerance to and physical dependence on morphine. *Eur. J. Pharmacol.* **213,** 91–97.
46. Fukagawa,Y., Katz, J. L., and Suzuki, T. (1989) Effects of a selective kappa-opioid agonist, U-50,488H, on morphine dependence in rats. *Eur. J. Pharmacol.* **170,** 47–51.
47. Miyamoto, Y., Portoghese, P. S., and Takemori, A. E. (1993) Involvement of delta$_2$ opioid receptors in the development of morphine dependence in mice. *J. Pharmacol. Exp. Ther.* **264,** 1141–1151.
48. Koob, G. F., Maldonado, R., and Negus, D. (1992) Precipitation of morphine withdrawal syndrome in rats by administration of mu-, delta- and kappa-selective opioid antagonists. *Neuropharmacology* **31,** 1231–1241.
49. Matthes, H. W. D., Maldonado, R., Simonin, F., Valverde, O., Slowe, S., Kitchen, I., Befort, K., Dierich, A., Le Meur, M., Dollé, P., Tzavara, E., Hanoune, J., Roques, B. P., and Kieffer, B. L. (1996) Loss of morphine-induced analgesia, reward effect and withdrawal symptoms in mice lacking the mu-opioid receptor gene. *Nature* **383,** 819–823.
49a. Kitchen, I., Slowe, S. J., Matthes, H. W. D., and Kieffer, B. (1997) Quantitative autoradiographic mapping of mu-, delta- and kappa-opioid receptors in knockout mice lacking the mu-opioid receptor gene. *Brain Res.* **778,** 73–88.
50. Matthes, H. W. D., Smadja, C., Valverde, O., Vonesch, J. L., Foutz, A. S., Boudinot, E., Denavit-Saubié, M., Severini, C., Negri, L., Roques, B. P.,

Maldonado, R., and Kieffer, B. L. (1998) Activity of the delta-opioid receptor is partially reduced, whereas activity of the kappa-receptor is maintained in mice lacking the mu-receptor. *J. Neurosci.* **18,** 7285–7295.

51. Stevens, C. W. and Yaksh, T. L. (1989) Magnitude of opioid dependence after continuous intrathecal infusion of mu-and delta-selective opioids in the rat. *Eur. J. Pharmacol.* **166,** 467–472.

52. Simonin, F., Valverde, O., Smadja, C., Slowe, S., Kitchen, I., Dierich, A., Le Meur, M., Roques, B. P., Madonado, R., and Kieffer, B. L. (1998) Disruption of the kappa-opioid receptor gene in mice enhances sensitivity to chemical visceral pain, impairs pharmacological actions of the selective kappa-agonist U-50,488H and attenuates morphine withdrawal. *EMBO J.* **17,** 886–897.

53. Bhargava, H. N. (1994) Diversity of agents that modify opioid tolerance physical dependence, abstinence syndrome, and self-administrative behavior. *Pharmacol. Rev.* **46,** 293–324.

54. Shippenberg, T. S. and Herz, A. (1987) Place preference conditioning reveals the involvement of D1-dopamine receptors in the motivational properties of mu- and kappa-opioid agonists. *Brain Res.* **436,** 169–172.

55. Funada, M., Suzuki, T., Narita, M., Misiwa, M., and Nagase, H. (1993) Blockade of morphine reward through the activation of kappa-opioid receptors in mice. *Neuropharmacology* **32,** 1315–1323.

56. Zhu, I., King, M., Schuller, A., Unterwald, E., Pasternak, G., and Pintar, J. E. (1997) Genetic disruption of the mouse Delta opioid gene. *1st European Opioid Conference,* Guildford, UK.

57. Pintar, J., Zhang, J., King, M., Schuller, A., Zhu, Y., Chen, Z. P., and Pasternak, G. (1998) Genetic analysis of opioid receptor function. *29th International Narcotic Research Conference,* Garmish-Partenkirchen, Germany.

58. Rothman, R. B, Holaday, J. W., and Porreca, F. (1993) Allosteric coupling among opioid receptors: evidence for an opioid receptor complex, in *Handbook of Experimental Pharmacology* (Herz, A., ed.), Springer-Verlag, Berlin, pp. 217–237.

59. Sora, I., Funada, M., and Uhl, G. R. (1997b) The mu-opioid receptor is necessary for [D-Pen2, D-Pen5] enkephalin-induced analgesia. *Eur. J. Pharmacol.* **324,** R1–R2.

60. Jiang, Q., Mosberg, H. I., and Porreca, F. (1990) Modulation of the potency and efficacy of mu-mediated antinociception by delta agonists in the mouse. *J. Pharmacol. Exp. Ther.* **254,** 683–689.

61. Rubinstein, M., Mogil, J. S., Japon, M., Chan, E. C., and Allen, R. G. (1996) Absence of opioid stress-induced analgesia in mice lacking β-endorphin by site-directed mutagenesis. *Proc. Natl. Acad. Sci. USA* **93,** 3995–4000.

62. König, M., Zimmer, A. M., Steiner, H., Holmes, P. V., Crawley, J. N., Brownstein, M. J., and Zimmer, A. (1996) Pain responses, anxiety and agression in mice deficient in pre-proenkephalin. *Nature* **383,** 535–539.

63. Sora, I., Takahashi, N., Funada, M., Ujike, H., Revay, R. S., Donovan, D. M., Miner, L. L., and Uhl, G. R. (1997a) Opiate receptor knockout mice define mu-receptor roles in endogenous nociceptive responses and morphine-induced analgesia. *Proc. Natl. Acad. Sci. USA* **94,** 1544–1549.

64. Tian, M., Broxmeyer, H. E., Fan, Y., Lai, Z., Zhang, S., Aronica, S., Cooper, S., Bigsby, R. M., Steinmetz, R., Engle, S. J., Mestek, A., Pollock, J. D., Lehman, M. N., Jensen, H. T., Ying, M., Stambrook, P. J., Tischfield, J. A., and Yu, L. (1997)

336 Maldonado and Valverde

Altered hematopoiesis, behavior and sexual function in mu opioid receptor deficient mice. *J. Exp. Med.* **185,** 1517–1522.

65. González, G. A. Yamamoto, K. K., Fischer, W. H., Karr, D., Menzel, P., Biggs, W., Vale, W. W., and Montminy, M. M. (1989) A cluster of phosphorylation sites on the cyclic AMP-regulated nuclear factor CREB predicted by its sequence. *Nature* **337,** 749–752.

66. Yamamoto, K. K., González, G. A., Biggs, W. H., III, and Montminy, M. R. (1988) Phosphorylation-induced binding and transcriptional efficacy of nuclear factor CREB. *Nature* **334,** 494–498.

67. González, G. A. and Montminy, M. R. (1989) Cyclic AMP stimulates somatostanin gene transcription by phosphorylation of CREB at serine 133. *Cell* **59,** 675–680.

68. Nichols, M., Weih, F., Schmid, W., De Vack, C., Kowenz-Leutz, E., Luckow, B., Boshart, M., and Schutz, G. (1992) Phosphorylation of CREB affects its binding to high and low affinity sites: implications for cAMP-induced gene transcription. *EMBO J.* **11,** 3337–3346.

69. Dash, P. K., Karl, K. A., Colicos, M. A., Prywes, R., and Kandel, E. R. (1991) cAMP response element-binding protein is activated by Ca^{2+}/calmodulin as well as cAMP-dependent protein kinase. *Proc. Natl. Acad. Sci. USA* **88,** 5061–5065.

70. Fiol, C. J., Williams, J. S., Chou, C., Wang, Q. M., Roach, P. J., and Andrisani, O. M. (1994) A secondary phosphorylation CREB341 at Ser 129 is required for the cAMP-mediated control of gene expression. A role for glycogen synthase kinase-3 in the control of gene expression. *J. Biol. Chem.* **268,** 32,187–32,193.

71. Xing, J., Ginty, D. D., and Greenberg, M. E. (1996) Coupling of the RAS–MAPK pathway to gene activation by RSK2, a growth factor-regulated CREB kinase. *Science* **273,** 959–963.

72. Hummler, E., Cole, T. J., Blendy, J. A., Ganss, R., Aguzzi, A., Schmid, W., Beermann, F., and Schutz, G. (1994) Targeted mutation of the CREB gene: compensation within the CREB/ATF family of transcription factors. *Proc. Natl. Acad. Sci. USA* **91,** 5647–5651.

73. Blendy, J. A., Kaestner, K. H., Schmid, W., Gass, P., and Schutz, G. (1996) Targeting of the CREB gene leads to up-regulation of a novel CREB mRNA isoform. *EMBO J.* **15,** 1098–1106.

74. Blendy, J. A. and Maldonado, R. (1998) Genetic analysis of drug addiciton: the role od cAMP response element binding protein. *J. Mol. Med.* **76,** 104–110.

75. Maldonado, R., Blendy, J. A., Tzavara, E., Gass, P., Roques, B. P., Hanoune, J., and Schutz, J. (1996) Reduction of morphine abstinence in mice with a mutation in the gene encoding CREB. *Science* **273,** 657–659.

76. Hayward, M. D., Duiman, R. S., and Nestler, E. J. (1990) Induction of the c-fos proto-oncogene during opiate withdrawal in the locus coeruleus and other regions of rat brain. *Brain Res.* **525,** 256–266.

77. Stornetta, R. L., Norton, F. E., and Guyenet, P. G. (1993) Autonomic areas of rat brain exhibit increased Fos-like immunoreactivity during opiate withdrawal in rats. *Brain Res.* **624,** 19–28.

78. Sheng, M. and Greenberg, M. E. (1990) The regulation and function of c-fos and other inmediate early genes in the nervous system. *Neuron* **4,** 571–582.

79. Curran, T. and Morgan, J. I. (1985) Superinduction of c-fos by nerve growth factor in the presence of peripherally active benzodiazepines. *Science* **229,** 1265–1268.

80. Lane-Ladd, S. B., Pineda, J., Boundy, V. A., Pfeuffer, T. Krupinski, J., Aghajanian, G. K., and Nestler, E. J. (1997) CREB (cAMP response element-binding protein) in the locus coeruleus: biochemical, physiological, and behavioral evidence for a role in opiate dependence. *J. Neurosci.* **17,** 7890–7901.

81. Carlezon, W. A., Jr., Thome, J., Olson, V. G., Lane-Ladd, S. B., Brodkin, E. S., Hiroi, N., Duman, R., Neve, R. L., and Nestler, E. J. (1998) Regulation of cocaine reward by CREB. *Science* **282,** 2272–2275.

82. Carlezon, W. A., Boundy, V. A., Haile, C. N., Kalb, R. G., Neve, R., and Nestler, E. J. (1997) Sensitization to morphine induced by viral-mediated gene transfer. *Science* **277,** 812–814.

83. Wise, R. A. (1996) Neurobiology of drug addiction. *Curr. Opin. Neurobiol.* **6,** 243–251.

84. Bozarth, M. A. and Wise, R. A. (1981) Heroin reward is dependent on a dopaminergic substrate. *Life Sci.* **29,** 1881–1886.

85. DiChiara, G. and North, A. (1992) Neurobiology of opiate abuse. *Trends Pharmacol. Sci.* **13,** 185–193.

86. Robbins, T. W. and Everitt, B. J. (1999) Drug addiction: bad habits add up. *Nature* **398,** 567–570.

87. Spyraki, C., Fibiger, H. C., and Phillips, A. G. (1983) Attenuation of heroin reward in rats by disruption of the mesolimbic dopamine system. *Psychopharmacology* **79,** 278–283.

88. Leone, P. and Di Chiara, G. (1987) Blockade of D_1 receptors by SCH23390 antagonizes morphine- and amphetamine-induced place preference conditioning. *Eur. J. Pharmacol.* **135,** 251–254.

89. Acquas, E., Carboni, E., Leone, P., and Di Chiara, G. (1989) SCH 23390 blocks drug-conditioned place preference and place-aversion; anhedonia (lack of reward) or apathy (lack of motivation) after dopamine-receptor blockade. *Psychopharmacology* **99,** 151–155.

90. Bischoff, S., Heinrich, M. Sonntag, J. M., and Krauss, J. (1986) The D-1 dopamine receptor antagonist SCH 23390 also interacts potently with brain serotonin (5-HT_2) receptors. *Eur. J. Pharmacol.* **129,** 367–370.

91. Skarsfeldt, T. and Larson, J. J. (1988) SCH 23390 a selective dopamine D-1 receptor antagonist with putative 5-HT_1 receptor agonist activity. *Eur. J. Pharmacol.* **148,** 389–395.

92. Koob, G. F (1992) Drugs of abuse: anatomy pharmacology and function of reward pathways. *Trends Pharmacol. Sci.* **13,** 177–184.

93. Pettit, H. O., Ettenberg, A., Bloom, F. E., and Koob, G. F. (1984) Destruction of dopamine in the nucleus accumbens selectively attenuates cocaine but not heroin self-administration in rats. *Psychopharmacology* **84,** 167–173.

94. Stinus, L., Nadaud, D., Deminière, J. M., Jauregui, J., Hand, T. T., and Le Moal, M. (1989) Chronic flupentixol treatment potentiates the reinforcing properties of systemic heroin administration. *Biol. Psychiatty* **26,** 363–371.

95. Harris, G. C. and Aston-Jones, G. (1994) Involvement of D_2 dopamine receptors in the nucleus accumbens in the opiate withdrawal syndrome. *Nature* **371,** 155–157.

96. Xu, M., Hu, H.-T., Cooper, D. C., Moratalla, R., Graybiel, A. M., White, F. J., and Tonegawa, S. (1994) Elimination of cocaine-induced hyperactivity and dopamine-mediated neurophysiological effects in dopamine D1 receptor mutant mice. *Cell* **79,** 945–955.

97. Rubinstein, M., Phillips, T. J., Bunzow, J. R., Falzone, T. L., Dziewczapolski, G., Ge, Zhang, Fang, Y., Larson, J. L., McDougall, J. A., Chester, J. A., Saez, C., Pugsley, T. A., Gershanik, O., Low, M. J., and Grandy, D. K. (1997) Mice lacking dopamine D4 receptor are supersensitive to ethanol, cocaine and methamphetamine. *Cell* **90,** 991–1001.

98. Baik, J. H., Picetti, R., Saiairdi, A., Thuiruet, G., Dierich, A., Depaulis, A., Le Meur, M., and Borrelli, E. (1995) Parkinson-like locomotor impairment in mice lacking dopamine D$_2$ receptors. *Nature* **377,** 424–428.

99. Maldonado, R., Saiardi, A., Valverde, O., Samad, T. A., Roques, B. P., and Borrelli, E. (1997) Absence of opiate rewarding effects in mice lacking dopamine D2 receptors. *Nature* **388,** 586–589.

100. Valverde, O., Saiardi, A., Samad, T., Roques, B.P., Maldonado, R., Borrelli, E. (1997) Increase of opioid analgesia in mice lacking dopamine D2 receptors. *28th International Narcotic Research Conference.* Hong-Kong, Chine, 1998.

101. Phillips, T. J., Brown, K. J., Burkhart-Kasch, S., Wenger, Ch. D., Kelly, M. A., Rubinstein, M., Grandy, D. K., and Low, M. J. (1998) Alcohol preference and sensitivity are markedly reduced in mice lacking dopamine D$_2$ receptors. *Nature Neurosci.* **1,** 610–615.

102. Sora, I., Wichems, C., Takahashi, N., Li, X. F., Zeng, Z., Revay, R., Lesch, K. P., Murphy, D. L., and Uhl, G. R. (1998) Cocaine reward models: conditioned place preference can be established in dopamine- and in serotonin-transporter knockout mice. *Proc. Natl. Acad. Sci. USA* **95**(13), 7699–7704.

103. Rocha, B. A., Fumagalli, F., Gainetdinov, R. R., Jones, S. R., Ator, R., Giros, B., Miller, G. W., and Caron, M. G. (1998) Cocaine self-administration in dopamine-transporter knockout mice. *Nature Neurosci.* **1,** 132–137.

104. Maldonado, R. and Koob, G. F. (1993) Destruction of the locus coeruleus decreases physical signs of opiate withdrawal. *Brain Res.* **605,** 128–138.

105. Rocha, B. A., Scerarce-Levie, K., Lucas, J. L., Hiroi, N., Castanon, N., Crabbe, J. C., Nestler, E. J., and Hen, R. (1998) Increased vulnerability to cocaine in mice lacking the serotonin-1B receptor. *Nature* **393,** 175–178.

106. Suzdak, P., Glowa, J., Crawley, J., Schwartz, R., Skolnick, P., and Paul, S. (1986) A selective imidazodiazepine antagonist of ethanol in rats. *Science* **234,** 1243–1247.

107. Maldonado, R., Micó, J. A., Valverde, O., Saavedra, M. C., Leonsegui, I., and Gibert-Rahola, J. (1991) Influence of different benzodiazepines on the experimental morphine abstinence syndrome. *Psychopharmacology* **105,** 197–203.

108. Navaratman, V. and Foong, K. (1990) Opiate dependence: the role of benzodiazepines. *Curr. Med. Res. Opin.* **11,** 620–630.

109. Valverde, O., Micó, J. A., Maldonado, R., and Gibert-Rahola, J. (1992) Changes in benzodiazepine-receptor activity modify morphine withdrawal syndrome in mice. *Drug Alcohol Depend.* **30,** 293–300.

110. Koob, G. F., Stinus, L., and Le Moal, M. (1989) Opponents process theory of motivation: neurobiological evidence from studies of opiate dependence. *Neurosci. Biobehav. Rev.* **13,** 135–140.

111. Zadina, J. E., Kastin, A. J., and Kersh, D. (1992) TYR-MIF and hemorphin can act as opiate agonist as well as antagonist in the guinea pig ileum. *Life Sci.* **51,** 869–885.

112. Rothman, R. B. (1992) A review of the role of anti-opioid peptides in morphine tolerance and dependence. *Synapse* **12,** 129–138.

113. Faris, P. K., Komisaruk, B. R., and Watkins, L. R. (1983) Evidence for the neuropeptide cholecystokinin as an antagonist of opiate analgesia. *Science* **219,** 310–312.

114. Galina, Z. H. and Kastin, A. J. (1986) Existence of antiopiate systems as illustrated by MIF-1/Tyr-MIF-1. *Life Sci.* **39,** 2153–2159.

115. Yang, H. Y. T., Fratta, W., and Majane, E. A. (1985) Isolation sequencing, synthesis and pharmacological characterization of two brain neuropeptides that modulates the action of morphine. *Proc. Natl. Acad. Sci. USA* **82,** 7757–7761.

116. Maldonado, R., Valverde, O., Ducos, B., Blommaert, A., Fournié-Zaluski, M. C., and Roques, B. P. (1995) Inhibition of morphine withdrawal by the association of RB 101, an inhibitor of enkephalin catabolism, and the CCK-B antagonist PD-134,308. *Br. J. Pharmacol.* **114,** 1031–1039.

117. Suzuki, T., Tsuji, M., Mori, T., Ikeda, H., Misawa, M., and Nagase, H. (1997) Involvement of dopamine-dependent and independent mechanisms in the rewarding effects mediated by Δ opioid receptor subtypes in mice. *Brain Res.* **744,** 327–334.

13

The 5-HT1B Knockout Mouse

An Animal Model of Vulnerability to Drugs of Abuse

Kimberly Scearce-Levie and René Hen

There is increasing evidence for the existence of genetic determinants that underlie the large individual differences in vulnerability to drugs of abuse. In human males, for example, 40–50% of the variance in alcohol preference appears to be genetically determined, according to twin studies, adoption studies, and familial studies *(1)*. In mice and rats, inbred strains display marked differences in responses to alcohol, cocaine, and other drugs of abuse *(2)*. Such responses include locomotor activation, drug self-administration, and conditioned place preference. Genetic studies have revealed that the differences in drug responsiveness between inbred strains are mediated by several genes that each contribute to a fraction of the phenotypic difference and have, therefore, been named quantitative trait loci (QTLs) *(2)*. In the case of two mouse inbred strains (C57BL/6 and DBA/2), the use of recombinant inbred strains and QTL analysis has allowed the identification of several genetic loci that are responsible for most of the strain differences in preference for different categories of drugs of abuse, including the psychostimulant cocaine, the sedative alcohol, and the opiate morphine *(2, 3)*. Therefore, such genetic studies suggest the existence of common molecular mechanisms that underlie vulnerabilities to different categories of drugs of abuse. This is in good agreement with the existence of brain reward pathways that appear to be involved in the hedonic effects of different classes of drugs. In particular, the mesolimbic dopaminergic pathways are involved in the rewarding properties of psychostimulants, opiates, alcohol, and nicotine *(4)*.

Although dopamine (DA) is a critical neurotransmitter in this pathway, serotonin (5-HT) appears to be able to modulate dopaminergic activity and, thereby, the reinforcing effects of various drugs of abuse *(5–9)*. We are interested in the nature of the 5-HT receptors that participate in these modulatory effects. Pharmacological studies have been difficult because of the large

From: *Cerebral Signal Transduction: From First to Fourth Messengers*
Edited by: M. E. A. Reith © Humana Press Inc., Totowa, NJ

number of 5-HT receptor subtypes and the lack of specific ligands *(10)*. Gene-targeting strategies, which allow alterations or deletions of a single gene, are a more specific alternative to classical pharmacological studies *(11,12)*. In addition, mice with targeted mutations in a particular gene are good models of genetic disorders because they undergo the same types of developmental and plastic changes as naturally occurring genetic variants. Therefore, we have used gene-targeting technology to test the involvement of a number of candidate genes in various responses to drugs of abuse.

The 5-HT_{1B} receptor is one such candidate gene. One or more QTLs associated with differential responses to cocaine, alcohol, and morphine contain the 5-HT_{1B} gene *(2)*. The 5-HT_{1B} receptors are abundant in the ventral tegmental area (VTA) and substantia nigra (SN) *(13)*, where they have been shown to modulate the activity of dopaminergic neurons *(5,14,15)*. The 5-HT_{1B} ligands have also been shown to modulate a number of responses to cocaine, including induction of immediate–early genes, self-administration, and drug discrimination *(8,16–18)*.

Therefore, we decided to study the effects of cocaine in mice lacking 5-HT_{1B} receptors. In this chapter, we demonstrate that the 5-HT_{1B} knockout (KO) mice display increased locomotor responses to psychostimulants such as cocaine and that they are more motivated to self-administer cocaine. Furthermore, using behavioral and biochemical measures, we suggest that even drug-naive KO mice are in a state that resembles wild-type mice that have been sensitized to cocaine by repeated exposure to the drug. This altered "state of mind" might be responsible for their increased vulnerability to drugs of abuse.

SELF-ADMINISTRATION OF COCAINE

Rationale

Intravenous self-administration of a drug is considered a good measure of the rewarding properties of that drug. There are several indications that the 5-HT_{1B} receptor might be able to modulate cocaine self-administration. The 5-HT_{1B} agonists are able to decrease the rate of self-administration of DA-releasing drugs on a fixed ratio schedule, indicating an increase in the rewarding efficacy of the DA-releasing drug *(19)*. Similarly, on a progressive ratio schedule of reinforcement, administration of a 5-HT_{1B} agonist raised the break point for self-administration of cocaine *(18)*.

To further investigate the role of the 5-HT_{1B} receptor in cocaine self-administration, we assessed the behavior of 5-HT_{1B} knockout mice under several different intravenous self-administration paradigms. We first measured how quickly they acquired stable self-administration behavior on a fixed-ratio reinforcement schedule. Then, we compared the dose-response curve for self-administration on this schedule between wild-type (WT) and KO mice. Next, we assessed their motivation to self-administer cocaine by switching the animals to a progressive ratio schedule of reinforcement. Finally, to confirm the role of the 5-HT_{1B} receptor in these behaviors, we evaluated the effect of the $5\text{-HT}_{1B/1D}$ antagonist GR127935 on the self-administration behavior of WT animals.

Methods

Self-administration experiments took place in mouse operant chambers (model ENV-300; Med Associates, Inc., Georgia, VT), as previously described *(20)*. Animals were maintained at 28 g body weight (± 5 g). Mice were trained to press a lever under a fixed-ratio 1–2 schedule, using sweetened condensed milk solution as a reinforcer. After responding was stable on both levers, each subject was implanted with a permanent indwelling jugular catheter in a modification of the method originally described by Weeks *(21)* and used previously in our laboratory for mice *(20)*. Briefly, mice were anesthetized using a combination of xylazine (20 mg/kg) and ketamine (65 mg/kg). Under an optical microscope, a catheter was inserted into the right external jugular and its tip advanced into the right atrium and secured to the vein using 5-0 Prolene sutures (Ethicon, Somerville, NJ). The free end of the catheter ran subcutaneously to an incision in the skin at the top of the skull and was then connected to a modified C313G cannula assembly (Plastic One, Roanoke, VA). The entire catheter unit was then embedded in dental cure (Geristore, DenMat, Inc., St. Maria, CA) and fixed to the skull. Catheter patency was maintained using the basic procedure described elsewhere *(20)*. At the end of each self-administration session, subjects received an injection of saline solution (2.5 mL) containing heparin (30 U/mL).

Two days after catheter implantation, mice started training under a fixed-ratio 1 (FR1) schedule of reinforcement, with cocaine (2.0 mg/kg/0.02 mL) as reinforcer. Sessions lasted until either 23 injections were self-administered or approx 3 h elapsed. At the start of each session, a priming injection of cocaine was given. Once mice met acquisition criteria for self-administration (at least 75% of active level pressings and at least 15 injections within 3 h for 3 consecutive sessions), they were switched to a progressive-ratio (PR) schedule. Daily PR sessions started with a 2.0-mg/kg priming infusion of cocaine and lasted up to 3 h. Each subsequent cocaine injection occurred upon completion of each of the ratios in the following exponential sequence: 2, 4, 6, 9, 12, 15, 20, 25, 32, 40, 50, 62, 77, 95, 118, 145, 178, 219, 268, 328, 402, 492, 603 *(22)*. There was a 1-h time limit to obtain each reinforcer, and failure to obtain the reinforcer terminated the session. The number of infusions earned is used as the dependent variable. Once mice reached a stable baseline of responding (number of injections not varying by more than 20% over 3 consecutive days), different doses of cocaine (0.5, 1.0, and 4.0 mg/kg/injection) were tested. Each dose was tested on 3 consecutive days in a random order. The number of injections earned was used as the dependent variable for statistical analysis [univariate and repeated measures ANOVA (analysis of variance)]. Subsequently, saline was substituted for cocaine to test extinction of drug self-administration. To confirm the specificity of cocaine's effects, independent groups of mice (seven WT and eight KO) were trained to respond for food under the same progressive-ratio schedule.

Two more independent groups of WT and KO mice were trained as described. Once mice reached a stable baseline of responding (number of injections not varying by more than 20% over 3 consecutive days), antagonist trials began. For these experiments, mice were given either a 10-mg/kg dose of GR127935 (n = 4 WT and 3 KO) or saline (n = 4 WT and 3 KO) 10 min before the cocaine self-administration session. Three consecutive sessions of cocaine self-administration after antagonist administration were conducted. The number of injections of cocaine earned on the third day of testing was used as the dependent variable for statistical analysis. Repeated measures ANOVA compared the number of cocaine reinforcements obtained by each mouse before (baseline) and after antagonist trials began. Treatment (saline or GR127935) and genotype were independent variables.

Results

Mice were initially trained to press a lever under FR1 and FR2 schedules using food as reinforcer. All mice successfully completed the food-shaping program in approx eight sessions (Fig. 1). There were no significant differences between WT and KO, indicating a similar ability of WT and KO to learn an operant behavior. Mice were then implanted with intravenous (iv) catheters and 2 d later trained to self-administer cocaine (2 mg/kg/0.2 mL/injection) through the catheter under a FR1 schedule. The KO mice acquired cocaine self-administration significantly faster than the WT (four versus eight sessions, respectively, Fig. 1). An unpaired t-test shows a significant difference in speed of acquisition of stable cocaine self-administration between genotypes ($t_{(17)}$ = −2.7, p < 0.05). However, the mean number of reinforcers obtained on the last day of the self-administration acquisition period was not significantly different (13.4 for the KO versus 15.5 for the WT, not shown). Once mice met cocaine self-administration acquisition criteria (at least 75% of active lever pressings and at least 15 injections within 3 h for 3 consecutive sessions), they were switched to a PR schedule. PR sessions started

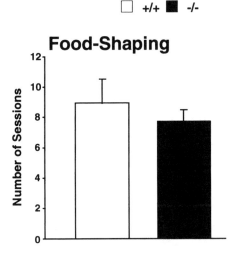

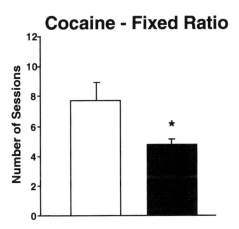

Fig. 1. Acquisition of self-administration. Each bar represents the mean number of training sessions ± SEM required to attain stable self-administration behavior (75% or active lever presses, and at least 15 rewards for 3 consecutive sessions) for 5 WT (white bars) and 10 KO (black bars) mice. **Top**: Acquisition of self-administration of food pellets. **Bottom**: Acquisition of self-administration of iv cocaine. The asterisk indicates a significant difference between WT and KO animals ($p < 0.05$).

with a 2.0-mg/kg priming infusion. In order to obtain each subsequent cocaine injection (2.0 mg/kg/injection), the mice had to complete an increasing number of responses, determined by the exponential sequence 2, 4, 6, 9, 12, 15, 20, 25, 32 ... . For all doses of cocaine tested (0.5–4.0 mg/kg/injection), KO mice ($n = 10$) showed significantly higher break points than WT mice

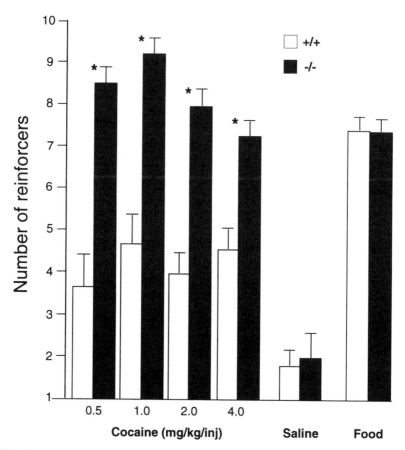

Fig. 2. Self-administration of cocaine on a PR schedule. Self-administration beha-vior was under a progressive ratio reinforcement schedule. Each bar represents the number of reinforcements (mean ± SEM) obtained by 10 KO (black bars) and 5 WT (white bars) mice in the cocaine experiment, or 8 KO and 7 WT in the food experi-ment. Significant differences between genotypes ($p < 0.05$) are marked by an asterisk.

($n = 5$; *see* Fig. 2). Repeated measures ANOVA confirms a significant effect of genotype ($F_{(1, 13)} = 7.10$, $p < 0.05$). In other words, the KO mice were willing to press the lever up to 25 times to obtain a single dose of cocaine, whereas the WT would press only 6 times. To test whether that differential response is specific to cocaine, saline was substituted for cocaine and the mice were again tested on a PR schedule. Both KO and WT showed extinc-tion of self-administration behavior, with the number of reinforcements obtained dropping to two for both genotypes. When responding for food pellets, WT and KO mice obtained an equal number of reinforcements

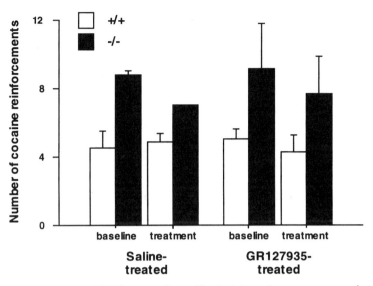

Fig. 3. Effect of GR127935 on cocaine self-administration on a progressive-ratio schedule. Each bar represents the mean ± SEM for the number of reinforcers obtained by independent groups of KO ($n = 3$ per group, dark bars) and WT ($n = 4$ per group, white bars).

(Fig. 2), indicating that the increased break point for cocaine self-administration in the KO does not result from nonspecific hyperactivity in the KO.

Pretreatment with GR127935 at the start of each daily self-administration schedule had no significant effect on the self-administration behavior of either genotype (Fig. 3). ANOVA confirmed our previous finding *(23)* that 5-HT$_{1B}$ KO mice have a higher break point than WT mice ($F_{(1, 10)} = 8.99$, $p < 0.05$). However, there was no effect or interaction of treatment ($p > 0.8$). For WT mice, the amount of cocaine self-administered did not change with time ($p > 0.7$). In other words, the mean number of reinforcements obtained after the antagonist sessions was not significantly different from the number of baseline reinforcements obtained. In KO mice, there was a tendency for the number of reinforcements obtained to decrease with time, but this effect was present in both groups pretreated with antagonist or saline. ANOVA shows a significant effect of time ($F_{(1, 4)} = 37.70$, $p < 0.005$), but no main effect ($p > 0.8$) or interaction ($p > 0.6$) with treatment for the KO mice. Therefore, GR127935 does not affect the self-administration behavior of either WT or KO mice. This suggests that whereas developmental absence of the 5-HT$_{1B}$ receptor can potentiate the rewarding effects of cocaine, acute blockade of this receptor in adult animals does not affect the rewarding properties of the drug.

Conclusions

We have shown that the KO mice pressed to a higher breaking point for cocaine than the WT on a progressive-ratio self-administration paradigm (Fig. 2). In other words, the 5-HT$_{1B}$ KO mice are willing to work harder to obtain cocaine, suggesting that the drug has a higher reinforcing efficacy for KO mice. Possibly related to this preference for cocaine is the fact that KO mice acquired iv self-administration of cocaine significantly faster than WT (Fig. 1). However, as previously reported (20), once stable responding was achieved on a FR1 or FR2 schedule, the two genotypes showed similar cocaine-taking behavior across a range of doses. These results might reflect the suggestion that progressive-ratio schedules better predict the reinforcing efficacy of drugs than low-value fixed-ratio paradigms (24). Alternatively, the biological substrates governing acquisition of self-administration may differ from those governing maintenance of self-administration.

These results suggest that the absence of the 5-HT$_{1B}$ receptor can enhance the rewarding properties of cocaine, as measured by progressive-ratio self-administration. However, acute administration of the 5-HT$_{1B/1D}$ antagonist GR127935 failed to affect self-administration behavior of wild-type mice. This accords well with recently published findings that GR127935 alone does not affect progressive-ratio self-administration behavior in rats (18). In sum, the genetic absence of the 5-HT$_{1B}$ receptor has a strong effect on cocaine self-administration, whereas pharmacological blockade of the same receptor fails to influence cocaine self-administration when measured by the same paradigm. This suggests that the enhanced self-administration behavior of 5-HT$_{1B}$ KO mice may be due to developmental compensations in other neural circuits.

LOCOMOTOR EFFECTS OF COCAINE

Rationale

Psychostimulants such as cocaine have been shown to increase loco-motion and, at high doses, to induce stereotyped movements (25,26). The ability of these substances to stimulate locomotion has been suggested to parallel their addictive potential (27). Because 5-HT$_{1B}$ KO mice appear to be more vulnerable to the reinforcing effects of cocaine, they may also be more sensitive to the stimulating effects of the drug. On the other hand, based on the role of the 5-HT$_{1B}$ receptor in neural circuitry, we might expect KO mice to be less sensitive to the locomotor effects of cocaine. Specifically, 5-HT$_{1B}$ receptor stimulation may enhance DA release by reducing GABA-ergic inhibition (5,14,15). Therefore, a knockout of the 5-HT$_{1B}$ receptor might be expected to remove this modulation of GABA release and thereby indirectly reduce DA neurotransmission. If increased DA release is essential to the locomotor activating effects of cocaine, we might expect 5-HT$_{1B}$ receptor knockout mice to have an attenuated locomotor response to cocaine.

In order to explore these conflicting hypotheses, we tested whether the locomotor responses to cocaine were altered in the KO mice. As before, we then compared the behavioral response of 5-HT$_{1B}$ KO mice to cocaine with the response of WT mice pretreated with a 5-HT$_{1B/1D}$ antagonist.

Methods

Cocaine (Sigma Pharmaceutical, St. Louis, MO) was dissolved in 0.9% saline on the day of the experiment. GR127935 (Glaxo Wellcome, UK) was dissolved in sterile distilled water by gently heating the solution for 20 min.

WT and KO mice (n = 13 mice per group) were introduced in an open field immediately after an ip injection of either saline or cocaine (10, 20, or 40 mg/kg). Locomotion was evaluated with a video-tracking device. In order to assess stereotyped behavior, the behavioral sessions were also videotaped and scored. In the acute cocaine experiment, animals were given a saline injection, then placed in the open field for 1 h on the 2 d preceding the drug challenge to allow them to habituate to the chamber. On the third day, animals were given an injection of either saline, 10 mg/kg cocaine, 20 mg/kg cocaine, or 40 mg/kg cocaine, placed in the open field, and monitored for 1 h.

To assess the effects of GR127935 on cocaine-induced locomotor activity, experimentally naive 129/Sv WT mice were habituated to the open field for 1 h. Then, they were randomly assigned to a treatment group: saline (n = 10), GR127935 (10 mg/kg, n = 10), cocaine (20 mg/kg, n = 10), or both cocaine and GR127935 (20 mg/kg and 10 mg/kg, respectively, n = 10). On the testing day, mice were habituated for 30 min, then given either an injection of saline or GR127935, returned to the open field for 30 min, and then given either an injection of saline or cocaine. After the cocaine injection, they were monitored for 60 min.

Results

Throughout the first hour after injection, KO mice displayed significantly more locomotion than WT in response to 10, 20, and 40 mg/kg cocaine (Fig. 4). Indeed, at the highest dose, KO mice showed nearly triple the amount of locomotion exhibited by WT mice. ANOVA shows a main effect of treatment ($F_{(3,95)}$ = 20.81, $p < 0.0001$), a main effect of genotype ($F_{(1,95)}$ = 20.54, $p < 0.0001$), a main effect of time ($F_{(11,1045)}$ = 40.25, $p < 0.0001$), a treatment by genotype interaction ($F_{(3,95)}$ = 4.98, $p < 0.005$), an interaction of time and treatment ($F_{(33,1045)}$ = 6.53, $p < 0.0001$), an interaction of time and genotype ($F_{(11,1045)}$ = 6.10, $p < 0.0001$), and an interaction of time, treatment, and

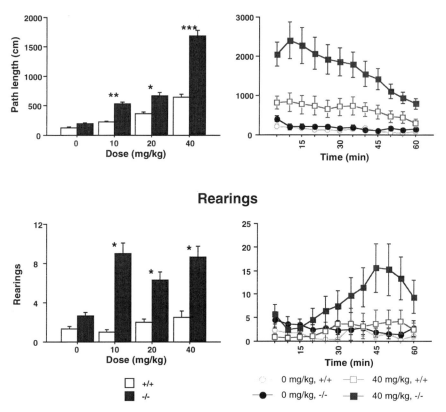

Fig. 4. Locomotor response to acute cocaine; effects of cocaine on open-field behavior ($n = 13$ per group). Significant differences between KO (dark bars) and WT (white bars) are marked by * ($p < 0.05$), ** ($p < 0.01$), or *** ($p < 0.001$). **Top, left**: Each bar represents mean path length + SEM in 5 min averaged over 1 hour. **Top, right**: Time course for 40 mg/kg and saline only. Each point represents mean path length ± SEM for each 5-min interval. **Bottom, left**: Each bar represents number of rearings ± SEM occurring in each 5-min interval averaged over 1 h. **Bottom, right**: Time course of rearings for 40 mg/kg and saline. Each point represents mean rearings ± SEM for each 5-min interval.

genotype ($F_{(33, 1045)} = 2.11$, $p < 0.0005$). Significant differences between genotypes, according to a *post hoc* Scheffé test, were found after treatments of 10 mg/kg ($p < 0.01$), 20 mg/kg ($p < 0.05$), and 40 mg/kg ($p < 0.001$). Similarly, KO mice showed an increase in the number of rearings after 10, 20, or 40 mg/kg cocaine, whereas cocaine did not stimulate rearing in WT mice (Fig. 4). Repeated measures ANOVA across time shows a main effect of genotype ($F_{(1, 94)} = 11.74$, $p < 0.001$), a main effect of time ($F_{(11, 1034)} = 5.38$,

$p < 0.0001$), an interaction of time and treatment ($F_{(33, 1034)} = 2.93, p < 0.0001$), and an interaction of time, treatment, and genotype ($F_{(33, 1034)} = 1.52, p < 0.05$). *Post hoc* Scheffé comparisons show significant differences between KO and WT mice after a cocaine injection of 10 mg/kg ($p < 0.05$), 20 mg/kg ($p < 0.05$), and 40 mg/kg ($p < 0.05$).

Cocaine can also increase the amount of repetitive, stereotyped behavior in rodents. Because stereotyped behavior can compete with horizontal loco-motion, the low stimulation of forward locomotion observed in WT mice after cocaine treatment may be due to increased stereotyped behavior. There-fore, we assessed the amount of stereotypy occurring in each genotype. In the KO mice, 10 mg/kg cocaine elicited a decrease in quiet behavior and an increase in repetitive movements and intense stereotypies, whereas this dose had no effect in the WT mice (Fig. 5). It is notable that at 10 mg/kg cocaine, all categories of behavior (except normal quiet) are increased in the KO mice compared to the WT mice, indicating an increased sensitivity to cocaine. Stereotypies became detectable in the WT mice at 20 mg/kg. At 40 mg/kg, the KO mice displayed predominantly rapid locomotor movements, whereas the WT mice displayed mostly repetitive movements and stereotypies. In general, the KO mice showed significantly less normal quiet behavior ($F_{(1,94)} = 20.47, p < 0.0001$) and this behavior decreased with increasing cocaine dose ($F_{(3, 94)} = 47.96, p < 0.0001$). There was also an interaction of genotype and treatment for normal quiet behavior ($F_{(3, 94)} = 3.10, p < 0.05$). Normal exploration also decreased with dose ($F_{(3,94)} = 3.14, p < 0.05$), but there were no effects or interactions of genotype. There was significantly more rapid locomotion in KO mice ($F_{(1, 94)} = 8.56, p < 0.005$), and this behavior increased with cocaine dose ($F_{(3, 94)} = 16.52, p < 0.0001$) and showed an interaction between treatment and genotype ($F_{(3, 94)} = 6.36, p < 0.001$). Similarly, repetitive movements also increased with dose ($F_{(3, 940)} = 7.99, p < 0.0001$), were more common in KO mice ($F_{(1, 94)} = 15.50, p < 0.0005$), and showed an interaction of genotype and treatment ($F_{(3, 94)} = 5.39, p < 0.005$). Intense stereotypy increased with treatment ($F_{(3, 94)} = 8.49, p < 0.0001$), but there were no significant genotype differences in this behavior. *Post hoc* Scheffé analysis revealed significant genotype differences in particular behaviors at specific doses, as indicated in Fig. 5.

To assess the overall intensity of the motor response to cocaine in a way that includes all types of behavior, numerical scores (0–4) were assigned to each category of behavior. Higher scores correspond to a stronger drug effect. When these scores are averaged and plotted against drug dosage (Fig. 5), it clearly shows that the dose response to cocaine is shifted to the left in the KO mice, suggesting that the KO mice are more sensitive to the motor effects of cocaine than the WT mice. ANOVA shows main

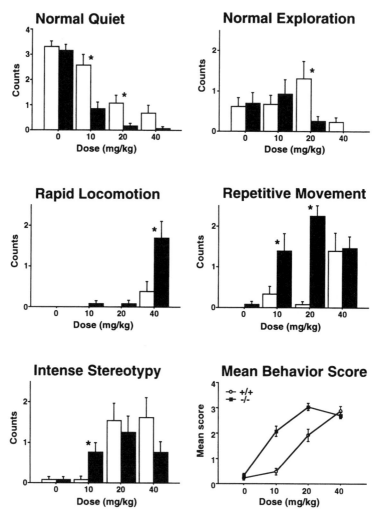

Fig. 5. Stereotypy in response to acute cocaine; mean number of times (out of a maximum of four) ± SEM that WT mice (white bars) and KO mice (black bars) are observed engaged in each behavior; $n = 13$ per group. Significant differences ($p < 0.05$) between genotype are marked with an asterisk. **Lower right**: Aggregate stereotypy score. Each point is the numerical mean ± SEM of behavioral scores (0–4) assigned to each group.

effects of treatment ($F_{(3,95)} = 51.289$, $p < 0.0001$), a main effect of genotype ($F_{(1,95)} = 15.206$, $p < 0.0005$), and an interaction of treatment and genotype ($F_{(3,95)} = 4.996$, $p < 0.0001$). *Post hoc* Scheffé analysis shows significant differences between KO and WT mice at 10 mg/kg ($p < 0.0001$) and 20 mg/ kg ($p < 0.05$). In summary, all measures of locomotor activity after acute

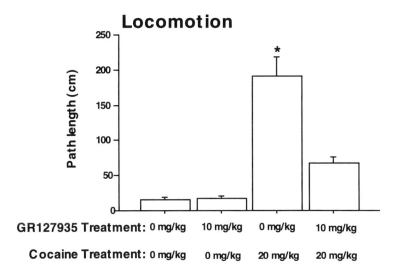

Fig. 6. Effect of GR127935 on locomotor response to cocaine; effect of GR127935 pretreatment on locomotor response to cocaine. Each bar represents the mean path length ± SEM for 10 WT mice per group. A significant ($p < 0.05$) difference from vehicle treament is indicated by an asterisk.

cocaine (horizontal locomotion, rearings, and behavioral scoring) suggest that 5-HT_{1B} KO mice are more sensitive to the locomotor effects of cocaine.

In order to determine if this increased sensitivity to cocaine is a direct result of the absence of the 5-HT_{1B} receptor, we pretreated WT mice with the 5-$HT_{1B/1D}$ antagonist GR127935 and then assessed their response to cocaine. The acute blockade of the 5-HT_{1B} receptor by GR127935 significantly reduced the locomotor response to cocaine in WT mice (Fig. 6). There was a main effect of treatment ($F_{(3,752)} = 24.53$, $p < 0.0001$). *Post hoc* Scheffé tests showed that mice given 20 mg/kg dose of cocaine alone were more active than mice pretreated with GR127935 ($p < 0.0001$), mice given GR127935 alone ($p < 0.0001$), or saline controls ($p < 0.0001$). Cocaine and/or GR127935 administration did not significantly alter rearings or nose-pokes (not shown). So WT mice pretreated with GR127935 not only show less response to cocaine than 5-HT_{1B} KO mice, they also show less locomotor response than WT mice given saline pretreatment.

Conclusions

Mice lacking the 5-HT_{1B} receptor show increased locomotion in response to all doses of cocaine tested (10, 20, and 40 mg/kg; *see* Fig. 4). Similarly, cocaine dose-dependently and significantly increases rearings in 5-HT_{1B} KO mice, although it does not strongly increase rearing in WT mice. These results cannot be explained by a simple trade-off between stereotyped behavior and forward locomotion, because at 10 mg/kg, KO mice display both more locomotion and more stereotyped behavior than WT mice (Fig. 5). Thus, all measures of locomotor response to acute cocaine indicate that 5-HT$_{1B}$ KO mice have a greater sensitivity to the drug. It is interesting that at high doses, WT mice tend to exhibit more intense stereotypy than KO mice, whereas the KO mice show much more rapid locomotion than WT mice at this dose. This suggests that there may be a shift in the balance between the neural circuits governing stereotypy and the neural circuits governing locomotor stimulation in the 5-HT$_{1B}$ KO mice.

Like the increased self-administration of cocaine, the increased locomotor sensitivity to cocaine in 5-HT$_{1B}$ KO mice probably cannot be explained by the simple absence of the 5-HT$_{1B}$ receptor. When WT mice were pretreated with the 5-HT$_{1B/1D}$ antagonist GR127935, their locomotor response to cocaine was actually reduced—the opposite of the effect of the genetic knockout of this same receptor. The observation than antagonist pretreatment has the opposite effect of the genetic knockout is a strong indication that developmental compensations are responsible for the heightened responsiveness to cocaine observed in 5-HT$_{1B}$ knockout mice.

BEHAVIORAL SENSITIZATION TO COCAINE

Rationale

Behavioral sensitization, or reverse tolerance, to the motor effects of cocaine has been shown to develop following repeated intermittent exposure to cocaine. This phenomenon has been suggested to reflect the addictive properties of cocaine *(28)*. It is possible that WT and KO mice sensitize differently to cocaine. It is also possible that the initial difference in locomotor response to cocaine reflects a difference in the neurochemical substrates underlying sensitization. We therefore studied the effects of repeated administration of cocaine on the locomotor behavior of KO and WT mice. In order to compare the effect of a genetic knockout of the 5-HT$_{1B}$ receptor with the effect of acute blockade of this behavior, we assessed sensitization in WT mice pretreated with GR127935, as well.

Methods

For the chronic cocaine experiment, animals were habituated during four daily 1-h open-field sessions in the absence of any drug treatment. On each day of chronic cocaine treatment, animals were given an injection of 20 mg/kg cocaine, then monitored in the open field for 1 h. All animals ($n = 16$ WT and 16 KO) received 5 total injections of cocaine, with a 2-d rest between each injection. This protocol has been shown to induce maximal behavioral sensitization *(25)*. At the end of the repeated cocaine regimen, animals were given a challenge injection of saline and monitored again for 1 h.

The WT 129/SV mice show little locomotor activation in response to cocaine *(see above)* and little sensitization after repeated injections *(see* Fig. 7*)*. Therefore, in order to assess the effect of GR127935 treatment on sensitization to cocaine, we used female C57/Bl6 mice, a strain that shows a robust locomotor response to cocaine and strong sensitization. After one habituation session, C57BL/6 mice were randomly assigned to the different test groups receiving either saline (*n* = 10), GR127935 (10 mg/kg, *n* = 10), cocaine (20 mg/kg, *n* = 14) or both GR127935 and cocaine (*n* = 13). After a week of rest, each mouse was reexposed to the same experimental paradigm, the only difference being the reduction of the GR127935 dose from 10 to 3 mg/kg, a dose shown to be sufficient, in this strain, to alter behaviors involving the 5-HT$_{1B}$ receptor *(29)*. These last test conditions were then repeated two times more, 1 wk apart, in order to test if 5-HT$_{1B}$ blockade could alter the development of behavioral sensitization to repeated cocaine treatment.

Results

As demonstrated (Fig. 5), 20 mg/kg of cocaine elicited more total locomotion from the KO than from the WT on each day of chronic treatment (Fig. 7). Repeated measures ANOVA shows main effects of genotype ($F_{(1, 27)} = 24.06$, $p < 0.0001$) and day ($F_{(4,108)} = 6.57$, $p < 0.0001$), with no interactions. Both genotypes showed evidence of behavioral sensitization. In WT mice, there was a main effect of day ($F_{(4,60)} = 4.26$, $p < 0.005$). According to Scheffé comparisons, WT mice were significantly more active ($p < 0.05$) on d 3, d 4, and d 5 than on d 1 of cocaine treatment. Similarly, for KO mice, there was a main effect of day ($F_{(4,48)} = 3.53$, $p < 0.05$), and KO mice were significantly more active on d 3 than on d 1 of cocaine treatment ($p < 0.05$).

Some studies have reported that most behavioral sensitization can be explained by context-dependent conditioning *(30)*. In order to see how much of the behavioral sensitization observed could be explained by conditioning to the open-field environment, after the sensitization paradigm was complete, we injected the mice with saline, and then monitored their locomotion in the open field. On this saline-treated day, both KO and WT mice were significantly more active than they were before beginning cocaine treatments ($F_{(1,27)} = 28.07$, $p < 0.0001$). However, their locomotion was much lower than after cocaine treatment, indicating most of the motor stimulation was due to the cocaine treatment. On the saline challenge day, KO mice were significantly more active than WT mice ($F_{(1, 27)} = 14.31$, $p < 0.001$).

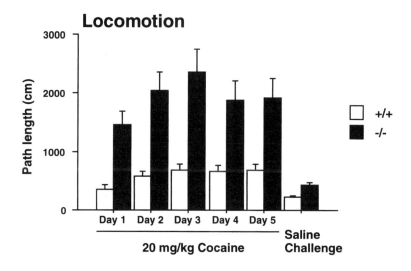

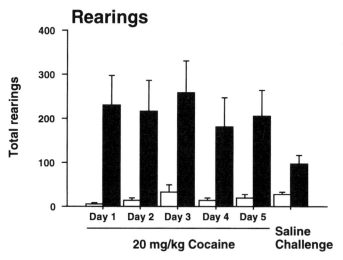

Fig. 7. Behavioral sensitization; behavioral response to repeated injections of 20 mg/kg of cocaine. **Top:** Bars (KO: white bars, $n = 16$; WT: black bars, $n = 16$) represent mean path length ± SEM covered in 5 min averaged over 1 h. **Bottom:** Bars represent total number of rearings ± SEM during 1 h.

Because KO mice were always more active than WT mice after cocaine treatment, it is not surprising that they show slightly more context-dependent sensitization after a saline challenge.

For rearings behavior, although KO mice again showed much more rearing than WT mice (Fig. 7), there was no sensitization of rearing. A repeated measures ANOVA across treatment days shows a main effect of genotype

$(F_{(1, 29)} = 10.47, p < 0.005)$, with no effects or interactions with treatment day. When the two genotypes are considered separately, however, WT animals did show a small increase in rearings after repeated cocaine treatments (a mean of 6.4 rearings on d 1 versus 34.4 on d 3). ANOVA shows a main effect of day in WT mice $(F_{(4, 56)} = 2.86, p < 0.05)$. KO mice show no significant change in the number of rearings over time. As was the case for locomotion, both groups of mice showed significantly more rearings after saline challenge at the end of the sensitization protocol compared to after saline habituation $(F_{(1, 30)} = 19.96, p < 0.0001)$. In general, rearing behavior appears to be more susceptible to context-dependent sensitization than horizontal locomotion. However, it increases less with repeated administration of cocaine.

Although the KO continued to have a larger increase in horizontal locomotion in response to cocaine throughout the experiment, the nature of the WT motor response changed as sensitization occurred. With repeated injections of cocaine, intense stereotypy increased progressively in the WT, which is a characteristic of sensitized behavior in many rodent strains (Fig. 8). In the KO, intense stereotypies reached a plateau after the second session which was significantly lower than for the WT. Behavioral differences between genotypes are most pronounced in scores for rapid locomotion (score = 2) and intense stereotypy (score = 4). KO mice show significantly more rapid locomotion scores $(p < 0.05)$ than WT mice on each day of cocaine administration. Repeated measures ANOVA show a main effect of genotype for rapid locomotion $(F_{(1,27)} = 20.23, p < 0.0001)$, with no interactions. Although both genotypes initially showed a similar amount of intense stereotypy, WT mice engaged in this behavior significantly more than KO after 4 and 5 d of cocaine treatment $(p < 0.05)$. For intense stereotypy, ANOVA shows a main effect of day $(F_{(4,108)} = 9.74, p < 0.0001)$ and an interaction of day and genotype $(F_{(4, 108)} = 4.42, p < 0.005)$.

Behavioral scores were aggregated as previously described (Fig. 8). The WT mice displayed a gradual increase in global behavioral scores, showing their progressive sensitization to cocaine. The KO, in contrast, started at a high level that is similar to the one reached by sensitized WT. The KO stayed at this high level throughout the experiment. ANOVA shows main effects of day $(F_{(4, 120)} = 23.30, p < 0.0001)$ and an interaction of day and genotype $(F_{(4, 120)} = 8.255, p < 0.0001)$.

We then tested whether the 5-HT$_{1B}$ antagonist GR127935 could affect the development or maintenance of sensitization in WT C57/Bl6 mice. GR127935 did not block the development of sensitization in these mice. Fig. 9 shows progressive increases in the locomotor response to cocaine across 4 d of cocaine treatment in mice treated with GR127935 and mice treated with saline. The locomotor activity of saline-treated WT mice (no cocaine treatment)

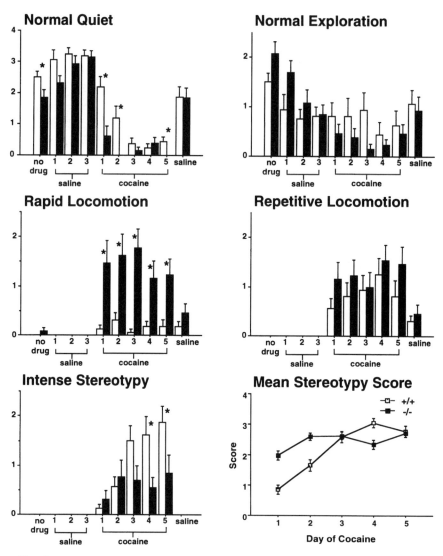

Fig. 8. Stereotypy after chronic cocaine; the mean number of times that WT mice (white bars) and KO mice (black bars) are observed engaged in each behavior ($n = 16$ WT and 16 KO mice tested repeatedly). Significant differences ($p < 0.05$) between genotype are marked with an asterisk. **Lower right**: Aggregate stereotypy score. Each point is numerical mean ± SEM of behavioral scores (0–4) assigned to each group.

did not significantly change over the course of the four injections (not shown). For cocaine-treated animals, the ANOVA showed a significant effect of the number of cocaine injections on locomotor activity ($F_{(3,24)} = 5.24$; $p < 0.01$). Indeed, there was a progressive increase of the locomotor response to a con-

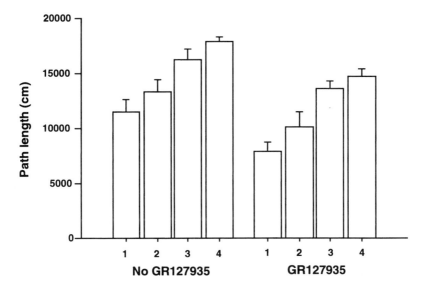

Fig. 9. Sensitization to cocaine in the presence and absence of GR127935; effect of GR127935 pretreatment on sensitization to cocaine in C57/Bl6 mice. Each bar is total path length ± SEM for 1 h of testing. **Left:** Locomotion across days for 14 animals treated with saline before cocaine. **Right:** Locomotion across days for 13 mice treated with 10 mg/kg GR127935 each day.

stant dose of cocaine. These results clearly showed the locomotor reactivity of C57BL/6 mice sensitized to repeated cocaine injections. Mice repeatedly injected with both cocaine and GR127935 also showed significant sensitization of their locomotor response to cocaine (injections: $F_{(3,33)} = 11.69; p < 0.0001$; Fig. 9). These results show that the 5-HT$_{1B}$ receptor is not essential for the development of locomotor sensitization to cocaine.

Conclusions

Behavioral sensitization to the locomotor effects of cocaine, as measured by increases in horizontal path length after repeated injections of a constant dose of cocaine, appears to occur in both WT and 5-HT$_{1B}$ KO mice (Fig. 7). Similarly, WT C57/Bl6 mice pretreated with the 5-HT$_{1B/1D}$ antagonist GR127935 show normal behavioral sensitization to cocaine (Fig. 9). These results indicate that the 5-HT$_{1B}$ receptor is not involved in the development of behavioral sensitization to cocaine.

There are some interesting behavioral differences between WT and KO mice that become apparent after a more detailed analysis of the behavioral response to repeated injections of cocaine. First, the increased behavioral sensitivity of 5-HT$_{1B}$ KO mice to cocaine was confirmed, as evidenced by their increased locomotion and rearings throughout 5 d of repeated cocaine injections (Fig. 7). It is notable that even sensitized WT mice do not show nearly as much locomotion as drug-naive KO mice, suggesting that there is a major qualitative difference between these two genotypes in their response to cocaine. The pattern of emergence of stereotyped behavior is also interesting (Fig. 8). KO mice show a small progressive increase in time spent engaged in rapid locomotion, whereas there is little change in the amount of time KO mice engage in repetitive or stereotyped movements over repeated cocaine injections. WT mice, in contrast, show little increase in rapid locomotion, but they show a very large progressive increase in intense stereotypy. In fact, the amount of intense stereotypy observed after four and five injections of 20 mg/kg of cocaine is the only measure of behavioral response to cocaine for which WT mice show significantly more activation than KO. This suggests that the neural circuitry governing the locomotor response to cocaine might be favored in KO mice, whereas the circuitry governing stereotypy might be favored in WT mice.

PHARMACOKINETICS OF COCAINE ABSORPTION AND METABOLISM

Rationale

To investigate whether pharmacokinetic differences might be responsible for the increased responsiveness to cocaine of the KO mice, we analyzed the levels of cocaine and its metabolites, benzoylecognine and norcocaine, in the brain and blood 15, 30, and 60 min after an acute injection of cocaine *(31)*.

Methods

Female mice 57–85 d old were given ip injections of cocaine (20 mg/kg). Separate groups of mice were decapitated 15, 30, or 60 min later. Trunk blood was collected and the brain was rapidly removed. Levels of cocaine, benzoylecognine, and norcocaine were assayed using extraction, derivitization, and gas chromatography–mass spectrometric (GC/MS) methods detailed below.

Brain Assay

Mouse brain was weighed and homogenized in 4X solution of $5N$ perchloric acid containing 1% NaF and internal standard (benzoylecogonine-d3, 1 ng/mg). The sample was then centrifuged at 16,500 rpm for 15 min and the supernatant was removed to a new tube and frozen at $-80°C$. Then, 800 µL of the supernatant was removed and the pH adjusted to pH 5.8–6.0, with a final volume of 1.5 mL. Bond-Elute Certified columns (3 mL, Varian Corp.) were placed in a Vac-Elute system. Each column was activated by washing with 2 mL of methanol, followed by 2 mL of phosphate buffer, 0.1 *M*, pH 6.0, without allowing the column to dry; the brain sample was applied and drawn through the column at the rate of 1 mL/2 min. The column was then washed with 6 mL of distilled water and 3 mL of 0.1 *M* HCl, and dried under full vacuum (15 in. of Hg) for 5 min. Next, 8 mL of methanol was passed through the column and dried for 30 s with full vacuum. The tips of the needles were wiped and labeled glass tubes placed in the Vac-Elut. Finally, 2 mL of methylene chloride:isopropyl alcohol (80:20) with 2% ammonium hydroxide was pulled through the column. The collected samples were dried at 40°C with a stream of nitrogen to half the original volume. The remaining liquid was transferred to a 2-mL crimp-top vial and the drying completed. To the dried sample, 50 µL each of pentafluoropropionic anhydride (PFPA) and 1,1,1,3,3,3-hexafluoroisopropanol (HFIP) were added. The vial was sealed with a Teflon-lined cap and heated for 10 min at 70°C. After evaporation of the mixture to dryness under a gentle stream of nitrogen at 50°C, the residue was taken up in 100 µL of butyl acetate, transferred to a 200-µL mini-glass insert, sealed in a glass vial, and analyzed by GC/MS. The GC/MS analysis was performed on a Hewlett-Packard Model 5890 gas chromatograph with a 13-m × 0.25-mm-inner diameter HP-5 capillary column. The flow of carrier gas (helium) through the column was 1.6 mL/min and the head pressure was maintained at 7.5 psi. The injection port temperature was 350°C, and the transfer line temperature was 285°C. After a splitless injection, the oven was maintained at 150°C for 1 min with the split flow turned off, and then the oven temperature was ramped to 285°C at 20°C/min. The mass analysis was done with a Hewlett-Packard MSD Model 5970 operated in electron impact mode at 70 eV with the electron multiplier voltage of the detector set at 200 volts above autotune. Data were collected in single-ion monitoring (SIM) mode and set to detect two ions of each compound: cocaine, 182 and 303; hexafluoroisopropyl derivative of benzoylecgonine, 318 and 439; and the hexafluoroisopropyl derivative of benzoylecoginine-d3, 321 and 442. The ion areas were summed and the ratios of drug to internal standard calculated and plotted on a standard curve, and final concentration was determined. Norcocaine and ecgonine were also detectable using this system.

Blood Assay

Trunk blood (200 µL) was removed and 200 µL of saline and 10 µL of 4% NaF was added. The sample was spun down at 13,000 rpm and the plasma removed to a new tube where 1 mL of 0.2*N* HCl was added along with BEG-D3 (10 µL of 2.5 ng/µL). Samples were frozen at 80°C. Microdisk columns (SPEC, 3 mL MP-1 from Ansys) were labeled and placed in

Vac-Elute holders. The columns were washed with 200 µL of MeOH, followed by 200 µL of 0.2N HCl. Aspiration was stopped and the blood sample loaded. The sample was drawn slowly through the column and then 500 µL of 20% acetone/water solution. The vacuum was increased to 15 in. of Hg and the column dried for 8 min. The vacuum was opened, the collection tips wiped, and collection tubes placed in the holder. Elution solvent (1 mL, 80%–20%–20% methylene chloride, isopropyl alcohol, and ammonia) was passed through the column and collected. Drying, derivatization, and detection by GC/MS were the same as for the brain assay.

Results

No significant differences were found between genotypes at any time period (data not shown). Brain levels were assayed by two-way between-subjects ANOVA. The effect of time was significant for both cocaine ($F_{(2,46)} = 24.8$, $p < 0.0001$) and benzoylecognine ($F_{(2,45)} = 43.4$, $p < 0.0001$). However, neither genotype nor the genotype by time interaction reached significant levels for either compound. Analysis of blood levels of cocaine and its metabolites revealed a similar lack of difference between the genotypes.

Conclusions

The increased behavioral sensitivity to cocaine observed in 5-HT$_{1B}$ KO mice cannot be explained by pharmacokinetic differences. KO mice showed normal brain levels of cocaine and its metabolites at several different time points after cocaine injection. This suggests that cocaine is absorbed, metabolized, and eliminated at the same rate in KO and WT mice.

LEVELS OF BASAL AND INDUCED FOS-RELATED ANTIGENS IN 5-HT$_{1B}$ MICE

Rationale

Acute administration of cocaine and other psychostimulants results in the induction of a number of immediate–early genes such as c-*fos* in the dorsal striatum and NAc (*17,32,33*). The induction of c-*fos* is short-lived and returns to baseline levels a few hours after the injection of cocaine (*33*). In response to repeated injections of cocaine, c-*fos* induction desensitizes while the levels of a different set of fos-related proteins termed chronic fos-related antigens (FRAs) become prominent, some of which are still expressed several days after the last injection of cocaine (*35*). The duration of that change in gene expression, as well as the fact that these "chronic FRAs" are specifically turned on after chronic exposure to psychostimulants, has led to speculation that these transcription factors might be responsible for the changes in gene expression that account for some long-lasting effects of drugs of abuse such as tolerance, sensitization and dependence *(35,36)*. It is possible that the difference in behavioral sensitivity to cocaine observed in KO mice will be reflected by a corresponding difference in striatal gene expression. We therefore analyzed the levels of fos-related antigens in both drug-naive KO mice and mice chronically treated with cocaine.

Methods

In order to isolate brain regions for Western blot and gel shift assays, following decapitation, whole brains were chilled in ice-cold physiological buffer as described previously *(34,35)*. Whole striatum was excised from 1-mm-thick coronal brain slices using a 12-gage syringe. Bilateral punches were pooled. In some experiments, striatum was obtained by gross dissection. Tissue was Dounce homogenized in approximately 20 volumes of electrophoretic mobility shift assay buffer (20 mM HEPES [pH 7.9], 0.4 M NaCl, 20% glycerol, 5 mM MgCl$_2$, 0.5 mM EDTA, 0.1 mM EGTA, 1% Nonidet P-40, 10 mg/mL leupeptin, 0.1 mM p-aminobenzamidine, 1 mg/mL pepstatin, 0.5 mM phenylmethylsulfonyl fluoride, 5 mM dithiothreitol). After incubating at 4°C for 30 min, samples were centrifuged at 15,000g for 20 min. Supernatants were then aliquoted and frozen for subsequent analysis. Protein concentrations were determined by the Lowry method.

The gel shift assay was performed as described previously *(35)*. The AP-1 probe was a double-stranded synthetic oligonucleotide derived from the promoter sequence of the human metallothionein (HMT) II gene and was identical to that employed by Sonnenberg et al. *(37)*. The probe was labeled with [α-^{32}P] dGTP and dTTP to a specific activity of $(2–9) \times 10^8$ cpm/mg by the Klenow DNA polymerase fill-in reaction. Following a 20-min incubation of the probe, binding buffer, and 10–40 mg of protein at room temperature, samples were electrophoresed at 150 V in a nondenaturing 6% acrylamide/0.24% N,N'-methylene-*bis*-acrylamide gel containing 25 mM Tris-borate (pH 8.3), 1 mM EDTA, and 1.6% glycerol. All gels were dried and exposed to X-ray film. Levels of binding were assessed by densitometry. Bands were shown to be specific for AP-1 binding by cold competition and supershift assays as reported previously *(34)*. Briefly, in the cold competition experiments, increasing concentrations (0, 3, 10, 30, and 100 ng) of unlabeled wild-type HMT AP-1 probe or mutant HMT AP-1 probe were added to samples to compete with the labeled HMT AP-1 probe for specific or nonspecific binding proteins.

For Western blotting, sodium dodecyl sulfate (SDS)-stop solution (2% SDS, 10% glycerol, 5% b-mercaptcethanol) was added to aliquots of electrophoretic mobility shift assay extracts (containing 30–50 mg of protein),

and samples were boiled and then subjected to one-dimensional SDS–polyacrylamide gel electrophoresis with 10% acrylamide/0.4% *bis*-acrylamide
in resolving gels. Proteins in resulting gels were transferred electrophoret-
ically onto nitrocellulose, blocked with 2% nonfat dry milk, and incubated
with an anti-M peptide primary antibody (anti-FRA, 1:4000, kindly provided by Dr. M. Iadarola) followed by horseradish peroxidase-conjugated
goat anti-rabbit IgG (1:4000). Immunoreactivity was visualized using enhanced
chemiluminescence techniques. Quantification of signals was achieved by
densitometry. The specificity of the resulting immunoreactive bands was confirmed by the observation that immunoreactivity was abolished when the
anti-Fra antibody was preadsorbed with purified M-peptide as established
previously *(35)*.

Results

We analyzed the expression of the FRAs in the brains of WT and KO
mice by Western blot and immunocytochemistry. Striatal protein extracts
from drug-naive WT and KO adult males were analyzed by Western blotting with an antibody directed against the DNA-binding domain of c-*fos*,
which recognizes a family of fos-related proteins (Fig. 10). Interestingly, a
number of fos-related proteins were more abundant in KO than in WT mice.
Among these are the 35–37-kDa proteins termed chronic FRAs that are
induced by chronic cocaine treatment. These chronic FRAs have recently
been shown to correspond to posttranscriptional modifications of ΔFosB (a
truncated form of FosB which results from alternative splicing of the *fosB*
gene) *(38,39)*. Optical density analysis of the three Western bands for each
genotype shows significantly greater density for KO ΔFosB bands than for
WT ($F_{(1,4)} = 44.97, p < 0.005$). The FosB band is unchanged in KO mice. In
addition, the level of FosB mRNA as assessed by Northern blot (not shown)
is the same in WT and KO mice. It is therefore likely that the increase in the
ΔFosB bands seen in the KO corresponds to a posttranscriptional event.

We analyzed more precisely the distribution of the FosB proteins in the
striatum by immunocytochemistry on brain sections, with an antibody raised
against an N-terminal sequence of FosB that has been shown to recognize
chronic FRAs *(34,35,39)* The number of immunoreactive cells was higher
in the KO than in the WT, which is in good agreement with the Western data
(Fig. 10). An automated count of FosB-positive nuclei from six WT and six
KO mice shows significant genotype differences in NAc ($F_{(1,10)} = 580.65$,
$p < 0.0001$) and in the CPu ($F_{(1,10)} = 67.76, p < 0.0001$). Interestingly, the
highest levels of expression of FosB proteins were found in the NAc, particularly in the shell subdivision. This structure appears to play a central role in
the rewarding effects of drugs of abuse.

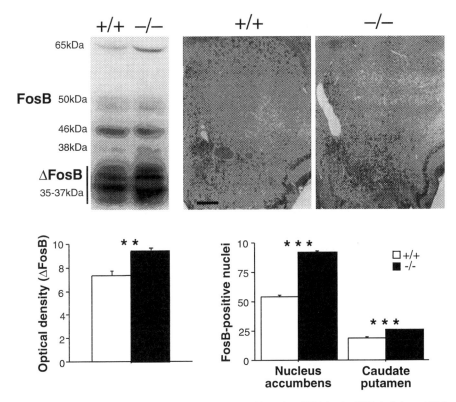

Fig. 10. Basal levels of striatal FRAs; basal levels of FRAs in WT (+/+) and KO (−/−) mice. **Top left**: Immunoblot shows ΔfosB protein in striatal extract. Results are representative of five independent experiments. **Lower left**: Mean optical density values of the ΔFosB immunoblot band from three KO (black bar) and three WT (white bar) mice. **Top right**: Immunohistochemistry showing FosB-positive nuclei in the NAc and CPu. Scale bar is 500 µm. **Lower right**: Count of mean number of FosB-positive nuclei from six WT and six KO mice.

FosB proteins were not the only ones to be upregulated in the KO. A 65-kDa band is three times more intense in the KO (Fig. 10). The other bands that have similar intensities in WT and KO mice could correspond to FRA1 and FRA2.

The band corresponding to c-*fos* (apparent molecular weight 58 kDa) is undetectable in striatal extracts of drug-naive mice. We were, however, able to detect c-*fos* on Western blots after acute exposure to cocaine (not shown). Interestingly, the induction of the c-*fos* band in the KO was reduced by 50% when compared to the WT. This agrees well with our previous immunohistochemical observations *(17)*.

Fos-related transcription factors form heterodimers with jun-related transcription factors. The resulting complex, termed AP-1, binds to a DNA motif called the AP-1 binding site that is found in a number of promoter regions and, as a result, modulates the transcription of downstream genes. In order to assess directly the presence of AP-1 transcriptional complexes in KO mice compared to WT mice, we performed gel retardation assays with striatal extracts from both strains and labeled oligonucleotides corresponding to the AP-1 binding site (Fig. 11). In baseline conditions, the specific bands were more abundant in KO than in WT, indicating an increase in AP-1 transcription complex in KO mice. Optical density analysis of AP-1 bands for three KO and three WT mice (basal) and six KO and five WT mice (chronic cocaine) shows significant differences between genotypes in basal conditions ($F_{(1,4)} = 7.55$, $p = 0.05$). The magnitude of the increase is similar to the one found in WT mice exposed chronically to cocaine.

In the supershift assay, an anti-FosB antibody was able to induce a supershift of the AP-1 band (Fig. 11), indicating that most of the AP-1 complex found in KO mice contains FosB.

EVIDENCE OF INCREASED VULNERABILITY
TO DRUGS OF ABUSE IN 5-HT$_{1B}$ KO MICE

Mice with a mutation in a single gene encoding the 5-HT$_{1B}$ receptor display an increased propensity to self-administer cocaine. We studied the reinforcing efficacy of cocaine by testing mice under a progressive ratio schedule *(40)*. In this schedule, the number of lever presses required to deliver an injection of cocaine is increased until a point at which the mice no longer respond. This "break point" reflects the reinforcing efficacy of the dose of cocaine tested. KO mice have a significantly higher break point than WT mice, indicating an increased motivation to self-administer cocaine (Fig. 2). They also acquire self-administration of cocaine more quickly than their WT counterparts (Fig. 1).

Our data about levels of fos-related transcription factors in the dorsal striatum and NAc indicate that postsynaptic changes (with respect to DA neurons) have occurred in the KO mice (Figs. 10 and 11). The increased levels of ΔFosB proteins in the NAc is particularly interesting because this structure appears to be critical for both the locomotor and rewarding effects of cocaine *(41)*. The increased levels of ΔFosB proteins in the KO are probably related to the observed increase in AP-1 binding complexes *(39,42)*. These transcription complexes are believed to regulate the expression of a number of genes that contain AP-1 motifs in their promoters.

The 5-HT$_{1B}$ receptors are expressed on the terminals of GABAergic striatal neurons that project to the SN and the VTA *(13)*. Activation of these

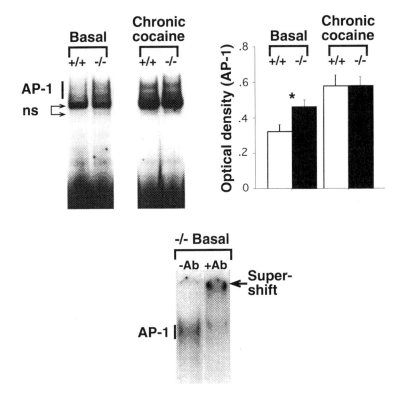

Fig. 11. AP-1 DNA-binding activity; basal AP-1 DNA-binding activity in striatal samples. **Top left**: Autoradiogram of a gel mobility-shift assay showing AP-1 DNA-bining activity in striatal extracts after eight daily injections of 20 mg/kg cocaine or saline. NS: nonspecifc binding. Results are representative of two independent experiments. **Top right**: Mean optical density of AP-1 bands for three KO (black bars) and three WT (white bars) mice for basal experiment, or six KO and five WT mice for chronic cocaine. **Bottom**: Autoradiogram of supershift assay. Left lane: no FosB antibody; right lane: after addition of FosB antibody. Results are representative of six independent experiments.

receptors has been suggested to inhibit GABA release onto dopaminergic neurons, thereby stimulating the activity of these neurons *(14)*. There are a number of independent indications that 5-HT$_{1B}$ agonists might have cocaine-like activities. The 5-HT$_{1B}$ agonist RU24969 induces, like cocaine, an increase in locomotor activity *(12)* and expression of immediate–early genes such as c-*fos* *(17)*. RU24969 can partially substitute for cocaine in drug discrimination studies *(16)*. In addition, RU24969 induces a leftward shift in doses of cocaine in an iv self-administration paradigm, suggesting that this compound might potentiate the effects of cocaine *(18)*. One might therefore expect that an antagonist of the 5-HT$_{1B}$ receptor would block or decrease the effects of cocaine. Indeed, the 5-HT$_{1B/1D}$ antagonist GR127935 decreased the effects of cocaine on locomotion (Fig. 6). We showed previously that GR127935 also decreased the ability of cocaine to induce c-*fos* in the striatum of WT mice and that this effect was mediated by the 5-HT$_{1B}$ receptor *(17)*. Because of these pharmacological studies, it was somewhat surprising that the locomotor response to cocaine in the KO mice is the opposite of what is found with the antagonist. Similarly, although KO mice appear more motivated to self-administer cocaine, GR127935 alone does not appear to affect cocaine self-administration (Fig. 3; ref. *18*). One possible explanation is the fact that, unlike an acute antagonist, the KO exerts its effect throughout the development and the life of the mouse. There is therefore plenty of time for compensatory mechanisms to come into play. One might, for example, hypothesize that in the absence of the 5-HT$_{1B}$ receptor, the dopaminergic tone is too low and that the organism compensates by boosting the activity of the dopaminergic system. Although this boost might generate normal DA activity in normal conditions, challenging situations such as a cocaine injection could reveal overactivity in the DA system.

In a number of ways, the 5-HT$_{1B}$ KO mice behave as if they had received a chronic regimen of cocaine and become sensitized to the drug. Behavioral sensitization, the increased responsiveness to cocaine following chronic administration, is long-lasting and has been suggested to underlie the addictive properties of cocaine *(25)*. Although behavioral sensitization usually refers to an increased motor response to cocaine, it is also accompanied by a number of biochemical and physiological changes such as increased levels of ΔFosB proteins and decrease of c-fos induction *(34–36)*, increased DA release in the NAc *(43)*, and increased responsiveness of DA D1 receptors *(44)*. The 5-HT$_{1B}$ KO mice display many of the characteristics of sensitized mice. They respond more vigorously to the locomotor effects of cocaine (Fig. 4). In a sensitization paradigm, they display little increase in stereotypy following repeated injections of cocaine (Fig. 7). They express high levels of the ΔFosB forms that are upregulated following chronic cocaine exposure while displaying a reduced induction of c-fos in response to acute cocaine exposure (Fig. 10; and ref. *17*). Furthermore, they express higher levels of transcription complexes that are specific for the AP-1 binding site and have also been shown to be increased by chronic cocaine treatment (Fig. 11). Finally, the KO's faster acquisition of cocaine self-administration (Fig. 1) and higher breaking point in the progressive ratio schedule (Fig. 2) are characteristics that are often associated with sensitization *(45)*. Because the DA system appears to be central to the effects of many categories of drugs of abuse, it is possible that the KO mice will respond more to other drugs that modulate the activity of DA neurons, such as opiates, cannabinoids, nicotine, or caffeine. Interestingly, the 5-HT$_{1B}$ KO mice also display increased self-administration of alcohol *(46)*.

There are a number of genetic arguments suggesting that common genes are involved in differential responsiveness to different categories of drugs such as alcohol, opiates, and psychostimulants. Several inbred strains of rodents such as the Fischer and Lewis rats and the C57BL/6 and DBA/2 mice differ in their responses to alcohol, opiates, and psychostimulants. Lewis rats self-administer cocaine, opiates, and alcohol at higher rates than Fischer rats *(47–49)*. Lewis rats also display greater conditioned place preference to these drugs and increased locomotor response to repeated cocaine injections *(47,50)*. Similarly, the C57BL/6 mice self-administer more cocaine, alcohol, and opiates in an oral self-administration paradigm when compared to the DBA/2 mice *(51–53)*. The analysis of a panel of 26 recombinant inbred strains has revealed that the differential responses to drugs of abuse of the C57BL/6 and DBA/2 strains are due to several QTLs, including one containing the gene encoding the 5-HT$_{1B}$ receptor *(26,54)*. Mutations in this gene might therefore be responsible for some of the differential responses to drugs of abuse of the C57BL/6 and DBA/2 mouse strains. Our preliminary results indicate that there are only silent mutations differentiating the C57BL/6 and the DBA/2 genes and therefore no change in amino acid sequence. However, we have also shown that the levels of 5-HT$_{1B}$ receptors are significantly lower in C57BL/6 than in DBA/2, suggesting the possibility of a mutation in the promoter sequence or in sequences contributing to RNA stability. We are currently investigating whether mutations in the 5-HT$_{1B}$ gene are associated with differential responses to drugs of abuse in rodents and humans. A recent study has found linkages to mutations in the human 5-HT$_{1B}$ gene in two independent populations diagnosed with antisocial alcoholism *(56)*. Because of their phenotype, the 5-HT$_{1B}$ KO mice are also a model for some of the biochemical changes that might be responsible for individual variations in vulnerabilities to drugs of abuse. The identification of proteins that are differentially expressed in the KO mice (such as the 65-kDa band in Fig. 10) might provide new candidate proteins associated with drug abuse.

REFERENCES

1. Plomin, R., Owen, M. J., and McGuffin, P. (1994) The genetic basis of complex human behaviors. *Science* **264,** 1733–1739.
2. Crabbe, J. C., Belknap, J. K., Buck, K. J., and Metten, P. (1994) Use of recombinant inbred strains for studying genetic determinants of responses to alcohol. *Alcohol Alcohol.* **2**(Suppl), 67–71.

3. Plomin, R., McClearn, G. E., Gora-Maslak, G., and Neiderhiser, J. M. (1991) Use of recombinant inbred strains to detect quantitative trait loci associated with behavior. *Behav. Genet.* **21,** 99–116.
4. Koob, G. F. (1992) Neurobiological mechanisms in cocaine and opiate dependence. *Res. Publ.—Assoc. Res. Nervous Mental Dis.* **70,** 79–92.
5. Benloucif, S., Keegan, M. J., and Galloway, M. P. (1993) Serotonin-facilitated dopamine release in vivo: pharmacological characterization. *J. Pharmacol. Exp. Therapeut.* **265,** 373–377.
6. Carroll, M. E., Lac, S. T., Asencio, M., and Kragh, R. (1990) Fluoxetine reduces intravenous cocaine self-administration in rats. *Pharmacol. Biochem. Behav.* **35,** 237–244.
7. Parsons, L. H. and Justice, J. B., Jr. (1993) Perfusate serotonin increases extracellular dopamine in the nucleus accumbens as measured by in vivo microdialysis. *Brain Res.* **606,** 195–199.
8. Parsons, L. H., Koob, G. F., and Weiss, F. (1996) Serotonin-1B receptor stimulation enhances cocaine reinforcement, cocaine-induced increases in nucleus accumbens dopamine, and cocaine-induced decreases in ventral tegmental area GABA. *Society for Neuroscience 26th Annual Meeting,* Washington, DC, p. 361.10.
9. Peltier, R. and Schenk, S. (1993) Effects of serotonergic manipulations on cocaine self-administration in rats. *Psychopharmacology* **110,** 390–394.
10. Hoyer, D., Clarke, D. E., Fozard, J. R., Hartig, P. R., Martin, G. R., Mylecharane, E. J., Saxena, P. R., and Humphrey, P. P. (1994) International Union of Pharmacology classification of receptors for 5-hydroxytryptamine (serotonin). *Pharmacol. Rev.* **46,** 157–203.
11. Lucas, J. J. and Hen, R. (1995) New players in the 5-HT receptor field: genes and knockouts. *Trends Pharmacol. Sci.* **16,** 246–252.
12. Saudou, F., Amara, D. A., Dierich, A., LeMeur, M., Ramboz, S., Segu, L., Buhot, M. C., and Hen, R. (1994) Enhanced aggressive behavior in mice lacking 5-HTIB receptor. *Science* **265,** 1875–1878.
13. Boschert, U., Amara, D. A., Segu, L., and Hen, R. (1994) The mouse 5-hydroxytryptaminel B receptor is localized predominantly on axon terminals. *Neuroscience* **58,** 167–182.
14. Cameron, D. L. and Williams, J. T. (1994) Cocaine inhibits GABA release in the VTA through endogenous 5-HT. *J. Neurosci.* **14,** 6763–6767.
15. Guan, X. M. and McBride, W. J. (1989) Serotonin microinfusion into the ventral tegmental area increases accumbens dopamine release. *Brain Res. Bull.* **23,** 541–547.
16. Callahan, P. M. and Cunningham, K. A. (1995) Modulation of the discriminative stimulus properties of cocaine by 5-HT1B and 5-HT2C receptors. *J. Pharmacol. Exp. Therapeut.* **274,** 1414–1424.
17. Lucas, J. J., Segu, L., and Hen, R. (1997) 5-HydroxytryptaminelB receptors modulate the effect of cocaine on cfos expression: converging evidence using 5-hydroxytryptaminelB knockout mice and the 5-hydroxytryptaminelB/1D antagonist GR127935. *Mol. Pharmacol.* **51,** 755–763.
18. Parsons, L. H., Weiss, F., and Koob, G. F. (1998) SerotoninIB receptor stimulation enhances cocaine reinforcement. *J. Neurosci.* **18,** 10,078–10,089.
19. Parsons, L. H., Weiss, F., and Koob, G. F. (1996) Serotonin1b receptor stimulation enhances dopamine-mediated reinforcement. *Psychopharmacology* **128,** 150–160.

20. Rocha, B. A., Ator, R., Emmett-Oglesby, M. W., and Hen, R. (1997) Intravenous cocaine self-administration in mice lacking 5-HT1B receptors. *Pharmacol. Biochem. Behav.* **57,** 407–412.
21. Weeks, J. (1962) Experimental morphine addiction: method of automatic intravenous injections in unrestrained rats. *Science* **138,** 143–144.
22. Depoortere, R. Y., Li, D. H., Lane, J. D., and Emmett-Oglesby, M. W. (1993) Parameters of self-administration of cocaine in rats under a progressive-ratio schedule. *Pharmacol. Biochem. Behav.* **45,** 539–548.
23. Rocha, B. A., Scearce-Levie, K. S., Lucas, J. J., Hiroi, N., Castanon, N., Crabbe, J. C., Nestler, E. J., and Hen, R. (1998) Increased vulnerability to cocaine in mice lacking the serotonin 1B receptor. *Nature* **393,** 175–178.
24. Kalivas, P. W. and Barnes, C. D. (1988) *Sensitization in the Nervous System.* Telford, Caldwell, NJ.
25. Tolliver, B. K., Belknap, J. K., Woods, W. E., and Carney, J. M. (1994) Genetic analysis of sensitization and tolerance to cocaine. *J. Pharmacol. Exp. Therapeut.* **270,** 1230–1238.
26. Wise, R. and Bozarth, M. (1987) A psychomotor theory of addiction. *Psychol. Rev.* **94,** 469–492.
27. Kalivas, P. (1995) Neural basis of behavioral sensitization to cocaine, in *The Neurobiology of Cocaine* (Hammer, R., ed.), CRC, Boca Raton, FL, pp. 81–98.
28. O'Neill, M. F., Fernandez, A. G., and Palacios, J. M. (1996) GR127935 blocks the locomotor and antidepressant-like effects of RU24969 and the action of antidepressants in the mouse tail suspension test. *Pharmacol. Biochem. Behav.* **53,** 535–539.
29. Post, R. M., Lockfield, A., Squillace, K. M., and Contel, N. R. (1981) Drug–environment interactions: context-dependency of cocaine-induced behavioral sensitization. *Life Sci.* **28,** 755.
30. Browne, S. P., Moore, C. M., Scheurer, J., Tebbett, 1. R., and Logan, B. K. (1991) A rapid method for the determination of cocaine in brain tissue. *J. Forensic Sci.* **36,** 1662–1665.
31. Graybiel, A. M., Moratalla, R., and Robertson, H. A. (1990) Amphetamine and cocaine induce drug-specific activation of the cfos gene in striosome-matrix compartments and limbic subdivisions of the striaturn. *Proc. Natl. Acad. Sci. USA* **87,** 6912–6919.
32. Young, S. T., Porrino, L. J., and Iadarola, M. J. (1991) Cocaine induces striatal c-fos-immunoreactive proteins via dopaminergic DI receptors. *Proc. Natl. Acad. Sci. USA* **88,** 1291–1295.
33. Hope, B. T., Kelz, M. B., Duman, R. S., and Nestler, E. J. (1994) Chronic electroconvulsive seizure (ECS) treatment results in expression of a long-lasting AP-1 complex in brain with altered composition and characteristics. *J. Neurosci.* **14,** 4318–4328.
34. Hope, B. T., Nye, H. E., Kelz, M. B., Self, D. W., ladarola, M. J., Nakabeppu, Y., Duman, R. S., and Nestler, E. J. (1996) Induction of a longlasting AP-1 complex composed of altered Fos-like proteins in brain by chronic cocaine and other chronic treatments. *Neuron* **13,** 1235–1244.
35. Nestler, E. J., Hope, B. T., and Widnell, K. L. (1993) Drug addiction: a model for the molecular basis of neural plasticity. *Neuron* **11,** 995–1006.

36. Sonnenberg, J. L., Macgregor-Leon, P. F., Curran, T., and Morgan, J. 1. (1989) Dynamic alterations occur in the the levels and composition of transcription factor AP-1 complexes after seizure. *Neuron* **3**, 359–365.
37. Chen, J., Kelz, M. B., Hope, B. T., Nakabeppu, Y., and Nestler, E. J. (1997) Chronic Fos-related antigens: stable variants of deltaFosB induced in brain by chronic treatments. *J. Neurosci.* **17**, 4933–4941.
38. Hiroi, N., Brown, J. B., Haile, C. N., Ye, H., Greenberg, M. E., and Nestler, E. J. (1997) FosB mutant mice: loss of chronic cocaine induction of Fos-related proteins and heightened sensitivity to cocaine's psychomotor and rewarding effects. *Proc. Natl. Acad. Sci. USA* **94**, 10,397–10,402.
39. Roberts, D. C., Loh, E. A., and Vickers, G. (1989) Self-administration of cocaine on a progressive ratio schedule in rats: dose-response relationship and effect of haloperidol pretreatment. *Psychopharmacology* **97**, 535–538.
40. Koob, G. F. (1992) Drugs of abuse: anatomy, pharmacology and function of reward pathways. *Trends Pharmacol. Sci.* **13**, 177–184.
41. Hiroi, N., Marek, G. J., Brown, J. R., Ye, H., Saudou, F., Vaidya, V. A., Duman, R. S., Greenberg, M. E., and Nestler, E. J. (1998) Essential role of the fosB gene in molecular, cellular and behavioral actions of chronic electroconvulsive seizures. *J. Neurosci.* **18**, 6952–6962.
42. Kalivas, P. W. and Stewart, J. (1991) Dopamine transmission in the initiation and expression of drug- and stress-induced sensitization of motor activity. *Brain Res. Rev.* **16**, 223.
43. White, F. J., Hu, X.-T., Henry, D., and Zhang, X.-F. (1995) Neurophysiological alterations in the mesocorticolimbic doparnine system with repeated cocaine administration, in *The Neurobiology of Cocaine* (Hammer, R. P., ed.), CRC, Boca Raton, FL, pp. 99–119.
44. Horger, B. A., Shelton, K., and Schenk, S. (1990) Preexposure sensitizes rats to the rewarding effects of cocaine. *Pharmacol. Biochem. Behav.* **37**, 707–711.
45. Crabbe, J. C., Phillips, T. J., Feller, D. J., Hen, R., Wenger, C. D., Lessov, C. N., and Schafer, G. L. (1996) Elevated alcohol consumption in null mutant mice lacking 5-HT1B serotonin receptors. *Nature Genetics* **14**, 98–101.
46. George, F. R. and Goldberg, S. R. (1988) Genetic differences in response to cocaine, in *Mechanisms of Cocaine Abuse and Toxicity* (Clouet, D., Asghar, K., and Brown, R., eds.), National Institute on Drug Abuse, Rockville, MD, pp. 239–249.
47. Li, A. A., Marek, G. J., Vosmer, G., and Seiden, L. S. (1989) Long-term central 5-HT depletions resulting from repeated administration of MDMA enhances the effects of single administration of MDMA on schedule-controlled behavior of rats. *Pharmacol. Biochem. Behav.* **33**, 641–648.
48. Suzuki, T., George, F. R., and Meisch, R. A. (1988) Differential establishment and maintenance of oral ethanol reinforced behavior in Lewis and Fischer 344 inbred rat strains. *J. Pharmacol. Exp. Therapeut.* **245**, 164–170.
49. Guitart, X., Beitner-Johnson, D., Marby, D. W., Kosten, T., and Nestler, E. J. (1992) Fischer and Lewis rat strains differ in basal levels of neurofilament proteins and their regulation by chronic morphine in the mesolimbic dopamine system. *Synapse* **12**, 242–253.
50. Belknap, J. K., Crabbe, J. C., Riggan, J., and O'Toole, L. A. (1993) Voluntary consumption of morphine in 15 inbred mouse strains. *Psychopharmacology* **112**, 352–358.

51. Belknap, J. K., Crabbe, J. C., and Young, E. R. (1993) Voluntary consumption of ethanol in 15 inbred mouse strains. *Psychopharmacology* **112,** 503–510.
52. Jones, B. C., Reed, C. L., Radcliffe, R. A., and Erwin, V. G. (1993) Pharmaco-genetics of cocaine: 1. Locomotor activity and self-selection. *Pharmacogenetics* **3,** 182–188.
53. Crabbe, J. C., Belknap, J. K., Mitchell, S. R., and Crawshaw, L. 1. (1994) Quanti-tative trait loci mapping of genes that influence the sensitivity and tolerance to ethanol-induced hypothermia in BXD recombinant inbred mice. *J. Pharmacol. Exp. Therapeut.* **269,**184–192.
54. Tolliver, B. K. and Carney, J. M. (1994) Comparison of cocaine and GBR 12935: effects on locomotor activity and stereotypy in two inbred mouse strains. *Phar-macol. Biochem. Behav.* **48,** 733–739.
55. Lappalainen, J., Long, J. C., Eggert, M., Ozaki, N., Robin, R. W., Brown, G. W., Naukkarinen, H., Virkkunen, M., Linnoila, M., and Goldman, D. (1998) Linkage of antisocial alcoholism to the serotonin 5-HT1B receptor gene in 2 populations. *Arch. Gen. Psychiatry* **55,** 989–994.

14

Regulation of Dopamine Transporter by Phosphorylation and Impact on Cocaine Action

Roxanne A. Vaughan

INTRODUCTION

The dopamine transporter (DAT) is the primary mechanism by which extracellular dopamine is cleared from the synaptic space. As such, it performs a key role in terminating synaptic transmission and in regulating the concentration of dopamine available for binding to presynaptic and postsynaptic dopamine receptors. DAT is also a major site of action for psychostimulants such as cocaine and amphetamine, and is believed to be involved with the reinforcing properties of these drugs. It has recently been found that activation of protein kinase C leads to reductions in dopamine transport and concomitant phosphorylation of DATs, suggesting that the protein undergoes functional regulation by phosphorylation. Other kinases may function similarly, providing neurons with a mechanism for fine temporal and spatial control of extracellular dopamine concentrations and subsequent neural activity. Therefore, phosphorylation of DAT has the potential to profoundly influence dopaminergic neurophysiology and may be related to mechanisms of cocaine abuse.

THE DOPAMINE TRANSPORTER

The availability of extracellular dopamine for dopaminergic neurotransmission is controlled to a great extent by the action of the dopamine transporter, which terminates dopaminergic nerve signaling by transporting dopamine (DA) from the extracellular space back into the presynaptic cell. Recent evidence from DAT knockout mice indicates that the vast majority of DA clearance can be accounted for by the action of the transporter *(1)*. DAT thus plays a primary role in controlling the intensity and duration of the synaptic response by regulating the concentration and availability of neurotransmitter to downstream receptors.

From: *Cerebral Signal Transduction: From First to Fourth Messengers*
Edited by: M. E. A. Reith © Humana Press Inc., Totowa, NJ

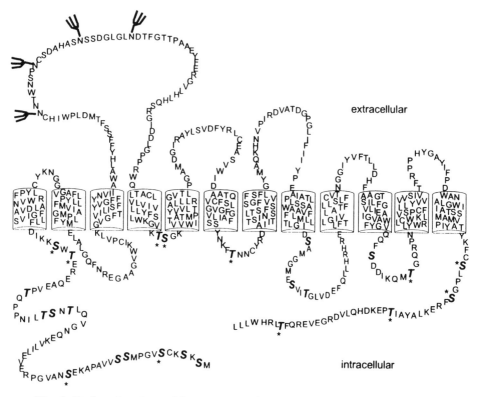

Fig. 1. Deduced amino acid sequence and predicted topological orientation of rDAT. Intracellular serines and threonines which represent potential sites of phosphorylation are shown in bold; residues found within consensus sequences for PKA, PKC, or CaM kinase are indicated by asterisks.

Single-copy genes encoding DAT have been cloned from human, rat, mouse, and bovine midbrain cDNA libraries *(2–8)*. The sequences indicate that the proteins contain 619–620 amino acids with 12 hydrophobic stretches suitable to be transmembrane-spanning domains (TMDs). A large loop between transmembrane helices 3 and 4 contains consensus sites for N-glycosylation, and the N- and C-termini are predicted to be oriented cytoplasmically (Fig. 1). Many of these properties predicted from sequence analysis have been confirmed biochemically. DAT has been shown to be an integral membrane protein requiring detergents for solubilization *(9–13)* and to contain N-linked carbohydrates as well as terminal sialic acids *(14–16)*. Its mass on sodium dodecyl sulfate (SDS) gels is 80 kDa, more than indicated by its nucleotide sequence, but glycosylation analysis demonstrates a core polypeptide 55 kDa with 25 kDa of mass contributed by carbohydrate *(15,16)*. Peptide mapping has verified that the carbohydrates are found in the TMD 3-4 loop, providing direct evidence for the extracellular orientation of this region *(17,18)*, whereas electron microscopic histochemistry is consistent with an intracellular N-terminal tail *(19,20)*. Other aspects of topological orientation and precise number of transmembrane domains are under continuing investigation, and we are far from understanding the three-dimensional structure of the protein and how this relates to function.

The dopamine transporter belongs to a large family of plasma membrane transporters for neurotransmitters, amino acids, and osmolytes, including carriers for norepinephrine (NE), serotonin (5-HT), γ-amino butyric acid (GABA), glycine, proline, taurine, and betaine, and the closely related vesicular monoamine transporters (VMATs) which package DA, NE, and 5-HT into synaptic vesicles (reviewed in refs. *21–23*). These transporters are all thought to have a similar 12-transmembrane domain structure and topological organization, and all display substrate translocation driven by cotransport of Na^+ and Cl^- down ionic electrochemical gradients. Although each transporter displays various unique properties, their considerable structural and functional similarities also make it probable that many of their basic molecular mechanisms will be conserved.

The physiological importance of the monoamine transporters is highlighted by conditions that perturb their normal action. The dopamine, norepinephrine, and serotonin transporters are all targets for the psychostimulant drug cocaine. The drug itself is not transported, but it binds to the protein and functions as a transport antagonist to block reuptake of the neurotransmitter. Animals treated with cocaine exhibit strongly elevated monoamine levels, which are thought to lead to downstream neuronal activation and the diverse physiological effects of the drug *(24)*. Although cocaine inhibits all the monoamine transporters, its reinforcing or addictive properties show the best pharmacological correlation with its activity at DAT *(25,26)* leading to the hypothesis that DAT represents a major site of drug action in reinforcement *(27,28)*.

Dopamine transporters also possess channellike properties, mediating ionic currents much larger than can be accounted for by the stoichiometric ion flux required to drive transport *(29)*. The physiological significance of these currents is not understood *(22)*, but in addition to DA transport and antagonist binding, they represent another measurable function of the protein.

Very little is known about the active sites on DAT for substrate transport or antagonist binding, and the molecular mechanisms underlying transport and inhibition are not understood. Uptake inhibitors such as mazindol, methylphenidate, and nomifensine, which are structurally dissimilar to cocaine, also bind to DAT with high affinity and inhibit DA uptake, but it is not clear if these blockers inhibit the same or different aspects of transport, or if antagonist actions are competitive or noncompetitive. Various residues and domains in the DAT primary sequence have been found to contribute to uptake and binding *(17,18,30–34)*, but the three-dimensional relationships of these sites are not known *(35)*. The multiple activities performed by DAT, their complex relationships, and the mechanistic unknowns may impact on the findings to be discussed.

REGULATION OF DAT FUNCTION BY PROTEIN KINASES

The presumed intracellular aspects of DAT and the other ion-coupled transporters contain numerous serine and threonine residues, many of which are found within canonical phosphorylation sites for protein kinase C (PKC), cyclic AMP-dependent protein kinase (PKA), and Ca^{2+}/calmodulin-dependent protein kinase (CaMK) (Fig. 1). Many of these potential phosphorylation sites are highly conserved throughout the transporter family, and each transporter also contains sites that are unique. The presence of these sites has raised widespread interest in the possibility that DAT and other neurotransmitter transporters undergo functional regulation by phosphorylation. Multiple approaches have been used to investigate this idea, including treatment of synaptosomes or DAT-expressing cells with activators or inhibitors of protein kinases followed by assessment of DA transport, ligand binding, or current flux, by direct assessment of DAT phosphorylation, and by phosphorylation site mutagenesis. The findings indicate that DAT may be subject to rapid functional regulation by phosphorylation, which would provide neurons with a previously unappreciated mechanism for fine-tuning synaptic dopamine concentrations. This chapter will present the evidence related to acute control of various DAT functions and possible relationship to actions of cocaine.

Effects of Protein Kinase Activators on Dopamine Transport

A variety of studies have examined the ability of protein kinase activators and other signal transduction compounds to regulate DA transport (Table 1). These studies have been performed in several types of DA uptake systems, including rat striatal synaptosomes or tissue *(36–41)*, mouse striatal synaptosomes and primary mesencephalic cultures *(42)*, COS7 and LLC-PK$_1$ cell lines transfected with rDAT *(43,44)*, and C6 glioma cells, Sf9 cells, and *Xenopus* oocytes expressing hDAT *(45–48)*. To date, the most effort has been devoted to investigating the effects of phorbol esters and other activators of PKC; mediators of other signal transduction pathways including cyclic

**Table 1
Second-Messengers That Affect DA Transport
and Ligand Binding to DAT**

DAT type	System	Treatment	V_{max}	K_m	B_{max}	K_d	Ref.
Rat	COS7	PMA	d20%	NC	NC[a]	i50%	44
Rat	LLC-PK$_1$	PMA	d35%	NC	—	—	43
Rat	Striatum	PMA	d15%	NC	—	—	41
Mouse	Striatum	PMA	d22%	NC	NC[b]	NC	42
Human	C6 glioma	PMA	d72%	d60%	d30%[a]	NC	46
Human	Sf9	PMA	d29%	NC	NC[a]	NC	45
Human	Oocyte	PMA	d69%	NC	d78%[b]	NC	48
Rat	Striatum	Ca^{2+}	i67%	NC	—	—	40
Human	C6 glioma	AA	d78%	d36%	d25%[a]	NC	47
Rat	Striatum	SNP	d35%	i500%	NC[a]	NC	39
Rat	Striatum	cAMP	*i33%	*NC	—	—	36

Note: V_{max} and K_m values were determined for [^{3}H]DA transport, except for * in which DA transport was assessed with RDE. B_{max} and K_d values were determined for binding of (a) [^{3}H]CFT or (b) [^{3}H]mazindol. Transport and binding values are shown as percent of control values.

Abbreviations: AA, arachidonic acid; PMA, phorbol 12 myristate, 13 acetate; SNP, sodium nitroprusside; d, decrease; i, increase; NC, no change; —, not done.

nucleotides, Ca^{2+}, arachidonic acid, and nitric oxide have also been examined, but to a much lesser extent.

Protein Kinase C

Several studies have now uniformly demonstrated that transport of dopamine is reduced after treatment with protein kinase C activators *(41–46,48)*. Application of phorbol 12-myristate 13-acetate (PMA) produces reductions in DA transport ranging from 15–20% in mouse and rat synaptosomes *(37, 41,42)* to over 80% in C6 glioma cells and oocytes *(46,48)*. These decreases were produced primarily by a reduction in V_{max}, and with one exception *(46)*, with no apparent change in the K_m for DA. Transport decreases produced by PMA were blocked or attenuated by PKC inhibitors such as staurosporine and bisindoylmaleimide *(41–46,48)*, were not produced by the inactive phorbol isomer 4αPDD (4αphorbol 12,13 didecanoate) *(41–43,45,48)*, and were mimicked by the physiological PKC activator diacylglycerol and the nonphorbol PKC activator (−) indolactam V *(41,42,44)*. These results (Table 2) provide compelling evidence that the observed reductions in DA transport are mediated by PKC, although it is not known if this occurs directly or by a downstream event.

Table 2
Effects of Protein Kinase C Activators on DA Transport

Treatment	Transport activity, % control
Vehicle	100
PMA	86.9 ± 3.0*
OA	86.5 ± 3.0*
PMA + OA	75.9 ± 2.5**
OAG	79.0 ± 2.6*
4αPDD	97.8 + 3.2
PMA	85.0 ± 4.4*
PMA + staurosporine	100.0 + 3.6

Note: Synaptosomes were treated with the indicated compounds for 20 min at 30°C prior to analysis for dopamine transport. Results are shown relative to control samples treated with vehicle (dimethylsulfoxide) ± SE. Upper group: Final concentations of PMA, OA, and 4αPDD were 10 μM, and OAG was 1 mM. Lower group: Final concentration of PMA was 1 μM and staurosporine was 3 μM. * and ** indicates statistically different groups, $p < 0.05$.
Abbreviations: OA, okadaic acid; OAG, 1-oleoyl-2-acetyl-*sn*-glycerol.
Source: Reprinted from ref. *41* with permission.

The PKC-induced inhibition is rapid, with uptake reductions detectable within 1–5 min, and maximum reductions reached by 20–30 min *(41–44, 46,48)*. These rapid response times are consistent with processes activated by second-messenger systems and, in conjunction with the inducement by diacylglycerol, indicate the potential for endogenous DAT regulation to occur in vivo by receptor-mediated processes (*see* Endogenous DAT Phosphorylation and Regulation).

While reduced DA uptake after PKC treatments might result from reduced DAT activity, other mechanisms such as enhanced DA release, loss of ionic gradients required to drive transport, loss of tissue viability, or effects on vesicular transport might also produce the observed effects. Many efforts have been made to address these possibilities, including direct measurement of membrane potential *(48)*, assessment of PMA effects on Na^+ characteristics *(44)*, measurement of activity of Na^+/K^+ ATPase, which maintains Na^+ electrochemical gradients *(42)*, measurement of Na^+-dependent [^{3}H]alanine transport, which would be altered by a Na^+ gradient reduction *(41,43,48)*, direct assay of [^{3}H]DA release *(43)*, assessment of synaptosomal health by LDH assay *(42)*, and reserpine pretreatment, which would indicate potential effects on vesicular DA utpake *(42)*. No evidence was found in any of these studies for PMA-induced loss of cell or synaptosomal viability, alteration of ionic gradients, increased release of DA, or effects on vesicular transport, leading to the belief that the reduced DA uptake observed in response to PMA is most likely due to decreased dopamine transporter activity.

Phosphatase Inhibitors

Protein phosphatase inhibitors such as okadaic acid (OA) also reduce DA uptake (Table 2), presumably mimicking the action of kinases by inhibiting substrate dephosphorylation *(41,42)*. These effects occur over times of 5–20 min, comparable to the time-course of PMA action *(41,42)*. OA inhibits protein phosphatase 1 (PP1) and protein phosphatase 2A (PP2A) at nano-molar levels, and other phosphatases such as PP2B (calcineurin) at much higher concentrations *(49)*. Inhibition of DA transport required OA doses of 0.5–10 mM, but it is not known if this indicates an action via calcineurin, or if this is a reflection of poor membrane permeability properties of the compound. The high-dose requirement may explain results of studies in which DA uptake in striatum or cells was not affected by lower doses of OA *(46, 50)*. Cotreatment of synaptosomes with OA and PKC activators were additive for reducing uptake (Table 2), suggesting a convergence of the affected pathways, possibly at DAT *(41,42)*.

Calcium

Potential upregulation of DA transport by Ca^{2+} is indicated by the finding that basal DA transport activity obtained in the absence of Ca^{2+} is stimulated more than twofold with increasing Ca^{2+} *(40)*. This increase is dose and time dependent, with strong increases observed after 5 min of Ca^{2+} pretreatment and with maximum effects obtained between 10 and 30 min. A 10-min treatment with 1–5 mM Ca^{2+} produced optimum transport activity, whereas higher Ca^{2+} concentrations were strongly inhibitory. The increased activity was produced by an increase in V_{max} with no effect on K_m (Table 1), but the mechanism underlying reduced transport was not pursued. The ability of Ca^{2+} to stimulate uptake was blocked by inhibitors of CaM kinase II, myosin light-chain kinase, and calmodulin, but not by PKC inhibitors, suggesting that the Ca^{2+}-induced stimulation involves activation of CaM kinase.

Independent confirmation of these exciting results has not yet appeared and there is a conflicting report that DA uptake was enhanced rather than reduced by omission of Ca^{2+} *(39)*. Nevertheless, these results have strong implications for other DAT regulation studies, because 1–2 mM Ca^{2+} is routinely added to uptake buffers and was present in most of the studies done with other compounds. These treatments may require re-evaluation in the absence of Ca^{2+}, as Ca^{2+}-induced upregulation of DA uptake may have counteracted downregulation produced by PMA or other inhibitors, or masked upregulation produced by other activators.

Arachidonic Acid

Receptor-mediated activation of phospholipase A2 catalyzes the formation of the phospholipid breakdown product arachidonic acid (AA), which can also activate PKC and has been related to many neurotransmitter effects. Investigation of possible AA involvement with DAT function has been examined in rat striatal synaptosomes *(38)* and in C6 glioma cells expressing hDAT *(47)*. In synaptosomes, AA profoundly reduced uptake (>90%) across a dose range of 10^{-5}–10^{-6} M, whereas the inactive AA analog, arachidic acid, had no effect.

More extensive characterizations were done in C6 glioma cells expressing hDAT, where application of exogenous AA resulted in complex, multidirectional effects on DA transport *(47)*. Treatment times of 30 min or less with either low or high doses stimulated DA uptake twofold, as did longer treatments at low doses. Maximum stimulation at all doses occurred by 15 min, and longer treatments (45–60 min) at higher doses reduced uptake as much as 90%. The inhibitory effect was produced by a decreased V_{max} counteracted by a small decrease in K_m (Table 1). Comparable reductions in DA transport were produced by receptor-mediated stimulation of endogenous arachidonic acid production, by preventing AA breakdown, and by inhibiting AA reincorporation into phospholipid. Transport inhibition was prevented by cotreatment with bovine serum albumin, which binds AA and prevents its action, but it was not blocked by staurosporine, suggesting a PKC-independent mechanism. Alternative possibilities for the AA effect include alterations in lipid microenvironment or effects on related proteins or membrane polarization *(47)*, but potential protein kinase involvement cannot yet be excluded.

Nitric Oxide

The nitric oxide (NO)-generating compounds sodium nitroprusside (SNP) and *s*-nitro-*N*-acetylpenicillamine (SNAP) also strongly inhibit DA transport in a dose-, time-, and temperature dependent manner *(39)*. Inhibition occurred with a $t_{1/2}$ of 10 min and was blocked by reduced hemoglobin, which binds NO and prevents its activity. One action of NO is to activate guanylate cyclase *(51)*, and 8-Br-cGMP inhibits DA uptake in mouse striatum *(42)*, presenting the possibility that the NO effect may be mediated through cGMP-dependent protein kinase. However, at present, we cannot exclude the possibility that other NO effects such as nitrosylation of amino acid side chains might provide the basis for reduced DAT function.

Cyclic Nucleotides

Regulation of DA transport has also been found in response to treatments with cGMP- and cAMP-related compounds. Treatment of striatal tissue with

8-Br-cGMP produced a small (10%) but statistically significant decrease in DA transport *(42)*. A similar result cited as data not shown was reported in relationship to NO-induced transport reductions *(39)*, but further work with this compound has not yet been done.

More effort has been devoted to investigating the effects of cAMP. Several studies have examined uptake of [^{3}H]DA in response to PKA activators such as cAMP analogs or forskolin, without obtaining evidence for transport regulation *(42,45,48,50)*. However, cAMP effects on DA uptake have been found using rotating-disk electrode (RDE) voltammetry, which measures zero-trans neurotransmitter uptake on a time-scale of seconds in contrast to the 1- to 10-min measurements typically performed for uptake of radiolabeled DA. When measured with RDE, striatal DA uptake increased by 33% after a 1-min exposure to 8-Br-cAMP, and it decreased by 20% after treatment with the PKA inhibitor H89 *(36)*. These effects were produced by changes in V_{max} with no change in K_m (Table 1). Enhanced activity was also produced by forskolin, but not the inactive dideoxyforskolin, and the increased activity induced by 8-Br-cAMP was blocked by H89, all strong indications for involvement of PKA. A significant characteristic of these findings was that the responses were rapidly transient. Pretreatment of the tissue for 12–15 min prior to uptake produced no effect, which may explain the lack of cAMP regulation seen in the studies that used longer pretreatments and assays.

An additional indication for cAMP involvement with DA transport regulation was found by treatment of DAT-expressing Sf9 cells with the PKA antagonist Rp-cAMPS *(45)*. Although the PKA agonist Sp-cAMPS and other cAMP analogs had no effect on transport, Rp-cAMPS elevated DA transport activity by 35%. This anomalous result was hypothesized to represent an indirect action mediated by signaling crosstalk rather than a direct effect of PKA *(45)*. Although the contradictory results of these two studies require futher clarification, these first indications of cAMP involvement with DA transport provide strong impetus for further investigations of this important second-messenger pathway.

Summary of Transport Regulation Studies

The studies examining protein kinase C have provided compelling evidence that DAT is regulated by this enzyme, particularly when considering the extent of the similarities that have been found using different species homologs of DAT and using transport systems as varied as mammalian synaptosomes, insect cells, and *Xenopus* oocytes. The studies have all shown reduced DA transport mediated by reduced V_{max}, and all found comparable time-courses and responses to inhibitors. Nevertheless, notable differences were also found, particularly the wide variation in extent of transport reduction, which ranged from 15% to 90%, and the variable sensitivities to OA and PMA, which were effective at 10-fold to 100-fold lower concentrations in cells and oocytes than in synaptosomes. These differences may be a reflection of enzymatic variations between the systems or indicate potential species-specific susceptibilities to regulation, and elucidating their basis may reveal important transport regulation characteristics.

The relationships of the other signal transduction compounds to DA transport have not been as well characterized, and many of the studies have produced conflicting results or are single reports needing further verification. Although this makes it premature to overgeneralize regarding pathway-specific properties, the mounting evidence that DA transport can increase or decrease in response to different effectors provides an emerging picture of multidirectional transport regulation, with the possibility for integrated control of extracellular DA by a wide variety of physiological signals.

Effects of Protein Kinase Activators on Antagonist Binding

The potential for binding of cocaine or other DA uptake inhibitors to be regulated by kinase activators has not been examined as thoroughly as for DA transport, and it is not clear how binding changes relate to phosphorylation conditions or to concomitant uptake changes. Studies examining kinase effects on antagonist binding have used [^{3}H]mazindol to monitor DAT presence/binding activity, whereas analysis of cocaine binding has been done with the cocaine analog [^{3}H]CFT (2β-carbomethoxy-3β-4-fluorotropane). Activity of other DAT ligands in response to kinase treatments has been less extensively investigated.

Considerable disparity exists among results of various studies that examined DAT binding after PMA treatment, with some studies reporting reduced binding and others showing no effects (Table 1). PMA treatment reduced the B_{max} of [^{3}H]CFT binding to hDAT expressed in C6 glioma cells by as much as 60% (46), and the B_{max} of [^{3}H]mazindol binding to hDAT in oocytes by over 90% (48), whereas a small PMA-induced decrease in [^{3}H]CFT binding to rDAT in COS cells was reported to be due to an increased K_d (44). Conversely, other studies have found no significant PMA-induced change in either binding parameter for [^{3}H]CFT or [^{3}H]mazindol in synaptosomes or Sf9 cells (42,45), and neither PMA nor staurosporine affected the K_i of a number of transport inhibitors assayed for inhibition of [^{3}H]DA uptake (45).

Effects of other second-messenger treatments on ligand binding are even less well characterized, with [^{3}H]CFT binding reported to be unaffected by NO treatments that inhibited uptake (39) but dose- and time-dependently reduced by all arachidonic acid treatments, even those that increased uptake

(47). Several of the studies that examined transport regulation did not include analysis of binding (Table 1).

The source of the discrepancies among the reports showing changes or no changes in binding is not known, making it difficult to assess how these findings relate to each other or to transport regulation. Many methodological variables between the studies prevent us from pinpointing mechanisms contributing to discrepancies, including species type and variables contributed by the system, such as enzymatic variations, different DAT expression levels, and differing ratios of cell surface to internal DAT pools. Variation among systems in proportion of total to surface DAT expression could alter the extent of transport or binding regulation, making fractional changes in activity more or less pronounced. In addition, although binding in these studies was assessed in whole cells or synaptosomes, the ligands used are thought to be able to cross membranes, therefore equilibrium binding may be measuring both internal and cell-surface DAT pools. Presumably, only DATs expressed at the cell surface contribute to uptake, leading to the possibility that functional regulation of a small pool of surface DATs might be undetectable in a larger total pool. Binding ligands capable of distinguishing surface from internal DAT pools will help address these issues.

Other potential contributions to differences between regulation of uptake and binding include methodological differences between uptake and binding assays (e.g., time-courses), as well as the possibilities that phosphorylation may have different effects on uptake and binding, that different kinases may differentially affect uptake and binding, or that binding of different ligands may be differentially affected by different phosphorylation activators. This wide variety of unknowns clearly indicates the need for additional study of these phenomena, but it also presents the potential for elucidating molecular and functional relationships between transport and binding.

Effects of Protein Kinase Activators on Currents

Transport-dependent and transport-independent currents mediated by DAT have also been shown to be regulated by PMA treatment *(48)*. In *Xenopus* oocytes expressing hDAT, PMA pretreatment reduced DAT-mediated ionic currents with a magnitude and time-course comparable to that of DA transport reduction. Because this property of transporters has only recently been identified, it has received the least study of any of the DAT properties with respect to regulation.

Effects of Protein Kinase Activators on Reverse Transport

Other more complex functions of DAT, such as reverse-transport-mediated release of DA induced by amphetamine, are also affected by phosphorylation conditions *(37,52,53)*. PKC-inhibiting treatments such as chelerythrine, Ro-31-8220, and calphostin C *(37)*, or H7 *(53)* reduced amphetamine-stimulated DA release from rat striatal tissue, although spontaneous DA release in the absence of amphetamine was not affected *(37)*. Conversely, treatment of bovine retina with OA produced an increase in amphetamine-stimulated DA release *(52)*. These effects were produced within time periods of 30–90 min, with shorter times not tested. These results may indicate an inverse relationship between the regulation of inward and reverse transport, and further contribute to the contention that DAT function is highly dependent on phosphorylation mechanisms.

DOPAMINE TRANSPORTER PHOSPHORYLATION

The above-described studies strongly implicate protein phosphorylation systems in the regulation of DAT function. Such involvement might occur by direct phosphorylation of the transporter or through phosphorylation of accessory regulatory proteins. Recent results show that DAT is phosphorylated in a PKC-dependent manner that correlates well with PKC-induced control of DAT function, providing strong evidence that phosphorylation of the transporter may be the mechanism underlying second-messenger-induced regulation of function.

In Vivo Phosphorylation

Metabolic labeling with $^{32}PO_4$ has identified DAT as an endogenous phosphoprotein both in rat striatal synaptosomes *(41)* and in cells expressing rat and human DATs *(43,54)*. These studies used highly specific DAT antisera to immunoprecipitate a $^{32}PO_4$-labeled protein that was identified as DAT based on its molecular mass, tissue distribution, and antibody specificity. These initial observations of DAT phosphorylation are also the first direct demonstration of phosphorylation of any of the neurotransmitter transporters in brain tissue.

In both cells and synaptosomes, DATs are phosphorylated in the absence of exogenous treatments, demonstrating endogenous basal turnover of phosphate *(41,43)*. Treatment with PMA induces threefold to fourfold increases in phosphate incorporation, which is detectable within 5 min and reaches apparent maximum by 15–20 min (Fig. 2). Basal and PMA-stimulated phosphorylation are blocked by the protein kinase inhibitors staurosporine and bisindoylmaleimide, and stimulation is not produced by 4αPDD *(41,43)*. Other PKC activators such as the diacylglycerol analog OAG (1-oleoyl-2-acetyl-*sn*-glycerol), and (−) indolactam V produced DAT phosphorylation increases comparable to PMA *(41)*. These data are strong evidence that activa-

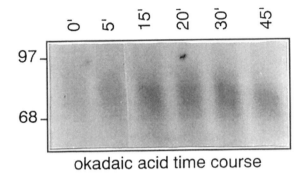

okadaic acid time course

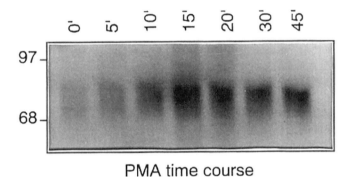

PMA time course

Fig. 2. In vivo phosphorylation of DAT. Rat striatal synaptosomes were metabolically labeled with $^{32}PO_4$, then treated with 10 µM OA or 10 µM PMA for the indicated times. PMA-treated samples received OA for the entire labeling period. Samples were subjected to immunoprecipitation with a DAT-specific antibody followed by SDS–polyacrylamide gel electrophoresis and autoradiography *(41)*. Molecular weight standards are shown in kDa. (Reproduced from ref. *41* with permission).

tion of PKC results in increased phosphorylation of DAT, although whether this occurs directly by PKC or by a kinase downstream of PKC is not known.

Phosphatase inhibitors also strongly increase DAT phosphorylation *(41)*. In fact, in synaptosomes, basal and PMA-stimulated phosphorylation of DAT are difficult to observe without inhibition of phosphatases. This suggests that in the brain, DAT phosphorylation levels are suppressed by robust dephosphorylation, and indicates that phosphatases are crucial elements in the regulation of the DAT phosphorylation state. The ability of PMA to stimulate phosphate incorporation in the presence of OA demonstrates that the effects of these compounds on DAT phosphorylation are additive (Fig. 2).

Okadaic acid increases DAT phosphorylation with rapid kinetics similar to those of PMA (Fig. 2), showing that endogenous dephosphorylation is rapid. Inhibition of DAT dephosphorylation requires micromolar concentrations of OA and calyculin, doses more compatible with inhibition of protein phosphatase 2B than PP1 or PP2A *(49)*. However, DAT phosphorylation was not affected by the PP2B inhibitor cyclosporin A *(41)*; therefore, these results are inconsistent with known patterns of phosphatase activity and/or could indicate the action of multiple phosphatases. Because phosphatases as well as kinases are subject to strong regulation *(49)*, determining the identity of the phosphatases which dephosphorylate DAT may be as critical to our understanding as the identification of the kinases that drive its phosphorylation.

Relationship of DAT Phosphorylation and DA Transport Regulation

Comparison of PKC-induced DAT phosphorylation and DA transport reduction shows the existence of a close correlation between the two and indicates the potential for DAT function to be regulated by its phosphorylation state. In both LLC-PK$_1$ cells and synaptosomes, all conditions that stimulate DAT phosphorylation lead to decreased DA transport, and conditions that are neutral for, or block, phosphorylation have no effect on uptake *(41,43)*. DAT phosphorylation and transport inhibition show good agreement with regard to time-courses, dose responses, inhibitor characteristics, and additivity of OA and PMA (Table 2 and Fig. 2). The finding that OA and PMA are additive for both transport inhibition and DAT phosphorylation is compatible with the signal for transport regulation being the phosphorylation state of DAT rather than that of an accessory protein.

The similarities found between cultured cells and brain in functional regulation and DAT phosphorylation suggest that cell lines will provide good model systems for examining these properties. However, some differences between these systems have been observed that may be indicators of important regulatory properties. As mentioned earlier, PMA reduces transport to a greater extent and at a lower dose in cells than in brain. This is paralleled by DAT phosphorylation, which was stimulated in cells at lower PMA concentrations than in the brain. The two systems also showed differential responses to okadaic acid, which produced a more pronounced stimulation of DAT phosphorylation in brain than in cells, possibly indicating that DATs undergo more robust dephosphorylation in the brain. This could explain the lesser extent of PMA-induced transport reduction obtained in brain compared to other systems, as stronger dephosphorylation of DAT may counteract the inhibitory effects of kinases. Other differences that might exist between cell systems and brain that could lead to regulation differences include phosphorylation sites and stoichiometry. These known and potential variables highlight the need to carefully characterize phosphorylation properties obtained in model systems prior to projecting mechanistic conclusions to the brain, as well as for the eventual purpose of using transfected cells to characterize DAT phosphorylation mutants.

At present, we do not know if DAT phosphorylation is involved with the other signaling pathways that regulate transport. Preliminary examination showed no enchancement of DAT phosphorylation after treatment of synaptosomes with forskolin or 8-Br-cAMP *(41)*, but this clearly needs additional evaluation in light of the recent cAMP findings *(36)*. Examination of DAT phosphorylation in response to the other compounds known to regulate uptake has not yet been reported, and is necessary to understand their mechanisms of action, keeping in mind that these other second messengers may alter DA transport in ways not involving DAT phosphorylation.

Phosphorylation Site Mutagenesis

Another important strategy for assessing involvement of phosphorylation in neurotransmitter transporter function is to mutate candidate phosphorylation sites to nonphosphate acceptor alanines and examine transport for loss or attenuation of protein-kinase-induced effects. This has been done for GABA transporter serines 24, 26, 562 *(55)* and for the GLYT1b glycine transporter residues T19, S239, T590, and S625 *(56)*, but none of these mutations performed singly or in combination altered the response of the transporters to PMA. Lack of effect may indicate that the sites are not involved in phosphorylation, or that multiple sites are involved and mutation of the indicated sites was insufficient to affect the overall response. However, these studies did not include phosphorylation analysis, so it is not known if the relevant residues were targeted.

There is only one preliminary report *(57)* describing mutation of all PKC consensus sites in the DAT. This mutant responded to PMA with DA uptake downregulation in the same manner as wild-type DAT, but phosphorylation analysis has not been performed yet to rule out other potential nonconsensus phosphorylation sites. The identification of DAT phosphorylation sites is a subject of current research. The rDAT primary sequence contains 14 serines and 10 threonines on putative intracellular aspects of the protein that could serve as phosphate acceptors for the serine/threonine kinases examined in these studies (Fig. 1). However, the large number of sites in conjunction with possible multisite phosphorylation makes it unlikely that a strictly molecular approach will succeed in identifying the appropriate sites. Phosphoamino acid analysis indicates that the vast majority of PMA-stimulated phosphate incorporation occurs on serines (R. Vaughan, unpublished result), potentially eliminating threonines as likely candidates for major regulatory effects for this pathway. In addition, although four tyrosines are present on the presumed intracellular aspects of DAT, phosphotyrosine was not detected.

Evidence for Endogenous DAT Phosphorylation

The above-described phosphorylation studies were done with exogenous treatments intended to drive maximal DAT phosphorylation, but such studies do not address the potential physiological control or set point of this process. The basal phosphorylation of DAT observed in synaptosomes shows that phosphate incorporation occurs in native tissue in the absence of exogenous treatments *(41)*, but the stoichiometry and regulation of phosphate levels by physiological stimuli is not understood.

Although this aspect of DAT phosphorylation has not yet been directly investigated, some observations may be relevant. Two lines of evidence have shown that DATs derived from brain are more acidic than predicted by amino acid sequence. Although the calculated isoelectric point (pI) of DAT is nearly neutral (7.05), we observed that isoelectric focusing (IEF) of photo-affinity-labeled rat striatal DAT resolves the protein into multiple acidic isoforms of pI 5.5–6.8 *(58)*. Removal of negatively charged sialic acids by deglycosylation did not restore the protein to a neutral form, compatible with the negative charges being a consequence of phosphorylation. Another indication that native DATs possess heterogenous negative charge is the ability of detergent-solubilized rat and primate striatal DATs to be adsorbed by anion-exchange resins *(10,13)*, and resolve into multiple components by graded elution *(10)*. Neither IEF nor chromatography identified DAT forms that were neutral, showing that all or most DAT polypeptides in the brain are negatively charged. Finally, DATs visualized on SDS gels by photoaffinity labeling and Western blotting also frequently display a pattern of multiple isoforms *(59)*. This is a common property of phosphoproteins, and phosphorylated DATs at times display an upward shift in migration on gels *(41,43)*, indicating the potential for the variable electrophoretic mobility of these DAT bands to reflect isoforms differentially phosphorylated in vivo.

Summary of Phosphorylation Studies

These data indicate the potential for robust and complex protein-kinase-mediated regulation of DA transport and possibly other DAT functions to be mediated by direct phosphorylation of the protein. Transporter phosphorylation may be a property common for transport control of many neurotransmitters, as serotonin transporters (SERTS) and VMATs expressed in cells are also phosphorylated in vivo *(60,61)*. Many aspects of this phenomenon remain to be investigated, including identification of the specific kinases and phosphatases that act on the protein, identification of sites and stoichiometry of phosphorylation, elucidation of endogenous control of phosphorylation/dephosphorylation, and analysis of potential crosstalk or convergence of various signal transduction pathways. However, it should be kept in mind that rigorous establishment of a link between DAT phosphorylation and DA transport regulation by site-directed mutagenesis has not yet been accomplished, leaving open the possibility that these phenomena are not directly related.

MECHANISM OF TRANSPORTER REGULATION AND RELATIONSHIP TO COCAINE

The mechanisms by which DAT phosphorylation is translated into functional upregulation or downregulation of DAT activities are not understood. Alteration of transporter activity can be hypothesized to occur by an increase or decrease in the number of transporters at the cell surface or by a change in DAT binding or transport efficiency. Evidence is accumulating that kinase-induced changes in activity of DAT *(45,48)*, as well as that of SERTs, norepinephrine transporters (NETs), and GATs *(55,62,63)*, are accompanied by redistribution of transporters between the plasma membrane and intracellular compartments. The evidence for direct change in DAT catalytic properties is less compelling, but the possibility still remains that transport may be controlled by multiple coincident processes in a manner similar to the complex events leading to desensitization and downregulation of G-protein-coupled receptors *(64)*. The molecular mechanisms underlying regulation of DAT functions have clear implications for various activities and actions of cocaine.

Regulation of Surface Expression

Two studies have specifically examined the relative distribution of cell-surface and internal DATs with respect to functional regulation. In oocytes, DAT surface expression response to PMA was examined by comparing [^{3}H]mazindol binding in whole cells and homogenates *(48)*. In intact oocytes, PMA treatment strongly reduced DA transport, transport-mediated currents, and [^{3}H]mazindol binding by about 90%. However, when binding was assessed in PMA-treated oocytes that were broken prior to assay, the binding activity detected displayed the same B_{max} and affinity as untreated controls. This suggests that the PMA-induced reductions in transport, current, and binding are all a consequence of DAT internalization rather than loss of binding affinity or alteration of transport catalytic properties *(48)*. Similar results were found when DAT surface expression in Sf9 cells was monitored with confocal microscopy *(45)*. PMA induced a 60% inhibition of uptake and a concomitant relocalization of DAT immunofluorescence from plasma membrane to intracellular compartments, consistent with internalization. Conversely, treatment with Rp-cAMPS, which increased DA transport activity, induced a recruitment of DAT immunostaining to the plasma membrane.

These results support the contention that uptake changes induced by kinases can be produced by membrane trafficking between intracellular compartments and plasma membrane. If comparable events are found to occur in the brain, they could have profound influences on the action of cocaine. Plasma membrane recruitment or internalization of DATs at the synapse would change the density of cocaine binding sites, which might alter the drug's efficacy and result in individual variations in sensitivity to the drug. Modulation of extracellular baseline DA levels by transport regulation could also magnify or diminish cocaine-induced DA concentration changes, thereby affecting associated neurophysiological events.

Alteration of Affinity

Regulation of DA transport by membrane trafficking does not preclude the possibility of concomitant regulation by other mechanisms such as changes in transporter catalytic efficiency. Although most second-messenger treatments have not been found to alter the affinity of DAT for DA or antagonists (Table 1), such changes have been found in some instances, including increased K_d for [^{3}H]CFT binding after PMA treatment *(44)*, and changes in the K_m for DA in esponse to PMA *(46)*, AA *(47)*, and SNP *(39)*. In addition, chromatographically resolved DAT charge isoforms, which may represent differentially phosphorylated DAT populations (*see* Evidence for Endogenous DAT Phosphorylation), display modest differences in cocaine and dopamine binding parameters *(10)*. Because these altered binding properties were detected in detergent-solubilized DAT samples *(10)*, membrane trafficking mechanisms presumably did not contribute to the results, and these findings may be consistent with direct alterations of transporter binding affinities induced by phosphorylation. Together, these results may be indications that changes in catalytic properties or mechanisms other than membrane trafficking contribute to regulation of DAT actions. The possible alteration of DAT affinity for cocaine by phosphorylation has direct implications for the drug's physiology. Acute and possibly chronic increases or decreases in DAT affinity for cocaine would affect the ability of the drug to block DA transport, leading to variabilities in extent of drug-induced DA overflow and subsequent downstream responses.

Reversibility of Regulation

Very little attention has been paid to the fate of upregulated or down-regulated DAT, although this point has great relevance to the physiology of the response. Two attempts to demonstrate reversibility of PMA regulation of DAT by washing PMA-treated synaptosomes or oocytes were negative *(42,48)*. However, because lipophilic phorbol esters are notoriously difficult to remove by washing out, we do not know if their use masks what would otherwise be a rapidly reversible inhibition of transport or if these findings are an accurate indication of slow recovery from functional down-regulation. On the other hand, the recent results showing occurrence of cAMP effects at treatment times of 1 min but not 15 min *(36)* indicates either a rapid reversibility or desensitization of cAMP-induced transport upreg-ulation. The time-course of reversibility is also an interesting point relative to the action of cocaine. If PKC-induced downregulation is only slowly reversible, reduction in DAT function produced by internalization or loss of affinity could produce effects longer-lasting than the metabolic persistence of the drug. Thus, if cocaine was administered at a time when DATs were already downregulated by a PKC-linked physiological stimulus (*see* the next subheading), the drug might be metabolically cleared before the downreg-ulation could reverse. This would potentially reduce interactions between DAT and cocaine and lead to suppression of drug effects.

Endogenous DAT Phosphorylation and Regulation

We currently have little information regarding the types of physiological stimuli that provide the endogenous control for DAT phosphorylation and/or functional regulation. This issue is of utmost importance to our under-standing of the relevance of these phenomena to normal neurophysiology, dopaminergic neurodegeneration, and mechanisms of drug abuse. Potential candidates for physiological control are receptors that couple to PKC and possibly to other effectors such adenylate cyclase, keeping in mind that, as yet, there is no direct evidence linking pathways other than PKC to phos-phorylation of DAT.

The potential involvement of DA receptors in regulating DAT phospho-rylation is of great interest, as this could provide a dopaminergic feedback mechanism for regulation of dopamine transport (Fig. 3). Evidence in sup-port of acute dopaminergic control of DAT function comes from studies showing that activation of D2 receptors increases striatal DA transport *(36,65)*, although the biochemical basis for this finding and potential relationship to DAT phosphorylation are not understood. Although it is well established that dopamine receptors are coupled to adenylate cyclase and PKA *(66)*, there is also evidence linking DA receptors to inhibition of PKC activity *(67)* and to activation of phospholipase C and inositol phosphate production *(66,68)*. The potential for the D2 receptor stimulation of DA transport to be mediated by PKA is not clear, as activation of D2 receptors inhibits adenylate cyclase and decreases cAMP levels *(66)*, but it produces the same transport increases as cAMP analogs and forskolin *(36)*. Speculation on possible PKC involvement with dopamine control of DAT function is also fraught with uncertainties. D_2 receptor stimulation of DA transport may be consistent with DA receptor-induced inhibition of PKC activity *(67)*, which would lead to decreased DAT phosphorylation and increased transport activity (Fig. 3B, center). However, if DA receptors increase inositol phosphate production *(66,68)*, this would activate PKC, leading to increased DAT phosphorylation and reduced transport activity (Fig. 3A, center).

At present, there is no clear biochemical explanation for dopaminergic regulation of DAT function, and further experiments will be required to clarify the relationships between DA receptors, PKA, PKC, and DAT phosphorylation. However, a DA receptor–DAT phosphorylation link would also present a potential mechanism for cocaine involvement in control of DAT activity via the drug's ability to alter DA levels (Fig. 3). DA overflow induced by cocaine would activate DA receptors and, depending on the direction of DA receptor–kinase coupling, would lead to increased or decreased DAT phosphorylation, and increased or decreased DA clearance. If DA receptors are positively coupled to PKC (Fig. 3A), the resulting increases in DAT phosphorylation might exacerbate the effects of cocaine by reducing the DA transport capability of DATs not occupied by the drug (Fig. 3A, right). Conversely, if DA receptors are negatively coupled to PKC, the subsequent reduction of DAT phosphorylation following cocaine-induced activation of DA receptors would dampen the drug effect by increasing the transport activity of non-cocaine-occupied transporters (Fig. 3B, right). Additional or alternative feedback scenarios for DAT function and cocaine action can be envisioned if DA receptor control of DAT phosphorylation is found to occur via PKA.

Aside from dopamine itself, there are many examples of neurotransmitter and drug abuse treatments which alter brain dopamine levels and/or DA transport activity. Although some of these effects may be mediated through activation of neural circuits rather than through second messengers, their potential relationship to DAT phosphorylation can be tested. DAT phosphorylation may also represent a point of convergence for second-messenger signaling initiated by multiple receptors and receptor types, resulting in highly integrated control of DA transport, and individual variations in activity of

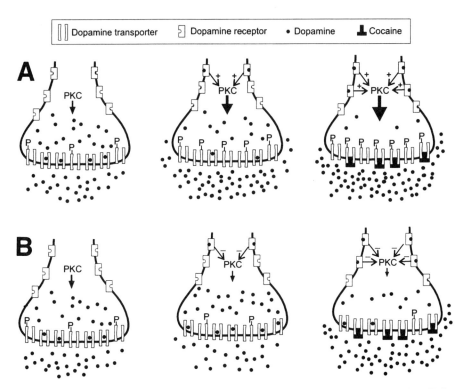

Fig. 3. Hypothetical dopaminergic feedback control of DA transport via PKC-induced DAT phosphorylation, and potential involvement of cocaine. Active and inactive DATs are represented with or without dopamine in the transport channel; phosphorylated DATs (P) are shown as being inactive. Relative activity of PKC is indicated by size of arrows. Not shown are synaptic vesicles or potential internalization of DATs. (**A**) Scenario in which activation of DA receptors increases PKC activity; *left*: neuron exhibiting a basal level of DAT phosphorylation and intermediate level of DA uptake; *center*: DA binding to presynaptic DA receptors activates PKC, leading to increased DAT phosphorylation and decreased DA uptake; *right*: presence of cocaine augments extracellular DA, leading to activation of additional DA receptors, further increases in PKC activity and DAT phosphorylation, and further reductions in transport activity of non-cocaine-occupied DATs. (**B**) Scenario in which activation of DA receptors decreases PKC activity; *left*: neuron exhibiting a basal level of DAT phosphorylation and intermediate level of DA uptake; *center*: DA binding to presynaptic DA receptors inhibits PKC, leading to decreased DAT phosphorylation and increased DA uptake; *right*: presence of cocaine augments extracellular DA, leading to activation of additional DA receptors, further reductions in PKC activity and DAT phosphorylation, and increased transport activity of non-cocaine-occupied DATs.

DAT kinases and phosphatases may be factors contributing to differential cocaine susceptibility.

SUMMARY

There is now considerable evidence that DA transport undergoes acute kinase-induced regulation in both positive and negative directions. For protein kinase C, phosphorylation studies are compatible with regulation of transport occuring by direct phosphorylation of DAT; for other second messengers, the relationships between DAT phosphorylation and transport regulation remain unknown. The rapid time-courses found for upregulation and downregulation of DA transport are compatible with receptor-mediated processes that would provide neurons with the capability for fine temporal and spatial control of synaptic dopamine concentrations.

The molecular events leading to regulation of transport and antagonist binding remain to be elucidated. The current evidence is complex, and even contradictory, possibly reflecting responses unique to specific transporter homologs, the particular activity assayed, actions of different kinases, or differential responses due to the type of system being utilized. These discrepancies highlight the rudimentary nature of our understanding of these processes and indicate the need for additional research to elucidate transporter phosphorylation characteristics and significance, both with relationship to normal neurophysiology and to mechanisms of drug abuse.

ACKNOWLEDGMENTS

Many thanks to Dr. Reith for his patience and encouragement during the preparation of this manuscript. Artwork was provided by Ms. Vickie Swift, UND Biomedical Communications Center.

REFERENCES

1. Giros, B., Jaber, M., Jones, S. R., Wightman, R. M., and Caron, M. G. (1996) Hyperlocomotion and indifference to cocaine and amphetamine in mice lacking the dopamine transporter. *Nature (London)* **379**, 606–612.
2. Donovan, D. M., Vandenbergh, D. J., Perry, M. P., Bird, G. S., Ingersoll, R., Nanthakumar, E., and Uhl, G. R. (1995) Human and mouse dopamine transporter genes: conservation of 5'-flanking sequence elements and gene structures. *Mol. Brain Res.* **30**, 327–335.
3. Giros, B., El Mestikawy, S., Bertrand, L., and Caron, M. G. (1991) Cloning and functional characterization of a cocaine-sensitive dopamine transporter. *FEBS Lett.* **295**, 149–154.
4. Giros, B., Mestikawy, S. E., Godinot, N., Zheng, K., Han, H., Yang-Feng, T., and Caron, M. G. (1992) Cloning, pharmacological characterization and chromosome assignment of the human dopamine transporter. *Mol. Pharmacol.* **42**, 383–390.

5. Kilty, J., Lorang, D., and Amara, S. G. (1991) Cloning and expression of a cocaine-sensitive dopamine transporter. *Science* **254**, 578–579.

6. Shimada, S., Kitayama, S., Lin, C.-L., Patel, A., Nanthakumar, E., Gregor, P., Kuhar, M., and Uhl, G. (1991) Cloning and expression of a cocaine-sensitive dopamine transporter complementary DNA. *Science* **254**, 576–578.

7. Usdin, T. B., Mezey, E., Chen, C., Brownstein, M. J., and Hoffman, B. J. (1991) Cloning of the cocaine-sensitive bovine dopamine transporter. *Proc. Natl. Acad. Sci. USA* **88**, 168–171.

8. Vandenbergh, D. J., Persico, A. M., and Uhl, G. R. (1992) A human dopamine transporter cDNA predicts reduced glycosylation, displays a novel repetitive element and provides racially-dimorphic *Taq I* RFLPs. *Mol. Brain Res.* **15**, 161–166.

9. Berger, S. P., Martenson, R., Laing, P., Thurcauf, A., DeCosta, B., Rice, K. C., and Paul, S. M. (1991) Photoaffinity labeling of the dopamine reuptake carrier protein with 3-azido[^{3}H]GBR-12935. *Mol. Pharmacol.* **39**, 429–435.

10. Gracz, L. M. and Madras, B. K. (1995) [^{3}H]WIN 35,428 ([^{3}H]CFT) binds to multiple charge-states of the solubilized dopamine transport in primate striatum. *J. Pharmacol. Exp. Ther.* **273**, 1224–1234.

11. Grigoriadis, D. E., Wilson, A. A., Lew, R., Sharkey, S., and Kuhar, M. J. (1989) Dopamine transport sites selectively labeled by a novel photoaffinity probe: ^{125}I DEEP. *J. Neurosci.* **9,** 2664–2670.

12. Lew, R., Grigoriadis, D. E., Sharkey, J., and Kuhar, M. J. (1989) Dopamine transporter: solubilization from dog caudate nucleus. *Synapse* **3**, 372–375.

13. Simantov, R., Vaughan, R., Lew, R., Wilson, A., and Kuhar, M. J. (1991) Dopamine transporter–cocaine receptor: characterization and purification. *Adv. Biosci.* **82,** 151–154.

14. Lew, R., Grigoriadis, D., Wilson, A., Boja, J. W., Simantov, R., and Kuhar, M. J. (1991) Dopamine transporter: deglycosylation with exo- and endoglycosidases. *Brain Res.* **539**, 239–246.

15. Patel, A. P. (1997) Neurotransmitter transporter proteins: posttranslational modifications, in *Neurotransmitter Transporters, Structure, Function, and Regulation* (Reith, M. E. A., ed.), Humana, Totowa, NJ, pp. 241–262.

16. Vaughan, R. A., Brown, V. L., McCoy M. T., and Kuhar, M. J. (1996) Species and brain-region specific dopamine transporters: immnological and glycosylaion characteristics. *J. Neurochem.* **66**, 2146–2152.

17. Vaughan, R. A. (1995) Photoaffinity-labeled ligand binding domains on dopamine transporters identified by peptide mapping. *Mol. Pharmacol.* **47**, 956–964.

18. Vaughan, R. A. and Kuhar, M. J. (1996) Dopamine transporter ligand binding domains: structural and functional properties revealed by limited proteolysis. *J. Biol. Chem.* **271**, 21,672–21,680.

19. Hersch, S. M., Yi, H., Heilman, C. J., Edwards, R. H., and Levey, A. I. (1997) Subcellular localization and molecular topology of the dopamine transporter in the striatum and substantia nigra. *J. Comp. Neurol.* **388**, 211–227.

20. Nirenberg, M. J., Vaughan, R. A., Uhl, G. R., Kuhar, M. J., and Pickel, V. M. (1996) The dopamine transporter is localized to dendritic and axonal plasma membranes of nirgostriatal dopaminergic neurons. *J. Neurosci.* **16**, 436–447.

21. Miller, J. W., Kleven, D. T., Domin, B. A., and Fremeau, R. T., Jr. (1997) Cloned sodium-(and chloride-) dependent high affinity transporters for GABA, glycine, proline, betaine, taurine, and creatine, in *Neurotransmitter Transporters, Struc-*

ture, Function, and Regulation (Reith, M. E. A., ed.), Humana, Totowa, NJ, pp. 101–150.

22. Povlock, S. L. and Amara, S. G. (1997) The structure and function of norepinephrine, dopamine, and serotonin transporters, in *Neurotransmitter Transporters, Structure, Function, and Regulation* (Reith, M. E. A., ed.), Humana, Totowa, NJ, pp. 1–28.

23. Schuldiner, S. (1997) Vesicular neurotransmitter transporters: pharmacology, biochemistry and molecular analysis, in *Neurotransmitter Transporters, Structure, Function, and Regulation* (Reith, M. E. A., ed.), Humana, Totowa, NJ, pp. 215–240.

24. Koob, G. F. and Bloom, F. E. (1988) Cellular and molecular mechanisms of drug dependence. *Science* **242**, 715–723.

25. Bergman, J., Madras, B. K., Johnson, S. E., and Spealman, R. D. (1989) Effects of cocaine and related drugs in nonhuman primates. III. Self-administration by squirrel monkeys. *J. Pharmacol. Exp. Therap.* **251**, 150–155.

26. Ritz, M. C., Lamb, R. J., Goldberg, S. R., and Kuhar, M. J. (1987) Cocaine receptors on dopamine transporters are related to self-administration of cocaine. *Science* **237,** 1219-1223.

27. Kuhar, M. J., Ritz, M. C., and Boja, J. W. (1991) The dopamine hypothesis of the reinforcing properties of cocaine. *Trends Neurosci.* **14**, 299–302.

28. Wise, R. A. and Bozarth, M. A. (1987) A psychomotorstimulant theory of addiction. *Psychol. Rev.* **94,** 469–492.

29. Sonders, M. S., Zhu, W.-J., Zahniser, N. R., Kavanaugh, M. P., and Amara, S. G. (1997) Multiple ionic conductances of the human dopamine transporter: the actions of dopamine and psychostimulants. *J. Neurosci.* **17**, 960–974.

30. Buck, K. and Amara, S. G. (1994) Chimeric dopamine-norepinephrine transporters delineate structural domains influencing selectivity for catecholamines and 1-methyl-4-phenylpyridinium. *Proc. Natl. Acad. Sci. USA* **91**, 12,584–12,588.

31. Giros, B., Wang, Y.-M., Suter, S., McLeskey, S. B., Pifl, C., and Caron, M. G. (1994) Delineation of discrete domains for subsrate, cocaine, and tricyclic antidepressant interactions using chimeric dopamine-norepinephrine transporters. *J. Biol. Chem.* **269**, 15,985–15,988.

32. Kitayama, S., Shimada, S., Xu, H., Markham, L., Donovan, D. M., and Uhl, G. R. (1992) Dopamine transporter site-directed mutations differentially alter substrate transport and cocaine binding. *Proc. Natl. Acad. Sci. USA* **89**, 7782–7785.

33. Kitayama, S., Wang, J. B., and Uhl, G. R. (1993) Dopamine transporter mutants selectively enhance MPP+ transport. *Synapse* **15**, 58–62.

34. Vaughan, R. A., Agoston, G. E., Lever, J., and Newman, A. H. (1999) Differential binding of tropane-based photoaffinity ligands on the dopamine transporters. *J. Neurosci.* **19**, 630–636.

35. Edvarsen, O. and Dahl, S. G. (1994) A putative model of the dopamine transporter. *Mol. Brain Res.* **27**, 265–274.

36. Batchelor, M. and Schenk, J. O. (1998) Protein kinase A activity may kinetically up-regulate the striatal transporter for dopamine. *J. Neurosci.* **18**, 10,304-10,309.

37. Kantor, L. and Gnegy, M. E. (1998) Protein kinase C inhibitors block amphetamine-mediated dopamine release in rat striatal slices. *J. Pharmacol. Exp. Ther.* **284**, 592–598.

38. L'hirondel, M., Cheramy, A., Godehueu, G., and Gowinski, J. (1995) Effects of arachidonic acid on dopamine synthesis, spontaneous release, and uptake in striatal synaptosomes from the rat. *J. Neurochem.* **64**, 1406–1409.

39. Pogun, S., Baumann, M. H., and Kuhar, M. J. (1994) Nitric oxide inhibits [^{3}H] dopamine uptake. *Brain Res.* **641,** 83–91.
40. Uchikawa, T., Kiuchi, Y., Akihiko, Y., Nakachim N., Yanazaki, Y., Yokomizo, C., and Oguchi, K. (1995) Ca^{2+}-dependent enhancement of [^{3}H] dopamine uptake in rat striatum: possible involvement of calmodulin-dependent kinases. *J. Neurochem.* **65**, 2065–2071.
41. Vaughan, R. A., Huff, R. A., Uhl, G. R., and Kuhar, M. J. (1997) Protein kinase C-mediated phosphorylation and functional regulation of dopamine transporters in striatal synaptosomes. *J. Biol. Chem.* **272**, 15,541–15,546.
42. Copeland, B. J., Neff, N. H., and Hadjjonstantinou, M. (1996) Protein kinase C activators decrease dopamine uptake into striatal synaptosomes. *J. Pharmacol. Exp. Therap.* **277**, 1527–1532.
43. Huff, R. A., Vaughan, R. A., Kuhar, M. J., and Uhl, G. R. (1997) Phorbol esters increase dopamine transporter phosphorylation and decrease transport V_{max}. *J. Neurochem.* **68**, 225–232.
44. Kitayama, S., Dohi, T., and Uhl, G. (1994) Phorbol esters alter functions of the expressed dopamine transporter. *Eur. J. Pharmacol.* **268**, 115–119.
45. Pristupa, Z. B., McConkey, F., Liu, F., Man, H. Y., Lee, F. J. S., Wang, Y. T., and Niznik, H. B. (1998) Protein kinase-mediated bidirectional trafficking and functional regulation of the human dopamine transporter. *Synapse* **30**,79–87.
46. Zhang, L., Coffey, L. L., and Reith, M. E. A. (1997) Regulation of the functional activity of the human dopamine transporter by protein kinase C. *Biochem. Pharmacol.* **53,** 677–688.
47. Zhang, L. and Reith, M. E. A. (1996) Regulation of the functional activity of the human dopamine transporter by the arachidonic acid pathway. *Eur. J. Pharmacol.* **315**, 345–354.
48. Zhu, S.-J., Kavanaugh, M. P., Sonders, M. S., Amara, S. G., and Zahniser, N. R. (1997) Activation of protein kinase C inhibits uptake, currents and binding associated with the human dopamine transporter expressed in *Xenopus* oocytes. *J. Pharmacol. Exp. Ther.* **282**, 1358–1365.
49. Cohen, P. (1989) The structure and regulation of protein phosphatases. *Ann. Rev. Biochem.* **58**, 453–508.
50. Tian, Y., Kapatos, G., Granneman, J. G., and Bannon, M. J. (1994) Dopamine and γ-aminobutyric acid transporters: differential regulation by agents that promote phosphorylation. *Neurosci. Lett.* **173**, 143–146.
51. Bredt, D. S. and Snyder, S. H. (1989) Nitric oxide mediates glutamate-linked enhancement of cGMP levels in the cerebellum. *Proc. Natl. Acad. Sci. USA* **86**, 9030–9033.
52. Bugnon, O., Ofori, S., and Schorderet, M. (1995) Okadaic acid modulates exocytotic and transporter-dependent release of dopamine in bovine retina in vitro. *Naunyn-Schmiedeberg's Arch. Pharmacol.* **351**, 53–59.
53. Giambalvo, C. T. (1992) Protein kinase C and dopamine transport—2. Effects of amphetatmine *in vitro*. *Neuropharmacology* **31**, 1211–1220.
54. Vrindavam, N. S., Arnaud, P., Ma, J. X., Altman-Hamamdzic, S., Parratto, N. P., and Sallee, F. R. (1996) The effects of phosphorylation on the functional regulation of an expressed recombinant human dopamine transporter. *Neurosci. Lett.* **216**, 133–136.
55. Corey, J., Davidson, N., Lester, H., Brecha, N., and Quick, M. (1994) Protein kinase C modulates the activity of a cloned γ-aminobutyric acid transporter

expressed in *Xenopus* oocytes via regulated subcellular redistribution of the transporter. *J. Biol. Chem.* **269**, 14,759–14,767.

56. Sato, K., Adams, R., Betz, H., and Schloss, P. (1995) Modulation of a recombinant glycine transporter (GLYT1b) by activation of protein kinase C. *J. Neurochem.* **654**, 1967–1973.

57. Lee, S. H., Son, H., Chang, M. Y., Kang, S. H., Chin, H., and Lee, Y. S. (1998) Protein kinase C-mediated regulation of dopamine transporter is not achieved by direct phosphorylation on the transporter. *Soc. Neurosci. Abstr.* **24**, 607.

58. Vaughan, R. A., McCoy, M. T., and Kuhar, M. J. (1992) Charge isoforms of dopamine transporters. *Soc. Neurosci. Abstr.* **18**, 1433.

59. Kokoshka, J. M., Vaughan, R. A., Hanson, G. R., and Fleckenstein, A. E. (1998) Nature of methamphetamine-induced rapid and reversible changes in dopamine transporters. *Eur. J. Pharmacol.* **20**, 269–275.

60. Krantz, D. E., Peter, D., Liu, Y., and Edwards, R. H. (1997) Phosphorylation of a vesicular monoamine transporter by casein kinase II. *J. Biol. Chem.* **272**, 6752–6759.

61. Ramamoorthy, S., Giovanetti, E., Qian, Y., and Blakely, R. D. (1998) Phosphorylation and regulation of antidepressant-sensitive serotonin transporters. *J. Biol. Chem.* **273**, 2458.

62. Apparsundaram, S., Schroeter, S., Giovanetti, E., and Blakely, R. D. (1998) Acute regulation of norepinephrine transport: II. PKC-modulated surface expression of human norepinephrine transporter proteins. *J. Pharmacol. Exp. Ther.* **287**, 744–751.

63. Qian, Y., Galli, A., Ramamoorthy, S., Risso, S., DeFelice, L. J., and Blakely, R. D. (1997) Protein kinase C activation regulates human serotonin transporters in HEK-293 cells via altered cell surface expression. *J. Neurosci.* **17**, 45–57.

64. Sibley, D. R. and Lefkowitz, R. J. (1987) Beta-adrenergic receptor-coupled adenylate cyclase. Biochemical mechanisms of regulation. *Mol. Neurobiol.* **1**, 121–154.

65. Meiergerd, S. M., Patterson, T. A., and Schenk, J. (1993) D_2 receptors may modulate the function of the striatal transporter for dopamine: kinetic evidence from studies *in vitro* and *in vivo*. *J. Neurochem.* **61**, 764–767.

66. Sibley, D. R., Monsma, F. J., and Shen, Y. (1993) Molecular neurobiology of dopaminergic receptors. *Int. Rev. Neurobiol.* **35**, 391–415.

67. Giambalvo, C. T. and Wagner, R. L. (1994) Activation of D1 and D2 dopamine receptors inhibits protein kinase activity in striatal synaptoneurosomes. *J. Neurochem.* **63**, 169–176.

68. Freidman, E., Jin, L. Q., Cai, G. P., Hollon, T. R., Drags, J., Sibley, D. R., and Wang, H.-Y. (1997) D_1-like dopaminergic activation of phosphoinositide hydrolysis is independent of D_{1A} dopamine receptors: evidence from D_{1A} knockout mice. *Mol. Pharmacol.* **51**, 6–11.

15

Dopamine Transporter mRNA in Human Brain

Distribution and Regulatory Effects
in Autopsy Studies of Cocaine Abusers

Deborah C. Mash, Li Chen Kramer, David Segal, and Sari Izenwasser

INTRODUCTION

Midbrain dopamine (DA)-containing neurons have effects on a number of distinct behavioral functions, including sensorimotor integration, motivation and affect, and cognition. Disruption of normal DA cell function occurs in Parkinson's disease, neuropsychiatric disorders, and psychostimulant dependence. Drugs of abuse and dependence, such as cocaine and amphetamines, act as psychostimulants in humans. Psychiatric complications of cocaine and amphetamine include paranoia, agitated delirium, delusional disorder, and the depressed mood and dysphoria associated with abrupt withdrawal. Amphetamine psychosis and cocaine arousal and reinforcement are presumed consequences of the stimulation of the mesolimbic DA pathway (for a review, *see* ref. *1*).

The mesolimbic DAergic system plays a primary role in mediating the rewarding effects of most abused drugs (for a review, *see* ref. *2*). Cocaine mediates its powerful reinforcing effects by binding to specific recognition sites on the DA transporter (DAT) protein *(3)*. The DAT is a cell-specific protein that controls extracellular levels of DA through the rapid reuptake of released DA into the presynaptic terminal. Identification and cloning of the gene for the DAT has provided insight into the molecular mechanism of DA reuptake inhibition by cocaine binding to the transport carrier *(4–7)*. The molecular cloning of monoamine transporters (for a review, *see* ref. *8*), along with the increased availability of highly specific radioactively tagged ligands have provided new tools to study the regulation of DAT in autopsy studies of cocaine abusers. The characterization of cDNAs and gene sequences from rat, bovine, and human brain has greatly facilitated studies of DAT structure–function relationships and gene transcript regulation. This chapter will

From: *Cerebral Signal Transduction: From First to Fourth Messengers*
Edited by: M. E. A. Reith © Humana Press Inc., Totowa, NJ

discuss the regulatory effects of cocaine on DAT mRNA content in human midbrain DA neurons from cocaine-overdose victims.

MOLECULAR CHARACTERIZATION OF THE HUMAN DOPAMINE TRANSPORTER

The recent cloning and expression of the human DAT gene has provided extensive information on structure–function relationships of the DAT protein *(4,5,9)*. The cloned DAT cDNA encodes a single polypeptide strand of 620 amino acids, which corresponds to a protein of 68,517 Daltons *(5)*. The cDNA encodes consensus sites for glycosylation and phosphorylation, suggesting that secondary processing may contribute to the regulation of the transport protein. Like other neurotransmitter carriers, the predicted structure of the DAT based on hydropathicity analysis suggests the presence of 12 transmembrane domains with an intracellular N-terminus and C-terminus *(5)*. Across transmembrane regions, DAT shares 68% identity with the norepinephrine transporter and 54% with the serotonin transporter. In addition, an aspartate residue and two serine residues, located in transmembrane regions, directly interact with dopamine's amino and hydroxyl groups, respectively *(10)*, and these residues are conserved across all monoamine transporters *(11)*. The high homologies in the regions forming the "uptake site" may be responsible for the inability of various drugs to distinguish among these three different transporters.

Expression of the cloned DAT in COS-7, mouse fibroblast or glioma cell lines *(4–6)* afforded biochemical characterization of saturable, Na^+- and Cl^--dependent DA transports. Psychotherapeutic drugs (mazindol, nomifensine, and tricyclic antidepressants), drugs of abuse (including cocaine, *d*-amphetamine, and phencyclidine [PCP]), and neurotoxins (6-hydroxydopamine and MPP^+, the active metabolite of MPTP) are known to bind with high affinity to DAT *(2,12,13)*. The pharmacological profile observed for inhibition of binding of the cocaine congeners in human striatum agrees with the values observed for inhibition of DA uptake and radiolabeled WIN 35,428 binding to the cloned human DAT *(14)*. Localization of DAT in brain tissue has been studied by functional uptake, radioligand binding to DAT protein, and *in situ* hybridization of DAT message; yet many questions remain about the precise location and role of transporters in DAergic circuits.

IN SITU HYBRIDIZATION OF DAT GENE EXPRESSION IN HUMAN MIDBRAIN

The DAergic systems in brain comprises three distinct pathways, including the nigrostriatal, mesocortical, and mesolimbic projections. The nigro-

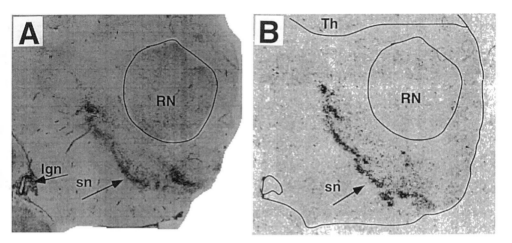

Fig. 1. DA transporter mRNA in human midbrain visualized by *in situ* hybridization. Panel (**A**) shows the Nissl-stained section for delineation of nigral DA-containing cell bodies. Panel (**B**) illustrates numerous cells within the pars compacta of the substantia nigra intensely labeled with the hDAT1-derived ^{35}S-labeled antisense RNA probe. Control subject, age 29 yr. Abbreviations: lgn, lateral geniculate nucleus; RN, red nucleus; sn, substantia nigra; th, thalamus.

striatal pathway originates in the substantia nigra pars compacta and terminates in the striatum. The mesolimbic pathway ascends from cells located in the ventral tegmental area (VTA) and projects to the limbic sectors of the striatum, amygdala, and olfactory tubercle. The mesocortical pathway originates in the VTA and terminates within particular sectors of the cerebral cortical mantle, including the prefrontal, cingulate, and entorhinal cortices *(15)*. Autoradiographic visualization of the distribution of DAT indicates that the topographic distribution of the transport carrier correlates well with DA innervation. The highest densities are localized to nigrostriatal terminals, with moderate densities in the mesolimbic terminals and low densities in the mesocortical terminals *(16,17)*.

The regional distribution of the DAT varies depending on the choice of probe and the target (message or protein). *In situ* hybridization studies have shown a restricted distribution of DAT mRNA. DAT mRNA is localized only in the DA-containing cell bodies of the substantia nigra, VTA, and the retrorubral cell groups *(18)*. Abundant DAT mRNA is found within the human substantia nigra pars compacta (Fig. 1), which contains the cell bodies that project primarily to the motor sectors of the striatum, corresponding to an increased gradient of DAT protein expression. The primate and human striatum is organized into distinct neurochemical compartments termed patch (striosome) and matrix *(19)*. The DAergic terminals within the patch arise from cell bodies localized to the substantia nigra pars compacta, whereas the DAergic projections to the matrix compartment originate from the VTA (rostrorubral area) and the substantia nigra pars compacta *(18,19)*. In the striatum, matrix-directed neurons have the lowest level of DAT mRNA, but the DA terminals in the matrix of the striatum maintain the highest density of DAT binding. These results suggest that there may not be a precise correlation between mRNA and ligand binding to the DAT protein *(18)*.

REGIONAL VARIABILITY AND THE EFFECT OF AGE ON THE HUMAN DAT mRNA

Reductions in DA and DA-related synaptic markers in the striatum are known to contribute to the cognitive and motor deficits associated with normal aging. Although cocaine, unlike methamphetamine, does not appear to be neurotoxic in experimental animals *(20)*, this question has not been resolved fully in the human. In the postmortem human striatum, a progressive decrease in DAT density with age has been demonstrated using [^{3}H]GBR 12935 *(13,21–23)*. Decreases in DAT density of 75% and 65% were reported for subjects ranging from 19 to 100 yr *(21)* and from 18 to 88 yr *(13)*, respectively. In vivo imaging of the DAT with cocaine and cocaine congeners ([^{11}C]cocaine *[24]* and [^{123}I]β-CIT *[25]*), and "classical" DAT inhibitors ([^{11}C]nomifensine *[26]*) also demonstrated a decline of DAT with increasing age. Using [^{11}C]cocaine, a gradual decline in the density of cocaine recognition sites was detected over an age range of 21–63 yr *(24)*. Using [^{123}I]β-CIT, a 51% decline in DAT density was observed over an age range of 18–83 yr *(25)*. A decrease in the [^{11}C]nomifensine striatum/ cerebellum ratios was observed over an age range of 24–81 yr *(26)*. Taken together, in vivo imaging with a variety of radiotracers demonstrates age-related declines in DAT density that occur at a rate of approx 10% per decade. It is unclear whether chronic cocaine abuse may cause accelerated senescence of DAergic neurons.

In keeping with the marked decline in DAT with normal aging, studies of the mRNA encoding the DAT demonstrated a profound loss of DAT gene expression in DA-containing substantia nigra neurons with increasing age *(11)*. The age-related decline in DAT mRNA was the result of both a decrease in the abundance of DAT mRNA per DA cell as well as a decrease in the total number of cells expressing DAT mRNA. The midbrain regions exhibiting the greatest cellular abundance of DAT mRNA in younger subjects had the greatest loss of DAT mRNA with age *(27)*. These results indicate that the changes in DA neurotransmission seen in normal aging may be related to altered DAT gene expression. Although a precipitous age-related decline (>95% in subjects >57 yr age) was reported for DAT mRNA, the mRNA for tyrosine hydroxylase (another phenotypic marker of DA neurons) decreased linearly with age *(28)*. Unbiased stereological cell counting using a monoclonal antibody against the DAT demonstrated an age-related reduction in the number of substantia nigra DAT immunoreactive neurons *(29)*. The lack of a correspondence between mRNA content and ligand binding to the DAT protein may indicate possible differential regulation of DAT mRNA and protein with normal aging.

At present, it is not known if this decline inDAT density corresponds to a loss of DA nerve terminals or to a decrease in the number of DAT expressed by aging DAergic neurons. Comparable changes in DAergic presynaptic and postsynaptic markers and DAT densities suggest that the decline in DAT labeling may be the result of reduced integrity of DAergic projections *(13)*. Age-related decreases have been shown for tyrosine hydroxylase *(28)*, striatal DA content *(30,31)*, and DA D_1 and D_2 receptors *(13)*. This decrease in DA synaptic markers suggests that the observed decrease in DAT density with normal aging may be the result of actual loss of DA neurons and terminals. Alternatively, if the DAT is regulated by synaptic DA content, the decrease in DAT density may be a compensatory response to the age-related decline in neurotransmitter content and turnover.

QUANTITATION OF DAT mRNA:
EFFECTS OF COCAINE TREATMENT AND
WITHDRAWAL IN RODENT MODELS OF COCAINE ABUSE

Chronic cocaine treatments do not appear to have the marked neurotoxic effects like those produced by amphetamine on DA and serotonin neurons (for a review, *see* ref. *32*). In fact, most preclinical studies have shown no reduction in radioligand binding to the DAT following chronic treatment of rats with cocaine, suggesting that DA terminals remain intact. Daily passive administration of cocaine for 10 d had no effect on binding to DA *(33)*, norepinephrine, or serotonin *(34)* uptake sites. Continuous infusion of cocaine for 7 d also had no effect on the number of DAT binding sites *(35)*. However, withdrawal from repeated administration of cocaine produces an decrease in transporter binding in the rat nucleus accumbens *(36)*. Because this decrease only occurs after withdrawal from the drug, it is likely to be a compensatory mechanism related to some other, earlier drug effect.

Many experimental cocaine treatment paradigms have been developed as animal models of human cocaine abuse and it has been reported that the dose, route, and frequency of administration influence the effects of chronic

Table 1
Effect of Cocaine Treatment
on the Regulation of DAT mRNA in Rodent Models

Cocaine treatment paradigm	Time after last dose	Method	Effect	Ref.
15 mg/kg ip, 2X daily 6.5 d	4 h	Ribonuclease protection assay	>40% ↓ in SN normal by 72 h	45
1 mg/kg iv, 2 wk (except weekends), 1 injection every 12 min for 2 h daily	10 d	*In situ* hybridization	22–36% ↓ in VTA	43
10 mg/kg daily, 8 d	1 h	*In situ* hybridization	No change	44
15 or 25 mg/kg daily, 8 d	1 h	*In situ* hybridization	20% ↓ in SN and VTA	44
15 mg/kg i.p, 3X daily (spaced 1 h apart), 14 d	4 h	*In situ* hybridization	9–12% ↓ in SNc and parabrachial pigmentosis	46
15 mg/kg ip, 3X daily (spaced 1 h apart), 14 d	10 d	Solution hybridization	No changes in SN or VTA	47,48

cocaine on DAT densities. Chronic treatment of rats with intermittent doses of cocaine demonstrated a twofold to fivefold increase in the apparent density of [^{3}H]cocaine and [^{3}H]BTCP binding in the striatum *(37)*. Rats allowed to self-administer cocaine in a chronic unlimited access paradigm had significant increases in [^{3}H]WIN 35,428 binding when the animals were sacrificed on the last day of cocaine access *(38)*. Rabbits and mice treated chronically with cocaine show an elevation in the density of [^{3}H]WIN 35,428 binding sites in the caudate *(39,40)*.

The inhibition of DA uptake by cocaine is altered in both the nucleus accumbens and caudate putamen following 7 d of chronic continuous cocaine administration *(35)*. Daily cocaine injections (15 mg/kg/d × 3 d) lead to a decrease in total dopamine uptake in the nucleus accumbens, with no change in the caudate putamen *(41)*. In contrast, a regimen of escalating doses over a 10-d period produces a transitory increase in DA uptake in the nucleus accumbens *(42)*.

There have been few reports of the effects of chronic cocaine on the expression of DAT mRNA in brain and the results, like those in binding studies, are variable. A number of different treatment paradigms have been used to study the effect of cocaine exposure on DAT mRNA content. These differences have included variable time courses for cocaine administration, doses, and withdrawal times (Table 1). Because of these differences across studies, it has proven difficult to draw relevant conclusions about DAT mRNA regulation with chronic cocaine based on these rodent studies.

One of the earlier reports of alterations in DAT mRNA was in rats treated with 1 mg/kg cocaine iv every 12 min over a 2-h period for 2 wk *(43)*. This regimen was chosen to mimic self-administration patterns; however, the animals were treated only during the work week (i.e., Monday through Friday), which meant that there was a 2-d withdrawal period in the middle of the study. However, 10 d after the last cocaine treatment, there was a decrease in DAT mRNA measured in the VTA *(43)*. Reduction in DAT mRNA was approx 20% in substantia nigra (SN) and VTA in rats 1 h after the last of eight daily ip injections of cocaine (either 15 or 25 mg/kg) *(44)*. This change was a dose-related effect with no significant change in mRNA levels in rats treated with passive injections of lower doses of cocaine (10 mg/kg/d). In this study, DAT binding was measured in parallel with mRNA and no changes were observed for [^{3}H]mazindol binding sites *(44)*. These findings demonstrated a lack of a correlation between message and radioligand binding sites. The most significant alteration in DAT mRNA was reported for rats injected with cocaine (15 mg/kg × 2 for 6.5 d) and killed 4 h after the last injection. In this study, a >40% decrease in DAT mRNA was observed in the substantia nigra pars compacta (SNc) *(45)*. This reduction appeared to be a transient effect, with mRNA levels returning to normal by 72 h after the last injection of cocaine. In contrast, a longer treatment period of 14 d of repeated cocaine administration (three injections daily of 15 mg/kg ip spaced 1 h apart) produced nominal reductions in mRNA content. *In situ* hybridization studies showed small, albeit significant decreases (9–12%) in DAT mRNA in SNc and parabrachial pigmentosis 4 h after the last injection *(46)*. Ten days after withdrawal from this same cocaine treatment paradigm, there were no significant alterations in DAT mRNA in SN or VTA *(47,48)*. It is not known whether these differences are dose related, or the result of the markedly different withdrawal times, or reflect the different methods used to measure mRNA content (Table 1). In the first study, *in situ* hybridization afforded the visualization of discrete regional changes *(46)*, whereas the use of solution hybridization and Northern blotting in the second study may not have detected subtle changes over subfields of DA cell groups *(47,48)* (for a review, *see* Table 1).

Although it is difficult to draw definitive conclusions from rodent studies on the effects of cocaine on DAT mRNA, decreases in DAT mRNA in DA cell body regions appear to occur as an early adaptation to cocaine exposure. This observation in based on the greater magnitude of this effect following shorter *(44,45)* rather than longer *(46)* cocaine treatments. In addition, this effect of cocaine exposure appears to be transient, as DAT mRNA levels return to normal within several days after cessation of the treatment *(45–48)*.

An interesting finding is that amphetamine appears to have an effect opposite to that of cocaine on DAT mRNA. Two studies have reported increases in DAT message following amphetamine treatment. In the first, rats were injected daily with amphetamine (2.5 mg/kg) for 5 d. Seven days later there was an increase in DAT mRNA in both the VTA and SN *(49)*. Similarly, after 3 d of withdrawal from 5 d of amphetamine injections (5 mg/kg ip), DAT mRNA was increased in rostral VTA and SN *(50)*. No changes in DAT mRNA content were seen after 14 d of withdrawal. These observations also suggest that changes in DAT mRNA in response to low-dose amphetamine exposure may not be long lasting.

DOPAMINE TRANSPORTER mRNA
IN AUTOPSY STUDIES OF CHRONIC COCAINE USERS

Although it is unclear exactly how the DAT protein regulates to alter the intensity and duration of DAergic neurotransmission, transcriptional regulation of transporter expression, increased membrane trafficking, and altered phosphorylation states of the DAT protein are possible mechanisms *(51)*. Cocaine congeners bind to two sites on the DAT, one of which mediates high-affinity DA uptake *(6)*. High-affinity cocaine binding sites on the DAT protein are increased in cocaine overdose victims *(52–54)*, although it is not known if this increase is correlated with a change in transporter function or gene product expression. In contrast to these findings, the apparent density of high-affinity cocaine binding sites in victims of agitated cocaine delirium were unchanged from control values, suggesting a defect in the ability of the transporter to regulate synaptic DA levels with a cocaine challenge *(14,55)*. Our group has used semiquantitative reverse transcriptase–polymerase chain reaction (RT-PCR) to assess whether chronic cocaine use leads to regulatory alterations in the expression of DAT mRNA in the substantia nigra from cocaine-overdose victims. These analyses were feasible because of the postmortem stability of human brain mRNAs *(56,57)*. In fact, human DAT mRNA has been shown to be translationally active postmortem *(58)*.

In these studies, cocaine fatalities were grouped into cocaine overdose (CO) deaths (a death attributed to the toxic effects of cocaine alone or cocaine in combination with alcohol) or CO cases presenting with preterminal excited delirium (ED; a syndrome marked by intense paranoia, increased body strength, bizarre and violent behavior, and hyperthermia) as described previously *(55,59)*. Postmortem human brain specimens were taken at autopsy from age-matched and drug-free control subjects and CO and ED victims. All subjects included in the study were matched for postmortem interval (range: 12–18 h). Medicolegal investigations of the deaths were conducted by forensic pathologists. Forensic pathologists evaluated the scene environment and circumstances of death and autopsied the victim in order to determine the cause and manner of death. The circumstances of death and toxicology data were reviewed carefully before classifying a death as a cocaine over-dose *(55,60)*. In a similar manner, controls were selected from deaths not caused by cocaine with no cocaine or metabolites detected in toxicology screens of blood or brain tissues. All cases were evaluated for common drugs of abuse, and alcohol and positive urine screens were confirmed by quantitative analysis of blood to exclude cases from the study based on evidence of polydrug or alcohol use prior to death. The cocaine toxicity cases selected for the present study had evidence of a number of different variables of chronic cocaine use based on review of the prior arrest records and treatment admissions, as well as on pathological signs (i.e., perforation of the nasal septum). Blood cocaine was quantified using gas–liquid chromatography with a nitrogen detector. Frozen brain regions were sampled for quantitation of cocaine and benzoylecgonine using gas chromatography–mass spectroscopy techniques *(61)*. Neuropathological analysis was done to verify the absence of any gross or histopathological abnormalities. All cocaine-overdose cases selected for study had cocaine and metabolite detected in the brain.

Total RNA was prepared from cryopreserved brain specimens of the substantia nigra *(62)* and quantified by using A_{260} values. The cDNA samples (20 μL) were prepared by using Superscript II reverse transcriptase, total RNA (5 μg), and oligo $(dT)_{12-18}$ primer. Two microliters of cDNA was amplified in a 100-μL polymerase chain reaction with 2.5 U of recombinant Taq DNA polymerase, a-$[^{32}P]$dCTP, and specific primers for human DAT, cyclophilin, and actin (Fig. 2 and Table 2). DAT cDNAs were amplified along with the housekeeping genes actin and cyclophilin, according to the method of ref. *63*. RT-PCR of serial dilutions of a pooled RNA sample was used to determine the optimum number of cycles and the linear range of detection. PCR reactions were performed on four dilutions of each cDNA sample so that at least three replicate values were within the linear range. PCR band intensities were obtained from a Molecular Dynamics (Sunnyvale, CA) Phospho-Imager by exposure of dried PCR gels. The measures were averaged to yield mean DAT, actin, and cyclophilin values for each sample (mean ± SE). The DAT/marker mRNA ratios were calculated and the differences between the group means were determined by using the Dunnett *t*-test.

Actin and cyclophilin, two constitutively expressed "housekeeping" genes in human brain, were used as marker genes to normalize the DAT mRNA level for each case (Fig. 3). The normalized DAT mRNA levels were represented by cDNA ratios of DAT and cyclophilin and DAT and actin

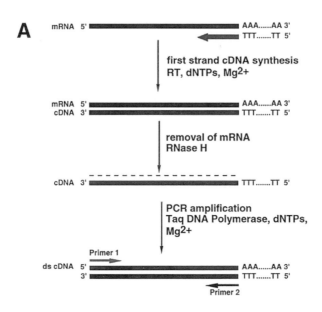

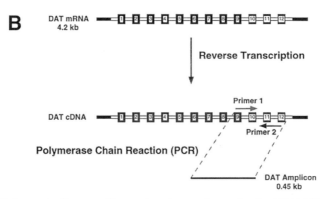

Fig. 2. Schematic diagram illustrating the semiquantitative RT-PCR method for measuring DAT mRNA in human midbrain. Target mRNA was reverse transcribed to the first-strand cDNA by Superscript II reverse transcriptase (RT), using oligo (dT) to hybridize to the 3' poly (A) tail. Template mRNA was removed by RNase H. The first-strand cDNA was then amplified by Taq DNA polymerase with a pair of gene-specific primers (Table 2).

compared across the brain samples and groups. These results demonstrate that the mRNA steady-state level of the cocaine-overdose group was not significantly different from the control group. In contrast, the mRNA levels of cocaine-overdose victims that exhibited preterminal excited delirium

Table 2
Primers for RT-PCR Amplication

Target gene	Primer sequence	Gene position	Amplicon
DAT	5'-TCCGGCTTCGTCGTCTTCTC-3' 5'-GATGTCGTCGCTGAACTGCC-3	nt#1078–1524	446 bp
Actin	5'-GCGGATGTCCACGTCACACTTCATG-3' 5'-CCACCGCCGCATCCTCCTCTTCTCT-3'	nt#394–580	186 bp
Cyclophilin	5'-TCCTAAAGCATACGGGTCCTGGCAT-3' 5'-CGCTCCATGGCCTCCACAATATTCA-3'	nt#280–445	165 bp

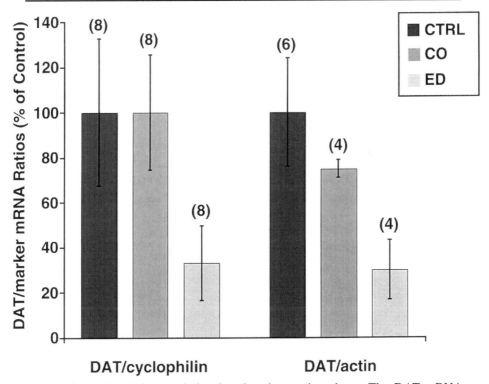

Fig. 3. DAT mRNA regulation by chronic cocaine abuse. The DAT mRNA levels were normalized to cyclophilin and actin mRNA levels (mean ± SE). The DAT/marker gene ratios were reduced significantly ($p < 0.05$) in the ED group, as compared to the age-matched and drug-free control (CTRL) group. The ratios were not statistically different between the CTRL and CO victims. DAT/cyclophilin and actin ratios were expressed as percentage (%) of the control.

was markedly reduced (67%, $p < 0.05$), as compared to age-matched and drug-free controls (Fig. 3). A decrease in DAT message was seen also when DAT mRNA values were normalized to actin for the ED group of cocaine fatalities ($p < 0.05$).

Although these studies failed to demonstrate a regulatory influence of chronic cocaine on DAT mRNAs in human cocaine abusers, DAT gene expression was altered significantly in cocaine abusers at risk for agitated cocaine delirium. If the cocaine-induced decrease in substantia nigra DAT mRNA leads to a decrease in the expression of functional DAT protein, then the behavioral effects of cocaine in this subgroup might increase with time, with the same dose of cocaine more efficiently blocking a reduced number of DA carrier sites. If this occurs within the mesolimbic DA neurons, it might explain the cocaine psychosis seen for these individuals. Paranoia in the context of cocaine abuse is common and several lines of evidence suggest that this phenomenon may be related to the function of the DAT protein and the phenomenon of cocaine-induced behavioral sensitization *(64)*. Genetic differences in the makeup of individuals who abuse cocaine may also underlie some of these differences in the susceptibility for adverse neuropsychiatric effects to develop with chronic cocaine abuse *(64)*.

The differential effects seen for DAT mRNAs in cocaine-overdose victims presenting with or without preterminal excited delirium may be a state or trait marker for these subgroups of cocaine abusers. The regulation of ligand binding to the DAT has been examined previously in fatal cocaine-overdose and excited cocaine-delirium victims using radioligand binding and in vitro autoradiography. Autoradiographic mapping with a single concentration of the cocaine congeners [^{3}H]WIN 35,428 and [^{125}I]RTI-55 failed to demonstrate an elevation in the apparent density of the DAT in the striatum of ED subjects as compared to drug-free and age-matched control subjects *(14,54)*. In addition, saturation binding with [^{3}H]WIN 35,428 and [^{125}I]RTI-55 demonstrated a significant decrease in the total number of DAT binding sites, but no change on the number of high-affinity binding sites, for the ED subgroup of cocaine-overdose victims *(14,54,65)*. In contrast to these results, cocaine-overdose victims had a marked increase in high-affinity cocaine binding sites as compared to drug-free and age-matched control subjects *(54)*. The lack of an upregulation of cocaine recognition sites on the DAT protein and mRNA in the ED subgroup of cocaine-overdose victims is consistent with the possibility of a diminished capacity for DA reuptake during a cocaine "binge." Because the concentration of synaptic DA is tightly regulated by reuptake mechanism(s), the decrease in DAT mRNA content could be a molecular defect that explains the paranoia and agitated delirium for this neuropsychiatric syndrome.

ACKNOWLEDGMENTS

This work was funded by grants from the National Institute on Drug Abuse (DA 06227 and DA 09484).

REFERENCES

1. Giros, B. and Caron, M. G. (1993) Molecular characterization of the dopamine transporter. *TIPS* **14**, 43–49.
2. Ding, Y.-S., Fowler, J. S., Volkow, N. D., Gatley, S. J., Logan, J., Dewey, S. L., Alexoff, D., Fazzini, E., and Wolf, A. P. (1994) Pharmacokinetics and in vivo specificity of [^{11}C]dl-threo-methylphenidate for the presynaptic dopaminergic neuron. *Synapse* **18**, 152–160.
3. Donnan, G. A., Kaczmarczyk, S. I., Paxinos, G., Chilco, P. J., Kalnins, R. M., Woodhouse, D. G., and Mendelsohn, F. A. (1991) Distribution of catecholamine uptake sites in the human brain as determined by quantitative [^{3}H]-mazindol autoradiography. *J. Comp. Neurol.* **304**, 419 –434.
4. Eshleman, A. J., Neve, R. L., Janowsky, A., and Neve, K. A. (1995) Characterization of a recombinant human dopamine transporter in multiple cell lines. *J. Pharmacol. Exp. Ther.* **274**, 276–283.
5. Giros, B., Mestikawy, S. E., Godinot, N., Zheng, K., Han, H., Yang-Feng, T., and Caron, M. G. (1992) Cloning, pharmacological characterization, and chromosome assignment of the human dopamine transporter. *Mol. Pharmacol.* **42**, 383–390.
6. Pristupa, Z. B., Wilson, J. M., Hoffman, B. J., Kish, S. J., and Niznik, H. B. (1994) Pharmacological heterogeneity of the cloned and native human dopamine transporter: disassociation of [^{3}H]WIN 35,428 and [3H]GBR 12935 binding. *Mol. Pharmacol.* **45**, 125–135.
7. Vandenbergh, D. J., Persico, A. M., and Uhl, G. R. (1992) A human dopamine transporter cDNA predicts reduced glycosylation, displays a novel repetitive element and provides racially-dimorphic Taq I RFLPS. *Mol. Brain Res.* **15**, 161–166.
8. Schloss, P., Mayser, W., and Betz, H. (1992) Neurotransmitter transporters. A novel family of integral plasma membrane proteins. *FEBS Lett.* **307**, 76–80.
9. Kitayama, S., Wang, J.-B., and Uhl, G. R. (1993) Dopamine transporter mutants selectively enhance MPP+ transport. *Synapse* **15**, 58–62.
10. Aquilonius, S.-M., Bergstrom, K., Eckernas, S. A., et al. *(1*987) In vivo evaluation of striatal dopamine reuptake sites using ^{11}C-nomifensine and positron emission tomography. *Acta Neurol. Scand.* **76**, 283–287.
11. Bannon, M. J., Poosch, M. S., Xia, Y., Goebel, D. J., Cassin, B., and Kapatos, G. (1992) Dopamine transporter mRNA content in human substantia nigra decreases precipitously with age. *Proc. Natl. Acad. Sci. USA* **89**, 7095–7099.
12. Dackis, C. A. and Gold, M. S. (1985) Bromocriptine as a treatment of cocaine abuse. *Lancet* **1**, 1151–1152.
13. DeKeyser, J., Ebinger, G., and Vauquelin, G. (1990) Age-related changes in the human nigrostriatal dopaminergic system. *Ann. Neurol.*, **27**, 157–161.
14. Mash, D. C. and Staley, J. K. (1996) The dopamine transporter in human brain: characterization and effect of cocaine exposure, in *Neurotransmitter Transporters: Structure, Function, and Regulation* (Reith, M. E. A., ed.), Humana, Totowa, NJ, pp. 315–343.
15. Bjorklund, A. and Lindvall, O. (1994) Dopamine-containing systems in the CNS, in *Handbook of Chemical Neuroanatomy,* Part I: vol 2 (Hockfelt, T. and Bjorklund, A., eds.), Elsevier, New York, pp. 55–122.
16. Graybiel, A. M. and Moratalla, R. (1989) Dopamine uptake sites in the striatum are distributed differentially in striosome and matrix compartments. *Proc. Natl. Acad. Sci. USA* **86**, 9020–9024.

17. Lowenstein, P. R., Joyce, J. N., Coyle, J. T., and Marshall, J. F. (1990) Striosomal organization of cholinergic and dopaminergic uptake sites and cholinergic M_1 receptors in the adult human striatum: a quantitative receptor autoradiographic study. *Brain Res.* **510**, 122–126.

18. Hurd, Y. L., Pristupa, Z. B., Herman, M. M., Niznik, H. B., and Kleinman, J. E. (1994) The dopamine transporter and dopamine D2 receptor messenger RNAs are differentially expressed in limbic- and motor-related subpopulations of human mesencephalic neurons. *Neuroscience* **63**, 357–362.

19. Graybiel, A. M. and Ragsdale, C. W. (1978) Histochemically distinct compartments in the striatum of human, monkey and cat demonstrated by acetylcholinesterase staining. *Proc. Natl. Acad. Sci. USA* **75**, 5723–5726.

20. Kleven, M. S., Woolverton, W. L., and Seiden, L. S. (1988) Lack of long-term monoamine depletions following repeated or continuous exposure to cocaine. *Brain Res. Bull.* **21**, 233–237.

21. Allard, P. and Marcusson, J. (1989) Age-correlated loss of dopamine uptake sites with [^{3}H]GBR 12935 in human putamen. *Neurobiol. Aging* **10**, 661–664.

22. Hitri, A., Casanove, M. F., Kleinman, J. E., Weinberger, D. R., and Wyatt, R. J. (1995) Age-related changes in [^{3}H]GBR 12935 binding site density in the prefrontal cortex of controls and schizophrenics. *Biol. Psychol.* **37**, 175–182.

23. Zelnik, N., Angel, I., Paul, S. M., and Kleinman, J. E. (1986) Decreased density of human striatal dopamine uptake sites with age. *Eur. J. Pharmacol.* **126**, 175–176.

24. Volkow, N. D., Fowler, J. S., Wang, G. J., Logan, J., Schyler, D., MacGregor, R., Hitzeman, R., and Wolf, A. P. (1994) Decreased dopamine transporters with age in healthy human subjects. *Ann. Neurol.* **36**, 237–239.

25. van Dyck, C. H., Seibyl, J. P., Malison, R. T., Laruelle, M., Wallace, E., Zoghbi, S. S., Zea-Ponse, Y., Baldwin, R. M., Charney, D. S., Hoffer, P. B., and Innis, R. B. (1995) Age-related decline in striatal dopamine transporter binding with iodine-123-β-CIT SPECT. *J. Nucl. Med.* **36**, 1175–1181.

26. Tedroff, J., Aquilonius, S., Hartvig, P., Lundquist, H., Gee, A. G., Uhlin, J., and Langstrom, B. (1988) Monoamine reuptake sites in the human brain evaluated in vivo by means of [^{11}C]nomifensine and positron emission tomography: the effects of age and Parkinson's disease. *Acta Neurol. Scand.* **77**, 192–201.

27. Bannon, M. and Whitty, C. (1997) Age-related and regional differences in dopamine transporter mRNA expression in human midbrain. *Neurology* **48**, 969–977.

28. McGeer, P. L., McGeer, E. G., and Suzuki, J. S. (1977) Aging and extrapyramidal function. *Arch. Neurol.* **34**, 33–35.

29. Ma, S. Y., Ciliax, B. J., Jaffar, S., Stebbins, G., Joyce, J., Kordower, J. H., Levey, A. I., Mash, D. C., and Mufson, E. J. (1999) Dopamine transporter immunoreactive neurons decrease with age in human substantia nigra: a stereologic analysis. *J. Comp. Neurol.* **409**(1), 25–37.

30. Adolfsson, R., Gottfries, C.-G., Roos, B.-E., Winblad, B. (1979) Post-mortem distribution of dopamine and homovanillic acid in human brain, variations related to age, and a review of the literature. *J. Neural Transmis.* **45**, 81–105.

31. Hornykiewicz, O. (1983) Dopamine changes in the aging human brain, in *Aging Brain and Ergot Alkaloids,* Series: Aging, Vol. 23 (Agnoli, A., Grepaldi, G., Spano, P. F., and Trabucchi, M. eds.), Raven, New York, pp. 9–14.

32. Seiden, L. S. and Ricaurte, G. A. (1987) Neurotoxicity of methamphetamine and related drugs, in *Psychopharmacology: The Third Generation of Progress* (Meltzer, H. Y., ed.), Raven, New York, pp. 359–366.

33. Kula, N. S. and Baldessarini, R. J. (1991) Lack of increase in dopamine transporter binding or function in rat brain tissue after treatment with blockers of neuronal uptake of dopamine. *Neuropharmacology* **30**, 89–92.

34. Benmansour, S., Tejani-Butt, S. M., Hauptmann, M., and Brunswick, D. J. (1992) Lack of effect of high dose cocaine on monoamine uptake sites in rat brain measured by quantitative autoradiography. *Psychopharmacology* **106**, 459–462.

35. Izenwasser, S. and Cox, B. M. (1992) Inhibition of dopamine uptake by cocaine and nicotine: tolerance to chronic treatments. *Brain Res.* **573**, 119–125.

36. Sharpe, L. G., Pilotte, N. S., Mitchell, W. M., and De Souza, E. B. (1991) Withdrawal of repeated cocaine decreases autoradiographic [^{3}H]mazindol-labeling of dopamine transporter in rat nucleus accumbens. *Eur. J. Pharmacol.* **203**, 141–144.

37. Alburges, M. E., Narang, N., and Wamsley, J. K. (1993) Alterations in the dopaminergic receptor system after chronic administration of cocaine. *Synapse* **14**, 314–323.

38. Wilson, J. M., Nobrega, J. N., Carroll, M. E., Niznik, H. B., Shannak, K., Lac, S. T., Pristupa, Z. B., Dixon, L. M., Kish, S. J. (1994) Heterogenous subregional binding patterns of ^{3}H-WIN 35,428 and ^{3}H-GBR 12935 are differentially regulated by chronic cocaine self-administration. *J. Neurosci.* **14**, 2966–2979.

39. Aloyo, V. J., Pazdalski, P. S., Kirifides, A. L., and Harvey, J. A. (1995) Behavioral sensitization, behavioral tolerance, and increased [^{3}H] WIN 35,428 binding in rabbit caudate nucleus after repeated injections of cocaine. *Pharmacol. Biochem. Behav.* **52**, 335–340.

40. Koff, J. M., Shuster, L., and Miller, L. G. (1994) Chronic cocaine administration is associated with behavioral sensitization and time-dependent changes in striatal dopamine transporter binding. *J. Pharmacol. Exp. Ther.* **268**, 277–282.

41. Izenwasser, S. and Cox, B. M. (1990) Daily cocaine treatment produces a persistent reduction of [^{3}H]dopamine uptake in vitro in rat nucleus accumbens but not in striatum. *Brain Res.* **531**, 338–341.

42. Ng, J. P., Hubert, G. W., and Justice, J. B., Jr. (1991) Increased stimulated release and uptake of dopamine in nucleus accumbens after repeated cocaine administration as measured by in vivo voltammetry. *J. Neurochem*, **56**, 1485–1492.

43. Cerruti, C., Pilotte, N. S., Uhl, G., and Kuhar, M. J. (1994) Reduction in dopamine transporter mRNA after cessation of repeated cocaine administration. *Mol. Brain Res.* **22**, 132–138.

44. Letchworth, S. R., Daunais, J. B., Hedgecock, A. A., and Porrino, L. J. (1997) Effects of chronic cocaine administration on dopamine transporter mRNA and protein in the rat. *Brain Res.* **750**, 214–222.

45. Xia, Y., Goebel, D. J., Kapatos, G., and Bannon, M. J. (1992) Quantitation of rat dopamine transporter mRNA: effects of cocaine treatment and withdrawal. *J. Neurochem.* **59**, 1179–1182.

46. Burchett, S. A. and Bannon, M. J. (1997) Serotonin, dopamine and norepinephrine transporter mRNA's: heterogeneity of distribution and response to "binge" cocaine administration. *Mol. Brain Res.* **49**, 95–102.

47. Maggos, C. E., Spangler, R., Zhou, Y., Schlussman, S. D., Ho, A., and Kreek, M. J. (1997) Quantitation of dopamine transporter mRNA in the rat brain: mapping, effects of "binge" cocaine administration and withdrawal. *Synapse* **26**, 55–61.

48. Spangler, R., Zhou, Y., Maggos, C. E., Zlobin, A., Ho, A., and Kreek, M. J. (1996) Dopamine antagonist and "binge" cocaine effects on rat opioid and dopamine transporter mRNAs. *NeuroReport* **7**, 2196–2200.

49. Shilling, P. D., Kelsoe, J. R., and Segal, D. S. (1997) Dopamine transporter mRNA is up-regulated in the substantia nigra and the ventral tegmental area of amphetamine-sensitized rats. *Neurosci. Lett.* **236**, 131–134.

50. Lu, W. and Wolf, M. E. (1997) Expression of dopamine transporter and vesicular monoamine transporter 2 mRNAs in rat midbrain after repeated amphetamine administration. *Mol. Brain Res.* **49**, 137–148.

51. Vaughan, R. A., Huff, R. A., Uhl, G. R., and Kuhar, M. J. (1997) Protein kinase C-mediated phosphorylation and functional regulation of dopamine transporters in striatal synaptosome. *J. Biol. Chem.* **272**, 15,541–15,546.

52. Little, K., McLaughlin, D., Zhang, L., McFinton, P., Dalack, G., Cook, E., Cassin, B., and Watson, S. (1998) Brain dopamine transporter messenger RNA and binding sites in cocaine users. *Arch. Gen. Psychiatry* **55**, 793–799.

53. Little, K. Y., Kirkman, J. A., Carroll, F. I., Clark, T. B., and Duncan, G. E. (1993) Cocaine use increases [^{3}H]WIN 35428 binding sites in human striatum. *Brain Res.* **628**, 17–25.

54. Staley, J. K., Hearn, W. L., Ruttenber, A. J., Wetli, C. V., and Mash, D. C. (1994) High affinity cocaine recognition sites on the dopamine transporter are elevated in fatal cocaine overdose victims. *J. Pharmacol. Exp. Ther.* **271**, 1678–1685.

55. Wetli, C. V., Mash, D. C., Karch, S. B. (1996) Cocaine-associated agitated delirium and the neuroleptic malignant syndrome. *Am. J. Emerg. Med.* **14**, 425–428.

56. Bannon, M. J., Poosch, M. S., Haverstick, D. M., Mandal, A., Xue, I. C., Shibata, K., and Dragovic, L. J. (1992) Preprotachykinin gene expression in the human basal ganglia: characterization of mRNAs and pre-mRNAs produced by alternate RNA splicing. *Brain Res. Mol. Brain Res.* **12**, 225–231.

57. Johnson, S. A., Morgan, D. G., and Finch, C. E. (1986) Extensive postmortem stability of RNA from rat and human brain. *J. Neurosci. Res.* **16**, 267–280.

58. Bannon, M. J., Xue, C.-H., Shibata, K., Dragovic, L. J., and Kapatos, G. (1990) Expression of a human cocaine-sensitive dopamine transporter in xenopus laevis oocytes. *J. Neurochem.* **54**, 706–708.

59. Staley, J. K., Boja, J. W., Carroll, F. I., Seltzman, H. H., Wyrick, C. D., Lewin, A. H., Abraham, P., and Mash, D. C. (1995) Mapping dopamine transporters in the human brain with novel selective cocaine analog [^{125}I]RTI-121. *Synapse* **21**, 364–372.

60. Escobedo, L. G., Ruttenber, A. J., Agocs, M. M., Anda, R. F., and Wetli, C. V. (1991) Emerging patterns of cocaine use and the epidemic of cocaine overdose deaths in Dade County, Florida. *Arch. Pathol. Lab. Med.* **115**, 900–905.

61. Hernandez, A., Andollo, W., and Hearn, W. L. (1994) Analysis of cocaine and metabolites in brain using solid phase extraction and full-scanning gas chromatography/ion trap mass spectrometry. *Forensic Sci. Int.* **65**, 149–156.

62. Chomczynski, P. and Sacchi, N. (1987) Single-step method of RNA isolation by acid guanidinium thiocyanate-phenol-chloroform extraction. *Anal. Biochem.* **162**, 156–159.

63. Chen, L., Segal, D., and Mash, D. C. (1999) Semi-quantitative reverse-transcriptase polymerase chain reaction: an approach for the measurement of target gene expression in human brain. *Brain Res. Protocols* **4,** 132–139.
64. Giros, B., Wang, Y.-M., Suter, S., McLeskey, S. B., Pifl, C., and Caron, M. G. (1994) Delineation of discrete domains for substrate, cocaine, and tricyclic antidepressant interactions using chimeric dopamine–norepinephrine transporters. *J. Biol. Chem.*, **269**, 15,985–15,988.
65. Staley, J. K., Wetli, C. V., Ruttenber, A. J., Hearn, W. L., and Mash, D. C. (1995) Altered dopaminergic synaptic markers in cocaine psychosis and sudden death, *NIDA Res. Monogr. Ser.* **153**, 491.

Index